Chemical Bonding at Surfaces and Interfaces

Edited by

Anders Nilsson

Stanford Synchrotron Radiation Laboratory, Menlo Park, California, USA

and

FYSIKUM, Stockholm University, Stockholm, Sweden

Lars G.M. Pettersson

FYSIKUM, Stockholm University, Stockholm, Sweden

and

Jens K. Nørskov

Technical University of Denmark, Lyngby, Denmark

ELSEVIER

Amsterdam • Boston • Heidelberg • London • New York • Oxford
Paris • San Diego • San Francisco • Singapore • Sydney • Tokyo

Chemical Bonding at Surfaces and Interfaces
Anders Nilsson, Lars G.M. Pettersson and Jens K. Nørskov
ISBN: 9780444528377

Authorized English reprint published by Beijing World Publishing Corporation, Ltd.

《表面和界面的化学键合》
ISBN: 9787523223451

Notice

Chemical Bonding at Surfaces and Interfaces

Contents

Preface

Molecular surface science has made enormous progress in the past 30 years. The development can be characterized by a revolution in fundamental knowledge obtained from simple model systems and by an explosion in the number of experimental techniques. Furthermore, the last 10 years has seen an equally rapid development of quantum mechanical modeling of surface processes using Density Functional Theory (DFT). The methods of surface science have been essential for the birth of nano-science and nano-technology, and more generally we are witnessing a rapid shift of the methods and concepts of surface science into a broad range of scientific disciplines where the interaction between a solid and the surrounding gas or liquid phase is an essential component. The purpose of the present book is to provide a broad overview of chemical bonding at surfaces, and to show how it can be applied in a range of scientific problems in heterogeneous catalysis, electrochemistry, environmental science and semiconductor processing.

We focus in the following on phenomena and concepts rather than on experimental or theoretical techniques, and the aim is to provide the common basis for describing the interaction of atoms and molecules with surfaces to be used very broadly in science and technology. The organization of the book reflects the general approach. We start with an overview of structural information on surface adsorbates and discuss the structure of a number of important chemisorption systems which will be further discussed in the subsequent chapters. In Chapter 2, we describe in detail the chemical bond between atoms or molecules and a metal surface in the observed surface structures. These two initial chapters set the stage for discussing chemical reactions at surfaces in the remaining parts of the book. We begin in Chapter 3 with a detailed description of experimental information on the dynamics of bond-formation and bond-breaking at surfaces. This is followed by an in-depth analysis of aspects of heterogeneous catalysis based on the d-band model, and examples are given of how modern theoretical DFT techniques can be used to actually design efficient heterogeneous catalysts. In Chapter 5, we turn our attention to adsorption and chemistry on the enormously important Si and Ge semiconductor surfaces. In the remaining two Chapters, we leave the solid-gas interface and turn our attention to solid-liquid interface processes by first studying the surface chemistry occurring on the electrodes in electrochemistry and in particular modern fuel cells for clean energy production. In the final Chapter, we give an overview of the environmentally important chemical processes occurring on mineral and oxide surfaces in contact with water and electrolytes.

It is the hope of the whole team of authors that the present effort will assist in providing a coherent and easily grasped picture of the fascinating chemistry occurring at the various surfaces that provide templates for wanted and unwanted catalysis in industry and in our environment.

Anders Nilsson
Lars G. M. Pettersson
Jens K. Nørskov

Chapter 1

Surface Structure

D. P. Woodruff

Physics Department, University of Warwick, Coventry CV4 7AL, UK

1. Why surface structure?

Quantifying and understanding the structure of surfaces, and particularly of adsorbates on surfaces, is a key step to understanding many aspects of the behaviour of surfaces including the electronic structure and the associated chemical properties. For example, any calculation of the electronic structure starts from the structure. Of course, it is now common to try to determine the structure of surfaces by *ab initio* methods, in which the structural model and the positions of the atoms are varied to find the lowest energy configuration which then forms the basis of the calculation of the electronic and chemical properties. Such methods have become increasingly powerful and effective in recent years, yet experimental tests of these optimised structures are crucial to ensure the integrity of such calculations, and there are certainly clear examples in the literature of the failure of these calculations to reproduce well-established experimental structural trends (e.g., CO on Pt(111) – see Section 4). A particular example of the significance of surface structure in surface chemistry is in the field of heterogeneous catalysis, in which one frequently reads references to 'the active site'. Underlying such statements is the belief that key steps in surface chemical reactions occur at specific geometrical sites on a surface, and that understanding the nature of these sites could greatly improve our understanding of how to make more efficient catalysts. In those cases in which a catalytic system is found to be 'structure sensitive' it seems likely that these active surface sites may be quite specific and thus their availability is dependent on the mode of catalyst preparation.

In this chapter, the objective is to illustrate some of the structural phenomena associated with adsorbate bonding at surfaces and to show how (experimental) quantitative surface structure determination can provide insight into the nature of adsorbate bonding at surfaces. To achieve this, a brief outline of the methods used for adsorbate structure determination is first given in Section 2. Details of these methods are not the focus of this chapter, yet it is important to understand the strengths and limitations of the various methods in order to evaluate the data that arise from them. In Section 3, are presented a few examples of the way that adsorbates may modify the structure of the outermost atomic layers of the surface onto which they are adsorbed, and the significance of such adsorbate-induced reconstruction. Section 4 includes illustrations of investigations of molecular adsorbates of varying size, while in Section 5 issues raised by careful quantitative measurements of chemisorption bondlengths, and the insight they give into bonding mechanisms, are discussed.

2. Methods of surface adsorbate structure determination

2.1. General comments

In this section, some key aspects of the various methods of surface adsorbate structure determination are described. Far more detailed descriptions of the individual methods may be found elsewhere (some relevant references are given), and the objective here is rather to highlight the particular strengths, limitations and special aspects of the techniques which need to be considered when evaluating and comparing the results of applications of these methods. One particular feature which is common to the great majority of these techniques is that the structure is extracted from the experiment through some kind of trial-and-error modelling. In this approach one 'guesses' a possible structure and then compares the results of the experiment with the results which would be expected from the guessed structure, through a computation based on the known physical phenomena that underlie the experiment. In many cases it is possible to refine the structural model in an automated and objective fashion by varying the structural parameter values in the model calculation and searching for the best agreement with experiment, typically identified as the minimum value of some kind of reliability- or *R*-factor. *R*-factors are commonly based on a sum of the squares of the differences of the experimentally measured and theoretically computed quantities. This type of optimisation, however, is only conducted within a specific structural model. For example, one may adjust the inter-layer spacings within the substrate, within a molecular adsorbate, and between the substrate and adsorbate, and may also adjust lateral positions of atoms, but typically within some applied symmetry constraints. It is then necessary to compare the results of such structural optimisations for different structural models. These models may only differ in the lateral registry of the adsorbate of the adsorbate – e.g., adsorption in atop, bridge or hollow sites – but may also include specific models of adsorbate-induced substrate reconstruction, such as changes in the atomic density of the outermost layer or layers of the substrate.

An important general limitation of this approach is that the ultimate structure determination is limited by the imagination of the researcher. If the correct structural model is not tested, the final solution will be the best structure tried, but not the correct one. Indeed, this best structure may differ fundamentally from the true structure. Notice, too, that this limitation also applies to *ab initio* total energy calculations to determine surface structures theoretically. Here, too, one must start from specific trial models of a structure which can then be optimised.

A second general issue in surface structure determination using the trial-and-error modelling approach is uniqueness. In any optimisation of a structural model one can find an optimal set of structural parameters which defines a minimum in the *R*-factor. This minimum value may represent a 'good fit' but is still not necessarily the correct structure. One can then compare the *R*-factor values associated with these local minima for different structural models, perhaps resulting in several 'good fits'. Ideally, one of the structural models gives a significantly lower *R*-factor. In some cases, however, the goodness-of-fit is similar for more than one best-fit structure. The risk of this problem arising can generally be greatly reduced by ensuring that the size of the data set being used for theory-experiment comparison is large. Large data sets not only reduce the likelihood of this type of ambiguity, but also reduce the size of the variance of the *R*-factor and thus render significant smaller differences in minimum *R*-factor values. For this reason the size of the data set is an important issue in determining the reliability of any experimental structure determination, as well as its precision. Of course, there are also situations in *ab initio* total energy calculations in which two structures have essentially the same lowest energy. In this case one must conclude either that the two structures really are energetically almost equivalent, in which case one expects coexistence of the two structures, or that the computation contains systematic errors in the accurate description of the underlying physics.

2.2. Electron scattering

In many ways the 'benchmark' method of quantitative surface structure determination is low energy electron diffraction (LEED) [1–3] This was the first method to be developed in the early 1970s and still accounts for the largest number of catalogued surface structure determinations [4]. A key feature of the technique is that, like conventional X-ray crystallography of bulk solids, it exploits the *long-range periodic order* of the sample to concentrate the elastically scattered low energy electrons into distinct diffracted beams. This can be both a strength and a limitation. In particular, the scattered electron intensity in the diffracted beams is dominated by contributions for those parts of a surface that have good long-range order, so the technique selectively provides information on these regions. If other regions lack this long-range order, the method is 'blind' to them, but also the information on the ordered parts is not distorted by the presence of the disordered regions. Because the elastic scattering cross-sections of atoms at the low energies (~30–300 eV) characteristic of LEED are very large, multiple scattering plays an important role and the structure can only be extracted through trial-and-error modelling. One further important feature of LEED is that it probes several atomic layers of the near-surface region, so getting a proper fit of experiment and theory requires not only a good description of the adsorbate geometry, but also of the substrate geometry including detailed layer relaxations and rumpling. Indeed, if these substrate relaxations are not well-described in the model, this may introduce systematic errors into the adsorbate geometry. In this sense, LEED gives the complete structure, but it is also important to describe *all* aspects to be confident of *any* of the conclusions.

Two rather different techniques that exploit the same underlying phenomenon of coherent interference of elastically scattered low energy electrons are photoelectron diffraction [5] and surface extended X-ray absorption fine structure (SEXAFS) [6,7]. Figure 1.1. shows schematically a comparison of the electron interference paths in LEED and in these two techniques. In both photoelectron diffraction and SEXAFS the source of electrons is not an electron beam from outside the surface, as in LEED, but photoelectrons emitted from a core level of an atom within the adsorbate. In photoelectron diffraction one detects the photoelectrons directly, outside the surface, as a function of direction or photoelectron energy (or both). The detected angle-resolved photoemission signal comprises a coherent sum of the directly emitted component of the outgoing photoelectron wavefield and other components of the same wavefield elastically scattered by atoms (especially in the substrate) close

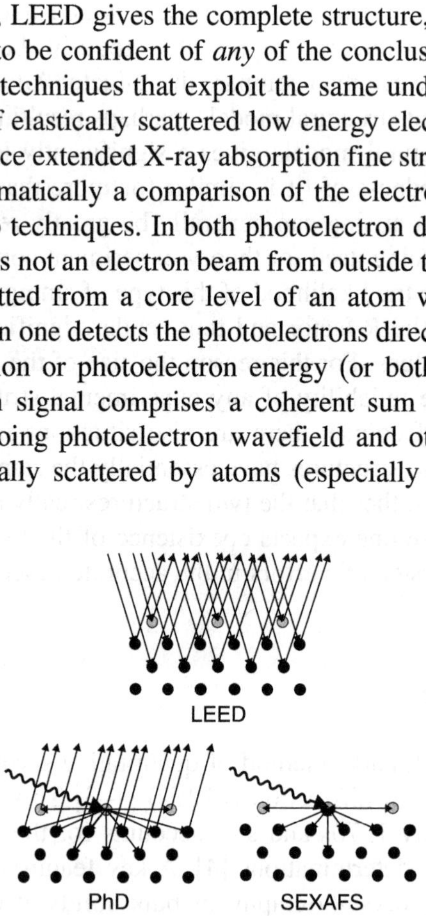

Figure 1.1. Schematic diagram showing the electron elastic scattering pathways contributing to the techniques of low energy electron diffraction (LEED), backscattering photoelectron diffraction (including the scanned-energy mode – PhD) and surface extended X-ray absorption fine structure (SEXAFS). Black disks represent substrate atoms, grey-shaded disks represent adsorbate atoms.

to the emitter. As one changes the collection angle, or the photoelectron energy (and thus the photoelectron wavelength), particular scattering paths switch in and out of phase with the directly emitted component, leading to intensity modulations. These modulations can be interpreted in terms of the structural environment of the emitter, through the use of multiple-scattering calculations for trial structures, in a fashion very similar to that used in the interpretation of LEED data. Indeed, insofar as the method of diffracted beam intensity collection in LEED involves the measurement of the intensity of these beams as a function of electron energy, the scanned-energy mode photoelectron diffraction (PhD) technique is closely similar to LEED, the photoemission intensity being measured as a function of photon, and thus photoelectron, energy. A key difference between LEED and PhD, however, is that PhD is *element specific* and *local*. The fact that the photoelectrons are emitted from an adsorbate core level with a characteristic binding energy means that the source of the photoelectron wavefield is known to be a specific elemental atom on the surface, and the structural information is centred on this atom. Furthermore, because the electron source is an outgoing spherical wave the structural information is local to the emitter atom, and does not depend on (or exploit) any long-range periodic order. This means that the technique determines the local structure independent of whether or not long-range order exists in the adsorbate layer; on the other hand, it also is unable to distinguish between those areas of the surface which have this long-range order and those which do not.

SEXAFS shares with photoelectron diffraction the elemental specificity and local character of the structural information content. The key difference between these two techniques is that while in photoelectron diffraction one measures the angle-derivative of the photoelectron emission cross-section, in SEXAFS one measures the total photoionisation cross-section (indirectly through the decay of the core holes created by the ionisation, leading to the emission of X-radiation or Auger electrons). SEXAFS exploits the fact that when the photoelectron wavefield emerges from the absorber atom a small fraction of this wavefield is elastically backscattered to the emitter, where it interferes with the outgoing component to modify the wavefield amplitude at the emitter; this wavefield amplitude enters the matrix element for the photoionisation cross-section as the final state. The total photoionisation cross-section is thus modulated with photon energy as the photoelectron energy changes, causing a change in the photoelectron wavelength, such that the back-scattering leads to alternately constructive and destructive interference. These modulations are the 'extended fine structure' of the technique's name. They provide information primarily on the distance of the emitter from the near-neighbour scattering atoms, although some limited directional information is contained in the way the amplitude of the modulations varies with the direction of the electric vector of the incident X-rays. Because both the source of electrons and the detector (in both cases the photoelectron emitting atom) are local in SEXAFS, the structural data is even more localised than in photoelectron diffraction, and for the same reason the amplitude of the modulations, and thus the ease of achieving good signal-to-noise ratios in the measured modulations, are about an order of magnitude lower. Typically SEXAFS provides accurate nearest-neighbour distances and limited information on the direction of these neighbours and the distances to other near neighbours. In photoelectron diffraction one is significantly more sensitive to non-nearest neighbours, and also obtains more specific directional information because the direction of photoelectron detection influences the scattering path-length differences explicitly. Full structural optimisation in SEXAFS also involves modelling through trial structures, although it is commonly possible to extract good nearest-neighbour distance information directly through Fourier transform methods providing these include corrections for the influence of phase shifts in the electron scattering events.

One final key feature of photoelectron diffraction which is not shared by LEED or SEXAFS is the ability to exploit so-called *chemical shifts* in photoelectron binding energies of atoms of the same element in different structural and electronic environments to obtain *chemical state specificity* in the local structural information.

All of these electron scattering techniques are typically capable of determining interatomic distances to a precision of ~0.02–0.05 Å, with specific cases in which somewhat worse, and occasionally even better, values are cited. For LEED and photoelectron diffraction one commonly finds the best precision for distances corresponding to atomic separations that are near-normal to the surface, with lower precision in locations parallel to the surface, a consequence of the fact that the scattered electrons are generally not detected at very grazing angles relative to the surface.

Because LEED typically involves incident beam currents of ~1 μA into an area of less than 1 mm^2, the problem of radiation damage can be severe for fragile adsorbed molecules and some surfaces. This problem can be substantially reduced by using channel-plate amplified systems in the measurement of the diffracted beam intensities to permit the use of incident currents of ~1 nA. By contrast, incident X-ray techniques have commonly been regarded as less of a problem for radiation damage. However, particularly when using modern third-generation synchrotron radiation sources that are capable of delivering high photon fluxes (~10^{11} photons/s) into highly focussed spots (~50 × 50 μm), there can also be significant damage problems in photoelectron diffraction and SEXAFS unless special precautions, such as defocusing of the incident radiation, are taken.

2.3. X-ray scattering

In contrast to low-energy electrons, X-rays are very weakly scattered by atoms, a property which leads to the success of X-ray diffraction as a means of determining the structure of bulk solids through scattering from atoms over a large depth into the crystals. While this property means that the X-ray scattering signal from surfaces is weak, surface X-ray diffraction (SXRD) [8,9] experiments can be performed experimentally by measuring the surface scattering at locations in momentum-transfer space far removed from those corresponding to diffraction from the underlying bulk. Like LEED, SXRD relies on good long-range periodic order, and indeed the quality of the order required for SXRD is typically higher than LEED in order to ensure that the weak surface diffraction beams are narrow and thus more easily detected above the diffuse scattering background. The benefit of performing these more demanding experiments is that because the scattering is weak, multiple scattering plays no significant role, and direct inversion Fourier transform methods are far more useful. Nevertheless, the final structural refinement generally still involves trial-and-error modelling. The simpler theoretical description also means that it is viable to tackle more complex surfaces involving much larger surface periodicity than in LEED. The intrinsically weak scattering, however, means that it is particularly demanding in SXRD to obtain precise structural information on the very-weakly-scattering low atomic number adsorbates (such as C, N and O) which comprise some of the most chemically interesting adsorbate molecules. We should also note that in many SXRD studies, measurements of the scattered intensities are made mainly at grazing angles (where the signals are largest) which allows one only to determine the relative lateral positions of surface atoms and not the spacing perpendicular to the surface. It is possible to extract such information from SXRD experiments, however, if measurements are made for a wider range of take-off angles (corresponding to so-called 'rod scans' in reciprocal space). While SXRD is capable of structural precision of ~0.01 Å, the actual precision depends strongly on the atoms being investigated and whether the position parallel or perpendicular to the surface is being determined. For low atomic number elements the location perpendicular to the surface may suffer from random errors of 0.1 Å or even significantly more.

A quite different surface structural technique which nevertheless exploits X-ray diffraction is X-ray standing wavefield (XSW) absorption [10–12]. In this technique one uses X-ray diffraction from the substrate to set up an X-ray standing wave with the same periodicity as the substrate scatterer-planes within, and outside, the crystal, due to the interference of the incident and diffracted X-rays. This standing wave can be scanned in a systematic way relative to the substrate scatterer-planes by scanning through the Bragg diffraction condition in either incidence angle or X-ray wavelength. If one measures the X-ray absorption at an adsorbate atom due to this standing wave, in such a scan, one can locate the absorber atom relative to the underlying substrate. Because the X-ray absorption is typically measured by core level photoemission, or by the X-ray fluorescence or Auger electron emission resulting from the refilling of the core hole, the energy of these emissions provides elemental specificity in the structure determination. Indeed, by exploiting chemical shifts in the core level photoemission this technique can provide chemical-state specific structural data. This added specificity is exploited at the lower photon energies typically associated with normal incidence to the Bragg scatterer-planes (NIXSW), and this variant of the technique is applicable to a wider range of materials due to its relative insensitivity to the mosaicity of the substrate crystal. Because the X-ray diffraction exploited in this technique relies only on the long-range periodicity of the substrate, there is no dependence on long-range order in the adsorbate. An important feature of XSW is that the adsorbate atom is determined relative to the extended bulk structure, because the standing wave is established in scattering from very many sub-surface layers. As such, the method provides no direct information regarding the position of the adsorbate atom relative to the nearest substrate atoms, and is completely blind to surface reconstruction (although such reconstruction may be inferred from a combination of the adsorbate location and plausible values of the chemisorption bond lengths). A further significant feature of the method is that the extraction of the basic structural parameters, the so-called coherent positions and coherent fractions, is model-independent. Moreover, in the simplest cases of single high-symmetry adsorption site occupation, the interpretation of these parameters in terms of the actual adsorbate location is trivial and unique. In more complex systems, however, simple modelling is still required to relate the measured structural parameters to a real structure. While precisions as high as 0.01 Å are sometimes claimed for this method, more typical values for adsorbates on surfaces are ~0.03–0.05 Å.

The radiation damage problems with these incident X-ray methods are similar to those described in the previous section for photoelectron diffraction and SEXAFS, namely that there are potential problems, but they can mostly be overcome by appropriate precautions.

2.4. Ion scattering

Ion scattering methods, covering a wide range of energies from ~1 keV to ~1 MeV, and mainly using low atomic number ions such as H^+, He^+ and Li^+, but also often including Ne^+ at low energies, have been used in a range of surface structural studies (e.g. Refs. [13,14]). The basic physical principle exploited is of elastic scattering *shadow cones*, such that atoms behind a scattering atom on the incident ion trajectory may be hidden from the incident beam within a certain range of relative lateral displacements but will scatter incident ions if this lateral displacement is exceeded. The visibility of scattering from these subsurface atoms as a function of incident direction thus provides information of the relative locations of the shadower (surface) atoms and shadowed (subsurface) atoms. Similar effects occur for the outgoing scattered ions, with surface atoms 'blocking' the scattered ions from subsurface atoms and preventing them from reaching the detector in certain directions. These methods formally exploit the well-defined crystallography of the surface but not explicitly the long-range order of an adsorbate. They have been used mainly to investigate a range of atomic adsorbate structures and have contributed little quantitative structural information on the local adsorption geometry of molecular species, although at the higher energies they can be particularly effective in investigating adsorbate-induced reconstructions of the outermost substrate layers. The precision of these methods is generally highest for higher energy ions (~100 keV – referred to as medium energy ion scattering or MEIS) for which the shadow cones are narrowest, when values of ~0.02–0.03 Å may be achieved. While each ion which scatters from a surface atom causes significant local damage due to the recoil of the scattering atoms, the information on this scattering atom relates to its position before the scattering event. For sufficiently low incident flux density, therefore, these methods can provide information on surfaces essentially devoid of damage induced by the incident beams.

2.5. Spectroscopic methods and scanning probe microscopy

While the methods summarised above are primarily directed to obtaining quantitative structural information on adsorbates on surfaces, a range of other methods may provide valuable qualitative information, yet much of this information must be treated with caution.

Perhaps the most obvious methods are the scanning probe microscopies, of which scanning tunnelling microscopy (STM) is the one most commonly able to offer atomic-scale resolution. Superficially, at least, STM provides a real-space mapping of surface atoms with sub-atomic resolution, so one might wonder why one needs the far more complex and indirect surface structural methods outlined above. The answer, of course, is that STM is a probe of the spatial variations of the surface *electronic* structure, not of the relative locations of the atomic centres on the surface. The electronic tunnelling probability depends on the overlap of the tails of the electron wavefunctions of the occupied and unoccupied valence states just outside the tip and the surface being scanned, and to a first approximation the surface corrugation obtained in STM is a contour of constant partial electronic density of states outside the surface. For an elemental surface this usually (but not invariably – e.g., [15]) leads to the peaks of the surface protrusions being located above the atom centres, but the amplitude of the surface corrugation has no simple relationship to the relative heights of atoms above the surface, except when comparing the height of symmetrically equivalent atoms (such as those defining the height of a surface step). Moreover, on compound surfaces or elemental surfaces in the presence of adsorbates, even the simple correlation between the lateral position of atoms and atomic-scale protrusions in STM ceases to be reliable. For example, adsorbate C and O atoms on metal surfaces are commonly imaged as dips rather than protrusions (e.g., [16]). Similarly, on the $TiO_2(110)$ surface, it is generally believed that the protrusions in the STM images correspond to the surface Ti atoms despite the fact that these atoms lie physically more than 1 Å lower in the surface than the O atoms [17]. Because of these electronic effects it is also not reliable to correlate apparent lateral shifts in atomic protrusions in STM images with real lateral shifts of the underlying atoms.

Despite these very real limitations, which certainly preclude the use of STM as a source of quantitative surface structural information, the technique can play a very valuable role in elucidating surface structural phenomena. For example, in low coverages of adsorbates on a surface (a situation in which other methods may lack sufficient sensitivity) it is often possible to determine the lateral registry of the adsorbate; at least in cases of high-symmetry adsorption sites, one may distinguish atop, hollow and bridge sites. STM images can also be helpful in the case of complex structures, as a source of possible structural models which may be tested in the trial-and-error modelling in quantitative structural methods, although there are clear pitfalls in interpreting such images too literally. However, the most valuable role of STM is in identifying inhomogeneity at surfaces such as step-site adsorption, island growth, coexistent surface structures, and in gaining information on the time-evolution of surface structural changes by such processes as nucleation and growth. For example, in early studies of the structure of the Cu(110)(2 × 1)–O surface phase in which the outermost Cu layer has only half the atom density of the underlying bulk layers there were often debates about 'where do all the Cu atoms go' in creating the missing-row structure. Sequential STM images during the evolution of the surface show [18,19] that the phase actually forms by the addition, rather than removal, of rows of surface Cu atoms, but the answer to the converse question which is then raised, namely 'where do all the Cu atoms come from', is surface steps.

Quite different information on surface structure arises from some spectroscopies. Most obvious are the vibrational spectroscopies, infra-red reflection absorption and electron energy loss. In these methods the comparison of the behaviour of molecular adsorbates on surfaces with the previously characterised behaviour in coordination compounds has led to the spectral fingerprint being used to infer local geometry. Much the best-known example of this is CO adsorption, the C−O stretching frequency being used to identify single, double and higher coordination adsorption sites (atop, bridge, hollow) by comparison with the considerable body of evidence on metal carbonyls (e.g., [20]). Even for these extremely well-characterised systems, however, this indirect approach to adsorption site determination has been found to be subject to misinterpretation, most conspicuously in the case of the c(4 × 2) phases formed by CO on Ni(111) and Pd(111). In both cases the vibrational spectroscopy was interpreted in terms of bridge site adsorption (Figure 1.2), in part because the vibrational frequency was deemed consistent with bridging sites, in part because the bridge site model leads to an appealing model with a periodic CO overlayer as seen in Figure 1.2. This view defined conventional wisdom for many

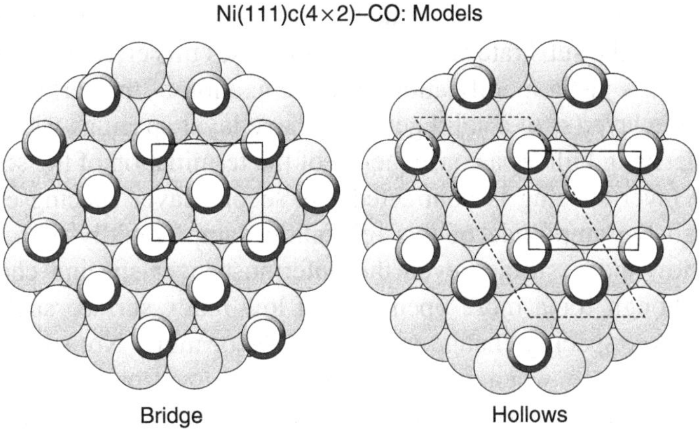

Figure 1.2. Plan view of the Ni(111)c(4 × 2)–CO surface phase showing the bridge site model favoured for many years on the basis of the interpretation of vibrational spectroscopy, and the mixed-hollow site model subsequently established through SEXAFS, PhD, and LEED measurements. To allow visibility of all surface atoms the C atoms are shown as the larger dark-shaded spheres and the O atoms as smaller white spheres. The full lines show the primitive unit mesh while the dashed lines show the centred (4 × 2) unit mesh.

years until quantitative structural studies by SEXAFS [21], PhD [22] and quantitative LEED [23] on Ni(111), and subsequently by PhD on Pd(111) [24], showed the true adsorption sites to be the (two inequivalent) three-fold coordinated hollows (Figure 1.2). Core level photoemission (X-ray photoelectron spectroscopy – XPS, albeit commonly performed with soft X-ray synchrotron radiation), may also show 'chemical shifts' in the photoelectron binding energy of adsorbates which depend on the coordination to the substrate. In most cases this is used only as a spectral fingerprint of the existence of multiple sites, but there has been some success in using these shifts to identify local coordination (e.g., Ref [25]). It is, however, in combination with a true quantitative technique that monitors photoemission, such as photoelectron diffraction and XSW, that these shifts have their greatest value in true surface structure determination.

3. Adsorbate-induced surface reconstruction

In early structural studies of adsorbates on surfaces there was an implicit assumption that the surface provided a rigid chequer board of identical sites into which atoms or molecules were adsorbed, the only structural parameters of interest being the lateral registry and the adsorbate–substrate chemisorption bondlength. It was, of course, understood that the surface could modify the adsorbate species, most obviously through partial dissociation, but also in more subtle ways, because this is the whole basis of heterogeneous catalysis. We now know, of course, that the adsorbate also induces changes in the substrate surface. In some cases this effect is quite subtle. The simplest example is just a modification of the relaxation of the surface layer(s). The outermost atomic layer(s) of a solid generally have layer spacings which differ from that of the underlying bulk as a consequence of the termination of the solid; typically the outermost layer spacing is contracted, the second layer spacing expanded and so on, although the amplitude of this relaxation damps rapidly with depth. For a close-packed low-index surface even the outermost layer spacing change may be only $\sim$1%, although for a more open-packed low-index surface such as fcc(110) the outermost layer spacing change may be $\sim$10% or more. Not surprisingly, these relaxations will change when material, including an adsorbate, is added to the surface. Typically the size of the clean surface relaxation is reduced, but in some cases larger changes may occur. This effect may also be local to the adsorbed atom. Consider, for example, the case of atomic O on Ni(100) [26,27]. At a coverage of 0.5 ML an ordered c(2 × 2) phase is formed in which the O atoms occupy alternate four-fold coordinated hollow sites in the surface (Figure 1.3). This means that in the second substrate layer half of the Ni atoms have an oxygen atom directly above them while the other half have no such O near-neighbour. This leads to a 'rumpling' of the second Ni layer, with the Ni atoms below the O adsorbates being 0.035 Å lower than those that are not covered in this way. This effect is marginal but detectable

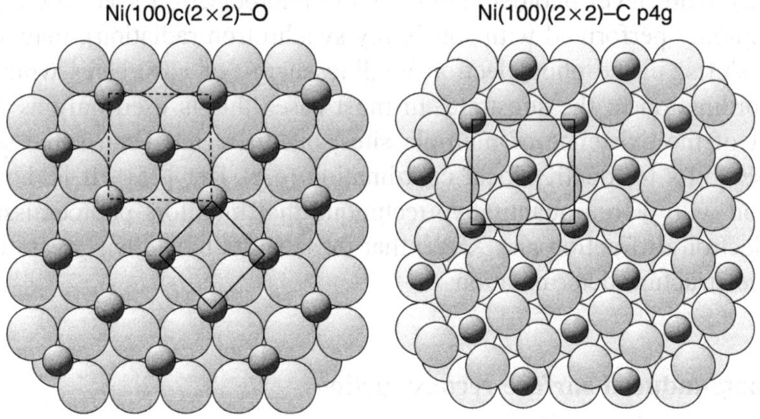

Figure 1.3. Plan view of the Ni(100)c(2 × 2)–O and Ni(100)(2 × 2)–C p4g surface structures. In each case the full lines show the primitive unit mesh while in the O-induced structure the dashed lines show the centred (2 × 2) mesh. In the case of the C-induced structure the outermost Ni atoms are shown as smaller more-darkly shaded spheres than those of the underlying substrate to see more clearly the relationship of this reconstructed layer to the substrate. Notice that this reconstruction also leads to some reduced Ni–Ni nearest-neighbour distances in the surface, so using the bulk atomic radii for these atoms would lead to some overlapping spheres.

with LEED. Interestingly, there is also a (2×2)-O surface phase of Ni(100) in which the O coverage is only 0.25 ML, so only every fourth second layer Ni atom is covered by an O atom and of the remaining $^3/_4$ of the second layer atoms there is $^1/_4$ of them that are more distant from the O adsorbates than is possible in the $c(2\times2)$ phase. In this case the overall rumpling amplitude of the second Ni layer is larger with a value of 0.100 Å. This leads to an interesting speculation: would the effect be even larger for isolated O atoms at low coverage? The answer is not known; the great majority of such subtle substrate distortion measurements have been made by LEED on surfaces with good long-range order in the adsorbate and with relatively small adsorbate–adsorbate distances.

A somewhat related question has been addressed, however, for the case of C on Ni(100). In this case at high coverage the influence of the adsorbate on the structure of the surface atomic layer of Ni is far more significant than these subtle changes in interlayer spacings. Figure 1.3 shows a schematic model of the structure formed by 0.5 ML of C on Ni(100). While the C atoms occupy alternate four-fold coordinated hollow sites just like O in Ni(100)c(2×2)–O, the Ni atoms of the outermost substrate layer are displaced tangentially within the surface plane relative to the C atoms such that the groups of four Ni nearest-neighbours of the C atom expand outwards and rotate alternately clock-wise and counter-clockwise to produce a larger (2×2) periodicity, but with characteristic glide symmetry lines which can be identified in the LEED pattern as being associated with a p4g space group [28]. The geometry of this structure, commonly referred to as the 'clock' reconstruction because of the clockwise and counter-clockwise rotations, is now rather well established by LEED [28,29], PhD [30,31], and SEXAFS [32]. The C atoms are almost coplanar with the reconstructed outermost Ni layer, and this layer expands significantly outwards away from the underlying bulk. However, the detailed mechanisms underlying this reconstruction have proved more controversial [33]. Early qualitative discussions centred on the idea that the C atom (unlike O in the unreconstructed Ni(100)c(2×2)–O phase) is almost small enough to penetrate the hollow site, and can do so when the hollow site is enlarged by the reconstruction. The idea that the C atom also forms a strong bond with the second layer Ni atom directly below as a result of this penetration has also been discussed, and at least one total energy calculation identifies this second layer bonding as very significant [34]. Experimental and theoretical studies of the vibrational properties of this surface led to the conclusion that the reconstruction can be regarded as a freezing of a specific nickel surface phonon mode [35–37], potentially clarifying the mechanism of the reconstruction, but this still does not provide a clear picture of the underlying driving force. It has also been shown that the reconstruction leads to a partial relief of adsorbate-induced compressive surface stress associated with the C adsorption [38]; the adsorbate-induced change in surface stress as a function of C coverage shows that the onset of the reconstruction around 0.3 ML halts the rise in compressive stress, presumably helping to lower the total energy.

One possible route to gaining more insight into the mechanism of this reconstruction is to compare the structure in this 0.5 ML ordered reconstructed surface with the *local* structure around C atoms adsorbed at lower coverage when the reconstruction does not occur. Notice that one feature of the higher coverage 'clock' reconstruction is that the local distortions of the Ni surface periodicity involve concerted interlocking movements that are only possible in a periodic structure; the same distortions applied locally around an isolated C atom would lead to much shorter Ni−Ni distances on the periphery of the distorted region. In this regard an STM study by Klink et al. [39] appeared to provide the key information. As expected, STM images at a C coverage of 0.5 ML, were found to show direct evidence for the lateral distortions of the (2×2) p4g-phase described above. At low coverage, however, images indicated that the C atoms adsorb in hollow sites without inducing a local clock reconstruction. On the basis of quantitative analysis of the STM images and associated line-scans, these authors deduced that there is a local lateral (radial) outward relaxation of the nearest-neighbour Ni atoms around these isolated adsorbed C atoms of 0.15 Å (albeit with an estimated precision of only ±0.15 Å). This local strain was interpreted as a signature of Ni−C near-neighbour repulsion and was taken as evidence that this repulsion is the key to the reconstruction and the compressive stress increase in the absence of reconstruction. A quantitative structure determination of the local environment of the C atoms at low coverage by PhD, however, showed that no such radial relaxation occurs in the Ni top-layer neighbours of the C atoms [31], implying that the apparent Ni atom displacements seen in the STM images are a result of the local electronic effects around the C atoms. Interestingly, the PhD structural study showed that the C−Ni nearest neighbour distances to both outermost and second layer Ni atoms were unchanged in the reconstruction; in the absence of the reconstruction the C atoms sit slightly higher above the (smaller) four-fold coordinated hollow site, while the Ni atom directly below the C atoms moves down to maintain a constant Ni−C bondlength when the reconstruction occurs. Indeed, the structural signature of the reconstruction in terms of bondlengths was found to be a reduction of the Ni−Ni distance within the outermost layer (which must be induced by the bonding to the C atoms), rather than any change in the Ni−C bondlengths.

Another example of adsorbate-induced surface reconstruction in which there have been studies of the structure of the precursor to the reconstruction at low coverages is the case of atomic oxygen on Cu(100). The ability of chemisorbed oxygen to form a missing row (or equivalently in terms of the equilibrium structure, an added row) reconstruction on Cu(110) has already been mentioned in Section 2.5 in discussing some of the benefits to be derived from the STM technique in structural studies. On Cu(110) the resulting (1×2) structure comprises −Cu−O−Cu−O− chains along the [001] direction with one 'missing' −Cu−Cu−Cu− chain between each of these, relative to the row spacing of the clean unreconstructed Cu(110) surface. On Cu(100) atomic oxygen also induces a missing row reconstruction which leads to similar [001] −Cu−O−Cu−O− chains although the detailed structure and mechanism of formation is rather different. The resulting ordered structure has a $(\sqrt{2} \times 2\sqrt{2})R45°$ unit mesh with every fourth [100] Cu surface atom row missing relative to the clean surface (Figure 1.4) and the O atoms occupy sites almost directly above second layer Cu atoms within the troughs produced by the missing Cu rows [40–44]. These O adsorption sites are essentially in the same location as four-fold coordinated hollow sites with respect to the underlying bulk solid, but the missing Cu rows means that the O atoms are only three-fold coordinated with respect to the (remaining) outermost layer Cu atoms. Notice that the fact that these erstwhile Cu neighbours are missing means that the O atoms can be located almost coplanar with the outermost layer Cu atoms; in this regard the structure shares with the clock-reconstructed Ni(100)(2 × 2)p4g–C surface the fact that the reconstruction allows the adsorbate atom to sit deeper in the surface. Coincidentally, STM dynamic imaging studies show that the Cu(100) $(\sqrt{2} \times 2\sqrt{2})R45°$–O structure is formed by the creation of missing rows, the rejected Cu atoms forming islands one atomic layer higher with the same reconstruction [45].

The Cu(100) $(\sqrt{2} \times 2\sqrt{2})R45°$–O surface reconstruction has an oxygen coverage of 0.5 ML, but there have also been many reports in the literature of a Cu(100)c(2 × 2)–O phase which would also involve a coverage of 0.5 ML. In truth, in quite a number of these earlier reports no ordered structure was actually observed experimentally and the adsorption phase studied was simply assumed to be this previously reported ordered phase. Reviewing this early literature thus requires some care! Indeed, a careful study of the relative intensities of the ½ and ¼ order

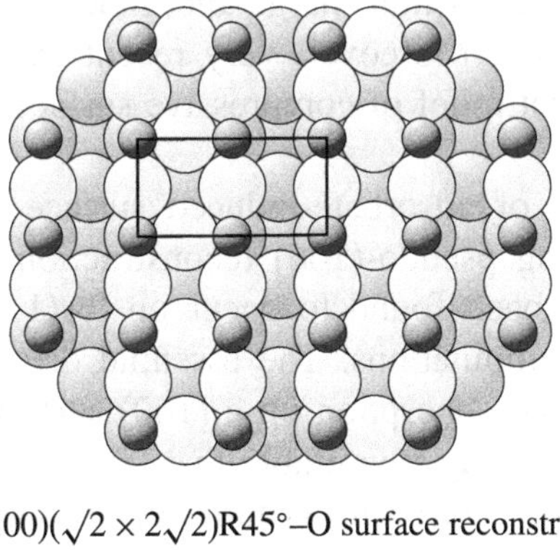

Figure 1.4. Plan view of Cu(100)$(\sqrt{2} \times 2\sqrt{2})R45°$–O surface reconstruction. The outermost layer Cu atoms are shown more lightly shaded than those of the underlying substrate to show more clearly the missing-row structure of this outermost layer. The full lines show the surface unit mesh.

LEED diffraction beam intensities (associated, respectively, with both possible surface phases and with the Cu(100) ($\sqrt{2} \times 2\sqrt{2}$)R45°–O phase alone) under different preparation condition led to the conclusion that no Cu(100)c(2 × 2)–O phase actually exists, but rather under certain conditions the additional diffracted beams of the Cu(100) ($\sqrt{2} \times 2\sqrt{2}$)R45°–O phase might be difficult to see if the long-range ordering is poor [46,47]. More recently, however, STM studies [48,49] appear to have resolved this issue, in that it appears that for low oxygen coverages the adsorbate forms extremely small ordered islands with a very high density of antiphase domain boundaries, such that there is little true long-range order but within the islands the periodicity is c(2 × 2). Such a surface phase is particularly easy to reconcile with earlier reports of a so-called 'four-spot' LEED pattern [50–52], with the general appearance of a c(2 × 2) phase but with splitting of the ½-order diffracted beams which could be attributed to antiphase domain boundaries.

Quantitative local structure determinations using the PhD technique do indicate a very significant difference between the local geometry of the chemisorbed O atoms at low and high coverage, consistent with the view that there is a low-coverage precursor state consistent with a 'local c(2 × 2)' phase. In particular, these experiments show that, while in the reconstructed ($\sqrt{2} \times 2\sqrt{2}$)R45°–O phase the O atoms lie only about 0.1 Å above the outermost layer Cu atoms, in the low coverage phase these atoms lie about 0.7 Å above (unreconstructed) four-fold coordinated hollow sites [43,44]. There have been a number of theoretical studies of the Cu(100)/O system aimed at elucidating the mechanism for the reconstruction; total energy calculations do show the reconstructed phase is energetically preferred, although the energy difference is surprisingly small (values of 0.3 eV [53] and 0.17 eV [54] have been found). One proposed mechanism relates to the large effective charge, anticipated to be on the adsorbed O atoms in the c(2 × 2) phase, which leads to a large surface dipole moment when the O atom is significantly above the surface. This may drive a phonon instability [55], but more recent work has cast some doubt on the true value of this charge and, indeed, even the proper definition of this effective charge [54]. Moreover, very recent experimental and theoretical work supports the view that relief of compressive surface stress plays a role in this reconstruction [56].

A quite different class of adsorbate-induced surface reconstruction is formed by those systems involving pseudo-(100) reconstruction of the outermost atomic layer; this behaviour has been found to occur on fcc(111) and (110) surfaces in several metal/adsorbate combinations. The essential driving force for such reconstructions appears to be that adsorption on a (100) surface (typically in a c(2 × 2) arrangement) is so energetically favourable that, even on a surface with a different lateral periodicity (and point-group symmetry), reconstruction of the outermost layer or layers to form this (100)-like geometry is favoured. This must occur despite the introduction of strain energy at the interface between the substrate and the reconstructed layer(s). Most of the examples of this phenomenon occur on fcc(111) surfaces, but atomic N adsorption on both Cu(110) and Ni(110) appears to involve essentially the same type of reconstruction [57]. In most cases these reconstructions are induced by atomic adsorbates, but Figure 1.5 shows a schematic diagram of the pseudo-(100) phase formed by methanethiolate, CH_3S–, on Cu(111). For simplicity the methanethiolate species, formed by reaction of the surface with methanethiol (CH_3SH) or dimethyl disulphide ($(CH_3S)_2$), are simply represented in the figure by the S atoms alone. The fact that this adsorbate induces a major density-lowering reconstruction of the Cu(111) surface was established more than 15 years ago by SEXAFS and NIXSW [58], but it is rather more recently that STM and qualitative LEED have shown that this reconstruction creates a near-square arrangement of surface Cu atoms [59], while MEIS has provided direct evidence of the existence of the reconstructed outermost single Cu layer [60].

Gaining a proper understanding of these pseudo-(100) reconstructions is difficult, particularly on the (111) surfaces, because there is a clear mismatch of symmetry (3-fold for the underlying bulk, nominal four-fold for the reconstructed layer) which, combined with the relatively atomically smooth character of the close-packed fcc(111) surface, means that these reconstructions are either incommensurate with the substrate or have a very long-range commensuration. In the case of the Cu(111)/CH_3S-system a commensurate $\begin{bmatrix} 4 & 3 \\ -1 & 3 \end{bmatrix}$ mesh has been proposed, but even if this structure is truly commensurate the surface unit mesh is 15 times the area of the substrate unit mesh, and has five different thiolate/substrate registry sites within the structure (Figure 1.5). Moreover, this suggested commensurate mesh is quite small

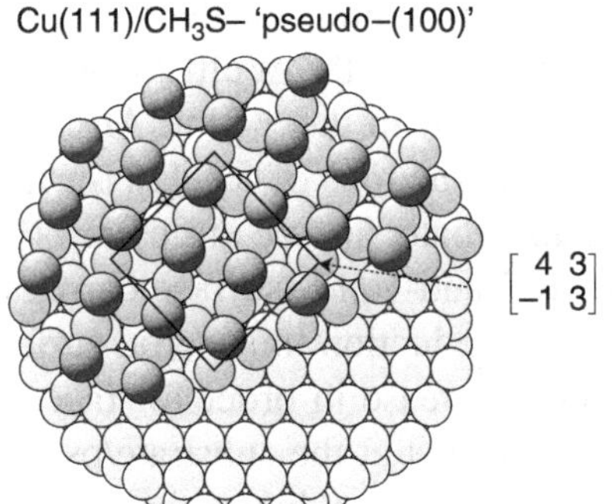

Figure 1.5. Plan view of Cu(111)/CH_3S- pseudo-(100) reconstruction assuming a commensurate $\begin{bmatrix} 4 & 3 \\ -1 & 3 \end{bmatrix}$ registry of the overlayer and substrate. The methylthiolate species are represented by the S head-group atoms alone, shown as the darkest spheres. The Cu atoms of the reconstructed pseudo-(100) layer are shown more darkly shaded than those of the underlying substrate. For clarity the reconstructed overlayer has been omitted from the lower right-hand side of the diagram, exposing the outermost unreconstructed Cu(111) layer.

compared with many other systems, in part because the Cu−Cu distance in this particular pseudo-(100) reconstructed layer is significantly larger (~15% in both principal directions) than the ideal (100) surface; one might then surmise that the actual spacing is less constrained within the layer and more susceptible to accommodate to the corrugations of the underlying substrate potential. By contrast, in the case of the pseudo-(100) reconstruction of Cu(111) induced by atomic nitrogen for which a $(25 \times 7\sqrt{3})$rect. commensurate phase has been proposed, the Cu–Cu distances are 4% larger than that in a Cu(100) surface in one direction and almost identical to the ideal (100) surface in the other [61,62]. Within this commensurate unit mesh are 144 different reconstructed Cu atoms and 72 N atoms, most of which could be in different registry sites. With this degree of complexity it is easy to see why there are no fully quantitative surface structure determinations of any of these systems, and indeed no *ab initio* total energy calculations. The rationale for the existence of these surface phases given above, however reasonable it may seem, must therefore formally remain as speculation.

More generally, the existence of these surface phases may be related to the superficially quite different adsorbate-induced surface restructuring phenomenon of faceting [57]. Here, too, the driving force is the energetic favourability of chemisorption onto one or more specific orientations of a substrate, but the restructuring involves a modification of the surface morphology into a 'hill-and-valley' corrugated structure in which at least one of the sides of the 'hills' comprises planar facets of the preferred orientation. This basic phenomenon of surface faceting has a long history and was traditionally observed on a microscopic scale with conventional optical microscopy. The surfaces present on the restructured faceted surfaces are those to be seen on the equilibrium shape of a small particle of the substrate material. This shape is related to the polar diagram of the surface free energy as a function of orientation (the so-called γ-plot) by the Wulff theorem, and facet planes occur at orientations corresponding to cusps in the γ-plot. An interesting unanswered question is why pseudo-(100) reconstructions of fcc(111)surfaces occur in preference to (100) faceting if the (100)/adsorbate surface structure is so energetically favoured. Of course, the creation of a hill-and-valley surface corrugation leads to a significant increase in surface area, so the reduction in specific surface free energy must more than compensate for this area increase in order for faceting to occur. Evidently, in these systems, the strain energy cost at the substrate/pseudo-(100) layer interface is less than the energy increase in faceting due to the increase in area.

Nevertheless, adsorbate-induced faceting of surfaces does occur under certain circumstances. An interesting and rather well-studied case is that of faceting of Cu(100) vicinal surfaces in an <010> zone to (410) as a result of atomic oxygen adsorption. The Cu(410) surface comprises (100) terraces just three atomic rows wide separated by a single atomic step, and it now seems rather well established that oxygen atoms decorate these steps in sites equivalent to those on the edges of the

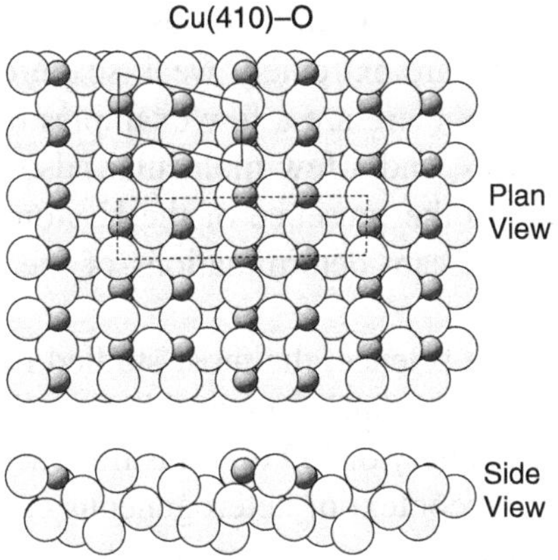

Figure 1.6. Plan and side views of the structure of the Cu(410)–O surface phase. The full and dashed lines show respectively the primitive and centred rectangular surface unit meshes (that are unchanged from those of the clean surface by the O adsorption).

missing rows in the Cu(100)($\sqrt{2} \times 2\sqrt{2}$)R45°–O surface reconstruction. Curiously, however, the Cu(410)-O surface also has a similar coverage of additional oxygen atoms coadsorbed in four-fold coordinated hollow sites in the middle of the (100) terraces (see Figure 1.6) – the very sites which appear to be unstable at high coverage on the extended Cu(100) surface. Notice, though, that if surface stress relief plays a role in destabilising the Cu(100) surface with these adsorption sites occupied, as has been suggested [56], the narrow terrace width of the (410) surface means that local lateral strain offers an alternative means of stress relief on this vicinal surface. Density functional theory (DFT) *ab initio* calculations and SXRD experiments [63] support the structure of Figure 1.6, but there has been no systematic theoretical study to understand why the (410) surface, with this particular terrace width, is so strongly favoured over other vicinal orientations for quite a wide range of average initial surface orientations.

4. Molecular adsorbates – local sites, orientations and intramolecular bondlengths

4.1. General issues and the case of CO on metals

Molecular adsorbates provide an increasingly complex set of challenges to surface structure determination, the complexity growing steeply with the number of atoms within the molecule. A special problem is also presented by hydrogen atoms which are present in the majority of chemically interesting molecules, but to which all truly quantitative experimental structural methods are, to a greater or lesser extent, 'blind'. In particular, H atoms are extremely weak scatterers of electrons and have no electronic core level. While there are a few examples of structure determination of atomic H adsorption phases, and a few molecular adsorbate structural studies in which marginal effects due to the presence of the H atoms have been established, there are no structurally significant determinations of the H atom positions within adsorbed molecules.

As in many areas of surface science, the most-studied molecular adsorbate structurally is CO; from a structural point of view this is a simple molecule (just two distinct atoms), while the chemistry of CO on transition metal surfaces, in particular, is chemically important in the context of heterogeneous catalysis. CO adsorption on such surfaces also shows some fascinating complexity. In particular, while it almost invariably bonds end-on to the surface through the C atom, it may adopt one-fold (atop), two-fold (bridge) or three- or four-fold (hollow) coordination sites on surfaces, with the relative energetics of these different sites being dependent on the substrate material, the substrate orientation, and the CO coverage. For example, the initial (lower coverage) adsorption site of CO on Ni(100) is atop, while on Ni(110) it is bridge and on Ni(111) it is hollow. Moreover, while CO forms a 0.5 ML coverage $c(4 \times 2)$ ordered phase on several fcc(111) metal surfaces, the local adsorption sites in this phase on Ni(111) and Pd(111) are the two inequivalent three-fold coordinated hollow sites (see the discussion in Section 2.5 and Figure 1.2 [21–24]), whereas on Pt(111) it is the atop and bridge sites which are co-occupied [64–66]. A further manifestation of the subtle energy changes involved in these different bonding sites is seen in the coverage dependence. For example, on Pt(111) and Rh(111), CO adsorbs at low coverage in atop sites only, but as the coverage increases both atop and bridge sites become occupied; indeed, on Ni(100) it appears that initial atop adsorption at low coverage changes to atop and bridge site occupation at intermediate coverage and then to pure bridge site occupation at the highest coverage [67]. Electronic structure measurements for this system have been interpreted as indicating that the energetic differences between these sites are very small, a consequence of increased π-bonding and increased σ-repulsion with increasing substrate-atom coordination number [68] (see also Chapter 2). It is perhaps not surprising, in view of this subtlety, that DFT ab initio calculations have been unable to reproduce all of these effects correctly, with hollow rather than atop sites being favoured at low coverage on both Pt(111) and Rh(111); this theoretical problem has attracted considerable debate and some proposed solutions (e.g., [69–72]).

One structural parameter of potential interest in the adsorption of simple diatomic molecules, such as CO, NO and N_2, is the intramolecular bondlength. Much of the motivation for studying such adsorbates is related to the way adsorption modifies the chemistry of these species as the basis for heterogeneous catalysis. Many such reactions involve scission of the intramolecular bond, and if the adsorption is of the molecular species, one might expect the relatively greater ease of bond scission to be reflected in the molecular precursor by a weakening, and hence a lengthening, of the intramolecular bond. Indeed, quite generally, the formation of the chemisorption bond might be expected to weaken the intramolecular bond. Unfortunately, it seems that the precision achievable in current surface structural techniques is inadequate to detect the small changes (typically only a few hundredths of an Ångström unit) that are predicted theoretically, and essentially all experimental reports indicate intramolecular bondlengths for these adsorbed species to be the same as in the gas phase to within the estimated experimental precision. Each method has its limitations in this regard. In photoelectron diffraction the determination of the intramolecular bondlength is most effective if the location of each constituent atom is determined separately by the photoelectron diffraction obtained from these atoms, but the intramolecular bondlength is then given by the difference in these positions; the difference measurement is then subject to greater errors than those in the location of each atom alone. In LEED, the weak scattering of the low atomic number constituent atoms reduces the precision in determining their location, a problem also present, in a much exaggerated form, for X-ray diffraction. One method that has been used to try to determine the intramolecular bondlength directly is NEXAFS (near-edge X-ray absorption fine structure). The energy of σ-symmetry final-state multiple-scattering shape resonances, relative to the absorption threshold, is related to the interatomic bondlength, and has been suggested to be the basis of a measurement of this bondlength [73]. Determining the exact energies of these broad resonances and the proper threshold energy are not without difficulties, however, nor is it clear that the spectral shape in the energy range of the resonance is wholly determined by the intramolecular scattering. This may be why rather significant ($\sim$0.13 Å) changes in the C$-$O bondlength for coadsorbed CO and alkali metals, indicated by this method, have not been supported by other methods, as described more fully in Section 5.

4.2. Simple hydrocarbons on metals

Perhaps the next simplest molecular adsorbates for which quantitative structural information exists are the unsaturated C_2 hydrocarbons, notably acetylene (ethyne, $HC{\equiv}CH$) and ethylene (ethene, $H_2C{=}CH_2$), adsorbed on a number of metal surfaces (especially, Cu, Ni and Pd), and also on Si(100), studied by LEED, SEXAFS, and PhD. In some systems adsorption of ethylene is accompanied by a surface reaction. In particular, on both Pt(111) [74] and Rh(111) [75] ethylene is converted to an ethylidyne species, $H_3C{-}C{-}$, which bonds to these surfaces through the C atom with the C$-$C axis essentially perpendicular to the surface, in three-fold coordinated hollow sites. In addition, ethylene adsorbed on Ni(111) at low temperature dehydrogenates to produce adsorbed acetylene as the surface is warmed towards room temperature; this particular system actually provided the first example of the use of a 'modern' surface science technique, namely ultraviolet photoelectron spectroscopy, to follow in situ a simple surface reaction on a single-crystal surface [76]. Subsequent use of vibrational spectroscopy [77] cast considerable light on this reaction, and on the adsorption of both C_2 hydrocarbons on several metal surfaces. In particular, these methods indicated not only that the C−C axis is essentially parallel to the surface for both molecules, but also that the adsorption is accompanied by a very significant reduction in the C−C bond order, as reflected by a softening of the C−C stretching vibrational mode. On Ni(111), e.g., this led to the suggestion that the C−C bond order in adsorbed ethylene is reduced from two to about one, and that in acetylene the reduction was from three to about 1.5. Vibrational spectroscopy was also interpreted as indicating that ethylene interacts with the surface by di-σ bonding, whereas acetylene interacts through π bonding [78]. Of course, these assignments do not formally define the adsorption geometry on the surface, although local qualitative structural models proposed did, for these cases, prove to be supported by true structural studies. In particular, acetylene on Ni(111) occupies a cross-bridge site such that the two C atoms are, respectively, in fcc and hcp hollow sites [79,80] (directly above third and second layer Ni atoms, respectively). Ethylene bonds in a parallel bridge geometry such that the two C atoms are in equivalent off-atop sites [80] (Figure 1.7). Notice that one consequence of the occupation of the two inequivalent hollow sites on the Ni(111) surface by the two C atoms in adsorbed acetylene is that the C−C bondlength is 1.44 Å, very much longer than the value of 1.21 Å in the gas phase molecule. Even allowing for the rather modest precision of this experimental structure determination by PhD (±0.15 Å), this bondlength expansion is significant. Indeed, the C−C bondlength for adsorbed ethylene (1.60 ± 0.18 Å) is also significantly larger than the gas phase value (1.34 Å).

These structural data therefore provide independent support for the reduced C−C bond order values of the adsorbed molecules indicated by vibrational spectroscopy.

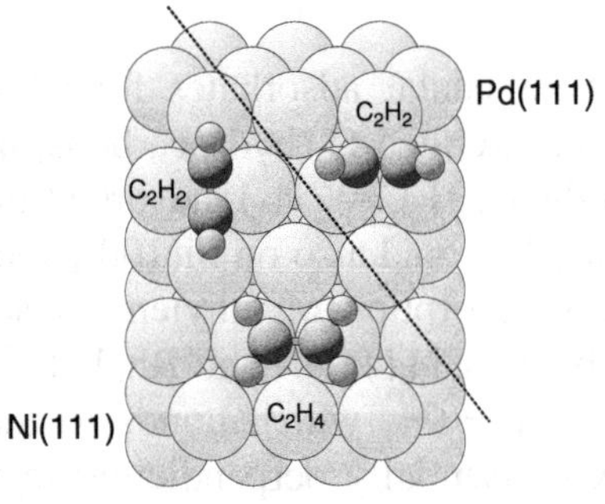

Figure 1.7. Plan view of a fcc(111) surface, showing the acetylene and ethylene adsorption geometries on Ni(111), and that of acetylene on Pd(111). The H atom positions are schematic only and have not been determined experimentally.

The structural behaviour of acetylene on Cu(111) is remarkably similar to that on Ni(111); here too, the results from vibrational spectroscopy [81] (and Auger electron spectroscopy [82]) were interpreted as indicating a rehybridisation of the C−C bonding to something more like the sp^3 bonding of singly bonded gas-phase ethane (H_3C-CH_3). Moreover, the adsorption geometry is found to be [83] the same cross-bridge site as on Ni(111), resulting in a C−C bondlength of 1.48 ± 0.10 Å, quite close to that in gas-phase ethane (1.54 Å), and much larger than in gas-phase acetylene (1.21 Å). The fact that this large extension of the C−C bond in adsorbed acetylene on Ni(111) and Cu(111) coincides with the separation of the two hollow sites on these surfaces initially led to the suggestion that the strong rehybridisation on these surfaces is, at least in part, structural in origin, but a wider survey of vibrational data from other systems [78] suggests the correct interpretation is more complex than this. Of course, Cu is generally far less reactive with hydrocarbons than Ni, and the dehydrogenation reaction of adsorbed ethylene on Ni(111) does not occur on Cu(111). Acetylene also has a rather low desorption temperature on Cu(111) (323 K), so at first sight the very strong rehybridisation and C−C bond lengthening on Cu(111) is puzzling. Why should a weakly adsorbed molecule be strongly modified by this adsorption? The solution to this puzzle appears to be found in the results of some *ab initio* cluster calculations of this adsorption system by Hermann and Witko [84]. These calculations confirm that the minimum energy adsorption structure corresponds to that found experimentally, with the long C−C bond, but also show that the low desorption energy is a consequence of quite strong bonding, offset by the energy cost of the intramolecular rehybridisation. In effect, therefore, the desorption temperature is a rather misleading indicator of the strength of the molecule-substrate electronic interaction. This conclusion also emerges from a more recent theoretical investigation of this system [85], although the detailed interpretation is somewhat different, as discussed in Chapter 2. This theme, of competing effects leading to a potentially misleading desorption temperature, is one to which we will return in a different guise in Section 5.

While the cross-bridge local adsorption site of acetylene on Cu(111) and Ni(111) is essentially identical, on the structurally similar Pd(111) surface the molecule adsorbs in a hollow site, as illustrated in Figure 1.7. This different adsorption site, first proposed on the basis of quantitative evaluation of NEXAFS spectra [86] and subsequently confirmed by PhD [87], provides a rationale for the significantly different behaviour seen in vibrational spectroscopy [78] for this system.

Following along the route of increasing complexity, or at least molecular weight, the other hydrocarbon that has attracted the largest number of structural studies is that of benzene, C_6H_6. We have remarked above on the first use of a modern UHV single crystal surface science study, using ultra-violet photoemission, to follow the dehydrogenation of adsorbed ethylene to acetylene on Ni(111). In fact this same paper, by Demuth and Eastman [76], also used the same method to infer, through the relative shift of the energy of the occupied π-orbitals, that benzene must adsorb on this surface with the molecular plane essentially parallel to the surface. Subsequently quite a number of quantitative structural studies using LEED and PhD have arrived at this same conclusion for adsorption on a range of metal surfaces. These quantitative studies have also addressed the interesting question of whether there is any change in the C−C bondlengths resulting from adsorption. Specifically, is there an increase of the average C−C distances, similar to that seen for the unsaturated C_2 species, and are there variations in the C−C distances around the benzene ring to reflect the symmetry of the adsorption site?

From a chemical view these are reasonably distinct questions, but from the point of view of structure determination they are somewhat related. If one allows the full range of symmetrically plausible distortions to be included in the modelling, there is a significant increase in the number of structural variables. The effect of these extra parameters on the quality of the theory/experiment fit may be partially coupled: i.e., a degradation of the fit, by changing one parameter, may be rectified by adjusting a second parameter. The net effect of this is that the estimated precision of each of several parameters is typically worse than the single parameter of a more symmetric adsorbate. For example, in a study of benzene adsorption on Ni(110) (without long-range order), analysis of PhD data on the assumption that the benzene ring is fully symmetric led to an optimum value of the C−C bondlengths of 1.45 ± 0.03 Å, significantly larger than the value in the gas phase molecule (1.39 Å) [88]. However, in the same study two different types of ring distortion were considered, in one of which a pair of opposite C−C bonds are of a different length from the remaining four C−C bonds. This is a feature of both the 'quinoid' distortion and a less-distorted 'H-flip' structure favoured by theoretical total energy calculations for benzene adsorbed on Cu(110) [85], although it is the less-distorted structure which seems to be supported by X-ray emission and absorption spectra on this surface [89]. However, on Ni(110) the PhD experiment found zero difference in the C−C bondlengths to be optimal, but the precision of the individual values was degraded to ± 0.05 Å, meaning that even the ring expansion became marginally significant. This precision is also inadequate to exclude either of the distortions considered in the theoretical and electron spectroscopy studies on Cu(110).

Of course, the most interesting type of ring distortion is the three-fold symmetric Kekulé distortion with alternating C−C bondlengths, a structure most readily reconciled with three-fold symmetric substrates. In fact, the great majority of experimental structural studies of adsorbed benzene have been on such surfaces (fcc(111) and hcp(0001)), and generally the best-fit structures have shown Kekulé distortion. However, the precision estimates of these studies are such that even the ring expansions of the best-fit structures are not formally significant. For example, LEED studies of benzene adsorption found ring radii in the distorted ring of 1.43 and 1.46 Å on Ru(0001) [90] and 1.48 and 1.50 Å on Ni(111) [91], but the error estimates of ±0.10 and ±0.15 Å, respectively, mean that neither the distortions, nor the expansions, are formally significant. Similarly, a PhD study of benzene adsorbed on Ni(111) [92] found alternating C−C bondlengths of 1.40 and 1.44 Å at low coverage and 1.40 and 1.46 Å at high coverage, but again with estimated errors of ±0.10 Å or larger. The reality of such distortions is thus as yet unproved. On the other hand, while all of these earlier measured expansions (in analyses which allow ring distortion) lie within the error estimates, the consistent trend to find best-fit structures with a net expansion strongly suggests that a real adsorption-induced ring expansion does occur, and indeed perhaps implies that the experimental error estimates may be unnecessarily pessimistic.

We may note that expansion of the benzene ring associated with such adsorption is fully compatible with simple molecular orbital considerations. On the assumption that the bonding scheme is similar to that for benzene ligands in organometallic complexes, the synergic Dewar-Chatt-Duncanson model [93] predicts that there is σ-donation from the filled π-bonding orbital of benzene to the metal and π-backbonding from filled metal d-states into the antibonding π^*-orbital of benzene. This lowers the C−C bond order and increases the C−C bondlength. For the 'classic' benzene complex $Cr(C_6H_6)_2$ this increase in C−C distance is 0.03 Å [94], somewhat smaller than the changes indicated in the various surface studies, but perhaps highlighting the need for extreme precision to solve this problem convincingly.

One interesting feature of the Ni(111)/benzene adsorption system is that both the azimuthal orientation of the benzene ring and the adsorption site change with increasing coverage, accompanied by the formation of an ordered $(\sqrt{7} \times \sqrt{7})R19°$ phase at the 'high' coverage limit of 1/7 ML. The evidence for the change in azimuthal orientation was first inferred from angle-resolved (valence) photoemission measurements [95,96], but the full structure determination of the two local geometries emerged from the PhD study mentioned above [92]. Figure 1.8 shows the resulting structures in a diagram that includes schematically the (expected) location of the H atoms. This simple picture gives some basis for understanding the preferred azimuthal orientation in the high-coverage ordered phase that is simply based on minimising steric hindrance as the coverage is increased. If the azimuthal orientation of the low-coverage adsorption geometry were preserved in the ordered $(\sqrt{7} \times \sqrt{7})R19°$ phase, it seems that the H atoms of adjacent molecules would overlap, while this is no longer true in the modified orientation. Of course, this rationale provides no help in understanding why this is not the preferred orientation at low coverage, nor why the change in local adsorption site between bridge and hollow occurs. As remarked above, coverage-dependent changes in adsorption site are well-known for CO on metal surfaces, but this seems to be the only example found so far for adsorbed benzene.

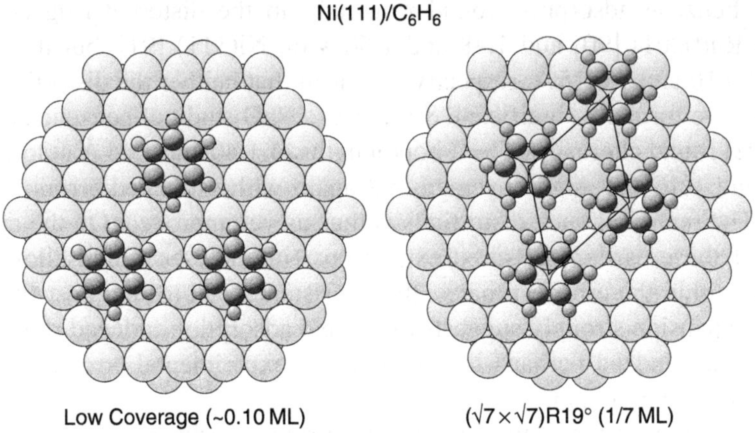

Figure 1.8. Plan view of the structure of the Ni(111) surface with adsorbed benzene, showing the local adsorption geometry at low (~0.10) coverage, and the local and long-range ordering geometry in the slightly higher coverage $(\sqrt{7} \times \sqrt{7})R19°$ phase. The H atom positions are schematic only and have not been determined experimentally.

4.3. Carboxylates on metals

Of the various other molecular adsorbates that have been subjected to rather complete structural studies, there are quite a number that have in common the property of bonding to the surface through a carboxylate, COO−, group. The simplest of these is the formate species, HCOO−, formed when a surface is exposed to formic acid (HCOOH) and the acid hydrogen atom is detached through interaction with the surface, generally combining with a second H atom from another molecule and desorbing from the surface as H_2. The formate species was first studied structurally on Cu(100) using O K-edge NEXAFS and SEXAFS [97,98]; the NEXAFS data showed clearly that the molecular plane is perpendicular to the surface, consistent with expectation that bonding to the surface is achieved through the two equivalent carboxylate O atoms. However, the interpretation of the SEXAFS data led to the surprising conclusion that the molecule occupies a cross-bridging site (with the O−O direction perpendicular to the Cu−Cu nearest neighbour direction) such that the two O atoms occupy off-hollow sites on the surface (see Figure 1.9). This geometry, with a surprisingly large Cu−O nearest neighbour distance of 2.30 Å, was heralded as the first example of a new type of molecule-surface bond; i.e., that the bonding to an extended metal surface was fundamentally different from the bonding to metal atoms in metal coordination compounds. Subsequent studies, first using SEXAFS to investigate the formate species on Cu(110) and reconsidering the interpretation of the initial SEXAFS data [99,100], but then using PhD to investigate both of these adsorption systems and reconsidering the SEXAFS data yet again [101], finally

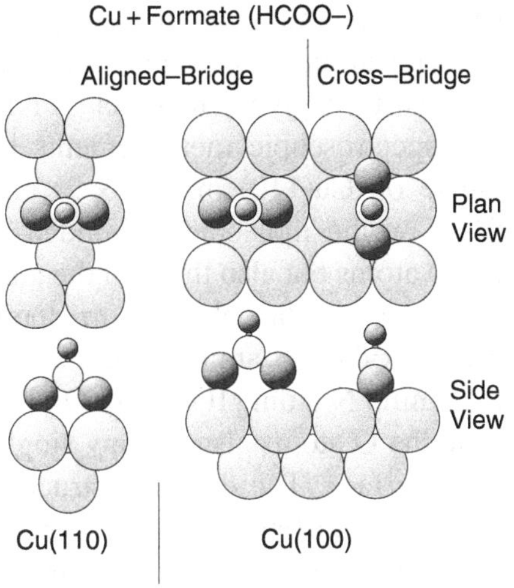

Figure 1.9. The local aligned-bridge adsorption sites of the formate (HCOO–) species on Cu(110) and Cu(100). Also shown is the cross-bridge site on Cu(100) originally proposed as a 'new type of surface bond' but subsequently shown to be incorrect.

resolved the structures. Specifically, on both surfaces the formate adopts an aligned bridge geometry such that each carboxylate O atom is near atop one of the two nearest neighbour surface Cu atoms, with Cu−O bondlengths of a little under 2.0 Å (Figure 1.9). The error in the SEXAFS analysis appears to have stemmed from mistaking intramolecular scattering from O atoms as due to substrate scattering from Cu atoms. The final structural conclusion, of course, is that the geometry of the formate-Cu surface bonding on both (100) and (110) surfaces is actually very much like that one might expect from coordination chemistry, and is *not* some new type of bonding. This interpretation of the adsorption geometry of formate on Cu(100) was further supported by the results of cluster calculations performed in the same year [102].

This local adsorption geometry for formate on Cu(100) and Cu(110), with the molecular plane perpendicular to the surface and bonding through the two carboxylate O atoms in near-atop sites, is also seen in other simple carboxylate species adsorbed on Cu(110), notably acetate (CH_3COO-) [103] and benzoate (C_6H_5COO-) [104] formed, respectively, by exposure to acetic acid and benzoic acid. Relatively recent X-ray spectroscopy measurements combined with theoretical calculations provide further information on the bonding of formate and acetate on Cu(110) [105].

A significantly more complex group of molecular carboxylic acids are the amino acids, and there have recently been some detailed structural studies of the simplest of these, glycine and alanine, on Cu surfaces. Glycine, NH_2CH_2COOH, is the simplest amino acid, containing the two key ingredients, namely the carboxylic acid (COOH) and amino (NH_2) groups. On deposition onto Cu surfaces the acid hydrogen is detached and a surface glycinate species, NH_2CH_2COO, is formed. As originally inferred from spectroscopic measurements [106,107], detailed PhD structural studies of glycinate on Cu(110) in a (3 × 2) phase and on Cu(100) in a (2 × 4) phase [108,109] show the molecule is bonded to the surface not only through the two carboxylate O atoms but also through the amino N atom. Vibrational spectroscopy provides evidence that other phases correspond to different bonding, in one phase through the carboxylate O atoms alone, and in another through one carboxylate O atom and the amino N atom [106,110]. On both surfaces the N atoms occupy near-atop sites, while the O atoms also occupy singly coordinated sites with respect to the underlying Cu surface. The different structures of the two surfaces, however, leads to some inevitable differences in the local bonding geometry; for carboxylate O bonding alone, as in formate, the local site on Cu(110) and Cu(100) is the same, bonding to two nearest-neighbour Cu atoms aligned along a <110>-type direction that is available in both surfaces. When bonding occurs through three points in the glycinate species, on the other hand, there is an inevitable mismatch of the intramolecular distances with the Cu–Cu spacings on the two different surfaces with square and rectangular surface unit meshes. The local structures of glycinate on these two substrates are shown in Figure 1.10. This figure, which includes the locations of the H atoms, is actually based on the results of DFT total energy calculations [1]. The experimental PhD studies of these systems [108,109] (which predate these calculations) determined only the locations of the N and O atoms (through N 1s and O 1s PhD spectra) and not the locations of the C (or H) atoms. Moreover, while these PhD studies allow one to determine the size and azimuthal directions of off-atop positions, it is not possible to distinguish between symmetrically equivalent directions, so the relative locations of each of the N and

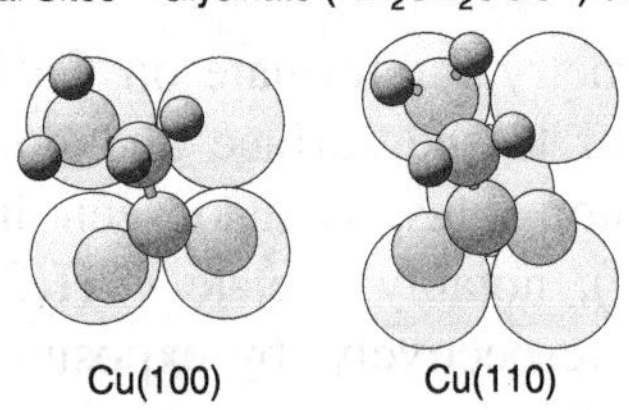

Figure 1.10. The local adsorption geometries of glycinate on Cu(100) and Cu(110) showing the near atop bonding of the amino N atom (top left in each diagram), and the near-atop or off-atop bonding of the two carboxylate O atoms at the bottom. The complete geometries shown here are as obtained from one particular set of DFT calculations [111], but are consistent with earlier experimental PhD studies of the N and O local adsorption sites and with earlier DFT calculations (e.g., [116] and earlier work by this group).

two O atoms within the molecule are less certain. Nevertheless, the basic geometry adopted by the adsorbed molecule is clear from the experiments, and is generally consistent with the more complete proposed local structure of Figure 1.10 offered by the DFT calculations. On Cu(100) not only the N atoms, but also the two O atoms are very close to atop sites. On the Cu(110) surface, with the larger Cu−Cu spacing in the <100>-type azimuth (by a factor of $\sqrt{2}$), the O atoms are far more significantly displaced from atop sites, but retain one-fold coordination.

While the PhD measurements determine only the *local* adsorption geometry, the fact that both the Cu(110)(3 × 2) and Cu(100)(2 × 4) glycinate phases show missing diffracted beams in LEED characteristic of glide symmetry means that one can infer the probable long-range ordering as shown in Figure 1.11 [109,112]. An interesting feature of these structures is that there are two inequivalent forms of the glycinate species on the surface that are related by mirror symmetry (or, within the ordered structure, by the glide symmetry); in Figure 1.11 one species has the amino group on the right-hand side, the other on the left. This is a manifestation of the general phenomenon of chirality – the existence of left- and right-handed forms of a molecule that are inequivalent and may have distinctly different chemical properties. In fact glycine itself (and the glycinate species too) is not intrinsically chiral, but when attached to the Cu surface in this way the two different handed forms – the two different enantiomers – are distinct. Of course, because the starting molecule is not chiral, the resulting surface must, on average, be covered with equal numbers of each enantiomer – a so-called racemic mixture. In Figure 1.11, this mixing occurs within each ordered domain of the adsorbate – i.e., the domains are heterochiral. This model is consistent with the space group symmetry inferred from the characteristic

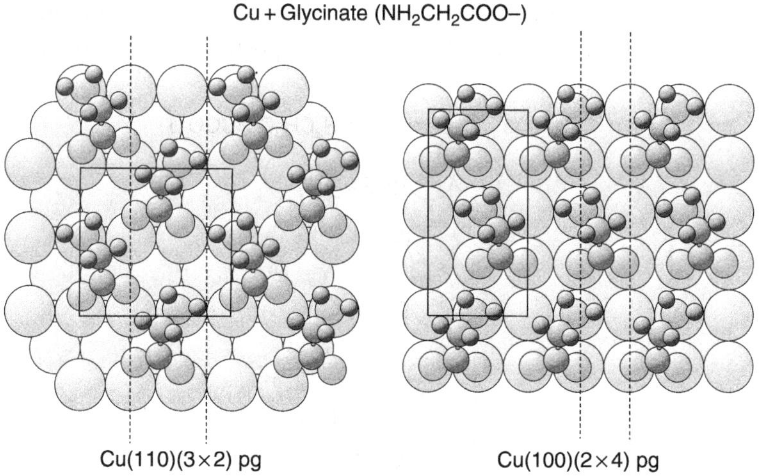

Figure 1.11. Heterochiral ordered structures of glycinate on Cu(110) and Cu(100). The full lines show the primitive unit mesh in each case, while dashed lines show the location of glide symmetry lines.

beam absences in LEED, although there have been suggestions, mainly based on STM studies [113–115], that homochiral domains (that lack the glide symmetry) may coexist on the surface; on Cu(100), coexistence of homochiral domains of a c(2 × 4) cannot be excluded on the basis of the LEED patterns alone [112]. The nature of the intermolecular hydrogen bonding associated with the formation of these ordered phases on Cu(110) has been investigated theoretically, including detailed comparisons with the results of near-edge X-ray absorption spectroscopy, but it seems the differences in these spectra to be expected for the heterochiral and homochiral domains are too small to allow them to be distinguished [116].

Of course, from the point of view of practical surface chemistry, the adsorption of a species that is intrinsically chiral is of far greater interest. In the pharmaceutical industry, in particular, there is a great need to produce specific chiral molecules of a single enantiomer, yet there is a dearth of enantioselective heterogeneous catalysts and most of this production uses the less attractive homogeneous catalysts. The next simplest amino acid beyond glycine, alanine, $NH_2CH_3C^*HCOOH$, is chiral, the chiral centre being at the C^* atom that is bonded to four different species, the amino NH_2, the methyl CH_3, the carboxylic acid COOH and a single H atom; one of the two equivalent H atoms of glycine is replaced by the methyl group in alanine. Interestingly, a single enantiomer of alanine(at)e also appears to form a (3 × 2) phase on Cu(110), although in this case the structure cannot be the same as the (3 × 2) heterochiral phase shown on Figure 1.11 for glycinate, because the structure *must* be homochiral. This means that the two molecules per unit mesh must adopt somewhat different local geometries. The solution appears to be as shown in Figure 1.12. For this system too, N 1s and O 1s PhD data [117] provided information on the local near-atop and off-atop locations of the N and O atoms, respectively, but in view of the large number of structural parameters involved in the four inequivalent O adsorption sites the final experimental structural optimisation involved adjustments

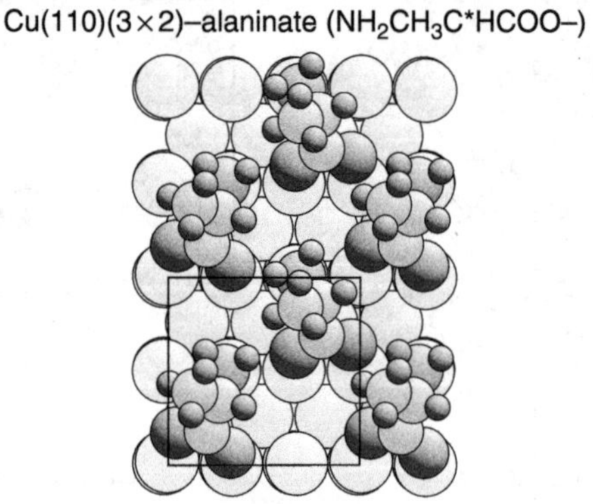

Figure 1.12. The (3 × 2) ordered structure of alaninate on Cu(110) as given by DFT calculations [118] and found to be largely consistent with PhD experiments [117].

of the structural parameter values yielded by an independent DFT study [111,118]. In particular, this optimisation of the PhD best-fit structure led to changes in the chemisorption bondlengths, discussed more fully in Section 5.

Another chiral adsorbate, also a carboxylic acid, that has attracted particular attention, is that of tartaric acid, HOOCHC*OHHC*OHCOOH, especially on Cu(110). This contains two chiral centres at the C* atoms that are each bonded to the carboxylic acid COOH group, a hydroxyl OH species, a H atom, and the other C* atom. Notice that because there are two chiral centres with the same bonding species, there are actually three different enantiomers of the molecule, namely, *R,R*, *S,S* and *S,R* where the *R* and *S* labels define the handedness of each chiral centre (see, e.g., Ref. [110] which also reviews work on this and other chiral species on surfaces). Because the *S,R* enantiomer contains identical elements of opposite chirality, this species is achiral, behaving like a racemic mixture of the two chiral enantiomers of a molecule with a single chiral centre; this enantiomer is also referred to as *m*- or *meso*-tartaric acid. So far there have been no experimental structure determinations of the adsorption site, although it is generally assumed that adsorption is again achieved through the (deprotonated) carboxylic acid O atoms in near-atop sites on the Cu substrate.

Vibrational spectroscopy has identified distinct phases as a function of coverage and preparation temperature that involve bonding through either one or both of the (deprotonated) carboxylic acid groups, leading to so-called monotartrate and bitartrate phases [110,119]. Figure 1.13 shows schematically the two different chiral enantiomers of the resulting bitartrate species, together with two different possible

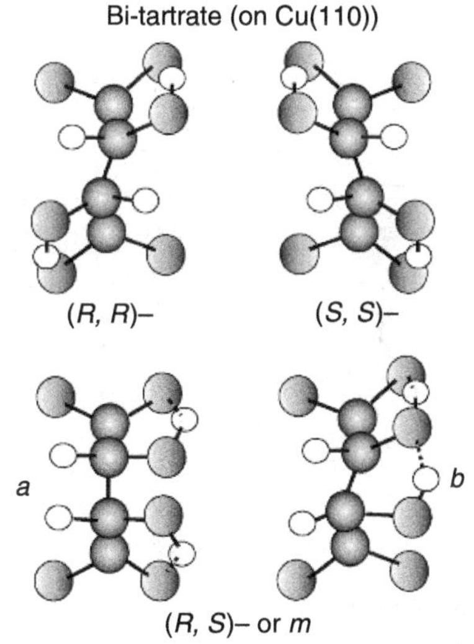

Figure 1.13. Schematic diagram of different forms of the bitartrate species as adsorbed on Cu(110).

representations of the achiral *R,S*-bitartrate to be discussed further below. In the case of the bitartrate phase of the chiral enantiomers an interesting finding is that the long-range ordering of the surface is itself chiral; i.e., the surface unit mesh lacks a mirror plane and, because no mirror-equivalent domains are possible, the whole surface diffraction pattern also displays this chirality. The upper part of Figure 1.14 shows the implied ordering into a $\begin{pmatrix} 9 & 0 \\ 1 & 2 \end{pmatrix}$ unit mesh of the *R,R* form of the molecule. Notice that the long-range ordering is clearly defined by the LEED pattern, but the detailed ordering of the molecules within the surface unit mesh is inferred from STM images (while the local adsorption site is simply assumed to involve carboxylate O atoms in near atop sites, as established experimentally for simpler carboxylate species). However, the qualitative character of distortions in the internal conformation of the molecule on the surface as predicted by DFT calculations [120], which influences the local intermolecular ordering, has received experimental confirmation from a (forward-scattering) X-ray photoelectron diffraction (XPD) study [121]. Of especial relevance for surface chemistry is the implication that there are bare chiral stripes on the surface between the triple rows of bitartrate species, and such regions could allow enantioselective chemistry of other molecules to take place. This result thus provides a potential rationale for the known ability of tartaric acid to produce chiral modification of real catalysis [110]. Evidently the chiral nature of the bitartrate on the surface is generating local ordering that reflects this chirality, a process generally believed to occur through the influence of intermolecular hydrogen bonding (as is also implicated in the ordering of glycinate and alaninate on these surfaces). It is therefore not surprising that the *S,S* enantiomer (which is a mirror image of the *R,R*

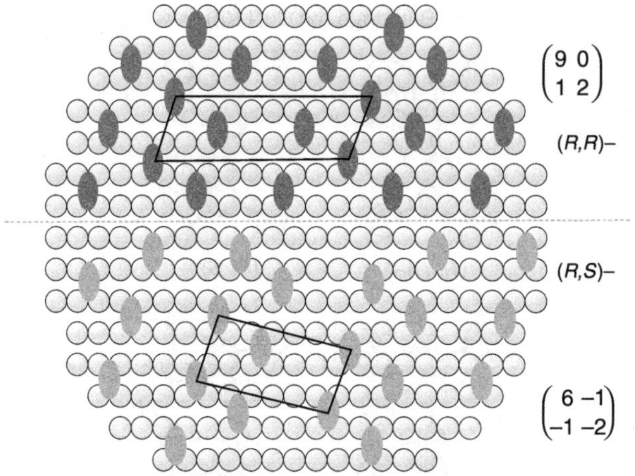

Figure 1.14. Schematic plan view of the outermost layer of Cu(110) showing the proposed adsorption structures for the ordered bitartrate phases, $\begin{pmatrix} 9 & 0 \\ 1 & 2 \end{pmatrix}$ and $\begin{pmatrix} 6 & -1 \\ -1 & -2 \end{pmatrix}$, derived from (*R,R*)- and (*R,S*)-forms of tartaric acid, respectively. The molecular adsorbates are represented by shaped ellipses.

species) leads to a $\begin{pmatrix} 9 & 0 \\ -1 & 2 \end{pmatrix}$ periodic structure, the mirror image of the $\begin{pmatrix} 9 & 0 \\ 1 & 2 \end{pmatrix}$ ordering. More recently, however, the ordering of the *R,S*-bitartrate phase on Cu(110) has been investigated by qualitative LEED [122]. As expected, the LEED pattern is not chiral. However, it appears to be described by the coexistence of two mirror-related domains of periodicity $\begin{pmatrix} 6 & -1 \\ -1 & -2 \end{pmatrix}$ and $\begin{pmatrix} 6 & 1 \\ -1 & 2 \end{pmatrix}$, each of which *is* chiral. The implication is that, like the glycinate species, adsorption of the achiral species on the surface has produced a chiral adsorbate. However, unlike glycinate, in which the domains appear to be heterochiral, and the (3×2) ordering certainly does not reflect the chirality, in the case of *R,S*-bitartrate the domains are homochiral and *do* reflect the local chirality. The inferred structure of one of these domains is shown schematically in the lower half of Figure 1.14. What is superficially surprising is that the *R,S*-bitartrate does reflect the chirality in this way, as one might anticipate that the intramolecular hydrogen bonding in this species, shown by the dashed lines in Figure 1.13 for this species, would retain the full symmetry as shown in representation (a). It appears, however, that this hydrogen bonding is asymmetric (representation (b), Figure 1.13) [120] and it is this asymmetry which influences the ordering.

Clearly these more complex molecular adsorbates represent a very considerable challenge for quantitative experimental structure determination, particularly if subtle molecular distortions are important, as appears to be the case in the tartaric acid system. PhD has the ability to determine the location of many of the constituent atoms in a largely independent fashion, yet this ability also leads to ambiguity in the relative positions of the different atoms within the molecule. XPD can provide more (but still incomplete) information on this matter of internal conformation. In the case of long-range ordered phases, as for the tartaric acid systems, conventional diffraction methods such as LEED and SXRD have the potential to obtain the complete structure, yet the number of structural variables to be solved independently is large, and even these methods are essentially insensitive to the location of the weakly-scattering H atoms. Nevertheless, SXRD, at least, does seem to have met with some success in the study of complex systems of this kind, such as thiouracil on Ag(111) [123]. A rather detailed picture of the structure of the much larger molecule quaterthiophene, on the same surface, has also been obtained using a combination of SXRD to gain information on the internal conformation [124] combined with NIXSW to gain information on the chemisorption bonding distance [125]. It would certainly be of considerable interest to be able to put some of the structural inferences for these chiral adsorbate systems on a firmer footing through further experimental structural studies of this kind.

4.4. Other substrates: molecules on Si

So far in this discussion of molecular adsorbates all the examples described have been on metal surfaces. There is no doubt that this reflects the balance of the available data, due, in large part, to the motivation to understand the surface chemistry underpinning heterogeneous catalysis. However, this motivation is equally relevant to oxide surfaces, while increasing interest in the possible integration of conventional (mainly silicon-based) and organic semiconductors [126] means that a few molecular adsorbates on semiconductors have also been studies. The case of oxide surfaces has long been something of a ‘Cinderella’ topic in surface science, in part because of the problems of preparing well-characterised surfaces, in part because of the problems of charging when using charged particles (electrons or ions) that form the bulk of surface spectroscopic and structural methods. Nevertheless, a small number of structure determinations have been performed for molecular adsorbates on oxide surfaces. These will be discussed in more detail in the following section in the context of the measurement of chemisorption bondlengths.

In general (and especially in group IV elemental semiconductors), semiconductor solids comprise a covalently bonded network, and creating a surface produces dangling bonds. Chapter 3 provides a far more extensive discussion of semiconductor surfaces and their properties. The dominance of covalent bonding in the bulk of these materials mean that we may anticipate that molecular adsorption at the surface will lead to attachment to these local dangling bonds. The situation is in contrast to metal surfaces, where it is not immediately obvious that metal atoms embedded in an extended (metallic) solid can form local bonds with adsorbed molecules in the same way as isolated metal atoms, although the evidence presented above clearly shows that they can. Almost all structural studies of molecular adsorbates on semiconductor surfaces have been performed on Si(100). In its bulk-terminated state there are two dangling bonds per surface Si atom, these atoms being bonded through two intact bonds to the layer below. It has long been known that the true situation is that adjacent pairs of surface Si atoms move together to form dimers, thus reducing the number of dangling bonds per surface atom to one, albeit with some energy cost in modifying the bond angles from those of the bulk solid (see Chapter 5). These dimers, however, are not symmetric, with their Si–Si axis parallel to the surface, but are asymmetric, with tilted dimer bonds. At low temperatures complex ordering of these asymmetric components occurs, but at room temperature only an average (2×1) ordering is seen (Figure 1.15), with the dimers flipping between the two different orientations dynamically. In the case of interaction with molecular water, H_2O, and ammonia, NH_3, (of potential relevance to oxidation and nitridation of the Si surface), a range of spectroscopies have established that dissociation at the surface leads to OH and H, and NH_2 and H, respectively, being attached to the dangling bonds, probably at either end of the same dimer (Figure 1.15). O 1s and N 1s PhD studies, respectively, have confirmed these local geometries for the OH [127] and NH_2 species [128,129], and provided quantitative details. One interesting conclusion of this is that the dimer to which these species are bonding becomes essentially symmetric, indirect evidence

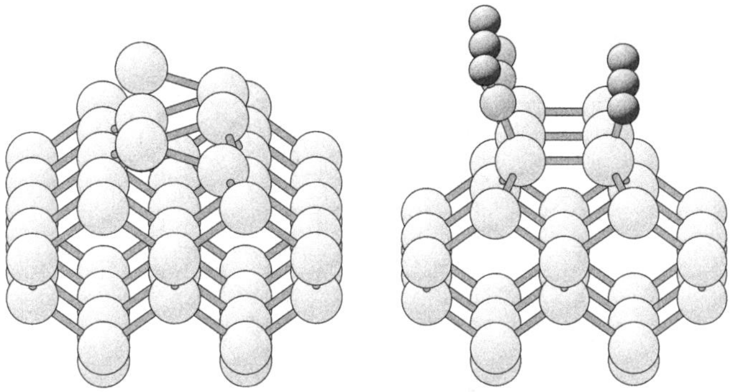

Figure 1.15. Pespective views of the clean Si(100) '(2 × 1)' surface with average (2 × 1) periodicity but a random mixture of the two orientations of the surface asymmetric dimers, and of the (2 × 1) phase of the same surface formed by the coadsorption of OH and H from the dissociation of H_2O. Note that only the local geometry of the OH species is known from experiments, and in reality it is likely that which end of the dimers contain the OH species will also be random, so this phase may also be properly described as '(2 × 1)'. The Si dimers, however, are now symmetric.

that both dangling bonds of the dimer are saturated by adsorbates, presumably the second species in each case being the ('invisible') H atoms. The structure of Ge(100)(2 × 1)–OH + H is found to be very similar to that of the equivalent Si surface [130]. The same technique has also been applied to study the interaction of ammonia with Si(111) [131]. This is a substantially more complex problem because of the nature of the Si(111) surface, with its (7 × 7) surface reconstruction, containing not only surface dimers but also adatoms and stacking faults (leading to the so-called DAS model) as a mechanism for reduction of the number of dangling bonds. However, the results of the N 1s PhD study appear to show rather clearly that the resulting NH_x species (probably NH_2) are adsorbed almost exclusively on the dangling bonds atop the so-called rest atoms of this surface reconstruction.

One perceived feature of the Si surface dimers on Si(100) is that they provide a means of attachment of molecular species incorporating a C=C double bond, a more direct kind of di$-\sigma$ bonding than that mentioned in discussing hydrocarbon adsorption on metal surfaces. With this in mind, quite a lot of attention has been focussed on the properties of the simplest unsaturated C_2 hydrocarbons on this surface, acetylene and ethylene. From a structural point of view there are two questions that have been discussed in the literature, namely, do these molecules adsorb above surface dimers in bridging sites aligned along the dimer direction (labelled as the bridge site in Figure 1.16), and does this lead to scission of the dimer bond? Extensive studies have been conducted with a range of spectroscopies, with STM and with theoretical total energy calculations by a range of methods,

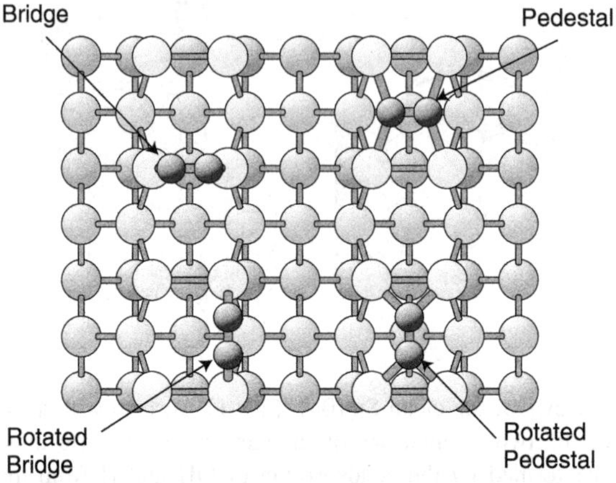

Figure 1.16. Plan view of the Si(100)(2 × 1) surface (with symmetric dimers) showing four possible sites for the adsorption of C_2H_2 and C_2H_4. Only the C atoms of the adsorbate are shown as small dark spheres. The Si atoms are shown with successively darker shading for deeper layers.

but the first quantitative experimental structure determination was achieved for ethylene adsorption using C 1s PhD. This study [132] concluded that, indeed, this species does adopt the bridging site above an intact dimer, the Si-Si separation being 2.36 ± 0.21 Å, only slightly larger than that found for the clean surface. The intramolecular C−C distance of 1.60 ± 0.08 Å also indicates a bond order of approximately one (or less), also consistent with the di−σ bonding.

Following this work a similar study of adsorbed acetylene was conducted by the same methods and led to rather similar conclusions [133], namely occupation of the bridging site above a Si dimer. The experimental precision in the measured Si−Si dimer bondlength was much worse (2.44 ± 0.58 Å) than it was for the C−C distance (1.36 ± 0.19 Å), but these results are still consistent with a lowering of the C−C bond order from the gas phase value. However, at a similar time another experimental structural study was published [134] investigating these same two adsorption systems. This led to similar results for adsorbed ethylene, but identified a quite different adsorption geometry (the pedestal site of Figure 1.16) for adsorbed acetylene. In this geometry, the molecule is located midway between two dimers and was suggested to have 'tetra-σ' bonding. The structure determination in this study was also based on C 1s PhD data, but instead of using multiple-scattering simulations to analyse the data, a simple 'holographic' inversion of the data was used to generate an 'image' of the adsorption site, from which the local geometry was extracted directly. There has been considerable interest in the last few years in the possibility of developing a viable method of direct inversion of PhD data to obtain surface structures, and several different algorithms have been discussed and tested, mainly on simulated data, but this application of one such method seems to be the first in which the analysis relied wholly on this approach and made no use of simulations of the data. It is therefore rather difficult to establish the reliability of the conclusions, particularly as it did not prove possible to gain access to the raw data. However, while an early STM study had favoured bridge site occupation [135], Wolkow [136] later suggested that the STM images shown in this same publication showed evidence for occupation mainly of sites midway between two dimers (i.e., pedestal or rotated pedestal – Figure 1.16) but also minority occupation of a single-dimer site, presumably the bridge site. A subsequent STM study [137] actually provided evidence for co-occupation of *three* different local geometries. While interpretation of STM images for adsorbate systems of this type is certainly not free from ambiguities, it does seem possible that multiple site occupation may occur, and this possibility was not explored in either of the PhD-based studies described above. In this context, it was also notable that one key difference in these two experimental studies was that the PhD (simulation) study was conducted at ~100 K whereas the holographic inversion study was performed at room temperature.

In an attempt to resolve these problems, a new PhD study (using multiple-scattering simulations for analysis) was undertaken [138]. In this study new experimental data were taken at room temperature and both the low temperature and room temperature data were subjected to a reanalysis that considered the possibility of multiple site occupation, considering as possible all four sites shown in Figure 1.16. These new experiments also paid particular attention to the possible role of radiation damage from the incident X-rays during the experiment, which is a further possible source of the original discrepancy between the conclusions from the two earlier experiments. A key finding of this new study is that the experimental data recorded at the two adsorption temperatures *do* differ, indicating some associated structural differences. At low temperature the bridge site remained strongly favoured as the only or majority site. At room temperature, on the other hand, while partial bridge site occupation remains, the results indicate that more than 50% of the adsorbed molecules occupied either the pedestal or rotated pedestal sites. The specific geometry (bondlength values) of the pedestal site identified in the earlier holographic study was found, however, to be inconsistent with the new data. Interestingly, these new experimental data showed no evidence for occupation of the rotated bridge site of Figure 1.16, despite this being thought to be the majority site in the more recent STM study [137].

Clearly the situation of acetylene adsorption on Si(100) is a complex problem, presumably related to the higher potential reactivity and capacity for bonding of the C≡C species. Fortunately, the situation for ethylene appears to be more straightforward, and it seems to be the interaction of Si(100) with C=C species that underpins most of the potential applications in creating suitable inorganic/organic electronic interfaces [126].

5. Chemisorption bondlengths

5.1. Metal surfaces

So far the discussion above has focussed on adsorption sites and the intramolecular bondlengths and conformation. However, as remarked earlier, for the case of metal surfaces in particular, the nature of the chemisorption bond might be expected to differ from that in metal coordination compounds, due to the influence of the highly itinerant valence electrons that create the metallic bonding within the solid. In fact the experimentally determined adsorption geometries, particularly of molecular adsorbates, do point to relatively local atomic bonds at the molecule/surface interface. The situation is less clear for atomic adsorbates, which commonly adopt the highest coordination sites on metal surfaces, a geometry that could correspond to high local atomic coordination, or to maximum embedding in the metallic electron gas. One piece of structural information that casts some light on this is the chemisorption bondlengths – i.e., the nearest-neighbour distance between the surface metal atoms and the bonding atom of the atomic or molecular adsorbate.

In the case of atomic adsorbates, this question seems to have last been addressed in detail some 20 years ago, based on the 38 available quantitative adsorbate structure determinations for the chalcogens O, S, Se, and Te and the halogens Cl and I adsorbed on various surfaces of Ag, Co, Cu, Fe, Ir, Ni, Pd, Rh, Ta, W and Zr [139,140]. The objectives of these early attempts to establish clear systematics for atomic chemisorption bondlengths were both practical and fundamental. The practical issue was simply that if the likely bondlengths were known, optimisation of structural searches using quantitative LEED could be made more quickly and efficiently; LEED calculations are far less demanding now, than they were in the 1980s, through the advance in fast and low-cost computing, so even for simple systems this potential to constrain possible solutions was quite important. The fundamental issue, of course, is the one that concerns us here – do the systematics of chemisorption bondlengths give any insight into the nature of chemisorption bonds?

The basic finding was that for chemisorption bonds, as for those in inorganic materials generally, bondlengths, D, and bond order, n, appear to be related by a Pauling-like expression [141] between an adsorbate X and the metal surface atoms M of the type

$$D(n) = D(1) - b \log_{10} n \qquad (1)$$

with the favoured value of b being around 0.80–0.85 Å, rather more than the value of 0.6 Å used by Pauling for C−C bonds of fractional bond order. One issue of debate was the appropriate value of the parameter $D(1)$, the bondlength for a bond order of unity; Mitchell [139,140] favoured the use of values obtained from the bulk structures of known inorganic X−M compounds, although there remains an issue over the definition of the valency. With the surface structural data base for atomic chemisorption systems now more than an order of magnitude larger than in 1986, there is the potential to revisit this issue far more thoroughly, but the task is a major one that will certainly not be attempted in this chapter!

At the time that this survey of chemisorption bondlengths was undertaken, the database of molecular adsorption structures [139] was even more sparse, inadequate for any clear pattern to be established. More recently, we have returned to this problem through a detailed study of one specific chemisorption system, namely CO on Ni surfaces [142,143]. The objective was two-fold. First, by conducting PhD experimental structure determinations for different phases of CO on Ni(100) and Ni(111) it was possible to obtain Ni−CO chemisorption bondlengths for 1-, 2-, and 3-fold coordinated surface sites; by using a common methodology (and thus common systematic errors, if any), one may expect to achieve a more reliable comparison of the associated bondlength changes. Secondly, by also studying the CO/H coadsorption phases on Ni(100), data was obtained on the influence on the bondlengths for different chemisorption energies at the same local bond order. The results of these experiments are summarised in Table 1.1, together with some related measured or calculated parameters. As remarked earlier, early LEED structural studies of CO adsorption on Ni(100) established that the atop (1-fold coordinated) geometry is favoured at low coverages, notably for the 0.5 ML c(2 × 2) phase [144–148], while spectroscopic studies have indicated that bridging (2-fold coordinated) sites are increasingly occupied at higher coverages [149–151]. In principle the 0.5 ML c(2 × 2) ordered phase corresponds to the coverage beyond which this transition occurs, but photoelectron and vibrational spectroscopic studies show that there is typically co-occupation of both states when the well-ordered c(2 × 2) LEED pattern is seen. The use of the photoelectron binding energy shifts in the PhD

Table 1.1
Summary of Ni−C bondlengths in several Ni/CO adsorption systems. Also included are data for CO adsorption on NiO. References for the various values are cited in the main text.

Surface	Coordination	Ni−C bondlength (expt.) Å	Ni−C bondlength (theory) Å	adsorption energy (eV)
Ni(100)/CO	1	1.73 ± 0.03	1.74	1.2
	2	1.89 ± 0.02		1.1
Ni(100)/CO+H	1	1.79 ± 0.02	1.79	~0.4−0.6
	2	1.88 ± 0.03		~1.0
Ni(111)/CO	3	1.93 ± 0.03		1.2
NiO(100)/CO	1	2.07 ± 0.02	2.49, 2.86, 2.03−2.05	~0.30

studies, however, allowed the local geometry of the distinct coadsorption sites to be determined independently. On Ni(111) the two inequivalent hollow (3-fold coordinated) sites are occupied at both low coverage [152] and in the 0.5 ML c(4 × 2) phase [153] described in Section 2. For all of these states, the adsorption energies for CO are similar in the range 1.0–1.2 eV as implied by early desorption studies [154,155] and more recently by novel calorimetric measurements [156].

In addition, the PhD technique was applied to the interesting case of CO adsorbed onto a hydrogen-saturated Ni(100) surface [150,157–159], which leads to a much more weakly adsorbed CO species at low temperatures [160,161], also believed to be atop a surface Ni atom. Warming this coadsorption phase causes restructuring at around 140 K, and partial CO desorption at 210 K, with associated spectral changes which have been associated with both hollow site and bridge site adsorption states. Based on these transition temperatures one may estimate the adsorption energy of the CO on the coadsorbed c(2 × 2) phase to be around 0.4–0.6 eV, a factor of 2–3 times less than in the absence of the coadsorbed hydrogen.

If we first consider the Ni−C bondlengths for the pure CO overlayers on Ni(100) and Ni(111), all with closely similar adsorption energies (Table 1.1), there is a clear trend with changing coordination number and corresponding bond order. Thus a bond order of 1 gives a Ni−C bondlength of 1.73 Å, that increases to 1.89 Å for a bond order of 0.50, and to 1.93 Å for a bond order of 0.33. This trend is entirely consistent with the bond-order/bondlength equation (1). The optimum value of b for these data is 0.50 Å, but based on such a small data set it is difficult to draw general conclusions as to the significance of this value, although it is perhaps notable that this value is much closer to 0.60 given by Pauling for C−C bonds than to 0.80–0.85 found by Mitchell for atomic chemisorption. The higher-temperature CO/H coadsorption phase, involving CO in bridging sites, that has a similar adsorption energy, is also consistent with this trend. For the low-temperature c(2 × 2)-CO+H phase, however, which is confirmed to involve atop adsorption, the Ni−C bondlength of 1.79 Å is some 0.06 Å longer than that for the same singly coordinated site in the absence of coadsorbed H. However, this bond extension is much less that the 0.16 and 0.20 Å associated with reducing the bond order by a factor of 2 or 3, despite the fact that we estimate the bond strength in this coadsorption phase to be a factor of 2 to 3 lower than in the absence of the coadsorbed H. At first sight this seems surprising, but the results of DFT calculations of the two c(2 × 2) atop phases (CO and CO/H) [142,143] proved revealing. These calculations reproduced almost exactly the experimentally determined Ni−C bondlengths, as shown in Table 1.1. In addition, however, the DFT results provide some insight into the possible reasons for the modest change in this bondlength, despite the large change in the inferred adsorption energy. Figure 1.17 shows the calculated electron charge density difference contour maps for the CO adsorbed in the atop sites with and without preadsorbed atomic H. These maps show the difference between the charge density in the actual minimum energy structures

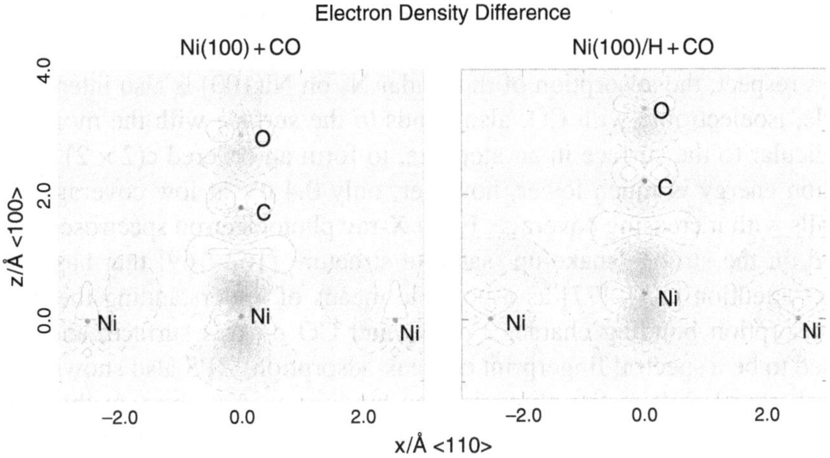

Figure 1.17. Electron charge density difference contour map for CO on Ni(100) and CO on Ni(100)/H in atop sites, derived from DFT calculations.

and in the sum of the isolated CO, and isolated surface, with the geometry of the chemisorption structures. The charge density differences thus represent the consequences of the local CO-metal bonding. The striking feature of these contour maps are the similarities in the two systems; despite the large difference in the adsorption energy, the electronic redistribution around the molecule and the nearest neighbour Ni atom, associated with the bonding, is much the same, suggesting the local bond strength is similar. Also notable in Figure 1.17, however, is the very strong rumpling of the outermost Ni layer in the coadsorption system, with the Ni atom bonded to the CO being pulled out of the surface layer by some 0.4 Å (the experimental results show the same effect but with a somewhat smaller value of 0.24 ± 0.08 Å). A further consequence of the CO adsorption on this surface is that the H atoms, above the Ni surface layer prior to the CO adsorption, are displaced downwards into the crystal surface in the presence of the coadsorbed CO; these H atoms are in the four-fold coordinated hollow sites on the surface and thus fall out of the plane of the contour map of Figure 1.17. As may be expected, the DFT calculations show there is a large energy cost in these CO-induced distortions of the Ni(100)/H surface. This accounts for the low overall adsorption energy, that is the sum of the strong local bonding energy and the energy cost associated with these distortions. Under these circumstances, the modest extension of the local Ni−C bondlength is understandable, because the local bond is not much weaker than in the absence of the preadsorbed H. This situation has some similarities to that of acetylene adsorption on Cu(111), discussed in Section 4.2. In that case the low adsorption energy was attributed to a combination of a strong chemisorption bond and a pronounced intramolecular bond extension. Here it is the distortion of the substrate required in the coadsorption that provides the energy cost. In both cases, therefore, the adsorption energy or desorption temperature fails to reflect the true strength of the local chemisorption bond.

In this respect, the adsorption of molecular N_2 on Ni(100) is also interesting. The molecule, isoelectronic with CO, also bonds to the surface with the molecular axis perpendicular to the surface in an atop site, to form an ordered c(2 × 2) phase. The adsorption energy is much lower, however; only 0.4 eV at low coverage, and this value falls with increasing coverage. Early X-ray photoelectron spectroscopy studies focussed on the strong 'shake-up' satellite structure [162–169] that has continued to attract attention [170–177] as a possible means of understanding the difference in chemisorption bonding character of N_2 and CO on this surface, and has been suggested to be a spectral fingerprint of weak adsorption. XPS also shows, however, a clear chemical shift in the photoelectron binding energy between the N bonded to the nearest-neighbour Ni atom and the N atom furthest from the surface, clearly reflecting a distinct difference in the electronic environment of the two N atoms as a result of the adsorption. For these reasons this has tended to be regarded as a case of weak chemisorption (rather than physisorption, as might be inferred from the adsorption energy). Indeed, an early and rather restricted PhD study of this adsorption system [178] had led to the conclusion that the Ni−N nearest-neighbour distance was 2.25 Å, this large value certainly implying a very weak bond. However, one advantage of the chemical shift in the N 1s photoelectron binding energy is that it means that the chemical-state specificity of the PhD technique can be applied to obtain quantitative structural information on the location of each of the N atoms in the molecule on the surface. Doing so [142,179] led to confirmation of the atop adsorption site and perpendicular orientation of the N−N axis to the surface, but also provided a revised value of the Ni−N chemisorption bondlength of 1.81±0.02 Å, this much shorter distance indicative of a strong local bond. The very different conclusions of the earlier study can be attributed to the limited data set that lacked the chemical-state specificity. The shorter bondlength is also confirmed by DFT calculations [142,179] that gave a value of 1.79 Å. However, while the DFT calculations also yielded a rather low adsorption energy (0.55 eV), they did not show any large adsorbate-induced modifications of the molecular or substrate structure that provided an obvious rationale for the apparent inconsistency of a strong local bond (as reflected by the bondlength and electronic modification) but a low adsorption energy. The origins of this effect have been the subject of a detailed combined experimental spectroscopic and theoretical study [180] and are discussed fully in Chapter 2.

One further example of a structural study of CO adsorption on Ni relates to the rather general issue of the role of coadsorbed CO and alkali metals on metal surfaces. Alkali atoms are known to be promoters of certain catalytic processes, improving turnover rates and/or selectivity in certain reactions, and are used as additives in some commercial catalysts. Indeed, in the specific case of CO on Ni, it is known that coadsorbed K does promote the dissociation of the CO [181,182]. One rationalisation for this effect has been that the coadsorbed alkali leads to increased occupation of the antibonding $2\pi^*$ orbitals of the CO, thus weakening the C−O bond. NEXAFS data for several CO/alkali coadsorption systems [183–185] have been suggested to indicate a lengthening of the C−O bond by about 0.13 Å, relative to its gas-phase value of 1.13 Å. However, this interpretation of the NEXAFS relies on the exact energy of rather broad σ-resonances in the spectra, but these occur in an energy range in which the coadsorbed alkalis may produce significant electronic modification. The structural interpretation of these changes is therefore not without controversy. In an attempt to resolve this using a more conventional structural probe, the PhD method was used to investigate the local geometry of the K, C and O atoms in the K/CO coadsorption phase on Ni(111) [186,187], comparing these data with those for pure CO adsorption and pure K adsorption on the same surface. The results were both interesting and disappointing. Specifically, no significant change in the C−O bondlength was found within the estimated precision of about 0.08 Å, and the Ni−C CO chemisorption bondlength also appeared to be unchanged by the coadsorbed K. On the other hand, the presence of the coadsorbed CO led to an increase in the Ni−K bondlength of 0.15 Å relative to the pure K layer. Interestingly, very similar conclusions were reached in a quantitative LEED study of CO and K coadsorption on Co$(10\bar{1}0)$ [188]. Thus, while there is no doubt that coadsorbed alkali atoms do modify the surface chemistry of CO on various surfaces, the main structural implication of coadsorption on these to Ni and Co surfaces is actually found to be a modification of the alkali chemisorption bondlength.

As described above, the adsorption systems of N_2, CO and CO/H on Ni(100) all show excellent agreement between DFT calculations and experiment in chemisorption bondlengths, but this is not always the case. One recent example of a significant discrepancy is that of alanine on Cu(110), already presented in Section 4.3. As described there, the experimental PhD study of this adsorbate system [117] agreed well with the DFT calculations [118] over the adsorption site, and indeed, because of the complexity of this system, the final optimisation of the experimental best-fit structure was based on the lateral offsets and distortions fond in the DFT calculations. However, using *all* the DFT parameter values actually gave a poor fit to the experimental data, but adjusting the height of the molecules above the surface by approximately 0.1 Å led to the final excellent fit. The implication is that the DFT calculations overestimate the Cu−O and Cu−N chemisorption bondlengths by about 0.1 Å. This discrepancy is very significantly larger than is typically found in such experiment/theory comparisons for chemisorption systems on metal surfaces, for which DFT is usually regarded as being rather accurate. Whether this problem reflects the complexity of this particular system is not yet known.

5.2. Oxide surfaces

As remarked in Section 4.4, there are currently very few experimental quantitative structural studies of adsorbates on oxide surfaces, but the few that there are have highlighted some potentially more serious issues in theory/experiment comparisons. The most significant of these is reflected in the data shown in the final row of Table 1.1. In a first exploration of the use of the PhD technique to investigate adsorption on oxide surfaces, NiO(100) epitaxial films were grown on Ni(100) and the adsorption of NO [189,190], CO and NH_3 [191,192] investigated structurally. All three molecular species were found to adsorb atop surface Ni atoms, but for the NO and CO adsorbates for which corresponding theoretical total energy calculations were available, the experiments show very much shorter Ni−N and Ni−C bondlengths. For the case of CO, shown in Table 1.1, the discrepancies relative to the first two theoretical values [193,194] (the only ones available at the time) are huge –0.42 and 0.79 Å; the situation is qualitatively similar for NO, with a discrepancy of 0.21 Å between experiment [189] and theory [195]. With hindsight, NiO was probably a bad system to choose as a first test of this approach, because this is known to be a difficult material to model theoretically, due to its highly correlated electronic behaviour. Nevertheless, the full scale of these problems only becomes evident through the availability of experimentally determined adsorption bondlengths. These results highlight, in a very clear way, the challenge for theoretical descriptions of the surface chemical properties of this material, and have led to further work to address this problem [196–199]. Much the most successful new work has been based on the use of the 'DFT+U' method in which gradient-corrected DFT methods are supplemented by a Hubbard Hamiltonian for the Coulomb repulsion and exchange interaction. For the NiO(100)/CO system [198] this leads to Ni−C bondlengths in the range 2.03–2.05 Å depending on coverage, in excellent agreement with experiment; other structural parameters and the adsorption energy are also in good agreement with experiment. Interestingly, though, the same method applied to the NiO(100)/NO system [199] leads to a good description of most of the physical properties but significantly overestimates the Ni−N bondlength relative to the experimental value.

The rutile $TiO_2(110)$ surface is probably the most investigated of all oxide surfaces in experimental surface science [200], and has also been subjected to a large number of theoretical investigations. From an experimental point of view, this surface has the advantage that good single crystals are readily available and that modest annealing in vacuum introduces a low level of colour centres associated with oxygen vacancies that renders the crystal semiconducting, thus avoiding the problems of charging in any technique involving electrons or ions. The surfaces of TiO_2 are also of considerable practical importance, not least because of its ability to catalyse the photochemical production of hydrogen from water [201]. Despite this, even the detailed relaxations of the outermost layer atoms of the clean unreconstructed (1×1) surface have proved controversial, although some consensus now seems to be emerging regarding the sign and approximate magnitude of these, particularly through recent new experimental studies [202,203].

Of the few adsorbate systems studied structurally on $TiO_2(110)$, there has been significant interest in its interaction with formic acid which produces a (2×1) surface phase [204] associated with deprotonation of the acid to create adsorbed formate, (HCOO) [205]. While there is a general consensus that the formate is bonded to the surface through the carboxylate O atoms with the molecular plane essentially perpendicular to the surface, the azimuthal orientation has proved more controversial. XPD measurements, exploiting high-energy intramolecular forward scattering of O 1s photoelectrons, concluded that the molecular plane lies in the [001] azimuth (see upper-right panel of Figure 1.18), corresponding to the direction of the rows of 5-fold-coordinated Ti atoms and the rows of bridging oxygen on the clean surface [206,207]. However, RAIRS (reflection-absorption infrared spectroscopy) [208] and NEXAFS (near-edge X-ray absorption spectroscopy) [209] studies provided support for a second (minority) formate species with its molecular plane rotated azimuthally by 90° (see lower left panel of Figure 1.18), that may account for approximately 30% of the surface coverage. None of these experiments identified the

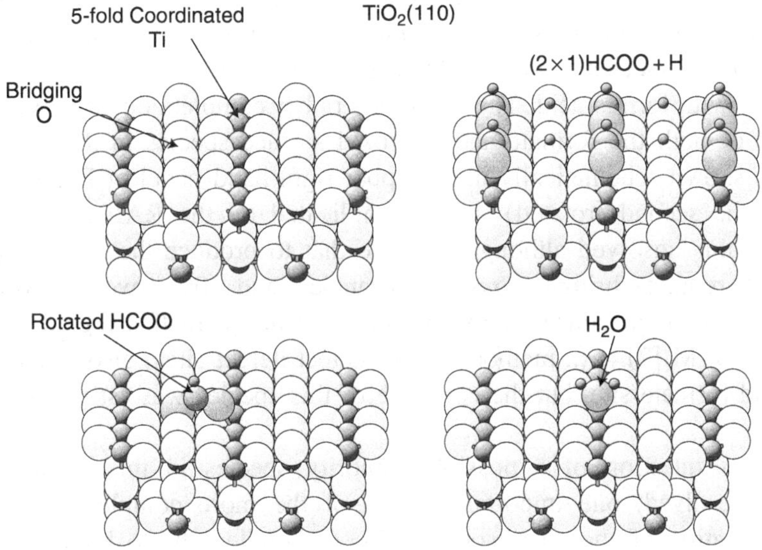

Figure 1.18. Structural models of the clean $TiO_2(110)$ surface (top left) and of various possible and actual adsorbate structures. In the TiO_2 substrate the Ti atoms are shown as smaller and more darkly shaded spheres than the O atoms.

local adsorption site, but an early chemical-state specific O 1s PhD study measured in a single emission direction, complemented by scanned-angle mode photoelectron diffraction data [206] was found to support the results of Hartree–Fock total energy calculations that the formate species, aligned along [001], is adsorbed with the O atoms in equivalent sites near-atop 5-fold coordinated Ti atoms in the surface. This adsorption site is also favoured by other theoretical total energy calculations [210–212].

More recently a new and far more extensive PhD study has been conducted, using both O 1s and C 1s spectra [213]. As in the earlier study, chemical-shift specific data was used for the O 1s signal to separate the adsorbate contribution from that of the substrate. The principal conclusion regarding the adsorption site, bridging adjacent 5-fold coordinated surface Ti atoms, is the same as that described above. However, the possible role of partial occupation of the rotated formate geometry (Figure 1.18) was also explored. Notice that in the preferred site, bridging two surface Ti atoms, the carboxylate O atoms are close to atop these Ti atoms, but in the rotated geometry one of the carboxylate O atoms occupies a bridging site, and indeed is assumed to arise through interaction with a bridging oxygen vacancy site. In fact the O 1s PhD data do indicate partial occupation of such bridging sites by the adsorbate O species, but the C 1s PhD shows clearly that any contribution from rotated formate species must be small (much less than 30%). However, it was noted that the O 1s photoelectron binding energy shift relative to the oxidic O atoms of the TiO_2 is essentially identical for adsorbed formate and hydroxyl, OH, species. Thus, the bridging O atoms contributing to the PhD modulations can be attributed to OH species formed by hydrogenation of the bridging O atoms during the reaction with formic acid. The result is a rather complete quantitative structural evaluation of both the formate and hydroxyl surface species in this coadsorption system, as shown schematically in the upper right-hand panel of Figure 1.18. The puzzle of the second (rotated) species implicated by the RAIRS and NEXAFS studies remains unsolved although we note that to produce the rotated species of Figure 1.18 requires bridging oxygen vacancies, and 30% coverage would seem to imply a very high concentration of such vacancies. In this regard, one recent study using STM [214] (a method able to detect defect sites) also found evidence of a second formate site, but the coverage of this species was estimated to be less than 1%.

While this question of a possible minority species in this surface reaction remains unresolved, one important aspect of the more complete PhD study is the determination of the chemisorption Ti–O bondlengths to both the carboxylate (2.08 ± 0.03 Å) and hydroxyl (2.02 ± 0.05 Å) species, which turn out to be in excellent agreement with the values (2.08 and 2.01 Å, respectively) for the one theoretical (DFT) calculation that considers this coadsorption situation explicitly [212]. Other theoretical calculations of the bondlength for the formate species alone [206,215] give closely similar values. The clear implication is therefore that the current theoretical descriptions of chemisorption on $TiO_2(110)$ are effective and reliable.

However, a recent investigation of the local structure of molecular water on this surface, using the PhD technique [216,217], reopens this question. As remarked above, there is a special interest in the interaction of water with this surface due to the potential importance of its ability to achieve photochemical decomposition. Describing the interaction of water with $TiO_2(110)$ theoretically has, however, proved a significant challenge, as testified by a large body of literature [218–232]. The basic problem is that experimentally it seems to be accepted that exposure of a perfect $TiO_2(110)$ surface to molecular water under surface science conditions does not lead to dissociation, although such dissociation can occur at surface (bridging) oxygen vacancy sites. By contrast, the great majority of theoretical calculations lead to the conclusion that dissociation should be facile, even on the perfectly ordered surface. A few calculations do reproduce the observed stability of adsorbed molecular water, but the appropriateness of the methods used to achieve this remains a subject of debate [231,232]. Nevertheless, there does appear to be a consensus between theory and experiment that the preferred site for adsorption of the molecule on the perfect surface is atop one of the five-fold coordinated Ti atoms in the surface layer. This site was first identified experimentally through STM studies [233,234], but has been confirmed in the recent PhD study [216,217]. Of course, the PhD study provides quantitative information on the structure and in particular the Ti–O chemisorption bondlength for the H_2O. The value for this parameter obtained was 2.21 ± 0.02 Å which may be compared with the theoretical values of 2.41 Å [221], 2.25 Å [224], 2.28 Å [227] and 2.32 Å [228], extracted from the few theoretical papers on the subject that actually quote their values of this bondlength. Clearly, the experimental bondlength is shorter than any of these theoretical values, and very significantly shorter than most of them. Of course, the adsorption energy of the water is relatively low, so this mismatch of experimental and theoretical chemisorption bondlengths may reflect some weakness of the theory to provide an accurate description of the weak bond. This interpretation would be consistent with the fact that the strong chemisorption bond of the formate species adsorbed on this surface appears to be well-described by similar theoretical treatments. However, we note that this same problem in the theory may lie at the heart of the problem of correctly describing the ability of the perfect TiO_2 surface to dissociate molecular water; indeed, the weakly chemisorbed state of the intact molecule is one point on the reaction path. In this context it is perhaps noteworthy that the theoretical chemisorption bondlength value closest to that of the experiment (2.25 Å) comes from the only theory paper [224], also quoting a value for the bondlength, that finds molecular water to be stable on the surface.

6. Conclusions

In this short chapter, I have attempted to review some of the main trends in results that have emerged from quantitative structural investigations of adsorption on surfaces, some of which provide significant insight into the nature of the adsorbate-substrate bonding. As in other areas of surface science, recent trends have been to tackle problems of increasing complexity, such as the adsorption of larger molecules (beyond diatomics!) and coadsorption systems. One rather general feature to emerge from these structural studies is that a surface is not a rigid atomic chequer-board on which adsorbates sit, but is a flexible structure that may even undergo radical restructuring as a consequence of the chemisorption. Insofar as the basis of heterogeneous catalysis is the modification of the properties of molecules in the presence of the surface, it is perhaps not surprising that such adsorbed molecules and atoms also modify the properties of the underlying surface. One further general trend to emerge, particularly for molecular adsorbates on metal surfaces, is that despite the fact that the surface atoms of the substrate are bound together by metallic bonds based on the highly delocalised valence electrons, the molecule appears to form quite local bonds with surface atoms similar to those in metal coordination compounds. At least for the model system of CO adsorption on Ni, it also seems that chemisorption bondlengths are related to bond order in a fashion very similar to that of free molecules. On the other hand, there do seem to be examples in which the total adsorption energy is not a reliable indicator of the chemisorption bond strength, because chemisorption may lead to intramolecular or substrate distortions that have a high energy cost and thus lower the adsorption energy. In these cases, experimental values of the chemisorption bondlengths can give further insight into the character of the bonding.

One further general issue concerns the reliability of theoretical (commonly *ab initio*) total energy calculations (mainly using DFT) as a means of establishing optimum values of chemisorption bondlengths. With ever-increasing use of such methods as a complement to a range of experimental methods that do not, themselves, provide quantitative structural information, it is important to understand the limitations of such theories. There is actually a dangerous trend to *replace* experiments by these theoretical methods, yet quantitative experimental structure determinations are essential to 'keep the methods honest'. For very many adsorbate systems there is excellent agreement between theoretical and experimental determinations of chemisorption bondlengths (typically, to within about 0.03 Å), but there are also examples where this is not the case. As we have seen, chemisorption of simple molecules on NiO represents an extreme case which has proved a major challenge to theory, but there other examples, numerically smaller but not necessarily less serious. The case of water on TiO_2 is potentially one such example, perhaps highlighting a problem in describing weak bonding that is nevertheless central to understanding the associated surface chemistry. We have also noted a rather significant discrepancy between theory and experiment in the chemisorption bondlength of alanine on Cu(110), perhaps reflecting a problem in dealing with a relatively complex molecular adsorbate. Clearly there is a continuing need for such experiment/theory comparisons, especially with the trend to study systems of increasing complexity. Theory can undoubtedly contribute in an important way to our understanding when placed alongside quantitative experiments, but should not replace such experiments, simply because they may be technically demanding.

References

[1] J.B. Pendry, *Low Energy Electron Diffraction* (Academic Press, New York, 1974).
[2] M.A. Van Hove, W.H. Weinberg, C.-M. Chan *Low-energy electron diffraction:* experiment, theory and surface structure. (Springer, Berlin, 1986).
[3] K. Heinz, Rep. Prog. Phys. 58 (1995) 637.
[4] P. R. Watson, M. A. Van Hove, K. Hermann. *NIST Surface Structure Database Ver. 5.0,* NIST Gaithersburg, MD. (2003).
[5] D.P. Woodruff, A.M. Bradshaw Rep. Prog. Phys. 57 (1994) 1029.
[6] J. Stöhr in X-ray Absorption, Principles, Techniques, Applications of EXAFS, *SEXAFS and XANES*, ed. R. Prins and D.C. Koeningsberger (Wiley, New York, 1988) 443.
[7] D.P. Woodruff, Rep. Prog. Phys. 49 (1986) 683.
[8] R. Feidenhans'l, Surf. Sci. Reports 10 (1989) 105.
[9] I. K. Robinson, D. J. Tweet, Rep. Prog. Phys. 55 (1992) 599.
[10] J. Zegenhagen, Surf. Sci. Rep. 18 (1993) 199.
[11] D. P. Woodruff, Prog. Surf. Sci. 57 (1998) 1.
[12] D.P. Woodruff, Rep. Prog. Phys. 68 (2005) 743.
[13] H. Niehus, W. Heiland, E. Taglauer, Surf. Sci. Reports 17 (1993) 213.
[14] J.F. van der Veen, Surf. Sci. Rep. 5 (1985) 199.
[15] S. Heinze, S. Blügel, R. Pascal, M. Bode, R. Weisendanger, Phys. Rev. B 58 (1998) 16432.
[16] P. Sautet, Surf. Sci. 374 (1997) 406.
[17] U. Diebold, Surf. Sci. Rep. 48 (2003) 53.
[18] D.J. Coulman, J. Winterlin, R.J. Behm, G. Ertl, Phys. Rev. Lett. 64 (1990) 1761.
[19] F. Jensen, F. Besenbacher, E Lægsgaard, I. Stensgaard, Phys. Rev. B 41 (1990) 10233.
[20] N. Sheppard, N.T. Nguyen, Adv. IR Raman Spectr. 5 (1978) 67.
[21] L. Becker, S. Aminpirooz, B. Hillert, M. Pedio, J. Haase, D. L. Adams, Phys. Rev.B 47 (1993) 9710.
[22] K.-M. Schindler, Ph. Hofmann, K.-U. Weiss, R. Dippel, V. Fritzsche, A. M. Bradshaw, D. P. Woodruff, M. E. Davila, M. C. Asensio, J. C. Conesa, A. R. González-Elipe, J. Electr. Spectros. Rel. Phenom. 64/65 (1993) 75.
[23] L. D. Mapledoram, M. P. Bessent, A. D. Wander, D. A. King, Chem. Phys. Lett. 228 (1994) 527.
[24] T. Gießel, O. Schaff, C. J. Hirschmugl, V. Fernandez, K.-M. Schindler, A. Theobald, S. Bao, W. Berndt, A. M. Bradshaw, C. Baddeley, A. F. Lee, R. M. Lambert, D. P. Woodruff, Surf. Sci. 406 (1998) 90.
[25] J. N. Andersen, C. O. Almbladh, J. Phys.: Condens. Matter 13 (2001) 11267.
[26] W. Oed, H. Lindner, U. Starke, K. Heinz, K. Müller, J. B. Pendry, Surf. Sci. 224 (1989) 179.

[27] W. Oed, H. Lindner, U. Starke, K. Heinz, K. Müller, D. K. Saldin, P. de Andres, J. B. Pendry, Surf. Sci. 225 (1990) 242.
[28] J. H. Onuferko, D. P. Woodruff, B. W. Holland, Surf. Sci. 87 (1979) 357.
[29] Y. Gauthier, R. Baudoing-Savois, K. Heinz, H. Landskron, Surf. Sci. 251/252 (1991) 493.
[30] A. Kilcoyne, D. P. Woodruff, A. W. Robinson, T. Lindner, J. S. Somers, A. M. Bradshaw, Surf. Sci. 253 (1991) 107.
[31] R. Terborg, J. T. Hoeft, M. Polcik, R. Lindsay, O. Schaff, A. M. Bradshaw, R. L. Toomes, N. A. Booth, D. P. Woodruff, E. Rotenberg, J. Denlinger, Surf. Sci. 446 (2000) 301.
[32] M. Bader, C. Ocal, B. Hillert, J. Haase, A. M. Bradshaw, Phys. Rev.B 35 (1987) 5900.
[33] D. P. Woodruff, in The Chemical Physics of Solid Surfaces vol. 7, Phase Transitions and Adsorbate Restructuring at Metal Surfaces eds. D. A. King and D. P. Woodruff (Elsevier, Amsterdam, 1994) 465.
[34] S. Reindl, A. A. Aligia, K. H. Bennemann, Surf. Sci. 206 (1988) 20.
[35] T. S. Rahman, H. Ibach, Phys. Rev. Lett. 54 (1985) 1933.
[36] W. Daum, S. Lehwald, H. Ibach, Surf. Sci. 178 (1986) 528.
[37] M. Rocca, S. Lehwald, H. Ibach, T. S. Rahman, Phys. Rev.B 35 (1987) 9510.
[38] D. Sander, U. Linke, H. Ibach, Surf. Sci. 272 (1992) 318.
[39] C. Klink, L. Oelsen, F. Besenbacher, I. Stensgaard, E. Laegsgaard, N. D. Lang, Phys. Rev. Lett. 71 (1993) 4350.
[40] H. C. Zeng, R. A. McFarlane, K. A. R. Mitchell, Surf. Sci. 208 (1998) L571.
[41] A. Atrie, U. Bardi, G. Casalone, G. Rovida, E. Zanazzi, Vacuum 41 (1990) 333.
[42] I. K. Robinson, E. Vlieg, S. Ferrer, Phys. Rev. B 41 (1990) 6954.
[43] M. C. Asensio, M. J. Ashwin, A. L. D. Kilcoyne, D. P. Woodruff, A. W. Robinson, Th. Lindner, J. S. Somers, D. E. Ricken, A. M. Bradshaw, Surf. Sci. 126 (1990) 1.
[44] M. Kittel, M. Polcik, R. Terborg, J. T. Hoeft, P. Baumgärtel, A. M. Bradshaw, R. L. Toomes, J.-H. Kang, D. P. Woodruff, M. Pascal, C. L. A. Lamont, E. Rotenberg, Surf. Sci.470 (2001) 311.
[45] F. Jensen, F. Besenbacher, E. Lægsgaard, I. Stensgaard, Phys. Rev. B 42 (1990) 9206.
[46] R. Meyer, C.-S. Zheng, K. G. Lynn, Phys. Rev. B 31 (1986) 8899.
[47] M. Wuttig, R. Franchy, H. Ibach, Surf. Sci. 213 (1989) 103.
[48] K. Tanaka, T. Fujita, Y. Okawa, Surf. Sci. 410 (1998) L407.
[49] T. Fujita, Y. Okawa, Y. Matsumoto, K. Tanaka, Phys. Rev. B 54 (1996) 2167.
[50] R. N. Lee, H. E. Farnsworth, Surf. Sci. 3 (1965) 461.
[51] G. Ertl, Surf.S ci. 6 (1967) 208.
[52] P. Hofmann, R. Unwin, W. Wyrobisch, A. M. Bradshaw, Surf. Sci. 72 (1972) 635.
[53] K. W. Jacobsen, J. K. Nørskov, Phys. Rev. Lett. 65 (1990) 1788.
[54] I. Merrick, J. E. Inglesfield, H. Ishida, Surf. Sci. 551 (2004) 158.
[55] J. E. Inglesfield, E. A. Colbourn, Phys. Rev. Lett. 66 (1991) 206.
[56] M. J. Harrison, D. P. Woodruff, J. Robinson, D. Sander, W. Pan, J. Kirschner, Phys. Rev. B (submitted).
[57] D. P. Woodruff J. Phys.: Condens. Matter 6 (1994) 6067.
[58] N. P. Prince, D. L. Seymour, D. P. Woodruff, R. G. Jones, W. Walter, Surf. Sci. 215 (1989) 566.
[59] S. M. Driver, D. P. Woodruff, Surf. Sci. 457 (2000) 11.
[60] G. S. Parkinson, M. A. Muñoz-Márquez, P. D. Quinn, M. Gladys, D. P. Woodruff, P. Bailey, T. C. Q. Noakes, Surf. Sci. 598 (2005) 209.
[61] V. Higgs, P. Hollins, M. E. Pemble, J. Pritchard, J. Electron Spectrosc. Relat. Phenom. 39 (1986) 137.
[62] S. M. Driver, D. P. Woodruff, Surf. Sci. 442 (1999) 1.

[63] E. Vlieg, S. M. Driver, P. Goedtkindt, P. J. Knight, W. Liu, J. Luedecke, K. A. R. Mitchell, V. Murashov, I. K. Robinson, S. A. de Vries, D. P. Woodruff, Surf. Sci 516 (2002) 16.
[64] D. F. Ogletree, M. A. Van Hove, G. A. Somorjai, Surf. Sci. 173 (1986) 351.
[65] I. Zasada, M. A. Van Hove, Surf. Rev. Lett. 7 (2000) 15.
[66] F. Bondini, G. Comelli, F. Esch, A. Locatelli, A. Baraldi, S. Lizzit, G. Paolucci, R. Rosei, Surf. Sci. 459 (2000) L467.
[67] J. T. Hoeft, M. Polcik, D. Sayago, M. Kittel, R. Terborg, R.L. Toomes, J. Robinson, D. P. Woodruff, M. Pascal, G. Nisbet, C. L. A. Lamont, Surf. Sci. 540 (2003) 441.
[68] A. Föhlisch, M. Nyberg, J. Hasselström, O. Karis, L. G. M. Pettersson, A. Nilsson, Phys. Rev. Lett. 85 (2000) 3309.
[69] P. Feibelmann, B. Hammer, J. K. Nørskov, F. Wagner, M. Scheffler, R. Stumpf, R. Watwe, J. Dumesic, J. Phys. Chem. B. 105 (2001) 4018.
[70] A. Gil, A. Clotet, J. M. Ricart, G. Kresse, M. Garcia-Hernández, N. Rösch and P. Sautet, Surf. Sci. 530 (2003) 71.
[71] G. Kresse, A. Gil, P. Sautet, Phys. Rev. B 68 (2003) 073401.
[72] L. Köhler, G. Kresse, Phys. Rev. B 70 (2004) 165405.
[73] J. Stöhr, J. L. Gland, W. Eberhardt, D. Outka, R. J. Madix, F. Sette, R. J. Koestner, U. Doebler, Phys. Rev. Lett. 51 (1983) 2414.
[74] U. Starke, A. Barbieri, N. Materer, M. A. Van Hove, G. A. Somorjai, Surf. Sci. 286 (1993) 1.
[75] A. Wander, M. A. Van Hove, G. A. Somorjai, Phys. Rev. Lett. 67 (1991) 626.
[76] J. E. Demuth, D. E. Eastman, Phys. Rev. Lett. 32 (1974) 1123.
[77] S. Lehwald, H. Ibach, Surf. Sci. 89 (1979) 425.
[78] N. Sheppard, Ann. Rev. Phys. Chem. 39 (1988) 589.
[79] S. Bao, Ph. Hofmann, K. -M. Schindler, V. Fritzsche, A. M. Bradshaw, D. P. Woodruff, C. Casado, M. C. Asensio, Surf. Sci. 307–309 (1994) 722.
[80] S. Bao, Ph. Hofmann, K. -M. Schindler, V. Fritzsche, A. M. Bradshaw, D. P. Woodruff, C. Casado, M. C. Asensio, Surf. Sci. 323 (1995) 19.
[81] B. J. Bandy, M. A. Chesters, M. E. Pemble, G. S. McDougall, N. Sheppard, Surf. Sci. 139 (1984) 87.
[82] N. D. S. Canning, M. D. Baker, M. A. Chesters, Surf. Sci. 111 (1981) 441.
[83] S. Bao, K. -M. Schindler, Ph. Hofmann, V. Fritzsche, A. M. Bradshaw, D. P. Woodruff, Surf. Sci. 291 (1993) 295.
[84] K. Hermann, M. Witko, Surf. Sci. 337 (1995) 205.
[85] L. Triguero, L. G. M. Pettersson, B. Minhaev, H. Ågren, J. Chem. Phys. 108 (1998) 1193.
[86] H. Hoffmann, F. Zaera, R. M. Ormerod, R. M. Lambert, J. M. Yao, D. K. Saldin, L. P. Wang, D. W. Bennett, W. T. Tysoe, Surf. Sci. 268 (1992) 1.
[87] C. J. Baddeley, A. F. Lee, R. M. Lambert, T. Gieβel, O. Schaff, V. Fernandez, K.-M. Schindler, A. Theobald, C. J. Hirschmugl, R. Lindsay, A. M. Bradshaw, D. P. Woodruff, Surf. Sci. 400 (1998) 166.
[88] J.-H. Kang, R. L. Toomes, J. Robinson, D. P. Woodruff, O. Schaff, R. Terborg, R. Lindsay, P Baumgärtel, A. M. Bradshaw, Surf. Sci. 448 (2000) 23.
[89] L. G. M. Pettersson, H. Ågren, Y. Luo, L. Triguero, Surf. Sci. 408 (1998) 1.
[90] C. Stellwag, G. Held, D. Menzel, Surf. Sci. 325 (1995) L379.
[91] G. Held, M. P. Bessent, S. Titmuss, D. A. King, J. Chem. Phys. 105 (1996) 11305.
[92] O. Schaff, V. Fernandez, Ph. Hofmann, K. -M. Schindler, A. Theobald, V. Fritzsche, A. M. Bradshaw, R. Davis, D. P. Woodruff, Surf. Sci. 348 (1996) 89.
[93] I. S. Butler, J. F. Harrod, Inorganic Chemistry: Principles and Applications (Benjamin/Cummings, Redwood City, CA 1989) pp. 679–689.
[94] E. Uhlig, *Organometallics*, 12 (1993) 4751.

[95] H.-P. Steinrück, W. Huber, T. Pache, D. Menzel, Surf. Sci. 218 (1989) 293.
[96] W. Huber, P. Zebisch, T. Bornemann, H. –P. Steinrück, Surf. Sci. 258 (1991) 16.
[97] J. Stöhr, D. Outka, R. J. Madix, U. Döbler, Phys. Rev. Lett. 54 (1985) 1256.
[98] D. Outka, R. J. Madix, J. Stöhr, Surf. Sci. 164 (1985) 235.
[99] A. Puschmann, J. Haase, M. D. Crapper, C. E. Riley, D. P. Woodruff, Phys. Rev. Lett. 54 (1985) 2250.
[100] M. D. Crapper, C. E. Riley, D. P. Woodruff, A. Puschmann, J. Haase, Surf. Sci. 171 (1986) 1.
[101] D. P. Woodruff, C. F. McConville, A. L. D. Kilcoyne, Th. Lindner, J. Somers, M. Surman, G. Paolucci, A. M. Bradshaw, Surf. Sci. 201 (1988), 228.
[102] A. Wander, B. W. Holland, Surf. Sci. 199 (1988) L403.
[103] K.-U. Weiss, R. Dippel, K. -M. Schindler, P. Gardner, V. Fritzsche, A. M Bradshaw, A. L. D. Kilcoyne, D. P. Woodruff, Phys. Rev. Lett. 69 (1992) 3196.
[104] M. Pascal, C. L. A. Lamont, M. Kittel, J. T. Hoeft, R. Terborg, M. Polcik, J. H. Kang, R. Toomes, D. P. Woodruff, Surf. Sci. 492 (2001) 285.
[105] O. Karis, J. Hasselström, N. Wassdahl, M. Weinelt, A. Nilsson, M. Nyberg, L. G. M. Pettersson, J. Stöhr, M. G. Savant, J. Chem. Phys. 112 (2000) 8146.
[106] S. M. Barlow, K. J. Kitching, S. Haq, N. V. Richardson, Surf. Sci. 401 (1998) 322.
[107] J. Hasselström, O. Karis, M. Weinelt, N. Wassdahl, A. Nilsson, M. Nyberg, L. G. M. Pettersson, M.G. Samant, J. Stöhr, Surf. Sci. 407, 221 (1998).
[108] N. A. Booth, D. P. Woodruff, O. Schaff, T. Gieβel, R. Lindsay, P. Baumgartel, A. M. Bradshaw, Surf. Sci. 397 (1998) 258.
[109] J.-H. Kang, R. L. Toomes, M. Polcik, M. Kittel, J. T. Hoeft, V. Efstathiou, D. P. Woodruff and A. M. Bradshaw, J. Chem. Phys. 118 (2003) 6059.
[110] S. M. Barlow, R. Raval, Surf. Sci. Rep. 50 (2003) 201.
[111] R. B. Rankin, D. S. Sholl, J. Phys. Chem. B 109 (2005) 16764.
[112] R. L. Toomes, J.-H. Kang, D. P. Woodruff, M. Polcik, M. Kittel, J.-T Hoeft, Surf. Sci. 522 (2003) L9.
[113] Q. Chen, D. J. Frankel, N. V. Richardson, Surf. Sci. 497 (2002) 37.
[114] X. Zhao, Z. Gai, R. G. Zhao, W. S. Yang, T. Sakurai, Surf. Sci. 424 (1999) L347.
[115] X. Zhao, H. Wang, R. G. Zhao, W. S. Yang, Mat. Sci. Eng. C 16 (2001) 41.
[116] M. Nyberg, M. Odelius, A. Nilsson, L. G. M. Pettersson, J. Chem. Phys. 119 (2003) 12577.
[117] D. I. Sayago, M. Polcik, G. Nisbet, C. L. A. Lamont, D. P. Woodruff, Surf. Sci. 590 (2005) 76.
[118] R. B. Rankin, D. S. Sholl, Surf. Sci. 574 (2005) L1.
[119] M. Ortega Lorenzo, S. Haq, P. Murray, R. Raval, C. J. Baddeley, J. Phys. Chem. B 103 (1999) 10661.
[120] L. A. M. M. Barbosa, P. Sautet, J. Am. Chem. Soc. 123 (2001) 6639.
[121] R. Fasel, J. Wilder, C. Quitmann, K.-H. Ernst, T. Greber, Angew. Chem. Int. Ed. 43 (2004) 2853.
[122] M. Parschau, T. Kampen, K.-H. Ernst, Chem. Phys. Lett. 407 (2005) 433.
[123] H. L. Meyerheim, Th. Gloege, H. Maltor, Surf. Sci. 442 (1999) L1029.
[124] H. L. Meyerheim, Th. Gloege, M. Sokolowski, E. Umbach, P. Bäuerle, Europhys. Lett. 52 (2000) 144.
[125] L. Kilian, W. Weigand, E. Umbach, A. Langner, M. Sokolowski, H. L. Meyerheim, H. Maltor, B. C. C. Cowie, T. Lee, P. Bäuerle, Phys. Rev. B 66 (2002) 075412.
[126] J. T. Yates, Jr. Science 279 (1998) 335.
[127] S. Bengió, H. Ascolani, N. Franco, J. Avila, M. C. Asensio, E. Dudzik, I. T. McGovern, T. Gieβel, R. Lindsay, A. M. Bradshaw, D. P. Woodruff, Phys. Rev. B 66 (2002) 195322.
[128] N. Franco, J. Avila, M. E. Davila, M. C. Asensio, D. P. Woodruff, O. Schaff, V. Fernandez, K.-M. Schindler, V. Fritzsche, A. M. Bradshaw, Phys. Rev. Lett. 79 (1997) 673.

[129] N. Franco, J. Avila, M. E. Davila, M. C. Asensio, D. P. Woodruff, O. Schaff, V. Fernandez, K.-M. Schindler, V. Fritzsche, A. M. Bradshaw, J. Phys.: Condens. Matter 9 (1997) 8419.
[130] A. Koebbel, M. Polcik, D. R. Lloyd, I. T. McGovern, O. Schaff, R. Lindsay, A. J. Patchett, A. M. Bradshaw, D. P. Woodruff, Surf. Sci. 540 (2003) 246.
[131] S. Bengió, H. Ascolani, N. Franco, J. Avila, M. C. Asensio, A. M. Bradshaw, D. P. Woodruff, Phys. Rev. B 69 (2004) 125340.
[132] P. Baumgärtel, R. Lindsay, O. Schaff, T. Giessel, R. Terborg, J. T. Hoeft, M. Polcik, A. M. Bradshaw, M. Carbone, M. N. Piancastelli, R. Zanoni, R. L. Toomes, D. P. Woodruff, New. J. Phys 1 (1999) 20.
[133] R. Terborg, P. Baumgärtel, R. Lindsay, O. Schaff, T. Gießel, J. T. Hoeft, M. Polcik, R. L. Toomes, S. Kulkarni A. M. Bradshaw, D. P. Woodruff, Phys. Rev. B 61 (2000) 16697.
[134] S. H. Xu, M. Keeffe, Y. Yang, C. Chen, M. Yu, G. J. Lapeyre, E. Rotenberg, J. Denlinger, J. T. Yates, Jr. Phys. Rev. Lett. 84 (2000) 939.
[135] L. Li, C. Tindall, O. Takaoka, Y. Hasegawa, T. Sakurai, Phys. Rev. B 56 (1997) 4648.
[136] R. A. Wolkow, Ann. Rev. Phys. Chem. 50 (1999) 413.
[137] S. Mezhenny, I. Lyubinetsky, W. J. Choyke, R. A. Wolkow, J. T. Yates, Jr., Chem. Phys. Lett. 344 (2001) 7.
[138] R. Terborg, M. Polcik, J. T. Hoeft, M. Kittel, D. I. Sayago, R. L. Toomes, D. P. Woodruff, Phys. Rev. B 66 (2002) 085333.
[139] K. A. R. Mitchell, Surf. Sci. 149 (1985) 93.
[140] K. A. R. Mitchell, S. A. Schlatter, R. N. S. Sodhi, Can. J. Chem. 64 (1986) 1435.
[141] L. Pauling, *The Nature of the Chemical Bond* (Cornell University Press, Ithaca, NY, 1960).
[142] D. I. Sayago, J. T. Hoeft, M. Polcik, M. Kittel, R. L. Toomes, J. Robinson, D. P. Woodruff, M. Pascal, C. L. A. Lamont, G. Nisbet, Phys. Rev. Lett. 90 (2003) 116104.
[143] J. T. Hoeft, M. Polcik, D. I. Sayago, M. Kittel, R. Terborg, R. L. Toomes, J. Robinson, D. P. Woodruff, M. Pascal, G. Nisbet, C. L. A. Lamont, 'Surf. Sci. 540 (2003) 441.
[144] K. Heinz, E. Lang, K. Müller, Surf. Sci. 87 (1979) 595.
[145] M. Passler, A. Ignatiev, F. Jona, D. W. Jepsen, P. M. Marcus, Phys. Rev. Lett. 43 (1979) 360.
[146] S. Andersson, J. B. Pendry, J. Phys. C: Solid State Phys. 13 (1980) 3547.
[147] S. Y. Tong, A. Maldonado, C. H. Li, M. A. Van Hove, Surf. Sci. 94 (1980) 73.
[148] S. D. Kevan, R. F. Davis, D. H. Rosenblatt, J. G. Tobin, M. G. Mason, D. A. Shirley, C. H. Li, S. Y. Tong, Phys. Rev. Lett. 46 (1981) 1629.
[149] P. Uvdal, P.-A. Karlsson, C. Nyberg, S. Andersson, N. V. Richardson, Surf. Sci. 202 (1988) 167.
[150] H. Antonsson, A. Nilsson, N. Mårtensson, I. Panas, P. E. M. Siegbahn, J. Electron Spectros. Rel. Phenom. 54/55 (1990) 601.
[151] J. Lauterbach, M. Wittmann, J. Küppers, Surf. Sci. 279 (1992) 287.
[152] R. Davis, D. P. Woodruff, Ph. Hofmann. O. Schaff, V. Fernandez, K.-M. Schindler, V. Fritzsche, A. M. Bradshaw, J. Phys.: Condens. Matter 8 (1996) 1367.
[153] M. E. Davila, M. C. Asensio, D. P. Woodruff, K.-M. Schindler, Ph. Hofmann, K.-U. Weiss, R. Dippel, P. Gardner, V. Fritzsche, A. M. Bradshaw, J. C. Conesa, A. R. González-Elipe Surf. Sci. 311 (1994) 337.
[154] J. C. Tracey, J. Chem. Phys. 56 (1972) 2736.
[155] K. Christmann, O. Schober, G. Ertl, J. Chem. Phys. 60 (1974) 4719.
[156] J. T. Stuckless, N. Al-Sarraf, C. Wartnaby, D. A. King, J. Chem. Phys. 99 (1993) 2202.
[157] C. Nyberg, L. Westerlund, J. Jönsson, S. Andersson, J. Electron Spectros. Rel. Phenom. 54/55 (1990) 639.
[158] H. Tillborg, A. Nilsson, N. Mårtensson, Surf. Sci. 273 (1992) 47.
[159] H. Tillborg, A. Nilsson, N. Mårtensson, J. N. Andersen, Phys. Rev. B 47 (1993) 1699.
[160] H. C. Peebles, D. E. Peebles, J. M. White, Surf. Sci. 125 (1983) L87.

[161] B. E. Koel, D. E. Peebles, J. M. White, Surf. Sci. 125 (1983) 709.
[162] C. R. Brundle, A. F. Carley, Discuss. Faraday Soc. 60 (1975) 51.
[163] J. C. Fuggle, E. Umbach, D. Menzel, K. Wandelt, C. R. Brundle, Solid State Commun. 27 (1978) 65.
[164] P. S. Bagus, C. R. Brundle, K. Hermann, D. Menzel, J. Electr. Spectros. Rel. Phenom. 20 (1980).
[165] C. R. Brundle, P. S. Bagus, D. Menzel, K. Hermann Phys. Rev. B 24 (1981) 7041.
[166] J. Stöhr, R. Jaeger, Phys. Rev. B 26 (1982) 4111.
[167] K. Horn, J. DiNardo, W. Eberhardt, H.-J. Freund, E. W. Plummer, Surf. Sci. 118 (1982) 465.
[168] H.-J. Freund, R. P. Messmer, C. M. Kao, E. W. Plummer, Phys. Rev. B 31 (1985) 4848.
[169] C. M. Kao, R. P. Messmer, Phys. Rev. B 31 (1985) 4835.
[170] A. Nilsson, H. Tillborg, N. Mårtensson, Phys. Rev. Lett. 67 (1991) 1015.
[171] H. Tillborg, A. Nilsson, N. Mårtensson, J. Electron. Spectros. Rel. Phenom. 62 (1993) 73.
[172] P. Decleva, M. Ohno, J. Chem. Phys. 96 (1992) 8120.
[173] M. Ohno, P. Decleva, Surf. Sci. 269/270 (1992) 264.
[174] M. Ohno, P. Decleva, G. Fronzoni, Surf. Sci. 284 (1993) 372.
[175] M. Ohno, P. Decleva, Surf. Sci. 296 (1993) 87.
[176] N. V. Dobrodey, L. S. Cederbaum, F. Tarantelli, Phys. Rev. B 57 (1998) 7340.
[177] N. V. Dobrodey, L. S. Cederbaum, F. Tarantelli, Phys. Rev. B 58 (1998) 2316.
[178] E. J. Moler, S. A. Kellar, W. R. A. Huff, Z. Hussain, Y. Zheng, E. A. Hudson, Y. Chen, D. A. Shirley, Chem. Phys. Lett. 264 (1997) 502.
[179] D. I. Sayago, M. Kittel, J. T. Hoeft, M. Polcik, M. Pascal, C. L. A. Lamont, R. L. Toomes, J. Robinson, D. P. Woodruff, Surf. Sci. 538 (2003) 59.
[180] P. Bennich, A. Nilsson, T. Wiell, O. Karis, M. Weinelt, N. Wassdahl, M. Nyberg, L. G. M. Pettersson, J. Stöhr, M. Samant, Phys. Rev. B 57 (1998) 9275.
[181] D. W. Goodman, Appl. Surf. Sci. 19 (1984) 1.
[182] R. D. Kelley, D. W. Goodman, in: D. A. King, D. P. Woodruff (eds.), *The Chemical Physics of Solid Surfaces and Heterogeneous Catalysis, Vol. 4: Model Catalysis Reactions*, Elsevier, Amsterdam, 1982, 427.
[183] F. Sette, J. Stöhr, E. B. Kollin, D. J. Dwyer, J. L. Gland, J. L. Robbin, A. L. Johnson, Phys. Rev. Lett. 54 (1985) 935.
[184] G. Paolucci, M. Surman, K. C. Prince, L. Sorba, A. M. Bradshaw, C. F. McConville, D. P. Woodruff, Phys. Rev. B 34 (1986) 1340.
[185] W. Wurth, C. Schneider, E. Umbach, D. Menzel, Phys. Rev. B 34 (1986) 1336.
[186] R. Davis, D. P. Woodruff, O. Schaff, V. Fernandez, K.-M. Schindler, Ph. Hofmann, K.-U. Weiss, R. Dippel, V. Fritzsche, A. M. Bradshaw, Phys. Rev. Lett. 74 (1995) 1621.
[187] R. Davis, R. Toomes, D. P. Woodruff, O. Schaff, V. Fernandez, K.-M. Schindler, Ph. Hofmann, K.-U. Weiss, R. Dippel, V. Fritzsche, A. M. Bradshaw, Surf. Sci. 393 (1997) 12.
[188] P. Kaukasiona, M. Lindroos, P. Hu, D. A. King, C. J. Barnes, Phys. Rev. B 51 (1995) 17063.
[189] R. Lindsay, P. Baumgartel, R. Terborg, O. Schaff, A. M. Bradshaw, D. P. Woodruff, Surf. Sci. 425 (1999) L401.
[190] M. Polcik, R. Lindsay, P. Baumgärtel, R. Terborg, O. Schaff, A. M. Bradshaw, R. L. Toomes, D. P. Woodruff, Faraday Disc. 114 (1999) 141.
[191] J.-T. Hoeft, M. Kittel, M. Polcik, S. Bao, R. L. Toomes, J.-H. Kang, D. P. Woodruff, M. Pascal, C. L. A. Lamont Phys. Rev. Lett. 87 (2001) 086101.
[192] M. Kittel, J.-T. Hoeft, M. Polcik, S. Bao, R. L. Toomes, J.-H. Kang, D. P. Woodruff, M. Pascal, C. L. A. Lamont Surf. Sci. 499 (2002) 1.
[193] G. Pacchioni, G. Cogliandro, P. S. Bagus, Surf. Sci. 255 (1991) 344.
[194] M. Pöhlchen, V. Staemmler, J. Chem. Phys. 97 (1992) 2583.

[195] H. Kuhlenbeck, G. Odörfer, R. Jaeger, G. Illing, M. Menges, Th. Mull, H. -J. Freund, M. Pöhlchen, V. Staemmler, S. Witzel, C. Scharfschwerdt, K. Wennemann, T. Liedtke M. Naumann, Phys. Rev.B 43 (1991) 1969.
[196] T. Bredow, J. Phys. Chem. B 106 (2002) 7053.
[197] G. Pacchioni, C. Di Valentin, D. Dominguez-Ariza, F. Illas, T. Bredow, T. Klüner, V. Staemmler, J. Phys.: Condens. Matter 16 (2004) S2497.
[198] A. Rohrbach, J. Hafner, G. Kresse, Phys. Rev. B 69 (2004) 075413.
[199] A. Rohrbach, J. Hafner, Phys. Rev. B 71 (2005) 045405.
[200] U. Diebold, Surf. Sci. Rep. 48 (2003) 53.
[201] A. Fujishima, K. Honda, Nature 238 (1972) 37.
[202] R. Lindsay, A. Wander, A. Ernst, B. Montanari, G. Thornton, N. M. Harrison, Phys. Rev. Lett. 94 (2005) 246102.
[203] G. S. Parkinson, M. A. Muñoz-Márquez, P. D. Quinn, M. J. Gladys, R. E. Tanner, D. P. Woodruff, P. Bailey, T. C. Q. Noakes, Phys. Rev. B 73 (2006) 245409.
[204] H. Onishi, T. Aruga, C. Egawa, Y. Iwasawa, Surf. Sci. 193 (1998) 33.
[205] H. Onishi, T. Aruga, Y. Iwasawa, J. Catal. 146 (1994) 557.
[206] S. A. Chambers, S. Thevuthasan, Y. J. Kim, G. S. Herman, Z. Wang, E. D. Tober, R. X. Ynzunza, J. Morais, C. H. F. Peden, K. Ferris, C. S. Fadley, Chem. Phys. Lett. 267 (1997) 51.
[207] S. Thevuthasan, G. S. Herman, Y. J. Kim, S. A. Chambers, C. H. F. Peden, Z. Wang, R. X. Ynzunza, E. D. Tober, J. Morais, C. S. Fadley, Surf. Sci. 401 (1998) 261.
[208] B. E. Hayden, A. King, M. A. Newton, J. Phys. Chem. B 103 (1999) 203.
[209] A. Gutiérrez-Sosa, P. Martínez-Escolano, H. Raza, R. Lindsay, P. L. Wincott, G. Thornton, Surf. Sci. 471 (2001) 163.
[210] S. P. Bates, G. Kresse, M. J. Gillan, Surf. Sci. 409 (1998) 336.
[211] P. Käckell, K. Terakura, Appl. Surf. Sci. 166 (2000) 370.
[212] P. Käckell, K. Terakura, Surf. Sci. 461 (2000) 191.
[213] D. I. Sayago, M. Polcik, R. Lindsay, J. T. Hoeft, M. Kittel, R. L. Toomes, D. P. Woodruff, J. Phys. Chem. B 108 (2004) 14316.
[214] M. Bowker, P. Stone, R. Bennett, N. Perkins, Surf. Sci. 511 (2002) 435.
[215] L.-Q. Wang, K. F. Ferris, A. N. Shultz, D. R. Baer, M. H. Engelhard, Surf. Sci. 380 (1997) 352.
[216] F. Allegretti, S. O'Brien, M. Polcik, D. I. Sayago, D. P. Woodruff, Phys. Rev. Lett, 95 (2005) 226104.
[217] F. Allegretti, S. O'Brien, M. Polcik, D. I. Sayago, D. P. Woodruff, Surf. Sci. 600 (2006) 1487.
[218] R. Schaub, P. Thostrup, N. Lopez, E. Lœgsgaard, I. Stensgaard, J. K. Nørskov, F. Besenbacher, Phys. Rev. Lett. 87 (2001) 266104.
[219] J. Goniakowski, M. J. Gillan, Surf. Sci. 350 (1996) 145.
[220] P. J. D. Lindan, N. M. Harrison, J. M. Holender, M. J. Gillan, Chem. Phys. Lett. 261 (1996) 246.
[221] M. Casarin, C. Maccato, A. Vittadini, J. Phys. Chem. B 102 (1998) 10745.
[222] D. Vogtenhuber, R. Podloucky, J. Redinger, Surf. Sci. 402–404 (1998) 798.
[223] P. J. D. Lindan, N. M. Harrison, M. J. Gillan, Phys. Rev. Lett. 80 (1998) 762.
[224] E. V. Stefanovich, T. T. Truong, Chem. Phys. Lett. 299 (1999) 623.
[225] W. Langel, Surf. Sci. 496 (2002) 141.
[226] C. Zhang, P. J. D. Lindan, J. Chem. Phys. 118 (2003) 4620.
[227] M. Menetrey, A. Markovits, C. Minot, Surf. Sci. 524 (2003) 49.
[228] C. Zhang, P. J D. Lindan, J. Chem. Phys. 121 (2004) 3811.
[229] L. A. Harris, A. A. Quong, Phys. Rev. Lett. 93 (2004) 086105.
[230] A. V. Bandura, D. G. Sykes, V. Shapovalov, T. N. Troung, J. D. Kubicki, R. A. Evarestov, J. Phys. Chem. B 108 (2004) 7844.

[231] P. J. D. Lindan, C. Zhang, Phys. Rev. Lett. 95 (2005) 029601.
[232] L. A. Harris, A. A. Quong, Phys. Rev. Lett. 95 (2005) 029602.
[233] I. M. Brookes, C. A. Muryn, G. Thornton, Phys. Rev. Lett. 87 (2001) 266103.
[234] R. Schaub, P. Thostrup, N. Lopez, E. Lægsgaard, I. Stensgaard, J. K. Nørskov, F. Besenbacher, Phys. Rev. Lett. 87 (2001) 266104.

Chapter 2

Adsorbate Electronic Structure and Bonding on Metal Surfaces

Anders Nilsson[1,2] and Lars G. M. Pettersson[2]

[1]*Stanford Synchrotron Radiation Laboratory, Post Office Box 20450, Stanford CA 94309, USA*
[2]*FYSIKUM, Stockholm University, AlbaNova University Center, S-10691 Stockholm, Sweden*

1. Introduction

When an atom or molecule is adsorbed on a surface new electronic states are formed due to the bonding to the surface. The nature of the surface chemical bond will determine the properties and reactivity of the adsorbed molecule. In the case of *physisorption*, the bond is rather weak, of the order of 0.3 eV. The overlap of the wave functions of the molecule and the substrate is rather small and no major change in the electronic structure is usually observed. On the contrary, when the interaction energy is substantially higher, there are rearrangements of the valence levels of the molecule, a process often denoted *chemisorption*. The discrete molecular orbitals interact with the substrate to produce a new set of electronic levels, which are usually broadened and shifted with respect to the gas phase species. In some cases completely new electronic levels emerge which have no resemblance to the original orbitals of the free molecule.

It is essential to have tools that allow studies of the electronic structure of adsorbates in a molecular orbital picture. In the following, we will demonstrate how we can use X-ray and electron spectroscopies together with Density Functional Theory (DFT) calculations to obtain an understanding of the local electronic structure and chemical bonding of adsorbates on metal surfaces. The goal is to use molecular orbital theory and relate the chemical bond formation to perturbations of the orbital structure of the free molecule. This chapter is complementary to Chapter 4, which contains a phenomenological description of the surface chemical bond using a small number of parameters.

The electronic structure of the surface chemical bond is discussed in depth in the present chapter for a number of example systems taken from the five categories of bonding types: (i) atomic radical, (ii) diatomics with unsaturated π-systems (Blyholder model), (iii) unsaturated hydrocarbons (Dewar-Chatt-Duncanson model), (iv) lone pair interactions, and (v) saturated hydrocarbons (physisorption).

2. Probing the electronic structure

A large number of surface-sensitive spectroscopic techniques are available for the study of the electronic structure of adsorbed molecules on surfaces [1]. However, it is often important to enhance the local information around the adsorbed entity. X-ray emission spectroscopy (XES) provides a method to locally study the electronic properties centered around one atomic site [2–4]. This is particularly important when investigating complex systems such as molecular adsorbates with many different atomic sites. Figure 2.1 shows an N_2 molecule adsorbed on a Ni surface in a perpendicular geometry. The gray zone represents the charge density of the valence electrons extending outside the metal surface and we have made a cut around the adsorbate to see more deeply into the molecule. Inside we can see one particular molecular orbital overlapping both the nitrogen and Ni atoms and the core electrons that are localized to one atom. Since the inner and outer nitrogen

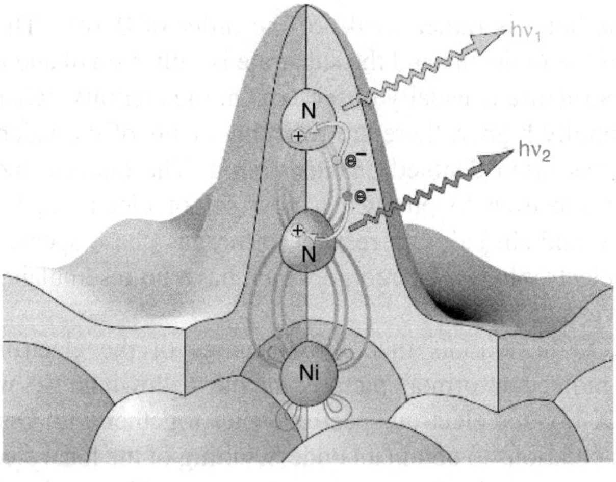

Figure 2.1. Schematic picture illustrating the local probe character in XES for N_2 adsorbed on a Ni surface. From the total charge density (gray envelope) valence electrons with *p*-angular momentum (contour lines) decay into the N 1*s* core hole. From Ref. [3].

atoms are non-equivalent the core-levels of the two atoms are shifted relative to each other [5]. We can create a core hole on either one of the nitrogen atoms through selective excitation using highly energy-resolved synchrotron radiation. The core holes can decay through transitions between core and valence levels with the emission of X-rays. Since the core electrons are localized to one atom we can, in an atom-specific way, study the valence electrons [3,6]. The decay process is governed by the dipole selection rule and with a N 1*s* core hole we will probe the N 2*p* projected valence states. Furthermore, angle-resolved measurements allow a separation of valence levels of different symmetries [3,6,7]. We thus have a tool to look into the nature of the surface chemical bond by disentangling the valence contributions from the different atoms.

The most common way to measure the occupied electronic structure is with valence band photoemission, also denoted Ultraviolet Photoelectron Spectroscopy (UPS), where the overall electronic structure is probed through ionization of the valence electrons [8]. Since the method is based on detection of electrons the technique is inherently surface sensitive and a significant fraction of the spectral contribution can be associated with the adsorbate. Figure 2.2 shows a comparison of UPS [9] and XES [6,10] spectra for N_2 adsorbed on Ni(100) (for a detailed discussion see Section 6). We can observe adsorbate-induced features in the region of 7–12 eV binding energy relative to the Fermi level in both spectroscopies. However, in the important regime around 0–5 eV, where the metal valence *d*-electrons reside, the UPS spectrum only shows strong substrate emission whereas the XE spectra reveal adsorbate-derived states. In addition, the XES provides a projection on the two different nitrogen atoms and a separation of π and σ states. However, there are special cases where the whole adsorbate electronic structure can be fully determined with UPS [11]. Using angular-resolved photoemission or UPS the k-dispersion and symmetry of the electronic states and thereby the band-structure can be measured [12]. If XES measures the local aspects of the electronic structure, UPS determines the more collective aspects. In this regard the two techniques become complementary to each other.

From a theoretical point of view XES furthermore provides a very strong basis for the evaluation of methods for population analysis, i.e., the decomposition of the molecular orbitals into atomic contributions [10]. Many different schemes subdividing the charge density into contributions assigned to respective atoms have been proposed, but the lack of means to directly measure the atomic populations in different orbitals has made all techniques somewhat arbitrary and a matter of taste. Due to the strongly localized character of the intermediate core-excited state, however, in combination with the direct dependence of the XES transition probability on the amount of local *p*-population (assuming a 1*s* core hole), XES provides a very sensitive tool to directly measure this atomic population. In all, we have an atom-specific tool, which can be used to address important questions regarding

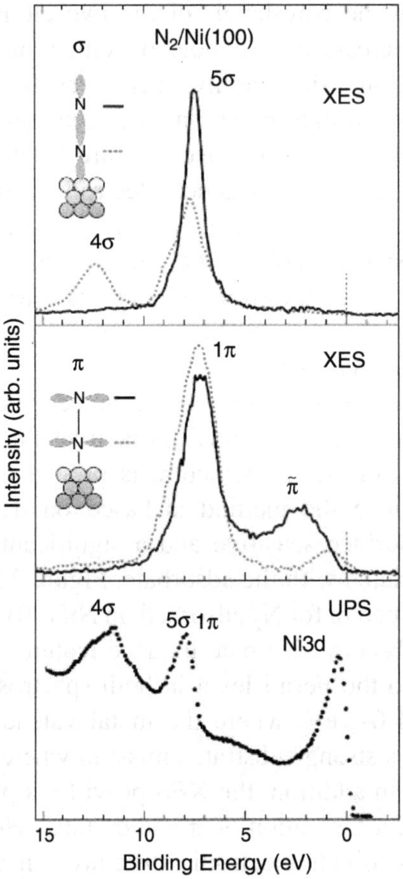

Figure 2.2. Comparison between atom-specific and symmetry-resolved XE spectra [6,10] with an UPS spectrum of N_2 adsorbed on Ni(100) measured at a photon energy of 35 eV [9]. From Ref. [3].

differences in the local electronic structure and surface bonding and which can be directly related to the molecular orbitals obtained from theoretical calculations.

Although the occupied orbitals are of main importance, since they are directly involved in the formation of the chemical bond, the unoccupied states also provide complementary information. In X-ray absorption spectroscopy (XAS), often denoted Near Edge X-ray Absorption Fine Structure (NEXAFS), we excite a core electron to the empty states above the Fermi level [3,4,13]. There is a close connection between XES and XAS where the former gives information on the occupied orbitals while the latter relates to the character and symmetry of the unoccupied levels. Both are governed by the dipole selection rule and the localized character of the core orbitals allows a simple atom-specific projection of the electronic structure; the major difference is in the final states. In XAS the empty electronic states are probed in the presence of the core hole which can lead to major deviations from a simple ground state picture [14]. One way to estimate the static influence of the core-hole, i.e., the shift in position and change of the shape of the spectrum, is to use the $Z+1$ approximation, which implies that the core-excited atom is replaced by the atom with nuclear charge increased by one $(Z+1)$ corresponding to the reduced screening due to the removal of a 1*s* electron [3,4,14]. In contrast, the XES final state contains no core hole and is similar to the valence hole state in UPS [3,14]. In analogy with the complementarity of XES and UPS, XAS thus gives a local projection of the unoccupied electronic states while the total unoccupied states can be probed using Inverse Photoemission Spectroscopy (IPS) [15].

The measured electronic structure, occupied or unoccupied, provides the fullest information when also combined with theory. Electronic structure calculations in surface chemistry have advanced immensely in the past decades and have now reached a level of accuracy and predictive power so as to provide a very strong complement to experiment. Indeed, the type of theoretical modeling that will be employed and presented here can be likened to computer experiments, where it can be assumed that spectra can be computed reliably and thus computed spectra for different models of the surface adsorption used to determine which structural model is the most likely. In the present chapter, we will thus consistently use the interplay between experiment and theory in our analysis of the interaction between adsorbate and substrate. Before discussing what quantities are of interest to compute in the analysis of the surface chemical bond, we will briefly discuss and justify our choice of Density Functional Theory (DFT) as approach to spectrum and chemisorption calculations.

DFT has emerged as a computational approach of comparable accuracy to the traditional correlated quantum chemical methods, but at a much lower computational cost. In relation to excited states it is sometimes said that DFT is only valid for the ground state. How can we then use DFT to obtain highly excited states as required for a treatment of XAS, XES and XPS? Here it should be noted that, although Hohenberg and Kohn in their seminal paper [16] specifically assumed the ground state, this is not a necessary constraint. The proof of the uniqueness of the energy functional relies on two assumptions: (1) that the density can be uniquely connected to a specific external potential and (2) the variational principle. Given that the external potential in chemistry is given by the nuclei this is already specified. The variational principle can be applied for any well-defined state that is bounded from below, i.e., the lowest excited triplet state, the lowest state with a specific occupation or the lowest state in a restricted variational space. Considering X-ray absorption we can variationally determine the lowest core-excited state using the constraint that the core-level should contain only one electron. Having determined the orbital corresponding to the lowest excitation we can now remove it from the variational space. The variational principle applied to the remaining orbital space, with the constraint on the core-occupation, will produce the best approximation to the second excited state. This procedure may be continued and leads to a well-defined set of orthogonal core-excited states [17].

Quantum chemical methods may be divided into two classes: wave function-based techniques and functionals of the density and its derivatives. In the former, a simple Hamiltonian describes the interactions while a hierarchy of wave functions of increasing complexity is used to improve the calculation. With this approach it is in principle possible to come arbitrarily close to the correct solution, but at the expense of interpretability of the wave function; the molecular orbital concept loses meaning for correlated wave functions. In DFT on the other hand, the complexity is built into the energy expression, rather than in the wave function which can still be written similar to a simple single-determinant Hartree-Fock wave function. We can thus still interpret our results in terms of a simple molecular orbital picture when using a cluster model of the metal substrate, i.e., the surface represented by a suitable number of metal atoms.

For interpretation of the spectroscopic data in terms of chemistry the molecular orbital picture is invaluable and we will extensively use this concept in our discussions of the surface chemical bond. The changes in the molecular orbitals as the adsorbate is brought in from gas phase to the surface provide a basis for the understanding of the bond-formation, but we also need to quantify the importance of the specific changes in terms of energetic contributions to the bond strength. This can be done using, e.g., Constrained Space Orbital Variation (CSOV) [18,19] analysis, in which the orbital space is subdivided and the fully relaxed electronic structure for the chemisorbed system is obtained by step-wise including specific orbital interactions. Starting with the adsorbate at infinite distance we can divide the orbital space into four subspaces: occupied orbitals on respectively adsorbate and substrate, and similar for the unoccupied states. Bringing the adsorbate into the equilibrium geometry on the cluster while disallowing all mixings between orbital classes gives the initial repulsion, which must be overcome through polarization, charge transfer and covalent bond-formation. The polarization effects are obtained by allowing mixing of occupied and unoccupied on adsorbate and cluster, respectively, while charge-transfer is obtained from allowing occupied on one mixing with the unoccupied of the other. Covalent bond-formation is then obtained when the wave function from these preceding steps is fully relaxed. At each step we can furthermore generate the density from the partially relaxed orbitals and use density differences between different steps to follow the flow of charge as the chemical bond is built up. The separation of the orbital spaces is exact at infinite distance, but at equilibrium the orbitals must necessarily mix due to the requirement of orthogonality. This energetic decomposition is thus not exact but does provide a reliable semi-quantitative picture of the importance of the different interactions [20]. Further decompositions, e.g., separating σ and π interactions, are of course possible and will be used in the following.

We should finally briefly discuss the calculation of spectra for the surface adsorbates which we will use to verify the theoretical models and to assign peaks in the spectra. The calculation of XES spectra has been discussed extensively previously [3]. Briefly we have shown that the ground state orbitals provide a balanced description of initial and final state and calculate the spectrum as the dipole transition between the valence orbitals and the selected $1s$ core level [21]. The success of this approach relies on similar charge transfer screening in the core-ionized initial (or intermediate) state as for the valence-ionized levels. XES thus reflects the ground state molecular orbitals.

XAS, on the other hand has a core-excited final state for which the effect of the core-hole must be taken into account. To obtain the full spectrum, i.e., valence, Rydberg and continuum excitations, we use the Slater transition-state approach [22,23] with a half-occupied core-hole. This provides a balanced description of both initial and final states allowing the same orbitals to be used to describe both initial and final states and all transitions are obtained in one calculation [23,24]. Details of the computational procedure can be found in the original papers as referenced in the following sections. In the present chapter, the focus is on the surface chemical bond and the spectra, measured or calculated, will mainly be used to obtain the required information on the electronic structure.

3. Adsorbate electronic structure and chemical bonding

Before going into details regarding different adsorbate systems on surfaces we will make some general remarks concerning the electronic structure of adsorbates on surfaces. In particular, we like to emphasize the uniqueness of the more localized *d*-electrons in transition metals, in comparison to the delocalized *s*- or *p*-electrons, in the formation of the chemical bond. Figure 2.3 shows a schematic illustration of the resulting electron density of states projected onto the adatom in the Newns-Anderson model [25,26] for two different cases. In this model, the interaction strength between the adatom wave function of one specific electronic level and the metal states is often denoted the hopping matrix element. In the case, when the hopping matrix element is much smaller than the band width of the metal states of interest the interaction leads to a broadened resonance-level of the projected states on the adatom (top Figure 2.3). The bottom of the resonance would reflect more bonding and the top more antibonding character of the wave function with respect to the adatom and the metal. If the band width is much smaller than the hopping matrix elements, then the bonding and antibonding states separate out as new distinct electronic levels, below and above the free adatom level. The latter is usually the case upon interaction

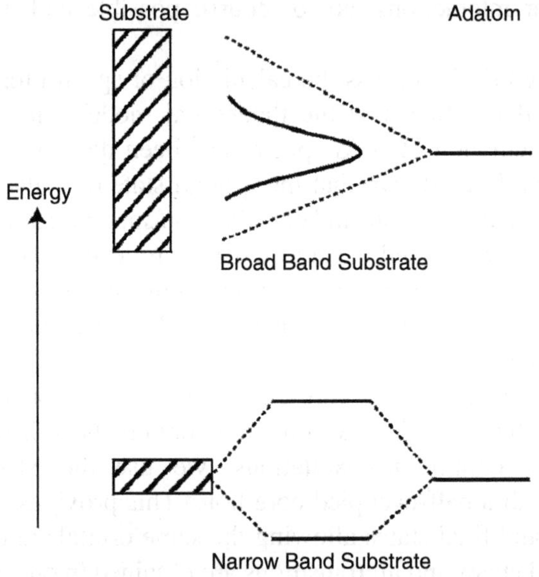

Figure 2.3. The projected adatom density of states in the Newns-Anderson model in two limiting cases: top) when the band width is larger and bottom) when the band width is smaller than the hopping matrix element.

with the more localized *d*-states and the former for the delocalized *s*- or *p*-states. In chapter 4, this model will be used in a DFT framework to predict trends in chemisorption energies of importance in catalysis.

We can directly observe the above two extreme cases in terms of adatom interactions using X-ray spectroscopy. Figure 2.4 shows XES measurements of atomic

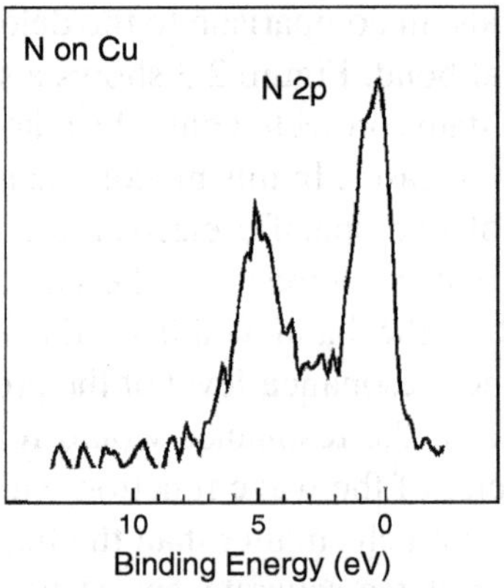

Figure 2.4. N K emission spectra of the N p_z component for atomic N adsorbed on Cu(100).

N adsorbed on Cu(100) [27], see Section 5 for more details. The experiment measures the N-2*p* density of states, i.e., the projection of the partial *p*-density of states onto the nitrogen adatom. The atomic level is split into two distinct levels which are bonding and antibonding, respectively, with respect to the Cu *d*-band. This represents the case when the hopping matrix element is larger than the width of the metal *d*-band.

The opposite case, i.e., when the band width is much larger than the hopping matrix element, can be seen in Figure 2.5 for the unoccupied K 4*s* states of K adsorbed on Ag. This has been measured using XAS of Ar adsorbed on Ag, since Ar using the $Z+1$ approximation becomes K as an effect of the final core hole state [28]. We can directly see that the K 4*s* level has become a broad asymmetric resonance. The adatom resonance has a tail towards lower energies with clear cut-off at the Fermi level. The 4*s* level mainly interacts with the delocalized unoccupied Ag *sp* electrons. Most of the 4*s* resonance is unoccupied which indicates that charge transfer has occurred from the adatom to the substrate.

It is essential to determine if the case where all electronic states are populated leads to a chemical bond. Let us consider some fundamental aspects of covalent bonding illustrated through the difference between the H_2 and He_2 molecules. The interaction of the 1*s* orbitals leads to bonding and antibonding orbitals. In the H_2 molecule only the bonding $1\sigma_g$ state is populated whereas in the He_2 molecule the antibonding $1\sigma_g$ orbital is also occupied. Figure 2.6 illustrates how charge has redistributed upon chemical bonding. The charge density difference plots are obtained by subtracting the frozen charge density of the two atoms at the correct bonding geometry from the fully relaxed molecular system. We directly see a large difference between the two molecular systems [29]. In the case of H_2 there is charge

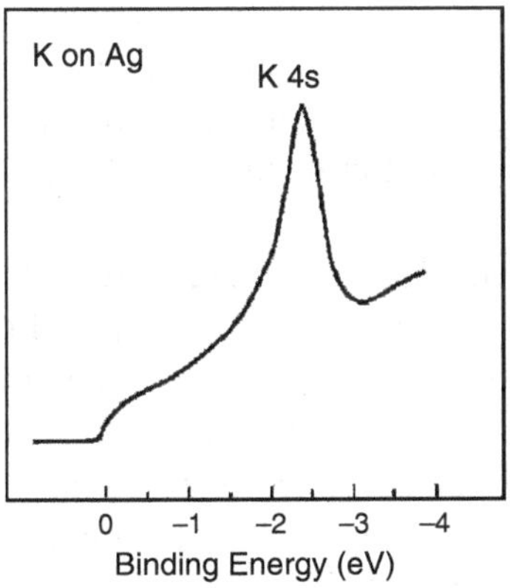

Figure 2.5. L-edge XAS measurements of Ar adsorbed on Ag(110). The projected Ar 4*s* states becomes the K 4*s* using the $Z+1$ approximation for the core hole state.

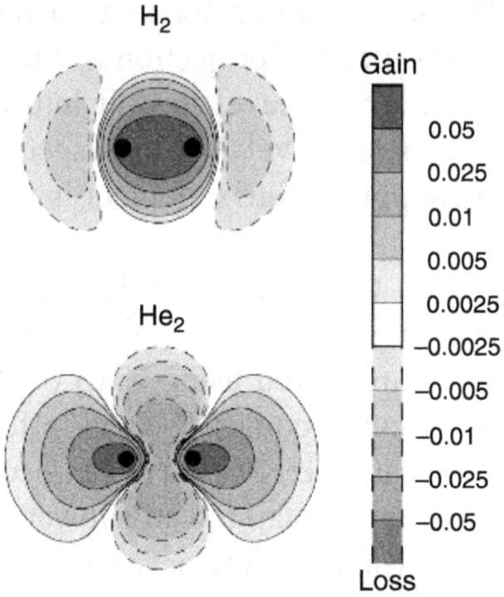

Figure 2.6. Molecule–atom charge density differences for H_2 and He_2 molecules. From Ref. [29].

building up between the two atoms forming an electron pair in the covalent bond. In He_2 there is instead a depletion of charge between the two connecting atoms due to repulsive interaction. Since both bonding and antibonding orbitals are occupied the resulting Pauli repulsion leads to a polarization of the charge away from the connection points in the charge distribution. For N adsorption on Cu as shown in Figure 2.4 both the bonding and antibonding states between N 2*p* and Cu 3*d* are occupied which would then lead to Pauli repulsion similar to the He_2 molecule. If we consider K on Ag and the 4*s* resonance, as shown in Figure 2.5, the bottom part of the resonance would imply bonding contribution and the top part antibonding. Since nearly the whole 4*s* resonance is unoccupied it would indicate a strong ionic contribution to the bonding. However, there is a small part of the tail that extends below the Fermi level indicative of a small covalent contribution. In this sense, the chemical bond of K on Ag would mainly be ionic with some small contribution from attractive covalent bonding. Let us take the discussion of ionic interactions on surfaces a little bit further.

Calculated adatom-induced densities of states for Cl, Si, and Li on jellium with an electron density corresponding to Al are shown in Figure 2.7 [30]. The location of the resonances relative to the Fermi level will determine the degree of electron occupation. The three different adatom cases represent atoms with a wide range of electronegativity. The Li 2*s* resonance is essentially above the Fermi level and resembles the K 4*s* resonance in Figure 2.5. It implies that charge has been taken away into the substrate. The opposite is seen for Cl where the 3*p* resonance is nearly fully occupied with transfer of charge from the metal to the adsorbate. These two cases provide clear examples of positive and negative ion adsorption as expected

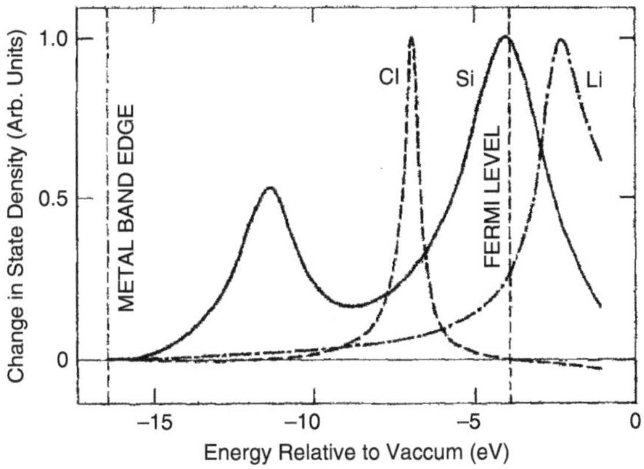

Figure 2.7. Theoretical adatom-projected density of states of different atoms on jellium with an electron density that corresponds to Al metal. Reproduced from [30].

based on the electronegativity with respect to Al metal. The density of states for Si adsorption shows two resonances which are related to Si 3*s* and 3*p* states. The 3*s* resonance is fully occupied meaning that the bonding and antibonding contributions counterbalance each other and cancel out any contributions to the bonding whereas the 3*p* resonance straddles the Fermi level. The occupied part of the Si 3*p* resonance would mainly correspond to bonding states resulting in covalent bond formation. It is interesting to note that the electronegativity of Si and Al are rather similar which implies that charge has been transferred in both directions to and from the adsorbed atom in the formation of the chemical bond. The covalent contribution in Cl adsorption is small since both bonding and antibonding contributions of the resonance are occupied. The same holds for Li adsorption since only a small part of the resonance is occupied.

Contours of constant charge density are shown in the top row in Figure 2.8 and the difference in charge density due to adsorption is shown in the middle row [30]. The total density contours in the vicinity of the Li and Cl atoms are more circular than those for Si, which show more of accumulation of charge into the bond region. We can notice that the substrate electrons are pulled towards the positively charged Li atom and pushed away from the negatively charged Cl. The difference in charge density shows clearly the displacements of charge resulting from chemical bond formation. Upon Li adsorption, charge has been transferred from the vacuum side of the adatom towards the metal. In the case of Cl adsorption the reverse is seen with transfer of charge from the substrate to the adatom. For Si, we see that the charge has decreased both in the central region of the adatom and in the substrate surface plane but increased in-between the substrate and the adatom. This is characteristic

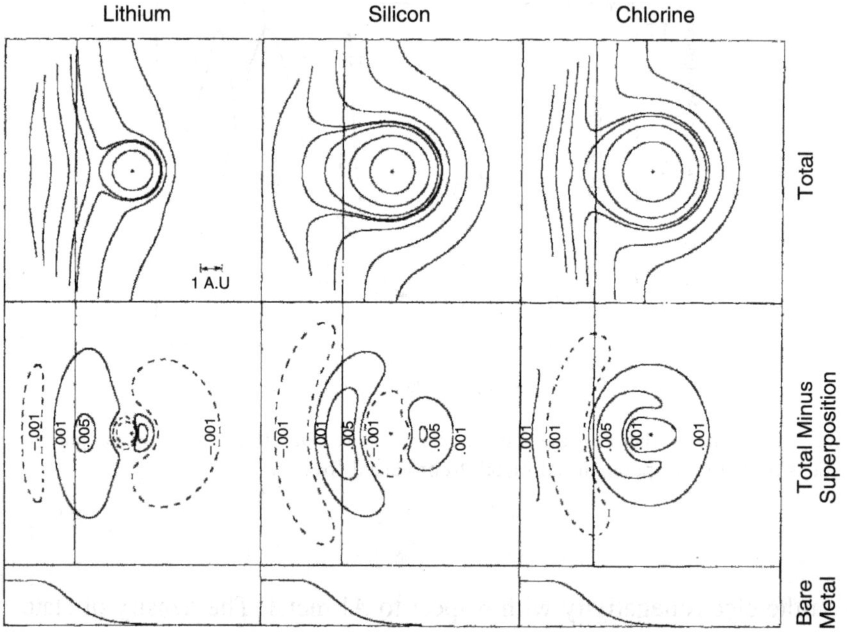

Figure 2.8. Electron density contours for atomic chemisorption on jellium with electron density that corresponds to Al metal. Upper row: Contours of constant electron density in the plane normal to the surface. Center row: Difference in charge density between isolated adatom and metal surface, full line gain and dashed line loss of charge density. Bottom row: Bare metal electron density profile. Reproduced from [30].

of covalent bonding and similar to the H_2 case, shown in Figure 2.6, but with the difference of involving *p*-orbitals instead of *s*-orbitals.

4. Adsorbate systems

In the following, we will discuss a number of different adsorption systems that have been studied in particular using X-ray emission spectroscopy and valence band photoelectron spectroscopy coupled with DFT calculations. The systems are presented with a goal to obtain an overview of different interactions of adsorbates on surfaces. The main focus will be on bonding to transition metal surfaces, which is of relevance in many different applications in catalysis and electrochemistry. We have classified the interactions into five different groups with decreasing adsorption bond strength; (1) radical chemisorption with a broken electron pair that is directly accessible for bond formation; (2) interactions with unsaturated π electrons in diatomic molecules; (3) interactions with unsaturated π electrons in hydrocarbons;

(4) lone pair interactions, and (5) saturated hydrocarbons which represent molecular systems without π or lone pair orbitals.

5. Radical atomic adsorption

These systems represent bonding to surfaces where the adsorbate atoms have unpaired electrons available for covalent interaction with unsaturated electronic states on the metal surface. We denote this bonding mechanism as radical adsorption where the open-shell electrons on the adsorbate atom can form electron pairs with the metal atoms at the surface. These radical atoms have in most cases been obtained through the dissociation of molecules on the surface. Let us make a simple picture of the electronic structure when a simple atomic adsorbate interacts with a transition metal, denoted the d-band model [31,32]. A similar description can also be found in Chapter 4.

We consider an adsorbate valence state interacting with the metal states on a transition metal as illustrated in Figure 2.9. The valence states of the transition metal surface atoms can be divided up into the free-electron-like *s*-electron states and the more localized *d*-electron states. Based on the discussion in Section 3 we would then expect that the interaction with the *s*-electrons would lead to a broad resonance whereas interaction with the *d*-states would lead to distinct new levels. We can imagine including the coupling between the adsorbate atomic level to the metal *s*-states first and then switching on the coupling to the metal *d*-states later. The coupling to the broad *s*-band then leads to a broadening and shift of the adsorbate state [26,30,32] (see Figure 2.9). There may be a large energy involved in this interaction, but since all the transition metals have a half-filled *s*-band in the metallic state and since the band is broad, there will only be small differences in this interaction from one metal to the next. The differences between the different transition metals must be associated primarily with the *d*-states. The interaction of the adsorbate states with a narrow distribution of states will give rise to the

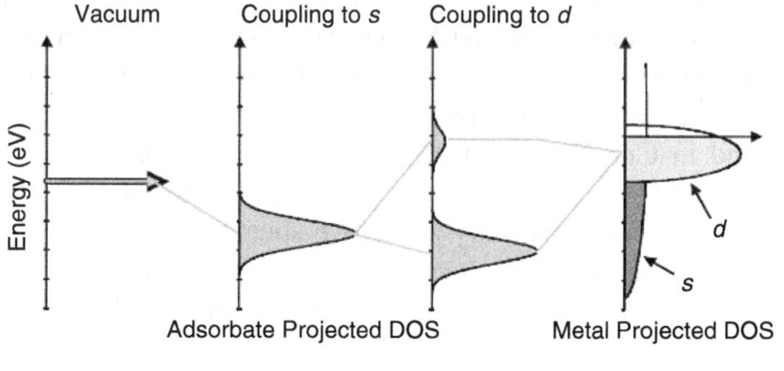

Figure 2.9. Schematic illustration of the formation of a chemical bond between an adsorbate valence level and the *s*- and *d*-states of a transition metal surface. Reproduced from [32].

formation of separate bonding and anti-bonding states as shown in the bottom of Figure 2.3. When bonding and anti-bonding states are formed, the strength of the bond will depend on the relative occupancy of these states. If only the bonding states are filled, there will be a strong bond, whereas if the anti-bonding states are also filled the bond becomes considerably weaker. This is essentially the same picture as the difference between H_2 and He_2 molecules as shown in Figure 2.6. We will in this chapter in particular discuss the appearance of these bonding and antibonding states in the electronic structure of the adsorbates on transition metals. Since the energy of the *d*-states relative to the Fermi level varies substantially from one metal to the next, the number of anti-bonding states that are above the Fermi level, and thus empty, will depend on the metal in question. As the *d*-states shift up in energy going from right (Cu) towards the left in the first transition-row, the strength of the adsorbate-metal bond should increase.

5.1. The electronic structure of N on Cu(100)

We will first demonstrate how we can observe the bonding and antibonding states due to the interaction of the *d*-band in atomic N adsorption on Cu(100). Since the 3*d* band in Cu is located a few eV below the Fermi level a significant fraction of the N 2*p*-Cu 3*d* states will be occupied and can be directly observed in XES and UPS. There is only a single phase of N adsorbed on Cu(100) showing a c(2 × 2) low-energy electron diffraction (LEED) pattern [33]. A surface extended X-ray absorption fine structure (SEXAFS) study [34] proposed that the N atoms are located in four-fold hollow (FFH) sites at a vertical height of 0.4 Å above the topmost Cu layer and thus being nearly coplanar with the surface making the geometry relatively simple.

Figure 2.10 shows a UPS spectra for clean and N covered Cu(100) in comparison with N 1*s* XES spectra [35]. The clean Cu UPS spectrum shows a strong spectral feature around 2–5 eV, which is due to the Cu *d*-band and a weak flat distribution close to the Fermi level which is related to the Cu *sp*-band. Upon adsorption there are some new weak features seen around 1 and 5 eV binding energy. The same structures are clearly seen in the XES spectrum, which has been put on a binding energy scale by subtracting the core-level binding energy from the emission energies [3]. We can directly note that the two components are located below and above the *d*-band in Cu. These can be attributed to the bonding and antibonding states with respect to the Cu *d*-orbitals. From the figure we can clearly see the difference in the adsorbate sensitivity in the two spectroscopies that probe the occupied electronic structure. In XES, we project the electronic states onto the adsorbate atom whereas in photoemission the joint electronic structure of the whole system is probed.

Figure 2.11 shows XES spectra of N adsorbed on Cu(100) symmetry-resolved in $2p_{xy}$ and $2p_z$ components [27] where xy is in the surface plane and z perpendicular

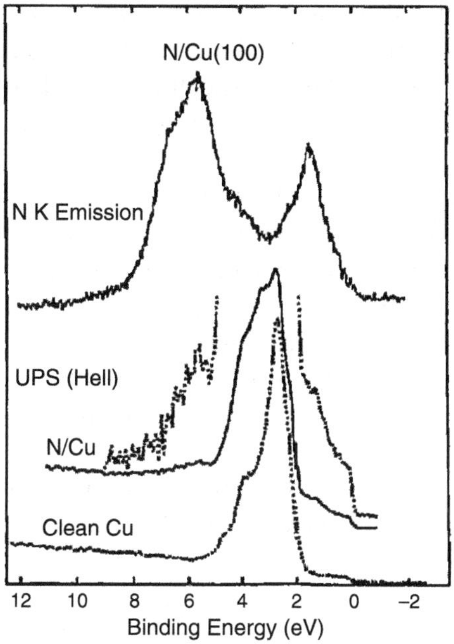

Figure 2.10. Comparison between XES and UPS spectra for N adsorbed on Cu(100). From Ref. [35]

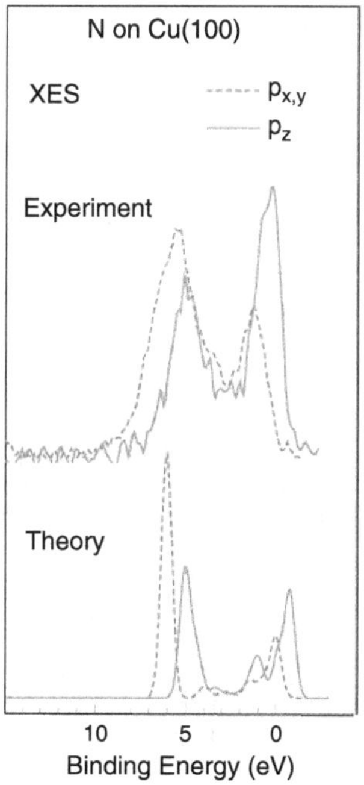

Figure 2.11. Experimental and theoretical symmetry-resolved XES spectra of N adsorbed on Cu(100). From Ref. [3].

to the surface. The spectra were obtained by assuming that half of the intensity at grazing emission comes from the $2p_{xy}$ and the other half from the $2p_z$ component [3]. It is interesting to observe that the $2p_z$ spectrum is shifted towards lower energy in comparison with the $2p_{xy}$ spectrum. In the lower part of Figure 2.11, we show the theoretical spectra obtained by DFT calculations on a $Cu_{61}N$ cluster to simulate the XES spectroscopic process [36]. There is in general a good agreement between experiment and theory. The theoretical spectra are much narrower in comparison with the experiment which partly can be related to the fact that only a single N atom is treated in the calculation. The partial N $2p_{xy}$ and N $2p_z$ densities of states obtained by band structure calculation of the c(2 × 2) overlayer show a much broader distribution [27,32]. We can also see in the experimental spectra substantial intensity between the two features corresponding to energies where the Cu *d*-band is located.

Figure 2.12 shows the atomic N 2*p* orbitals surrounded by atomic Cu 3*d* orbitals in an ideal, FFH adsorbate geometry where the N atom is located at the same vertical

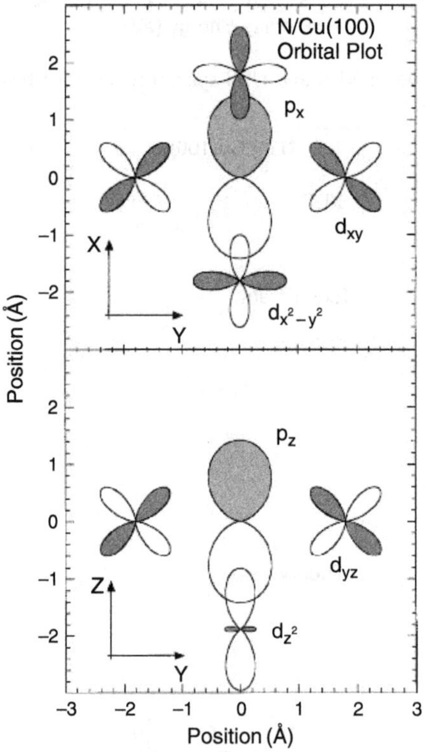

Figure 2.12. Schematic N 2*p* and Cu 3*d* orbital plots. The top figure shows the surface from above with the N sitting in a hollow adsorption site surrounded by four Cu atoms. The bottom picture shows a side-view with a Cu atom from the second layer located directly below the N atom. From Ref. [27].

height as the first-layer Cu atoms. The upper part of the figure shows the in-plane N $2p_{xy}$ atomic orbitals while the lower part shows the out-of-plane N $2p_z$ orbital. We have chosen to plot only the N $2p_x$ and $2p_z$ orbitals together with selected Cu $3d$ orbitals that have non-zero overlap with the N $2p$ orbitals. The simplified adsorbate geometry has C_{4v} symmetry and therefore the N $2p_y$ and Cu $3d_{xz}$ orbitals are obtained by a 90° rotation about the z-axis. The N $2p_x$ orbital forms both σ- and π-bonds with the Cu $3d$ orbitals. The σ-bonds result from the overlap with the first-layer Cu $3d_{x^2-y^2}$ and $4s$ orbitals, while the π-bonds arise from the overlap with the Cu $3d_{xy}$ orbitals in the first layer and the $3d_{xz}$ orbital on the second-layer Cu atom located directly below the adsorbate. For this idealized geometry the remaining nearest-neighbor Cu orbitals have zero overlap with the N $2p_x$ orbital. The lower portion of Figure 2.12 shows that the N $2p_z$ orbital forms π-bonds with the first-layer Cu $3d_{yz}$ and $3d_{xz}$ orbitals. The N $2p_z$ σ-bonds result from the overlap with the second-layer Cu $3d_{z^2}$ and $4s$ orbitals. The actual adsorbate geometry involves a displacement of the N atom outwards away from the first Cu layer by 0.4 Å. This displacement makes possible contributions from additional Cu orbitals other than those shown for the idealized geometry in Figure 2.12 and also introduces some mixing of the N $2p_{xy}$ and N $2p_z$ orbitals in the σ- and π-bonds described above.

Figure 2.13 shows the surface Brillouin zone (SBZ) of the c(2 × 2) overlayer on Cu(100). The high-symmetry points in the SBZ are denoted Γ, X, and M. Since there are one N and two Cu atoms per unit cell there should exist two bands for each N $2p$ atomic orbital. Thus there are six predominantly adsorbate-derived bands arising from the bonding and antibonding combinations of the N $2p_x$, $2p_y$ and $2p_z$ orbitals with the underlying Cu $3d$ and $4s$ orbitals. Each of these six bands contains both σ and π contributions at different points in the SBZ, as shown in Figure 2.13. A straightforward tight-binding analysis of the model problem [37] can directly indicate the bonding nature of the N-derived surface states at the different symmetry points. This is shown in Figure 2.13 for the N $2p_x$ and N $2p_y$ orbitals in the xy plane (parallel to the surface) in terms of bonding (or antibonding) interactions. There is a degeneracy for the N $2p_x$ and N $2p_y$ bands at the Γ and M points and therefore only the N $2p_x$ orbital is shown. At the Γ point the phases of the wave functions of the two Cu $3d_{x^2-y^2}$ orbitals and the two Cu $3d_{xy}$ orbitals have to be the same leading to both an in-phase and out-of-phase overlap with the N $2p_x$ orbital. This leads to σ and π non-bonding interactions. At the M point the phase of the Cu orbitals in both x and y directions have to change sign between neighboring sites which gives σ and π bonding interactions. We therefore expect that one of the adsorbate bands will disperse to higher binding energies from the zone center to the M edge since it becomes more bonding while the other band shows the opposite dispersion since it becomes more antibonding. At the X point the interactions are generally a mixture of non-bonding as well as bonding or antibonding combinations. The σ

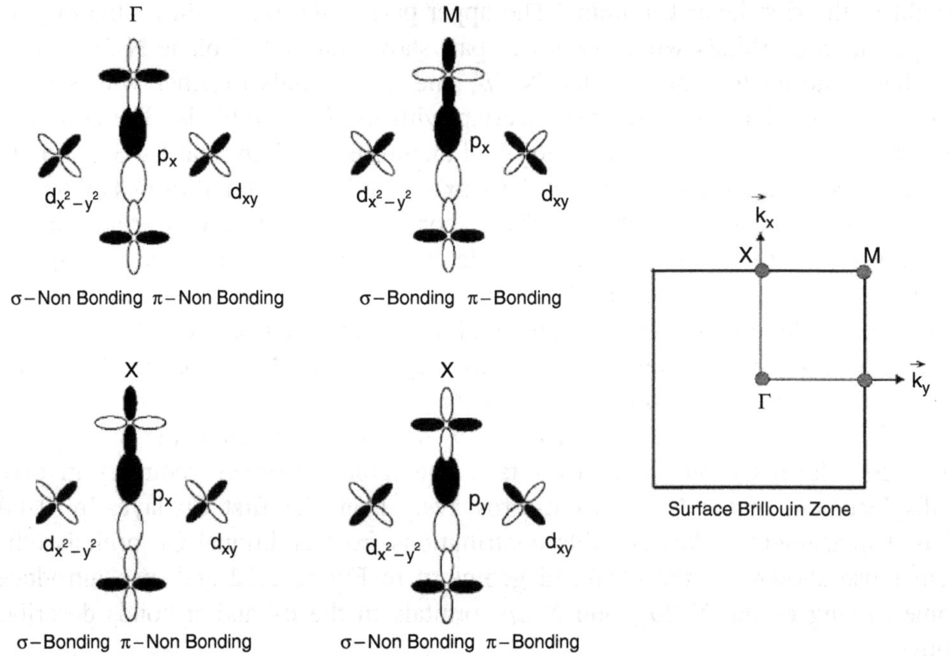

Figure 2.13. Schematic N $2p_x$, N $2p_y$ and Cu $3d$ orbital plots at the different high symmetry points in the surface Brillouin zone. Note that the degeneracy is lifted at the *X* point where both N $2p_x$ and N $2p_y$ orbitals are shown. From Ref. [3].

interactions between N $2p_x$ and the first layer Cu $3d_{x^2-y^2}$, $3d_{z^2}$, and $4s$ are bonding or antibonding while the π interaction with the first layer Cu $3d_{xy}$ is non-bonding. The reverse is true for N $2p_y$ with the σ interactions being non-bonding and the π interactions bonding or antibonding. Similar arguments can be made for the N $2p_z$ orbital, not shown. The π interaction of N $2p_z$ with the first layer Cu $3d_{yz}$ is bonding or antibonding while the π interaction with $3d_{xz}$ is non-bonding. In all cases the interactions of the N orbitals with those on the second-layer Cu are independent of *k*.

This provides the general picture with bonding bands below and antibonding bands above the bulk Cu *d*-band. The states with the largest bonding or antibonding character occur at the M point with a large N $2p$ contribution and are directly observed as experimental features at 6 and 1 eV in the XES spectra in Figure 2.11. The non-bonding states at the Γ point are predominantly of Cu $3d$ character with little contribution of N $2p$ and are observed between the bonding and antibonding spectral features around 2–4 eV. The implication of the occupancy of both the bonding and antibonding bands will be discussed in the next section.

5.2. Chemical bonding of atomic adsorbates

In the previous section, we have discussed the occupied electronic structure of N adsorbed on Cu. We now return to the picture shown in Figure 2.9 and address the difference in population of the antibonding states between transition and noble metals such as Ni and Cu. Let us compare atomic N adsorbed on the Ni(100) and Cu(100) surfaces [3,32]. The details of the adsorption geometry have been discussed in Chapter 1. Figure 2.14 shows the XES and XAS spectra on a common binding energy scale [3]. Spectra related to adsorbate (in-plane) p_{xy} states are drawn with solid lines while spectra due to (out-of-plane) p_z states are plotted with dotted lines. The position of the Fermi level in the spectra is indicated by dotted vertical lines. The XAS spectra are shown to the right of the vertical line and are all normalized to the same step height well above the Fermi level.

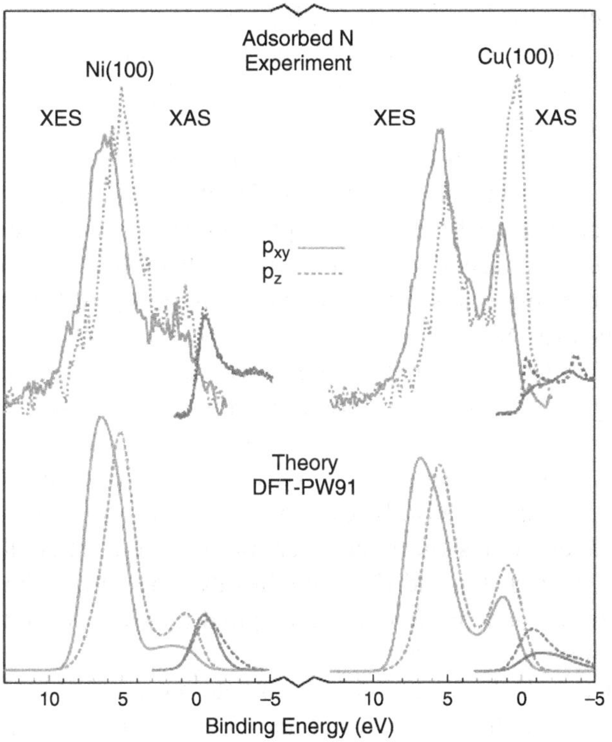

Figure 2.14. Top: Comparison of the XES (occupied states) and XAS (unoccupied states) spectra of atomic N adsorbed on Ni(100) and Cu(100) with separated p components. The intensity scaling between the XES and XAS spectra is arbitrary. Bottom: Calculated density of states projected onto the p_x, and p_z valence states of N chemisorbed onto a Cu(100) and Ni(100) surface in the c(2 × 2) structure. Reproduced from [32].

In the XES spectra of N adsorbed on Cu, both the p_{xy} and p_z components exhibit two strong peaks, representing the bonding and antibonding states, discussed above. In the XAS spectra, on the other hand, no strong peaks are observed. For N adsorbed on Ni we only observe one strong peak at high binding energy in the XES spectra, due to occupied bonding states. The antibonding states can now be seen in the XAS spectra directly above the Fermi level. The results clearly show how the antibonding states are shifted from below to above the Fermi level going from Cu to Ni. The population of the antibonding states will have a large influence on the total adsorption energetics. If the antibonding states are occupied the net bonding effect will cancel, resulting in Pauli repulsion. In Ni metal the Fermi level is located at the top of the *d*-band whereas in Cu it is located 2 eV above the *d*-band. This implies that the antibonding states are occupied for the adsorption on Cu and unoccupied in the case of Ni. For adsorption on Cu, the only net bonding effect would come from the remaining 4*sp* interaction.

The question is if this simple chemical bonding picture is general for other atomic adsorbates including also C and O atoms. Figure 2.15 shows XAS and XES spectra for C on Ni(100) and O on both Ni(100) and Cu(100) in comparison with the previously discussed atomic N adsorption system. Adsorption of C on Ni(100) leads to a nearly identical adsorption site as for N on Ni(100) and a similar substrate reconstruction, see discussion in Chapter 1. There exists for O on Cu(100) a low-coverage non-reconstructed phase with a FFH site [38] which is the phase used for the experiment in order to allow a simple comparison with the other adsorbate systems. In all cases, we see a very similar trend where the interaction gives rise to bonding and antibonding states, where the latter are occupied for the adsorption on Cu and unoccupied for Ni.

The effect of populating the antibonding states can directly be seen in the adsorption energies. Figure 2.16 shows the computed dissociative adsorption energies for N_2, O_2 and CO as a function of the center of the *d*-band with respect to the Fermi level on different transition metals [32]. We see a major change in energy between Ni and Cu for C, N, and O adsorbates confirming the picture that the population of antibonding states affects the bond strength. The bond strength increases also beyond Ni going towards Co and Fe. It implies that the antibonding states are still partly occupied on Ni and continue to be emptied as we move left in the periodic table of the transition metal series.

It is interesting to note the difference in adsorbate-projected intensity ratio between the bonding and antibonding states for N and O on Cu in Figure 2.15. The atomic 2*p* orbital is lower in energy for O compared to N. This results in a more polarized bond and as seen in Figure 2.15 the bonding states contain more adsorbate character than the antibonding states. Since the oxygen atom has one electron more compared to nitrogen we expect that the two unpaired electrons will be in the O $2p_{xy}$ states for a completely planar geometry. This would lead to the O $2p_z$ atomic orbital

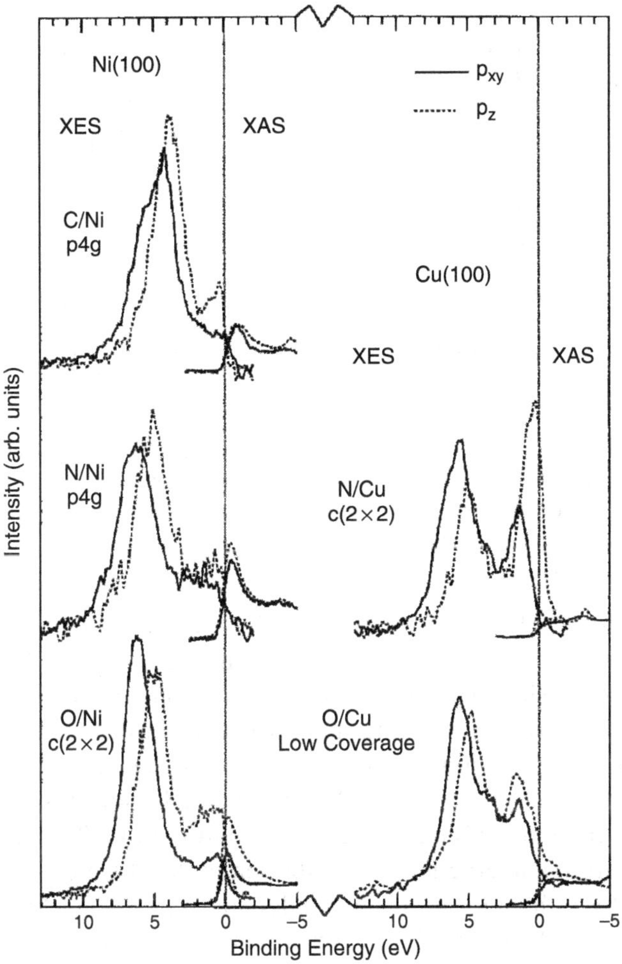

Figure 2.15. A comparison of the XES and XAS spectra of C, N, O/Ni(100) and N, O/Cu(100) with p_{xy} (solid lines) and p_z (dotted lines) separated. The intensity scaling between the XES and XAS is arbitrary. From Ref. [3].

being fully occupied implying Pauli repulsion with respect to interactions with the second layer below the O atoms. Therefore the O atomic adsorbate is located 0.5 and 0.8 Å above the first surface layer on Cu [39] and Ni [40], respectively, where in comparison N/Cu(100) is at 0.4 Å. Furthermore, a stronger adsorbate contribution in the bonding states would imply a larger ionic contribution to the chemical bond in comparison to adsorbed N. The adsorption of O on various metals and alloys relating the energetics to the center of the *d*-band is discussed in detail in Chapter 4.

We expect similar bonding configurations as discussed above for second-row elements such as Si, P, and S. Both F and Cl should become more extreme in ionic

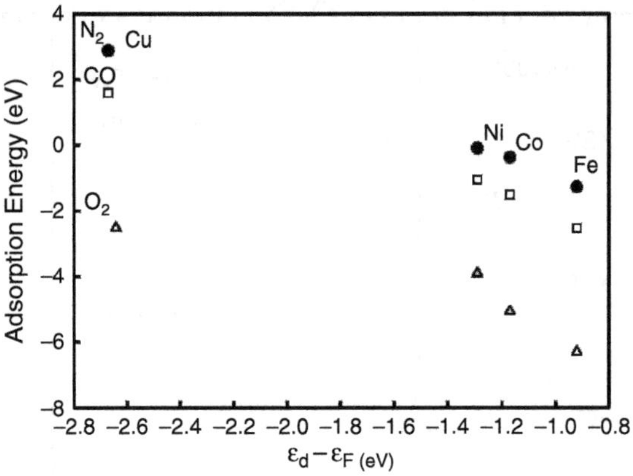

Figure 2.16. Calculated dissociative nitrogen (•), carbon monoxide (□), and oxygen (Δ) chemisorption energies over different 3*d* transition metals plotted as a function of the center of the transition metal *d*-bands. A more negative adsorption energy indicates a stronger adsorbate-metal bond. Reproduced from [32].

character where most of the adsorbate character will be in the bonding state. UPS studies of Cl adsorption on Cu(100) shows Cl induced states at 6 and 2 eV below the Fermi level [41]. These can be attributed to the bonding and antibonding states, respectively. In particular the bonding states are related to a broad and large spectral feature. The large degree of ionic character can also be seen through the lack of a tail towards the Fermi level in the antibonding state which indicates that they are fully occupied. There can thus only be a small contribution of Cl 3*p* character above the Fermi level confirming a large ionic contribution to the chemical bond. This picture was confirmed by a theoretical cluster model study of F/Cu(100) and Cl/Cu(100) in which a corresponding orbital analysis was used to analyze the adsorbate wave function [42]. Polarization effects in the substrate, however, affect the resulting surface dipole and lead to only a small change in the work function in spite of the fully ionic adsorbate [42]. The interactions of ions with surfaces in electrochemical systems are discussed in Chapter 6.

For atomic H adsorption on surfaces the electronic structure as obtained by UPS studies and DFT calculations on Ni, Pd, and Pt shows a similar picture. There is a strong bonding H-induced feature around 7–9 eV below the Fermi level observed both in UPS and band structure calculations [43]. This has been related to that the H 1*s* level interacts with both the metal *s*-and *d*-bands. Since the H 1*s* level is much lower in energy in comparison with the previously discussed adsorbates, for which the outer level was of *p* character, it is anticipated that the metal *s*-electrons will be more strongly mixed into the adsorbate bonding resonance. Since no X-ray spectroscopy measurements can be conducted on H it is difficult to derive how much H 1*s* character there is in the *d*-band region, respectively, above the Fermi level. The antibonding level has been observed in UPS for H adsorbed on Ti(0001) at 1.5 eV below the Fermi level as a surface state [44]. Since there is a band gap in the substrate at this region in the bulk-projected band structure there is no broadened interaction with the substrate and the spectral feature comes out rather sharp. We can expect that the antibonding H 1*s*-metal level is broadly distributed over the *d*-band region and again the amount of filling will depend on the position of the *d*-states with respect to the Fermi level.

6. Diatomic molecules

The CO and N_2 adsorbed on the late transition metals have become prototype systems regarding the general understanding of molecular adsorption. The adsorption energy is only 5–10% of the molecular dissociation energy. This has lead to the plausible assumption of a weak molecule–surface interaction where the chemisorption process causes only a small modification of the molecular orbital structure of the free molecule. However, it is important to differentiate between the contributions to the strength of the surface chemical bond and the orbital interactions leading to rehybridization of the electronic structure to minimize repulsion. The top of Figure 2.17 shows a description of the traditional Blyholder model [45–49] where a dative bond between the 5σ and metal states of σ symmetry is formed, leading to charge donation into the metal which is compensated by a back-donation into the molecular 2π^*. In this frontier-orbital picture a synergism between the π and σ bonds is achieved, where the internal molecular bond is weakened due to the increased population of the antibonding 2π^* in the back-donation.

We will in this section show how XES can provide atom-specific information to test if the assumptions leading to this simple picture are reasonable. What is the nature of the new electronic states formed upon adsorption and are there any changes in the remaining molecular orbitals, i.e., is a frontier-orbital picture sufficient? It will be shown below that the simple frontier orbital model needs to be modified: the σ-donation instead leads to repulsive interaction whereas it is the π-interaction that causes the attractive bonding contribution, see bottom of Figure 2.17. To describe the electronic structure of the π-system it is necessary to involve both the 1π and 2π^* orbitals in the interaction with the substrate. Both the σ-repulsion will be minimized and the π-bonding maximized as the *d*-electron states shift towards the Fermi level. This relates the bond description to trends in adsorption energies according to the Nørskov-Hammer *d*-band model [49,50]. The mixing of the 1π and 2π^* orbitals will partly break the internal π-bond and restore one part of the electron pair into a radical state on the inner atom to become available for bonding to the surface and the other part to become a lone pair on the outer atom. We will demonstrate how we can experimentally build a picture of the electronic structure

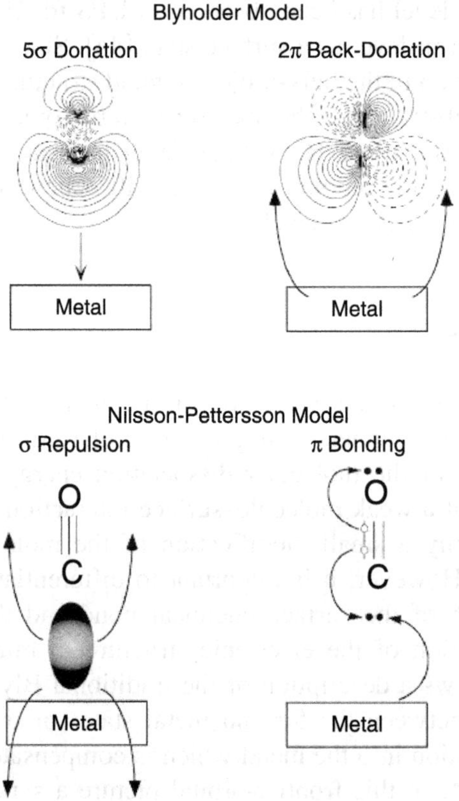

Figure 2.17. Schematic picture of the description of CO metal bonding via (top) the frontier-orbital picture, often denoted Blyholder model, with 5σ donation and $2\pi^*$ backdonation interactions with metal electron states and (bottom) the Nilsson-Pettersson model with σ-repulsion and π-bonding involving a rehybridization of both the 1π and $2\pi^*$ orbitals with the metal *d*-states.

and of the important contributions to the surface chemical bond for CO and N_2 adsorbed on Ni(100) and how this can be applied to understand adsorption in different sites, on different metals and also the coadsorption between CO and alkali metals. Furthermore, the resulting model is more generally applicable and applies not only to surface adsorbates but also to many inorganic systems involving CO coordination chemistry [51].

6.1. N_2 adsorbed on Ni(100)

The upright adsorption geometry of the N_2 molecule on Ni(100) in the on-top site leads to two chemically inequivalent N atoms. If a separation between the two atoms can be made, this homonuclear system would provide an ideal case to study how the electronic states redistribute in a molecule upon adsorption since any asymmetry between the two atoms can only be due to the bond-formation with the substrate. Furthermore, the resulting adsorption energy is very small, around 0.4 eV [52], and we might expect that the influence of the adsorption process on the electronic structure of the molecule should be minor. When chemisorbed on Ni(100) the molecular axis of N_2 has been found to be perpendicular to the surface [5,53] with a Ni–N bond distance of 1.81 Å [54]. For a more in-depth discussion about the structure of N_2 on Ni(100) see Chapter 1.

The XAS spectrum for N_2 on Ni(100) exhibits two 1*s* to $2\pi^*$ resonances at 400.6 and 401.0 eV, corresponding to the outer and inner N atoms, respectively [55]. Hence, by using different excitation energies, site-specific XES spectra can be recorded. The resulting XES spectra for the outer and inner N atoms are shown in the left part of Figure 2.18 [10]. The corresponding spectra, obtained from Density Functional Theory (DFT) calculations simulating the radiative decay process using a N_2/Ni_{13} cluster, are shown in the right part of Figure 2.18 [2] giving a generally good agreement with the experimental data. From the symmetry and binding energies of the spectral features, it is straightforward to assign all features above 5 eV binding energy in analogy with UPS measurements, see Figure 2.2. To facilitate the comparison with the much studied CO molecule we shall use $C_{\infty v}$ symmetry notation for the molecular orbitals.

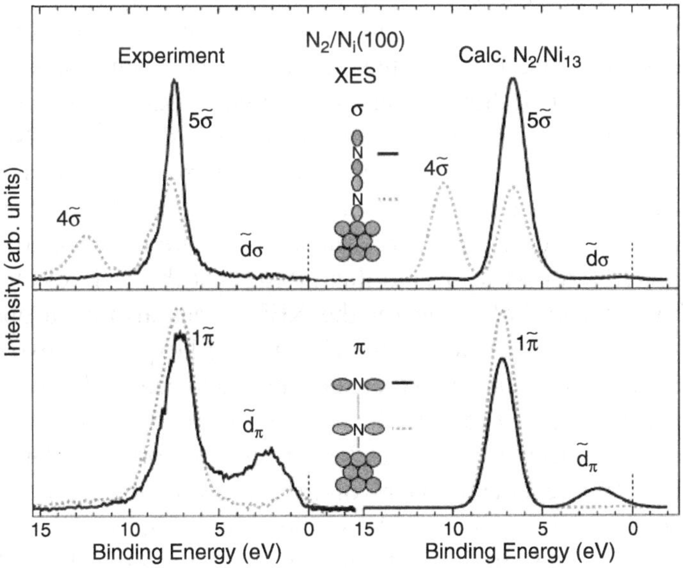

Figure 2.18. Experimental and theoretical atomically resolved XES spectra for N_2 adsorbed on Ni(100). The upper panel displays states of σ symmetry and the lower panel states of π symmetry. From Ref. [3].

The novel information contained in Figure 2.18 is the large difference in the states located on the inner and outer N atoms and the clearly resolved structures within 5 eV binding energy, i.e., in the Ni *d*-band region. All spectral peaks, representing the N 2*p*-derived and N-atom-projected molecular orbitals, exhibit different intensities or shapes for the inner and outer N atoms. Interesting findings are the localization of the 4σ state to the inner N atom, with no visible spectral intensity from the outer N atom, and the larger 5σ localization to the outer N atom. There is also a small polarization of the 1π orbital towards the inner nitrogen atom. Near the Fermi level we find new states that are formed as direct consequence of the surface chemical bond. These states arise from interaction of molecular π and π^* states, as discussed below, with the Ni *d*-states. There is a state located on the outer N atom centered at about 2.5 eV binding energy denoted $\tilde{d}_\pi$. In some sense we can label this state as a lone-pair orbital on the outer atom. Closer to the Fermi level there is also intensity on the inner nitrogen atom. An important question is how the π-system can be bonding when the new substrate-induced orbital essentially is a lone-pair, non-bonding orbital on the outer nitrogen atom.

Let us now interpret the adsorbate electronic structure by considering the orbital contour plots which have been obtained from the same calculation that led to good agreement with the experimental XES spectra. In Figure 2.19, orbital plots for gas phase and adsorbed N_2 for molecular orbitals (MO's) of π symmetry are shown. In the lower panel the gas phase 1π and the adsorbate $1\tilde{\pi}$ orbital are shown. As seen in the XES spectra, the 1π polarizes upon adsorption towards the inner nitrogen atom. It also mixes with the d_π of the interacting Ni atom. The amplitude of the $1\tilde{\pi}$ orbital has the same phase between the Ni and two nitrogen atoms, constituting a bonding orbital between all three centers. In the middle panel of Figure 2.19, orbital plots of the lone-pair $\tilde{d}_\pi$ states are shown. The orbital is mainly of Ni 3*d* character with a nodal plane centered at the inner nitrogen atom. It clearly shows the outer nitrogen lone-pair character of this state as seen experimentally. In the top panel of Figure 2.19, we show the $2\pi^*$ and $2\tilde{\pi}^*$ orbitals which are unoccupied and therefore cannot be observed with XES. It is to this orbital that the N 1*s* electron is excited when preparing the core hole state for the XES; it has also been observed using inverse photoemission at 4 eV above the Fermi level [15]. It has two nodal planes between the two nitrogen atoms and the metal atom and can be denoted antibonding.

To create the new adsorbate orbitals, we need to involve the whole original π system of the free molecule, i.e., both the 1π and $2\pi^*$ orbitals. This is described as a three-orbital allylic interaction involving the formation of a totally bonding ($1\tilde{\pi}$), a non-bonding ($\tilde{d}_\pi$) and a totally antibonding orbital ($2\tilde{\pi}^*$) as illustrated in Figure 2.20 [10,56]. The allylic configuration can be readily derived using first- and second-order perturbation theory [57]. First-order corresponds to charge transfer between the N_2 unit and the metal atom and second-order to polarization within the N_2 unit. If we only consider the Ni atom that is directly involved in the bonding,

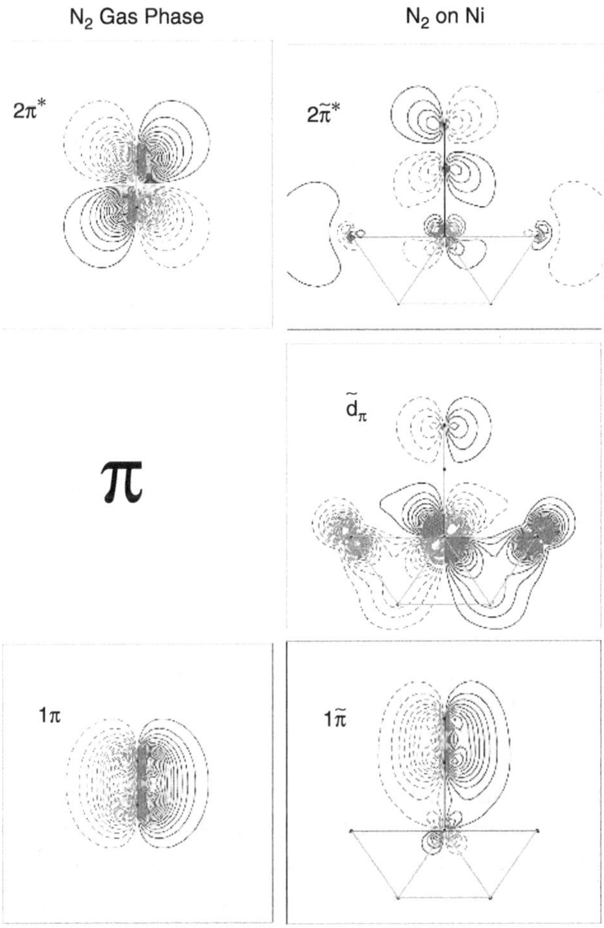

Figure 2.19. Contour plots of π orbitals for N_2 in the gas phase and adsorbed on Ni(100). Solid and dashed lines indicate different phases of the wave function. From Ref. [3].

the π-system will involve three atoms and three π orbitals will thus be generated. The lowest orbital will always be bonding between all three centers and the highest orbital will be antibonding. The intermediate orbital should be antibonding between the end-atoms, which for a symmetrical molecule results in no contribution on the center atom and this orbital can be denoted non-bonding. The bonding orbital is similar to the free molecule 1π orbital, but slightly polarized towards the inner nitrogen atom with a small but significant contribution from the Ni 3*d* orbital. Since the 1π population is smaller compared with the free molecule we have weakened the N–N bond and instead formed a covalent Ni–N interaction. The intermediate orbital is essentially non-bonding with a main contribution from the metal as seen from the low intensity in the spectrum.

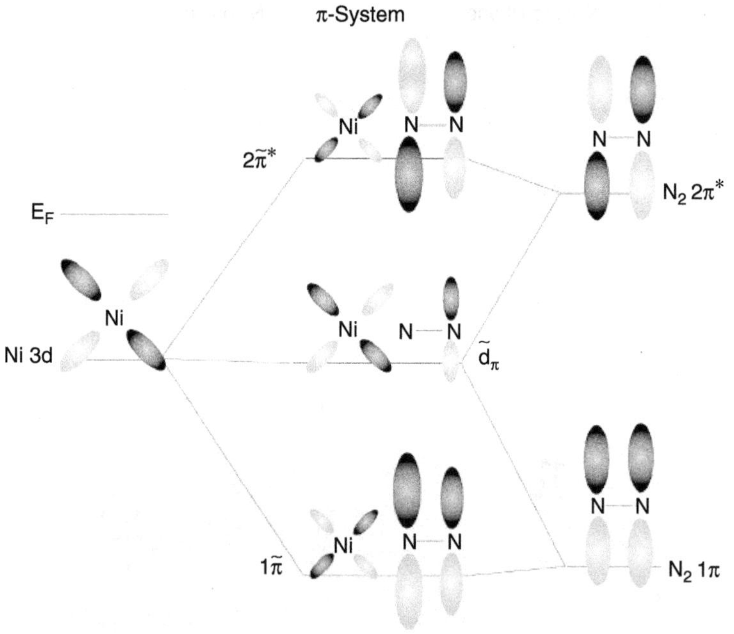

Figure 2.20. Schematic illustration of the π orbital interactions in the N_2−Ni adsorption system in terms of the atomic N $2p$ and Ni $3d$ orbitals. From Ref. [3].

We need to address how the 1π and $2\pi^*$ orbitals of the free molecules are perturbed to build up the allylic configuration. We illustrate this in Figure 2.21, where a mixing of N_2 1π and $2\pi^*$ orbitals can be viewed as a partial breaking of the internal π-bond preparing the molecule for bonding to the metal. As an extreme case let us consider first the complete breaking of the internal $N_2\pi$-bond by mixing the 1π and the $2\pi^*$ orbitals until the original atomic orbitals at each atomic center are obtained. A positive combination of the 1π and $2\pi^*$ molecular orbitals leads to an atomic orbital on the inner nitrogen atom, while a negative combination leads to an atomic orbital on the outer nitrogen atom. In chemical terms, the breaking of the internal $N_2\pi$-bond into atomic radicals, the original atomic orbitals, allows to make additional bonds at the expense of the internal N_2 bond. However, the initial bond breaking increases the total energy of the system and the energy gained from forming new bonds must compensate the first step for bonding to occur. Due to this energetic balance the internal π-bond is not broken completely for the molecule, but only polarized to increase the bonding interaction between the $N_2\pi$-system and the metal d_π-states. The 1π orbital is polarized towards the inner nitrogen atom, thereby making this atom more available for bonding to the Ni atom and simultaneously the internal N_2 bond is weakened; the outer nitrogen part of the broken π electron-pair then forms the lone pair $\tilde{d}_\pi$ -orbital. In a simple way we can regard the radical state

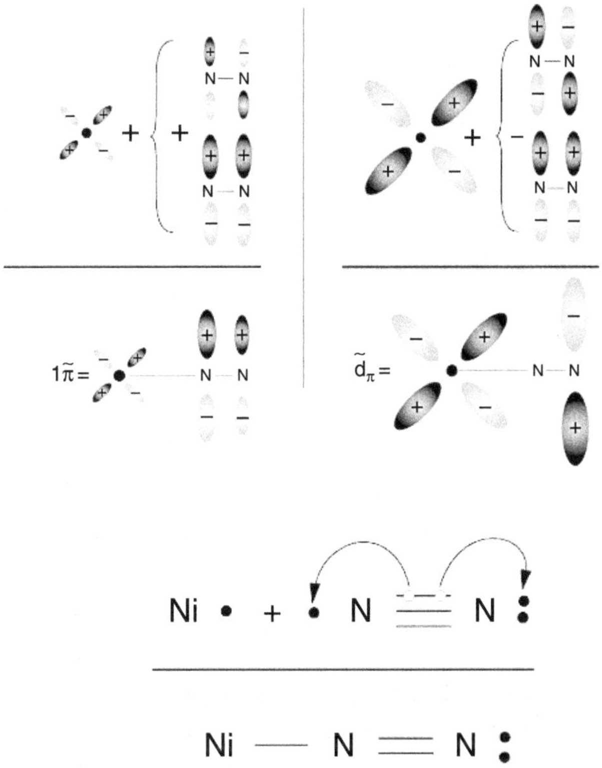

Figure 2.21. The composition of the adsorbate orbitals of π symmetry in a perturbational treatment for N_2 adsorbed on Ni. Polarization is achieved through intergroup and intragroup mixing. In terms of a valence bonding model, this can be seen as the partial breaking of the $N_2\pi$ bond, forming nitrogen radicals. From Ref. [3].

that is partly formed on the inner atom to interact with the *d*-states in the metal in a similar manner as in the radical atomic adsorption case, see previous section, forming bonding and antibonding states. The bond strength contribution from the π-states would then depend on the occupancy of the antibonding states and thereby also the position of the *d*-band in the metal. This would give rise to the well-known relationship of adsorption energy to the metal *d*-band center position [49,50]. We have to be aware that the bonding and antibonding states of the radical state is only a virtual picture and is in reality mixed with the internal N_2 molecular orbitals forming the allylic configuration.

So far we have reduced the chemisorbate electronic structure to the treatment of a triatomic model. For the real surface the orbitals illustrated in Figure 2.20 will undergo changes due to the interaction with the localized *d*-band and the free-electron-like 4*sp*-band, which form extended states in the metal. In Ni metal the

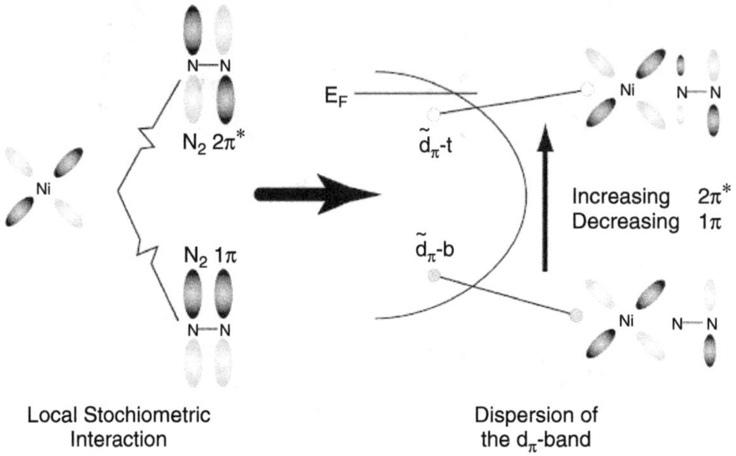

Figure 2.22. The modification of the non-bonding d_π orbital into a band due to the band character of the metal d-states. $\tilde{d}_\pi - b$ and $\tilde{d}_\pi - t$ refer to d-orbitals at the bottom and top of the d-band, respectively. From Ref. [3].

d-band dominates between 4 eV and the Fermi-level. The $\tilde{d}_\pi$ has mainly metal character whereas both the $1\tilde{\pi}$ and $2\tilde{\pi}^*$ orbitals are mainly localized on the N_2 molecule and largely preserve their molecular character. We therefore concentrate on the $\tilde{d}_\pi$-state as here the influence from the broad metal-band is the largest. In Figure 2.22, the variation of the $\tilde{d}_\pi$- state across the Ni d-band is shown schematically. The $\tilde{d}_\pi$-state always has a nodal plane between the two outermost atoms in the allylic configuration. At the bottom of the band this nodal plane coincides with the inner nitrogen atom forming the lone-pair state. Going towards the top of the band this nodal plane is located more between the inner and outer nitrogen atoms. In the perturbative description, the mixing of the 1π orbital into the metal d_π will change according to the variation of the relative orbital energies across the d-band. Towards the top of the band the relative $2\pi^*$ contribution increases and the 1π contribution decreases; this affects the location of the nodal plane in the molecule. We can clearly see the signature of this effect in the inner nitrogen spectrum shown in Figure 2.18 where additional intensity on the inner nitrogen atom close to the Fermi level is seen. We have more contribution of $2\pi^*$ character at these energies. Similar effects will also be seen for all the CO adsorption systems that will later be described.

It has been erroneously argued that the 1π orbital is mixing only within the occupied orbital space and therefore should not directly contribute to the adsorbate bond strength [58,59]. The argument has been based on considering the $1\tilde{\pi}$ as the bonding and the $\tilde{d}_\pi$ as the antibonding orbital formed from the 1π interaction with the metal d-states, i.e., a two-orbital picture. If, in this picture, the $\tilde{d}_\pi$ is fully occupied the net bonding contributions from the 1π with respect to the metal would be canceled which would require that all the 1π orbital contributions are within the occupied orbital space. However, this is contradicted by experiment. If we inspect the experimental part of Figure 2.18 we can easily see that the occupied part of the π-electron system does not sum to equal contributions on the two atoms. The contribution on the outer atom is larger and therefore the unoccupied $2\pi^*$ level should be polarized towards the inner atom as also observed near the Fermi level in the experimental spectrum in Figure 2.18 and illustrated in Figure 2.20. This is achieved by a small mixing of the 1π orbital with the $2\pi^*$ similar to what is shown for the $1\tilde{\pi}$ orbital in Figure 2.21. In this sense, there is a contribution of 1π to the chemical bond. It is interesting to note that the theoretical calculations in Figure 2.18 show a close to even occupation of both nitrogen atoms by adding the atom populations in the $1\tilde{\pi}$ and $\tilde{d}_\pi$ orbitals. This is also reflected in the orbital contour plot in Figure 2.19 where the $2\tilde{\pi}^*$ level is evenly populated on the two nitrogen atoms similar to the gas phase $2\pi^*$ orbital. Furthermore, the above described variation in the 1π and $2\pi^*$ contributions to the $\tilde{d}_\pi$-orbital, as observed in the experiment, is not seen in the theoretical calculations shown in Figure 2.18. In this sense, it looks like the details of the π-interaction are not fully described by the theoretical approaches [58,59]. However, the contribution to the adsorption strength of the 1π will not be large, albeit still large enough to overcome the σ-repulsion, which will be our next topic. In the case of CO, the involvement of the 1π in the chemical bond is also seen in the theoretical calculations, as will be described in the next section.

Let us now turn to the σ-system. The important adsorbate-orbitals of σ-symmetry are depicted in Figure 2.23. From the orbital plots we observe the same polarization as the experimental and theoretical XES spectra indicate, $4\tilde{\sigma}$ to the inner atom and $5\tilde{\sigma}$ to the outer nitrogen atom. The experimental XES spectra do not indicate any intensity in the $4\tilde{\sigma}$ orbital on the outer nitrogen atom. However, from inspection of the orbital plot we can attribute the component on the outer nitrogen to N 2*s* character. It is interesting to note that the charge redistribution of the 5σ orbital into the adsorbate $5\tilde{\sigma}$ orbital goes in opposite direction compared with a σ donation picture. We also show a $d\tilde{\sigma}$ orbital which mainly consists of metal character and can be seen in the XES spectra between the $5\tilde{\sigma}$ state and the Fermi level.

The σ-system can be summarized in a molecular orbital diagram, similar to the allylic configuration of the π-system, shown in Figure 2.24. However, for symmetry reasons more atomic orbitals are involved (N 2*s*, N 2*p*, and Ni bands), which makes the situation more complicated. We picture some of the highest orbitals as the $4\tilde{\sigma}$ and $5\tilde{\sigma}$ with mainly adsorbate character and the $d\tilde{\sigma}$ orbitals with mainly metal character, where we use, e.g., the $4\tilde{\sigma}$ to indicate the 4σ-derived orbital resulting from the interaction with the metal. The $4\tilde{\sigma}$ and $5\tilde{\sigma}$ are bonding orbitals with respect to the metal and will undergo a downwards shift in energy in comparison with the free molecule while the essentially metal $d\tilde{\sigma}$ is an antibonding orbital, which will therefore shift upwards in energy. In this orbital diagram, all orbitals, both bonding

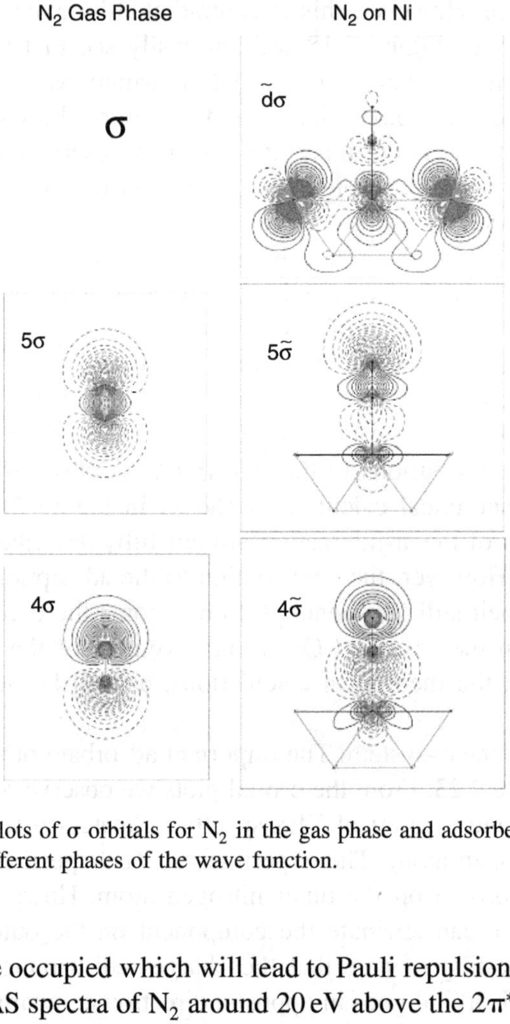

Figure 2.23. Contour plots of σ orbitals for N_2 in the gas phase and adsorbed on Ni(100). Solid and dashed lines indicate different phases of the wave function.

and antibonding, are occupied which will lead to Pauli repulsion. The N_2 $6\sigma^*$-orbital has been seen in XAS spectra of N_2 around 20 eV above the $2\pi^*$ resonance [60] and to mix with this orbital is very costly. This is an essential difference as compared to the π-system where the orbital mixing involves a low-lying unoccupied orbital. To minimize the Pauli repulsion the system will depopulate some of the $d\tilde{\sigma}$ orbitals through a polarization of both the 4*sp* and 3*d* density from the central Ni atom out towards the surrounding metal centers. However, the energy gain is not enough to overcome the Pauli repulsion including the cost for $6\sigma^*$-orbital mixing and metal polarization. The resulting picture is a repulsive σ-system. This interpretation is supported by calculations where the energetic gain of σ and π interactions can be separated based on a constrained space orbital variation (CSOV) theoretical method [10,19,61,62]. At a longer molecule-metal bond distance or for a transition metal with lower occupation in the *d*-shell the Pauli repulsion could become smaller and the σ-system net bonding. The lone-pair bonding as described in Section 9

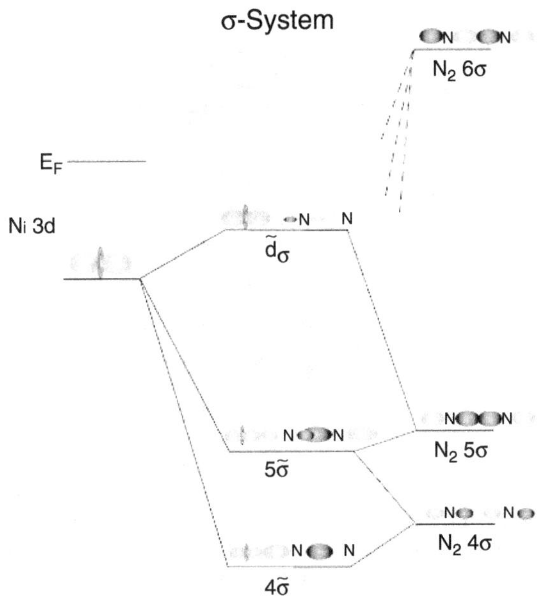

Figure 2.24. Schematic illustration of the σ orbital interactions in the N_2−Ni adsorption system in terms of the atomic N $2p$ and Ni $3d$ orbitals. From Ref. [3].

has a similar interaction as the σ-system with the surface and the bond distances are much further out to minimize the repulsion. However, for N_2 the energy gain through π bonding is large enough to overcome the σ repulsion at the equilibrium bond distance. It is interesting to note that this picture is rather different compared with the 5σ donation bonding scheme. As the *d*-band shifts towards the Fermi level more and more *d*-states will become unoccupied which leads to reduced Pauli repulsion. Far to the left in the periodic table the σ-system could even become bonding. The σ-system would in this respect also give the trend in the adsorption energy to the metal *d*-band center position according to the Nørskov-Hammer *d*-band model [49,50].

We can observe the signature of attractive π-interaction and repulsive σ-interaction from charge density difference plots shown in Figure 2.25. We observe loss of charge on all three atoms along the Ni−N−N axis, corresponding to σ interaction, whereas charge has been gained perpendicular to this axis. The latter is due to π-bonding. There seems to be charge building up between the inner nitrogen and Ni atom that could mistakenly be interpreted as σ bonding, however, this is attributed to the π bond, see discussion about CO in the next section. Furthermore, if the σ donation picture were valid we would expect to see a decrease of charge on the inner nitrogen atom along the bonding axis and an increase in charge on the Ni atom corresponding to the d_σ orbital. Since both lose charge we have a situation similar to

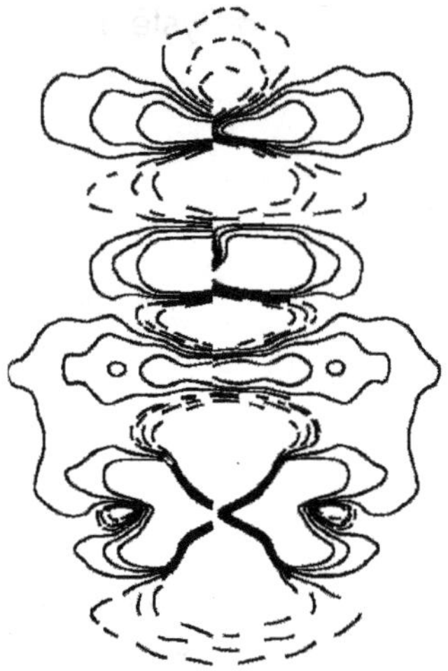

Figure 2.25. Charge density difference plot of N_2 adsorbed on Ni(100). Regions of electron loss are indicated with dashed outer line and increase with full line. We have chosen a plane containing the interacting metal atom with one N_2 molecule in the same plane. From Ref. [3].

the He_2 case as shown in Figure 2.6. There is a polarization of charge away from the coordinated Ni atom into the surrounding metal and a *d-d* rehybridization on the Ni atom to reduce the d_σ and increase the d_π character [10]. The latter could be argued to be essential for the π-bonding and therefore there is a strong synergism. However, from a careful analysis of the energy gain from different orbital interactions there is no large difference in the strength of the π-interaction if the σ-interaction is taken into account or not in accordance with earlier general conclusions about the CSOV approach [20]. We can note that the increase in charge density on both nitrogen atoms implies that more $2\pi^*$ character has become occupied compared to loss of 1π character into the empty orbital space as discussed above. Let us return to this issue in the discussion about CO adsorption on Ni.

From these results we can derive a model of the electronic structure and the surface chemical bond in that it involves all molecular orbitals and that the resulting binding energy is obtained from a balance between repulsion in the σ-system and bonding based on the π orbitals. Instead of bonding through σ donation the main effect from the σ-system is thus a repulsive interaction. The balance between attractive π-interaction and repulsive σ-interaction explains why the adsorption energy is low even though the adsorbate electronic structure is the result of a complete rehybridization. Likewise, the attractive adsorbate-metal interaction in the π-channel leads to a weakening of the internal molecular bond, which is countered to some extent by a strengthening of the internal molecular bond through the repulsive molecule-metal interaction in the σ-channel [10].

6.2. CO adsorbed on Ni(100)

We will now compare the N_2 system to the much more studied isoelectronic CO molecule adsorbed on Ni(100). Like N_2, CO adsorbs in a c(2 × 2) overlayer structure on Ni(100), occupying on-top sites with the carbon end down with a C−Ni distance of 1.73 Å, see Chapter 1 for details. However, the adsorption energy of 1.2 eV [63] is much higher in comparison to that of N_2. It is therefore very interesting to see how the difference in electronegativity of the carbon and oxygen atoms influences the surface-chemical bond in comparison to the isoelectronic N_2.

The top part of Figure 2.26 shows XES spectra of CO on Ni(100) which are compared with gas phase in the bottom part [3,62]. We observe the dominant $1\tilde{\pi}$ state for chemisorbed CO in both the carbon and oxygen XES spectra. Towards lower binding energy, new states are observed, which differ in the oxygen and carbon spectra. We denote these spectral features as the d_π-band, similar to N_2. At the bottom of the d_π-band (higher binding energy), intensity is only observed in the

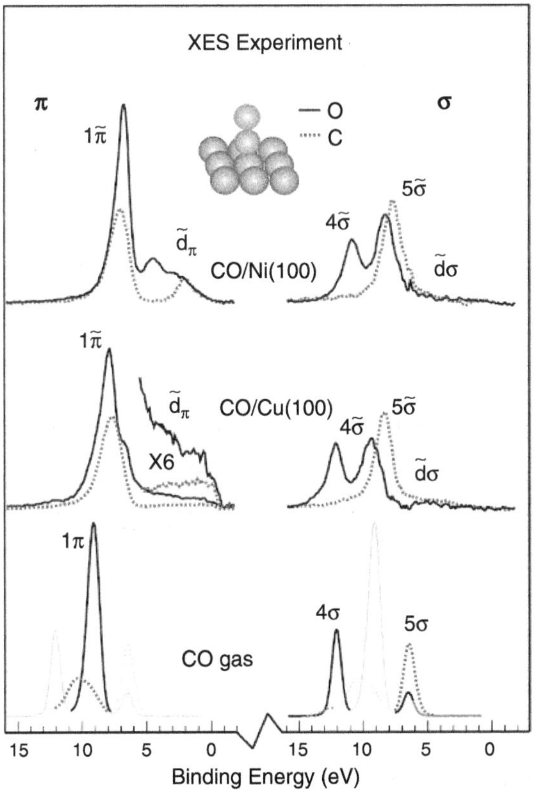

Figure 2.26. Experimental XE spectra for CO gas and adsorbed on Ni(100) and Cu(100). From Ref. [62].

oxygen spectrum. This state is a characteristic oxygen lone-pair state of π-symmetry at 4.5 eV binding energy. At the top of the band, close to the Fermi-level, intensity is present in both the carbon and oxygen XES data, as described in connection to Figure 2.22. Plots of π-symmetry orbitals are shown in Figure 2.27 [62]. Putting the CO π-electronic structure into the perspective of the isoelectronic N_2, we find that in principle these systems behave equivalently. This follows from the overall similarity of the XES spectra for the two adsorbates and their similar orbital character. Both systems can be described in terms of an allylic molecular orbital diagram as shown in Figure 2.20. For both molecules, the 1π forms a bonding combination to the metal *d*-states, maximizing overlap through internal polarization. In the XES spectra in Figure 2.26, we see that for chemisorbed CO the $1\tilde{\pi}$ orbital is more polarized towards carbon in comparison with the free molecule. At the same time a characteristic lone-pair state on the outer atom is formed with large Ni *d*-character. Due to the width of the Ni *d*-band, these states are rather broad and towards the Fermi-level contributions on both atoms can be seen in analogy with Figure 2.22. We show in the orbital plots in Figure 2.27 states both at the top (d_π−t) and bottom (d_π−b)

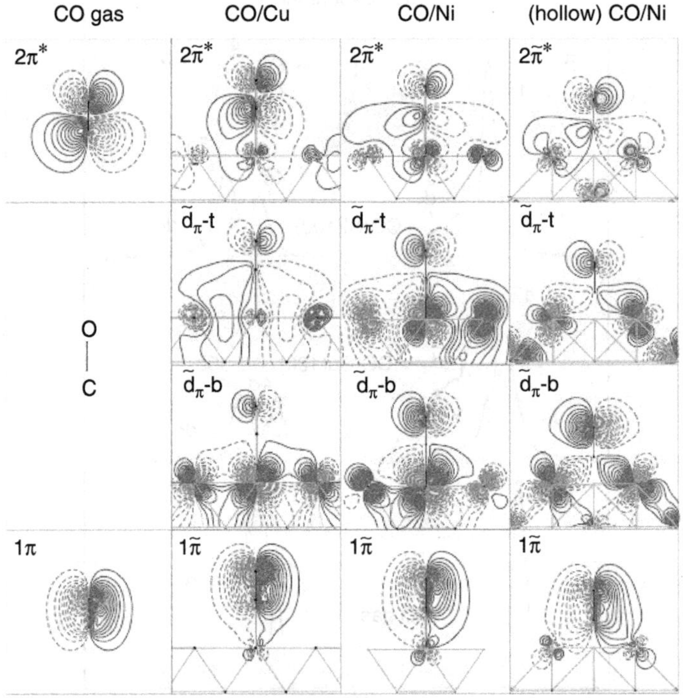

Figure 2.27. Contour plots of π orbitals for CO in the gas phase and adsorbed on Cu(100), Ni(100) on top and Ni(100) hollow adsorption sites. Solid and dashed lines indicate different phases of the wave function.

of the Ni *d*-band. From an inspection of the orbitals we note that there are small contributions close to the Fermi level also from the carbon atom in the d_π orbital. There is one significant difference compared to N_2 in terms of the orbital character. In the CO adsorbate the $2\tilde{\pi}^*$ level is mainly polarized towards the oxygen atom whereas it is the reverse for the free molecule. This means that the orbital rotation in the π orbital space has a large effect on the $2\tilde{\pi}^*$ and significant 1π character must thus have moved into the empty orbital space, i.e., significant π and π^* mixing has occurred to create the new $2\tilde{\pi}^*$ orbital.

In accordance with earlier UPS results [48,64], we identify the modified $5\tilde{\sigma}$- and $4\tilde{\sigma}$-states in the contour plots in Figure 2.28. In addition to these features, we find weak intensity tailing off to the Fermi-level. Similar to the N_2 case a dramatic redistribution of the σ molecular orbitals takes place upon adsorption. This is also seen in the orbital contour plots shown in the middle panel of Figure 2.28 [62]. The $5\tilde{\sigma}$ orbital polarizes towards the (outer) oxygen atom and the $4\tilde{\sigma}$ towards the (inner) carbon atom (mainly C 2*s* character) similar to adsorbed N_2. The relative strength of the $4\tilde{\sigma}$- and $5\tilde{\sigma}$-states in the oxygen XES spectrum is a measure of the degree of polarization upon adsorption. In the free CO molecule the intensity of the 4σ is 5 times larger than the 5σ, see Figure 2.26. On Ni, this ratio changes dramatically and the $4\tilde{\sigma}$ is even weaker than the $5\tilde{\sigma}$. This is the same trend as found for N_2. Upon adsorption the relative energy positions of the $4\tilde{\sigma}$ and $5\tilde{\sigma}$ orbitals also deviate from what is found for the free molecule. In the adsorbate, this difference

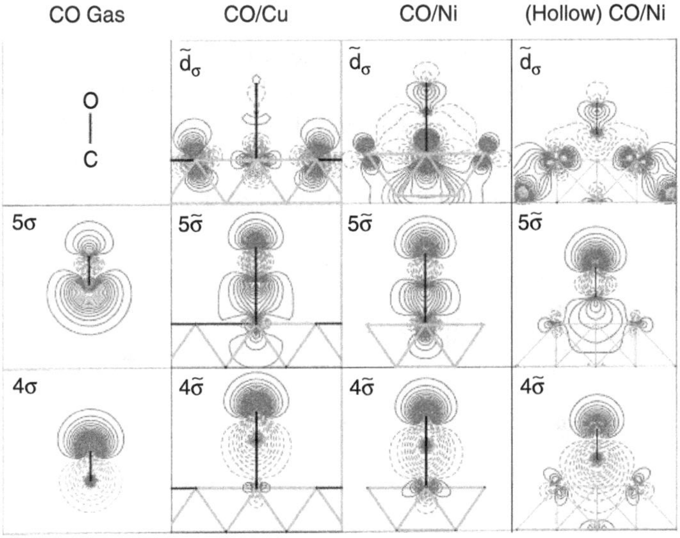

Figure 2.28. Contour plots of σ orbitals for CO in the gas phase and adsorbed on Cu(100), Ni(100) on top and Ni(100) hollow adsorption sites. Solid and dashed lines indicate different phases of the wave function.

is 2.5 eV in comparison to 5.5 eV for the free molecule. It is interesting to compare this to the N_2 on Ni(100) system, where the opposite trend, with an increased energy splitting between the two σ-orbitals, is observed; the $4\tilde{\sigma}$ orbital undergoes a larger shift to higher binding energy than the $5\tilde{\sigma}$ orbital [10]. The different behavior found for the two isoelectronic molecules can be directly related to the difference in orbital character of the $5\tilde{\sigma}$ and $4\tilde{\sigma}$ orbitals [62]. The 5σ is an internally antibonding orbital in the free CO molecule as evidenced, e.g., by the bond contraction found upon ionization from the 5σ level [65]. This is due to the large C 2*s* contribution that becomes antibonding with respect to the O 2*p* contribution. Upon adsorption the polarization away from the carbon atom towards oxygen results in the orbital changing character to bonding leading to higher binding energy. The opposite occurs for the $4\tilde{\sigma}$ orbital which is bonding for the free molecule and becomes antibonding in the adsorbate leading to a decrease in binding energy; the orbital splitting therefore becomes much smaller [62]. The 5σ orbital in N_2 is bonding and loses bonding character upon polarization since the atoms no longer give the same contribution to the orbital character leading to a shift closer to the Fermi level; the opposite is true for the antibonding 4σ orbital. These arguments show that the internal redistribution is essential for understanding the energetic positions of the orbitals. However, the bonding character towards the metal will also shift the orbital down in energy. It is interesting to note that the shift in the 5σ orbital to higher binding energies in comparison with the 4σ orbital in the past has been interpreted as a sign of attractive 5σ donation [45,66]. However, the overall interaction of the σ-system with the metal is similar to the N_2 case and leads to adsorbate-metal repulsion for the systems considered [62].

Figure 2.29 shows charge density differences for CO adsorbed on Ni compared to CO in gas phase [3]. The results for the total density are rather similar to what has been observed elsewhere [58,66,67] and there interpreted in terms of the simple frontier orbital picture illustrated in the upper part of Figure 2.17. The left part of Figure 2.29 shows the total difference in the charge density and it is very similar to

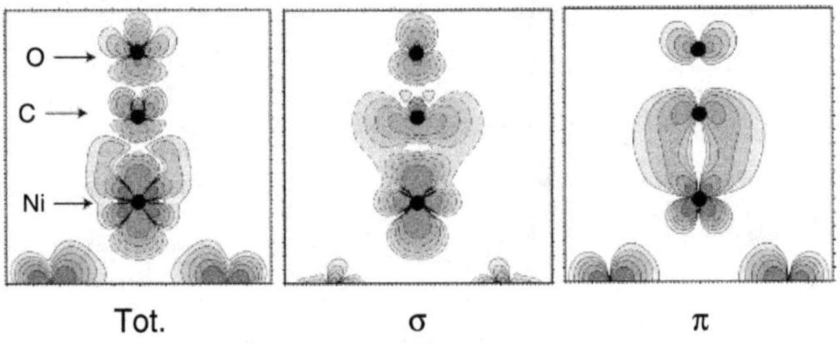

Figure 2.29. Charge density difference plots of CO adsorbed on a Ni_{13} cluster. Regions of electron loss are indicated with dashed outer line and increase with full line. We have chosen a plane containing the interacting metal atom with one CO molecule in the same plane. From Ref. [3].

N_2 on Ni shown in Figure 2.25. There seems to be a gain of charge of π-symmetry resembling backbonding into the $2\pi^*$ orbital. If we look at the σ-interaction on the axis between the O–C–Ni centers in the charge difference plot we observe similar changes as in the case for N_2 on Ni. There is a loss of charge on all atomic centers and there seems to be some small gain of charge between the C and Ni atoms. However, in the theoretical calculations it is possible to individually relax the σ- and π-systems, the result of which is also shown in Figure 2.29. We observe that the entire σ-system loses charge and the entire π-system gains charge, i.e., a σ to π charge transfer which, however, is not expressible in terms of a simple frontier orbital picture. The antibonding $d\tilde{\sigma}$ orbital is depopulated as much as possible to minimize the repulsion. Since it involves both Ni 3*d* and Ni 4*s* orbitals it will also lead to a weakening of the substrate Ni–Ni bonds. The π-interaction in the total density difference has in the past been interpreted as evidence of backdonation into the CO $2\pi^*$ from the Ni 3*d* orbitals [66]. However, investigation of the relationship between charge transfer to and polarization within the adsorbed CO unit in forming the allylic configuration reveals that, if we use the gas-phase molecular orbitals to decompose the wave function, the $2\pi^*$ becomes populated by 22% while the 1π is depopulated by 12%. The net effect is the observed 10% increase of the CO π-population for chemisorbed CO, which however, as seen in the rightmost panel in Figure 2.29, goes into the formation of a partial π-bond between the carbon and nickel as well as population of the lone-pair state on the oxygen. This result highlights the importance of orbital mixing within the π-system upon adsorption and supports the allylic configuration picture as shown in Figure 2.20. In this context, it is also interesting to note that the calculated energetic separation between the 1π and $2\pi^*$-orbitals is comparable for CO (9.60 eV) and N_2 (9.55 eV). Consequently, the internal polarization results from orbital mixing of similar magnitude for the two adsorbates on the metal surface. This is seen in the relative intensities in the XES spectra from the new lone-pair states of π-symmetry and the gain in charge of π-character in the charge density difference plots, which are all rather similar. However, breaking up the internal π-bond in homonuclear N_2 takes significantly more energy compared to heteronuclear CO, which could explain the large difference in adsorption energy between the two systems.

Let us address one fundamental concept that is essential to a complete understanding of CO adsorption on metal surfaces. CO adsorbs on nearly all systems with the carbon end down. This has in the past been attributed to the fact that the frontier orbitals 5σ and $2\pi^*$ are located on the carbon atom. However, the preference for adsorption through carbon can easily be understood based on the allylic configuration. Since the O 2*p* is much lower in energy compared to C 2*p* it is energetically much more favorable to form the lone pair state on the oxygen and the virtual radical state on the carbon atom. Such reasoning also explains why there is O lone-pair character in the non-bonding orbitals of CO_2 and H_2CO. From a simple chemical viewpoint, carbon atoms can make 4 bonds whereas oxygen can make 2. The possibility of carbon to rehybridize as *sp*, sp^2 or sp^3 in conjunction with the higher electron affinity of oxygen gives greater flexibility on the carbon side to form the required new bond between adsorbate and surface.

It has also been shown that the same electronic structure and allylic bonding mechanism is applicable to CO in general in inorganic chemistry and to metal carbonyls and to, e.g., CO coordination to the heme group in particular [51]. The spectra from the σ-orbitals in metal carbonyls seem to be split into several components, which can be attributed to the fact that there is no clear separation of σ and π orbitals but instead a number of orbitals formed based on the overall symmetry of the whole molecule. The relative intensity of the carbon and oxygen contributions is different in the different symmetry-adapted orbitals. We can make a connection to adsorbed CO where the σ-bands could have different contributions of carbon and oxygen character at different points in the Brillouin zone. From an inspection of Figure 2.26 we indeed observe that the $5\tilde{\sigma}$ orbital contributions from carbon and oxygen do not overlap completely in the measured carbon and oxygen spectra.

In summary, the electronic structure of N_2 and CO adsorbed on Ni is characterized by a complete rehybridization of the adsorbate valence-states in the presence of the metal-surface. In the π-system a characteristic allylic orbital structure is found both for N_2 and for CO, slightly modified due to the varying electronegativity of the involved atoms. The experimentally observed π-electronic structure is at odds with the usual frontier-orbital model where only the π* of the adsorbate molecule is considered. Often the simple picture in Figure 2.29 has been denoted the Blyholder model, but this has no real connection to the original paper [46]. In that study, there is no discussion of the σ-interaction while the π-system is described in a three-orbital model similar to the allylic configuration. In some sense the frontier-orbital picture could hold if we treat the π- and σ-systems separately, in which case the free molecule 1π and 2π* would be the frontier orbitals in the formation of the allylic configuration giving the π-bonding. In the electronic structure of local σ-symmetry, a full rehybridization of the participating molecular orbitals is observed. Energetically, the π- and σ-systems behave in opposite ways. The π-interaction is stabilizing the adsorbate-substrate complex, whereas the σ-interaction is destabilizing [62] in agreement with previous work by Bagus and Hermann [19] and Bagus et al. [61]. The attractive π-interaction is balanced by the repulsive σ-interaction. This balance explains why the total adsorption energy and the molecule-metal bond are weak even though the adsorbate electronic structure is the result of a complete rehybridization. Likewise, the attractive adsorbate metal interaction in the π-channel leads to a weakening of the internal molecular bond, which is to some extent countered by a strengthening of the internal molecular bond through the repulsive CO-metal interaction in the σ-channel. Therefore there are only small changes in the internal bond length upon adsorption of CO. A thorough discussion of these mechanisms is given in references [62,68].

6.3. CO adsorbed on Cu(100) and other metals

The adsorption of CO has been studied on a large number of different metals. It is important to understand the trends in chemical bonding and electronic structure and how they depend on the properties of the metal. It has been established that a linear relationship between the adsorption energy and the energetic position of the metal *d*-band exists where the adsorption energy increases for CO adsorption on transition metals going to the left in the periodic table [49,69]. In the following section, we will discuss this trend based on our current understanding of the ingredients for the CO chemisorption bond formation in terms of the allylic π-interaction and repulsive σ-interaction. We will make a comparison between CO adsorbed on Ni and Cu where the former corresponds to an open *d*-shell configuration ($d^{8.4}$) and the latter to a situation where the *d*-shell is more or less completely filled ($d^{9.6}$). The center of the *d*-band in Cu is lower by 2 eV in comparison with Ni and the question is how this is reflected in the electronic structure of the adsorbate.

CO forms a c(2 × 2) overlayer also on Cu(100) with an on-top geometry similar to Ni(100) [70]. The 0.7 eV adsorption energy is lower by 0.5 eV [71] in comparison with Ni(100) (1.2 eV) which is also manifested as a longer Cu−CO bond with a distance of 1.9 Å [72]. The experimental XES spectra of CO adsorbed on Cu [62] are shown in the middle row in Figure 2.26. The left panel of Figures 2.27 and 2.28 show contour plots of the dominant molecular orbitals as previously discussed for CO on Ni in the preceding section.

If we look at the π-symmetry spectra in Figure 2.26 we observe that there is quite a dramatic difference in the d_π states for CO on Cu compared to Ni. First of all there is much lower intensity in this spectral region corresponding to smaller adsorbate character in the case of Cu and there is a broad distribution extending from the $1\tilde{\pi}$ and all the way to the Fermi level. There is larger intensity in the oxygen spectrum compared to that of carbon illustrating that the allylic configuration is still a reasonable description, however. From the orbital plots we see that the $3d$ involvement in the $1\tilde{\pi}$ state is much smaller in Cu compared to Ni and that the lone-pair orbital contains more metal d_π orbital character. Also the $2\tilde{\pi}^*$ orbital resembles more the original gas phase $2\pi^*$ level indicating less mixing involving the 1π orbital. Furthermore, there is also an energy shift towards the Fermi level of the $1\tilde{\pi}$ state in adsorbed CO compared to the 1π orbital in gas phase CO. Since there is a mixing of $2\pi^*$ character involved in forming the $1\tilde{\pi}$ state, as shown in Figure 2.27, there is a destabilization of the orbital; the effect is larger for Ni than for Cu. All this evidence points clearly to a much smaller π-interaction with the surface for CO on Cu. The allylic configuration is still applicable, but with much less orbital mixing: the resulting orbitals retain more of the pure CO, respectively, metal *d*-orbital character. It is interesting to note that the orbital plots indicate that the nature of the contributions to the d_π state close to the Fermi level for adsorption on Cu is rather different from that on Ni. The wave function on the metal shows a clear 4*p* character on Cu due to the fact that the Cu *d*-band lies at much higher binding energy and the metal states close to the Fermi level correspond to the Cu 4*sp* band. In this sense, the width of the metal states in the d_π band is much larger in Cu than Ni since it contains both Cu 3*d* and 4*sp* states giving rise to the broad distribution in the spectra. The two situations correspond to the two extreme cases in the Newns-Anderson model described in Section 3 with narrow and broad bands of the substrate.

In the spectra of σ-symmetry and in the orbital plots we see a large similarity between CO on Ni and Cu. The polarization of the $5\tilde{\sigma}$ orbital is slightly smaller on Cu than on Ni which is also seen in the orbital plots. There is a smaller contribution of Cu 3*d* in the $5\tilde{\sigma}$ and $4\tilde{\sigma}$ orbitals in comparison with Ni. Since the *d*-band is further down in Cu we expect that the $d\tilde{\sigma}$ states will be more occupied for CO on Cu leading to increased σ-repulsion. Since both the π-interaction is much smaller for CO on Cu and the σ-repulsion larger it leads to lower adsorption energy and longer metal−carbon bond length. The longer bond will result in smaller overlap between the CO and metal orbitals as is also seen in the orbital contour plots [3]. We note essentially a rather similar change in the electron density as for CO on Ni with one exception. There is no increase in *d*-population of π-symmetry on the Cu atom as seen in the case of Ni. This can be understood based on that Cu has a full *d*-band and cannot modify its *d*-shell occupation through rehybridization within the 3*d*-shell.

Let us now address the relationship between the energetic position of the center of the *d*-band in different transition and noble metals and the CO adsorption bond strength as shown in Figure 2.30 [49,69]. Shifting the center of the *d*-band towards the Fermi level increases the adsorption energy of CO. This is also reflected in stronger electronic structure changes as the interaction strength increases. Recent XES studies of CO on Ru show stronger redistribution in the σ-system and more adsorbate intensity in the lone-pair d_π-states [73]. Following the discussion of N_2 on Ni in the previous section the mixing of the 1π and 2π* orbitals partly breaks the internal π-bond and leads to a radical virtual state on the carbon atom and a lone-pair state on the oxygen atom. The radical virtual state can form bonding and antibonding states with the metal where the bond strength will depend on the occupancy of the latter. This will lead to a dependence of the π-contribution to the adsorption energy on the energetic position of the *d*-band center as shown in Figure 2.30. The σ-repulsion will also depend on the *d*-band center. When the *d*-band becomes less filled the $d\tilde{\sigma}$ band will reside more above the Fermi level, resulting in a smaller Pauli repulsion. It could be that at the beginning of the transition metal series

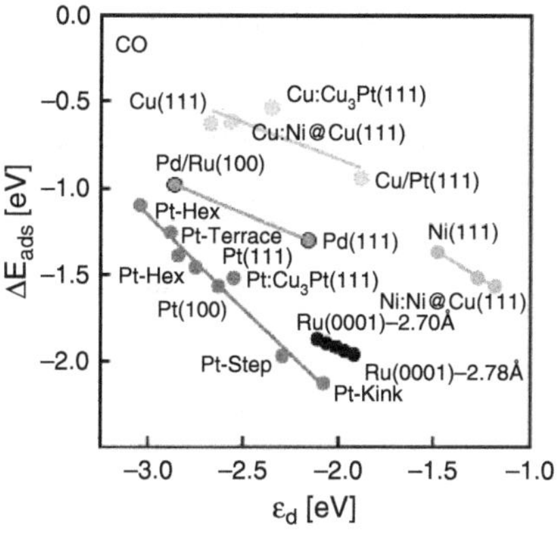

Figure 2.30. Computed CO chemisorption energies as a function of the *d*-band center (ε_d) of the metal surface. From Ref. [69].

the σ-interaction becomes attractive. In this sense, both the π- and σ-interactions contribute to the adsorption energy in such a manner that the CO bond strength increases when the *d*-band center shifts towards the Fermi level.

6.4. CO adsorbed in different sites

The interplay between the electronic and geometric structures of adsorbates is of fundamental importance for the understanding of many surface phenomena. Molecular CO is known for its ability to populate different adsorption sites, depending on the metal, substrate structure, coverage, temperature and influence from coadsorbate species. These often coexisting phases indicate only small energetic differences for different sites, which has been interpreted as indicative of rather similar bonding. The nature of CO bonding in different sites has been studied using the CO−H coadsorption system on Ni, where CO can be populated in on-top, bridge and hollow sites, see Chapter 1 for details regarding structural information.

Figure 2.31 shows C and OK- emission spectra of CO in different sites and orbital contour plots are shown in the middle panels for on-top and right panels for hollow sites in Figures 2.27 and 2.28 [68]. We note similar qualitative changes in the electronic structure upon adsorption for CO in bridge and hollow sites as observed in the previous case of the on-top site. The 5σ orbital polarizes towards the (outer) oxygen atom and the 4σ towards the (inner) carbon atom (mainly C 2*s* character). The relative strength of the 4σ- and 5σ-states in the oxygen XES spectra is a measure of the degree of polarization upon adsorption. The degree of polarization increases with increasing Ni coordination. In the π-system we observe the dominant $1\tilde{\pi}$ state in both the carbon and oxygen XES spectra. Towards lower

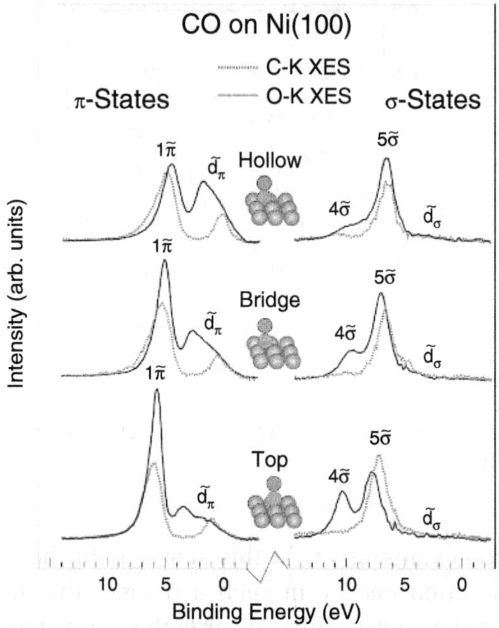

Figure 2.31. Experimental XES spectra for CO adsorbed on Ni(100) in on-top, bridge and hollow sites.

binding energy the d_π-band state can be observed with a dominant oxygen lone-pair character. Both the amount of $1\tilde{\pi}$ polarization and adsorbate character in the lone-pair state increases with increasing Ni coordination. We observe that the $1\tilde{\pi}$ orbital for CO in the hollow site has the same contribution on both atoms which is rather different compared with the free molecule. We can also see a dramatic polarization of the $2\tilde{\pi}^*$ towards the oxygen atom for CO in the hollow site indicating a major mixing with the 1π orbital.

Both the π–bonding and the σ-repulsion increase with increasing Ni coordination, such that the resulting adsorption energy is rather similar for on-top and bridge sites and slightly less favorable for hollow sites. The gain in π–bonding with increased Ni coordination is thus lost due to the increased σ-repulsion. The energy contributions, π–bonding and σ-repulsion, from the two symmetry channels partly compensate each other. However, the interaction is very different between the sites causing a dramatic change in the electronic structure which increases with increasing Ni-coordination. Based on these findings we can understand the change in vibrational frequencies and the different reactivity of CO in the different adsorption sites [68]. In adsorption systems where the π-interaction is weaker, such as CO on Cu and N_2 on Ni the σ repulsion will dominate at higher coordination, leading to population of only on-top sites. We anticipate that for NO with one additional electron the

π-interaction will increase while the σ-interaction remains quite similar to that of CO leading to a larger probability for adsorption in sites with higher coordination.

6.5. Coadsorption of CO and K on Ni(100)

The promoter action of alkali metals in heterogeneous catalysis has been an intriguing issue for a long time for which the co-adsorption of CO and K on single-crystal surfaces has become the prototype system [74,75]. There are three important changes in the chemisorption properties of CO on metal surfaces due to the presence of K: the heat of adsorption of CO increases, the internal C−O bond is weakened and the adsorption site shifts to maximum substrate coordination, see discussion in Chapter 1. The extra stabilization of the co-adsorbate overlayer has been related to the formation of an essentially ionic K^+−CO^- lattice [74,76,77]. The extra charge on the CO molecule has been attributed to an increased population of the CO $2\pi^*$ [78]. However, as discussed in Chapter 1 there seems not to be any significant changes in the internal C−O bond length.

In the saturated coadsorbed phase of CO and K on Ni(100) a c(2 × 2) overlayer is formed. The XPS spectra show C 1*s* and O 1*s* binding energies similar to CO adsorbed in the hollow site [64]. This indicates that the balance between π-bonding and σ-repulsion changes when K is present. Note that this interplay is also related to the intramolecular CO bond, i.e., an increased repulsive σ-interaction between the adsorbate and the substrate gives rise to a somewhat strengthened CO bond [62]. As discussed in Chapter 1 the Ni−K bond length increases by 0.15 Å due to coadsorption and the K atoms reside above the level of the neighboring O atoms in CO [79,80]. The lack of direct bonding between K and the metal surface indicates a strong interaction between the CO and K atoms in the first layer, covalent or ionic. It has been shown that the K 2*p* XAS spectrum for CO and K coadsorbed on Ni(100) exhibits crystal-field splitting demonstrating a large degree of ionic interaction between CO and K [77].

The spectra for the CO/K/Ni(100) system are presented in the bottom panel of Figure 2.32 [3,81]. In the upper part of Figure 2.32, we also show XES spectra of CO on Ni(100) in hollow sites on hydrogen-precovered Ni(100), discussed in the previous section. Starting with the $1\tilde{\pi}$ system a strong interaction is observed, similar to hollow site CO. A closer inspection of the d_π-distribution for the two hollow adsorption systems indicates a somewhat stronger π-interaction for the alkali system. By applying a curve-fitting procedure to the oxygen spectra, a similar ratio between the integrated d_π and the 1π features is obtained. The main spectral difference is instead found in the carbon spectrum at the upper part of the d_π band, which is significantly more intense for the alkali-modified system. As is evident from Figure 2.22, a larger $2\pi^*$ character in the d_π states gives rise to electronic states with an increased amplitude on the carbon atom, as compared to the oxygen

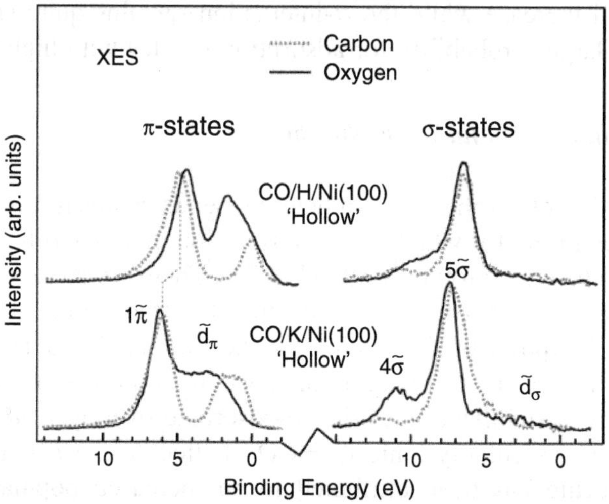

Figure 2.32. Experimental XES spectra for CO adsorbed on Ni(100) in hollow sites coadsorbed with hydrogen and potassium, respectively. From Ref. [3,81].

atom. This can be viewed as an upward shift of the Fermi level with respect to the d_π band seen in Figure 2.22 and, as a consequence, increased population of $2\pi^*$ character. The $1\tilde{\pi}$ state shifts to higher binding energy by 1 eV, compared with the hollow site CO on the hydrogen coadsorbed system, supporting the energy shift of the Fermi level. This clearly fits into the picture of a negatively charged CO through increased $2\pi^*$ population.

Turning to the σ-system, the amount of σ-polarization for the alkali system appears smaller than for the other hollow phase; in the oxygen spectrum a more distinct $4\tilde{\sigma}$ feature is found. The degree of internal polarization is somewhere in between the case of CO adsorption in bridge sites on Ni(100), and in hollow sites on hydrogen modified Ni(100), see Figure 2.31. Thus, in terms of the bonding properties, these findings indicate that the presence of alkali decreases the repulsive interaction of the σ-system, leading to a shorter adsorbate distance, in agreement with previous work on Ni(111) [79].

The driving force for the reduced σ-repulsion upon coadsorption of alkali can be understood by considering the interplay between the electrostatic properties of the CO-alkali overlayer and the local covalent interaction. For the c(2 × 2)CO/K/Ni(100) system, the formation of an essentially ionic K^+CO^- lattice has been established, seen also in the downward shift of the π-system. This gives rise to an image charge plane of opposite sign in the substrate. Locally, the Ni atom directly coordinated to the CO unit is depleted of charge whereas the CO unit carries additional charge. The charge depletion in the metal can be related to a decreased metal 4*sp* density. This means that the local σ-repulsion between the CO σ-system and the metal states is reduced, altering the energetic balance between π-attraction and σ-repulsion such that adsorption in more highly coordinated sites becomes favored. Consequently, the CO metal distance decreases leading to an increased spatial overlap and to a stronger, attractive π-interaction.

7. Unsaturated hydrocarbons

In the previous section, we have seen how we can understand the bonding of N_2 and CO to metals in terms of the allylic model with a partial mixing of π and π^*. Both N_2 and CO have unsaturated π-bonding character with a rather large experimental $\pi \rightarrow \pi^*$ excitation energy around 6 eV, which in the general case makes it unfavorable to fully populate the π^* orbital, as would be required for completely breaking an internal π-bond and instead forming two bonds to the substrate in a lying-down geometry. As a consequence N_2 and CO generally bond in an upright geometry and only partially involve the π^* orbital, as described in the allylic Nilsson-Pettersson bonding model. In the present section we will discuss the bonding to metals of unsaturated hydrocarbons: acetylene, ethylene and benzene. Here the relevant π to π^* excitation energy is much lower, 3.5–3.9 eV, which is low enough that it can be compensated by the formation of two covalent bonds to the surface in a lying-down geometry. Furthermore, the loss of π-character and mixing in of π^* both contribute to weaken the π-bond leading to large bond-elongations upon adsorption. This is in contrast to the case of CO and N_2 where the polarization and depopulation of the antibonding 5σ-orbital compensates for the bond-weakening in the π-system leading to smaller effects on the internal bond-length than expected from the π and π^* orbital-mixing as discussed in the previous section.

The description of the bonding of unsaturated hydrocarbons to metals was originally developed by Dewar, Chatt and Duncanson and is now known as the well-established DCD model based on a frontier-orbital concept [82]. In this model, the interaction is viewed in terms of a donation of charge from the highest occupied π-orbital into the metal and a subsequent backdonation from filled metal-states into the lowest unoccupied π^*-orbital, see Figure 2.33. Contrary to the case of the standard Blyholder model for CO and N_2 the DCD frontier-orbital model is supported by experimental XES measurements [83]. In the present section, we will show how we can experimentally identify and quantify the contributions of the different π-orbitals involved in the interaction with the surface. The DCD model will be shown to very well describe the chemical bonding of ethylene on Cu and Ni surfaces. Furthermore, the differences in bonding of benzene to Cu and Ni will be discussed.

The DCD model gives a good description of the final bonding to the surface, but does not describe the energetics and bond-formation. In order to address such questions we will use a different viewpoint which takes into consideration which

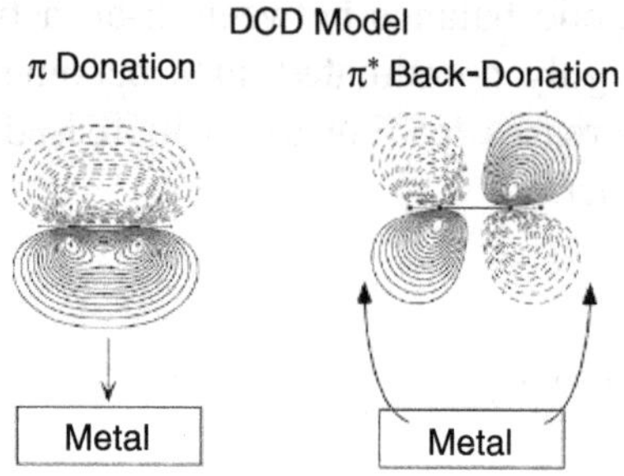

Figure 2.33. Schematic picture of the frontier-orbital description of ethylene-metal bonding via π-donation and π*-backdonation interactions with metal electron states. From Ref. [3].

molecular electronic states are involved in the formation of the bond. This is the spin-uncoupling, or bond-preparation, picture of Triguero et al. [84], based on the fact that for a closed-shell molecule an excitation is required to break the spin-coupling in an internally bonding orbital in order for two new bonds to be formed to the surface. Knowledge of this excited state allows an estimate of the rehybridization cost and thus of the different contributions to the total bond energy. In the present section, we will complement the discussion of the standard DCD model with discussing also the bond energetics according to the spin-uncoupling concept.

7.1. Ethylene (C_2H_4) adsorbed on Ni(110) and Cu(110)

We will begin the present section by demonstrating how we can use XAS to deduce the adsorption structure of ethylene on the two surfaces and draw conclusions on the chemical bond-formation. C 1*s* XAS probes the unoccupied *p*-density of states locally around the carbon atoms, and probes the symmetry selectively along the direction of the E-vector of the exciting light. The Cu(110) and Ni(110) surfaces have two-fold symmetry with metal rows in the first atomic layer. Using linearly polarized X-rays we can thus use XAS to project the unoccupied electronic structure in three different directions and determine if there is an alignment of the molecules on the surface. Figure 2.34 compares XAS spectra for ethylene adsorbed on Cu(110) (full lines) and Ni(110) (dashed lines) where the E-vector has been oriented in the three high-symmetry directions: out-of-plane (110), in-plane parallel with the surface atomic rows ($1\bar{1}0$) and in-plane perpendicular to the rows (001) [85]. For ethylene on Cu(110) we identify the strong π* resonance at 284.8 eV only in the out-of-plane spectrum directly showing that the molecule is lying down with the molecular plane parallel with the surface. The C−C σ* resonance is seen as a broad feature at 297 eV only in the ($1\bar{1}0$) spectrum demonstrating an alignment of the C−C axis parallel with the Cu rows on the (110) surface. XAS thus gives the orientation of the molecular axis relative to the surface and together with theoretical calculations it has been

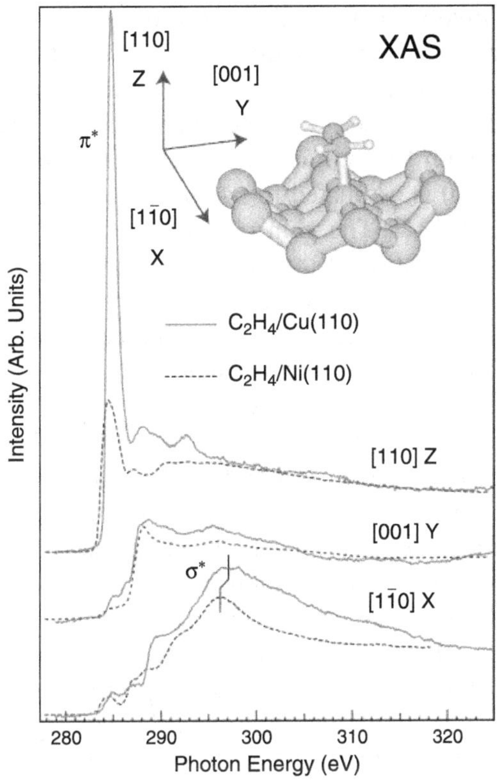

Figure 2.34. Symmetry-resolved experimental C 1*s* XAS of ethylene adsorbed on Cu(110) (solid lines) and Ni(110) (dashed lines). The coordinate system and molecular geometry used for the calculations are shown in the inset. From Ref. [85].

shown that the ethylene molecule is adsorbed lying down with the molecular axis parallel with the rows in a bridge site coordinating to two Cu atoms as indicated in Figure 2.34. This adsorption geometry implies that the ethylene molecule is bound in the so-called di-σ configuration, i.e., two σ-bonds have been formed between the molecule and the surface. A similar analysis can be performed for ethylene adsorbed on Ni(110) (dashed lines in Figure 2.34) where the π^* appears in the out-of-plane direction at 284.5 eV and the C−C σ^* shape-resonance shows states along the Ni rows at 296.3 eV. This indicates a similar molecular orientation for ethylene on Ni(110) as on Cu(110), but the spectra also indicate important differences in the interaction of ethylene with the two different metals. On Cu, the π^* resonance is much sharper and more intense than on Ni, and the C−C σ^* shape-resonance is found at a slightly higher energy. The shift of the shape-resonance towards lower energy as well as the lower intensity and broadening of the π^* resonance on Ni compared to Cu are consistent with a stronger adsorbate-surface interaction with greater adsorbate π^* involvement for the transition metal Ni than for Cu. In fact, the desorption temperature of ethylene on Cu(110) is less than 225 K [86], whereas on Ni(110) the molecules are adsorbed up to room temperature, at which point they start to decompose [87]. It is clear that the interaction of ethylene is significantly stronger with Ni than with Cu in spite of a similar heat of adsorption on the two metals, which on Ni(110) is estimated to 64 kJ/mol [88] while the computed value is 54 kJ/mol on Cu(110) [84]. The gain in bonding on Ni is thus balanced by an additional cost in rehybridization of the molecule upon adsorption, resulting in only a small net gain in the resulting chemisorption energy.

The shift in σ^* shape-resonance position can be correlated with changes in the C−C bond-length by the empirical 'bond-length with a ruler' approach based on the sensitivity to the bond-length of the energy position of this antibonding state [13,89]. By applying this simple method the C−C bond-length is estimated from the experiment to 1.41(2) Å for ethylene adsorbed on Cu(110) and 1.43(2) Å on Ni(110) [85]. Comparison with the bond-lengths of 1.34 and 1.54 Å for gas phase sp^2 hybridized ethylene and the saturated sp^3 hybridized alkane ethane shows that the carbon has gone a long ways towards rehybridization to sp^3, i.e., losing much of the C−C double-bond character. How can we describe the resulting surface chemical bond in a simple molecular orbital picture? For this we need to know the occupied orbitals and thus turn to XES for an experimental decomposition in terms of orbital symmetries and character. For simplicity, the orbital labels of gas phase ethylene (D_{2h} symmetry) will be used also for the adsorbate even though this is formally not correct since the symmetry is broken upon adsorption.

Gas phase ethylene belongs to the point group D_{2h} and has the ground state electronic configuration $(1a_g)^2(1b_{3u})^2(2a_g)^2(2b_{3u})^2(1b_{2u})^2(3a_g)^2(1b_{1g})^2(1b_{1u})^2$. The $1a_g$ and $1b_{3u}$ correspond to the bonding and anti-bonding combinations of the two C 1*s* orbitals. The $2a_g$ and $2b_{3u}$ orbitals correspond mainly to the C 2*s* orbitals in bonding and anti-bonding combinations. The $3a_g$ orbital is the C−C bonding σ-orbital, and the $1b_{1u}$ is the π-orbital. The lowest unoccupied molecular orbital is the π^* ($1b_{2g}$).

In Figure 2.35, we compare the symmetry-projected XES spectra for ethylene on Cu(110) and Ni(110) and show in the top part the measured orbital contributions along the atomic rows (($1\bar{1}0$) or X direction). Along this direction, there are mainly four states in the XES spectra of ethylene adsorbed on Cu(110). The main feature is the $3a_g$, which shows up at 8.7 eV binding energy. On the higher binding energy side, there is a feature corresponding to the $2b_{1u}$ at 14.0 eV, and a very weak peak at 18.4 eV corresponding to the $2a_g$. On the low binding-energy side, there is a broad feature, with its maximum at 5.6 eV, which stems from the gas phase π-state. This state appears due to a rehybridization of the molecule upon adsorption, which leads to an upward bending of the C−H bonds. For the molecule adsorbed on Ni(110) all the corresponding peaks are shifted to lower binding energy. This trend is true also

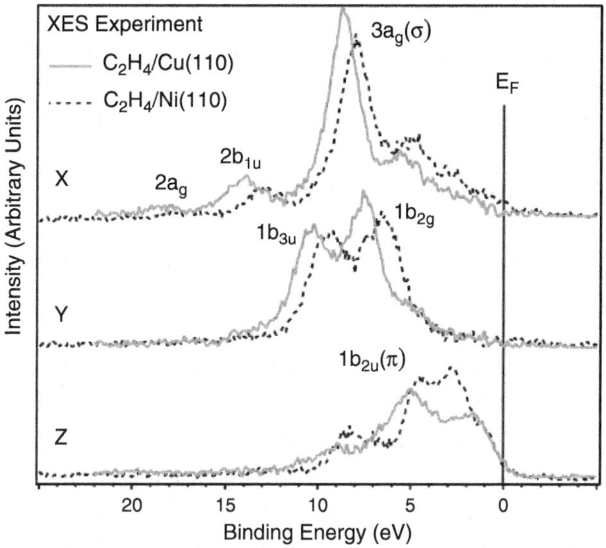

Figure 2.35. Symmetry-resolved XES spectra of ethylene on Cu(110) (full lines) and Ni(110) (dashed lines) [85]. From top to bottom the $(1\bar{1}0)$ (X), (001) (Y) and (110) (Z) projected spectra are shown.

for the spectral contributions along the other directions. This can be explained by the destabilization of the C−C bonding orbitals due to the interaction with the surface, in agreement with the stabilization of the anti-bonding σ^* in the XAS spectra. Along the X-direction, the peak assigned to the $3a_g$ can be seen at 8.0 eV, the $2a_g$ and $2b_{1u}$ at 12.9 and 17.2 eV, and the maximum of the π-derived states is at 4.8 eV. The larger intensity in this region in the case of Ni indicates a larger σ–π mixing than on Cu. This mixing cannot simply be described by a linear combination of σ and π resulting from an angle between the molecular axis and the metal surface. There is almost no contribution of the sharp π^* resonance along the X-direction in the XAS spectra shown in Figure 2.34, which would be the case if the molecules were tilted on the surface.

The reason that the π^*-orbital does not mix significantly with the σ^*, as it does with the occupied orbitals, can be related to the energy splitting of 10 eV between the π^* and σ^*, which is substantially larger than the energy separation of $\approx$4 eV from the occupied orbitals. In C_{2v} symmetry the σ^* cannot mix with the occupied σ-orbitals due to symmetry reasons.

Along the Y-direction (across the rows), which is shown in the middle of Fig. 2.35, there are two main features, assigned to the $1b_{3u}$ and $1b_{2g}$ at, respectively, 10.4 and 7.6 eV on Cu, and at 9.3 and 6.6 eV on Ni. For both metals there is also some intensity which can be assigned to the rehybridization and mixing with π-states starting at 4.8 eV.

The lower part of Figure 2.35 shows the spectra in the out-of-plane Z-direction which most directly reveals the adsorbate–substrate interaction. Along this direction the spectra are characterized mainly by three features. For ethylene adsorbed on Ni(110), there is a small peak at 8.3 eV, which is due to the rehybridization and mixing of σ and π, allowing σ ($3a_g$) intensity to show up in the out-of-plane Z-direction, as well as π intensity ($1b_{2u}$) in the surface plane, mainly along the molecular axis (X-direction). At lower binding energy there is also the remainder of the π states at 4.6 eV and some intensity, caused by adsorbate-metal interaction, starting at 3.5 eV and continuing all the way up to the Fermi level. For Cu the corresponding values are 8.7, 5.0, and 3.0 eV. The intensity of the metal-induced states is significantly larger on Ni than on Cu because of the stronger interaction. We would now like to interpret these experimental observations in terms of an orbital interaction diagram describing the adsorbate-substrate bond-formation, but to do this it is necessary to study the orbital character also based on theoretical DFT calculations. This is so, because the experimental information is projected only on the carbon atoms and the relative phases of adsorbate and substrate contributions, necessary to determine their bonding or antibonding characters, are thus not known. In Figures 2.36 and 2.37, we show representative π- and π*-derived orbitals for the different spectral regions from DFT calculations that also reproduce the spectroscopic features [85].

The orbital illustrated in the leftmost part of Figure 2.36 is the symmetrical $1b_{2u}$ (π) orbital of the free molecule; for the adsorbed molecule it becomes distorted with slightly larger distortion on Ni (right) than on Cu (middle). The unoccupied

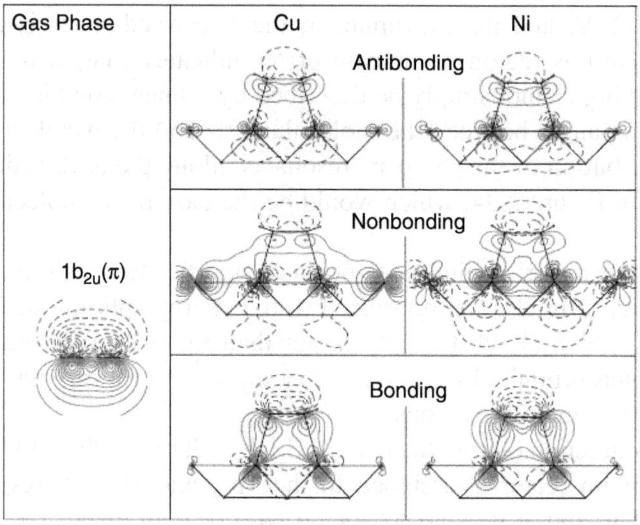

Figure 2.36. Calculated representative orbital characters for ethylene $1b_{2u}$ (π) interacting with the metal *d*-states. From Ref. [85].

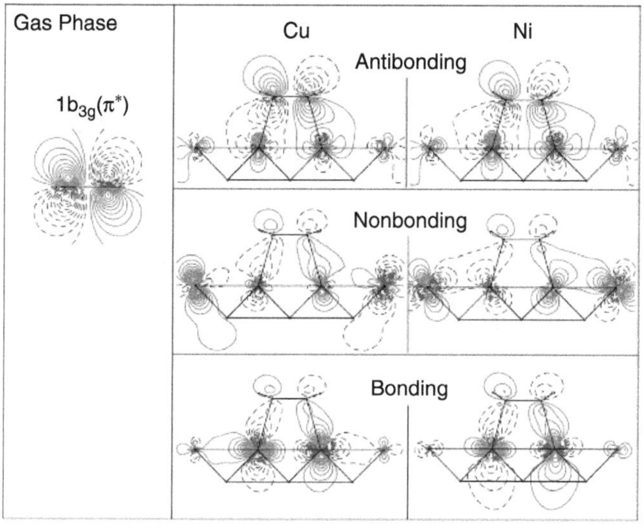

Figure 2.37. Calculated representative orbital characters for ethylene $1b_{3g}$ (π^*) interacting with the metal *d*-states. From Ref. [85].

π^*-orbital is shown for the gas phase molecule in the left part of Figure 2.37 and on Cu (middle) and Ni (right). The bonding and non-bonding combinations become occupied for the adsorbate system and we note a slightly greater involvement of the π^* in the orbitals from the adsorption on Ni than on Cu.

At lower binding energy, in the region from 4.8 eV to the Fermi level in Figure 2.35, we note that the occupied orbitals are of both π and π^* character in bonding and non-bonding combinations with the metal states. One of the relevant non-bonding orbitals of π-symmetry for each surface is shown in Figure 2.36 and one for each surface of π^*-symmetry is shown in Figure 2.37. Since we observe states of both π and π^*-character in XES we find agreement with the DCD bonding model of ethylene, where the bonding is described as a donation of molecular π-electrons into the metal and back-donation from the metal into the unoccupied molecular π^*-orbital [83]. We note that the characters of the metal *d*-orbitals go from pointing towards the carbon atoms for the bonding orbitals (d_{z^2}), to mainly pointing away from them for the non-bonding orbitals (d_{xz}), and again pointing towards them for the anti-bonding orbitals (d_{z^2}).

A simplified schematic molecular orbital diagram is shown in Figure 2.38. The figure shows the interaction between the metal 3*d* band and the occupied π-orbital (left part) and the unoccupied π^*-orbital (right part). The interaction with the π- and π^*-orbitals results in a large number of bonding, non-bonding and anti-bonding combinations in the figure; only three energy levels of each π and π^* symmetry are drawn in the figure. The anti-bonding orbitals remain unoccupied, the

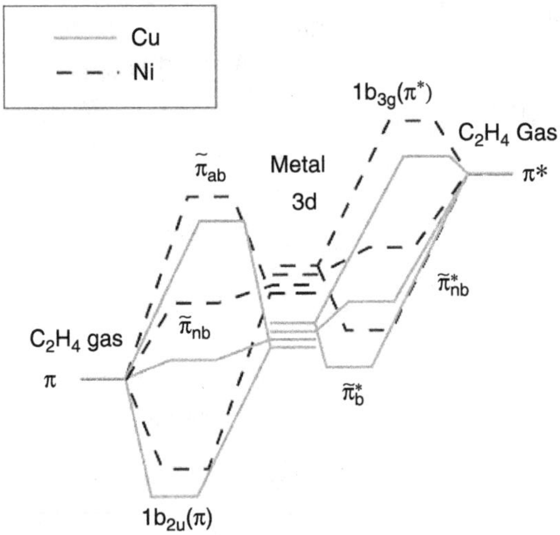

Figure 2.38. Schematic orbital interaction diagram illustrating the interaction of the ethylene $1b_{2u}$ (π) (left side) and $1b_{3g}$ (π^*) (right side) orbitals with the metal *d*-band where the *d*-band center of Cu lies at higher binding energy than that of Ni. From Ref. [85].

bonding occupied and the non-bonding, which are close to the Fermi level, may be more or less occupied depending on the metal *d*-band position. The strength of the bond to the surface is determined by the amount of π- and π^*-states being occupied. It is important here to realize that the π- and π^*-states cannot mix directly due to symmetry; any charge-transfer between these orbitals must be mediated by the surface, as in the established DCD model, but how large is the charge-transfer and how can it be evaluated in direct connection with experiment?

Establishing the amount of charge transfer between the π- and π^*-states requires a decomposition of the calculated wave functions, which can be done in a large number of more or less arbitrary ways. If we exploit the property of XES to give an atom-projected density of the *p*-contributions to the molecular orbitals we can calibrate calculated XES spectra of the adsorbate against the free molecule, for which the occupation in the π-orbital ($1b_{2u}$) is known, to quantify the amount of $2p$ character in each of the orbitals based on the intensity relative to the π-orbital of the gas phase. To separate the π and π^* contributions we can use the *gerade* and *ungerade* symmetry of the C 1*s* molecular orbitals, forming a bonding (*gerade)* combination without a node ($1a_g$) and an antibonding (*ungerade)* combination ($1b_{3u}$) with a node. Due to the dipole selection rule spectra projected on the $1a_g$ core level probe only the π-symmetry (bonding between the C atoms) while spectra projected onto the $1b_{3u}$ probe the π^* (antibonding between the C atoms).

Integrating the π- and π^*-contributions to the theoretical out-of-plane XES spectra and normalizing to the intensity of the gas phase π-orbital, which is assumed to have 1.0 π-electron/atom, yields occupations of 0.84 π and 0.16 π* electrons/atom for ethylene adsorbed on Cu(110) [83]. The corresponding numbers for ethylene adsorbed on Ni(110) are 0.78 π and 0.31 π* electrons/atom. This means that the π–π* donation/backdonation is larger in the case of Ni(110) than for Cu(110). This is reasonable considering that the incompletely filled *d*-band in the case of Ni will allow for smaller exchange repulsion as well as for a more efficient rehybridization to form bonds to the adsorbate. Furthermore we note a total increase of occupation in the π-system of ethylene on Ni(110) to 1.09 electrons/atom, whereas on the Cu(110) surface the total occupation stays constant at 1.0 electron/atom, as in the gas phase. The relatively larger back-donation for Ni can be understood from the fact that the *d*-band lies closer to the molecular π* in Ni than in Cu. All in all we find that the DCD model gives a good representation of the resulting bonding in terms of π–π* donation/back-donation [83] and that the strength of the interaction correlates well with the position of the center of the *d*-band for the two metals [49,50] (see Chapter 4).

7.2. Benzene on Ni and Cu surfaces

Benzene adsorbs weakly on Cu and strongly on Ni. It is interesting to study how the differences in adsorption strength are reflected in the electronic structure of the adsorbate-substrate complexes as determined based on the XAS and XES spectra for benzene on Cu(110) and Ni(100) shown in Figures 2.39 and 2.40, respectively [83,90].

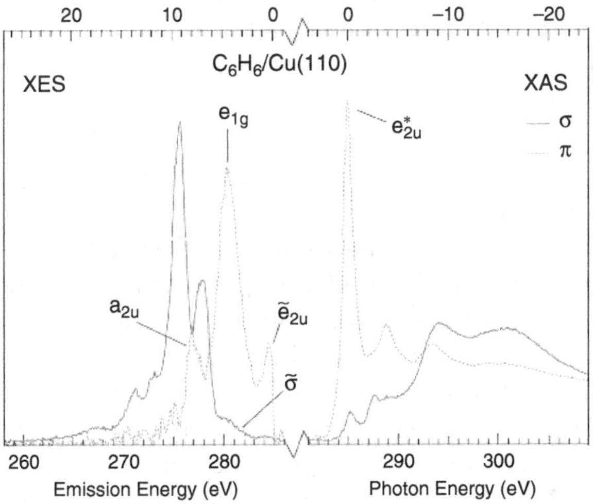

Figure 2.39. σ- (full line) and π- (dashed line) resolved XES (left) and XAS (right) symmetry-resolved spectra of benzene on Cu(110) shown using photon energy scale (bottom) and binding energy scale (top). From Ref. [90].

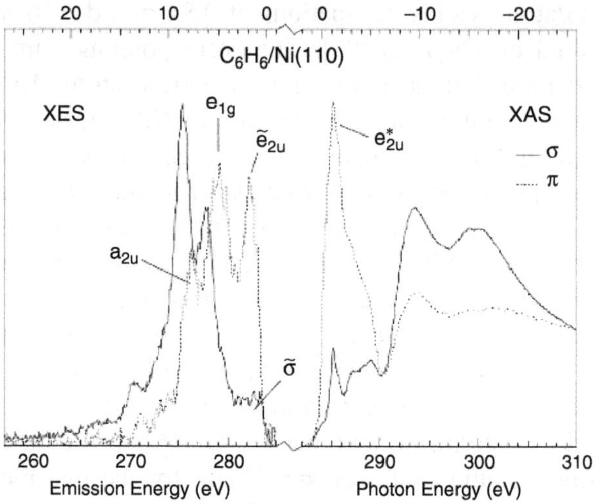

Figure 2.40. σ-(full line) and π-(dashed line) resolved XES (left) and XAS (right) symmetry-resolved spectra of benzene on Ni(110) shown using photon energy scale (bottom) and binding energy scale (top). From Ref. [90].

If we start with states of π-symmetry (dashed lines) we find three distinct peaks in the XES spectra reflecting the occupied states. The $1a_{2u}$ and $1e_{1g}$ π-like orbitals are essentially intact from the gas phase, while the third state, labeled $\tilde{e}_{2u}$, is not seen for the free molecule. Based on symmetry-selection rules, it can be shown that this state is derived from the lowest unoccupied molecular orbital (LUMO) $e_{2u}\pi^*$-orbital that becomes slightly occupied upon adsorption. We anticipate a similar bonding mechanism as discussed in the previous section for adsorbed ethylene with the exception of a weaker rehybridization due to the extra stability in the π-system from the aromatic character.

The new occupied $\tilde{e}_{2u}$ state appears differently in the spectra from benzene on Cu and Ni. In the case of adsorption on Ni (Figure 2.40), the new π-state is located 1.7 eV below the Fermi level and hence overlapping the Ni *d*-band region. The fraction of adsorbate character of this state has to be of similar magnitude as that of the higher binding energy π-orbitals since the observed emission intensity is of comparable strength. The formation of a single bonding state rather than a band suggests that the e_{2u} orbital interacts mainly with a narrow distribution of states, i.e., the Ni *d*-band. For benzene adsorbed on Cu (Figure 2.39), the new occupied $\tilde{e}_{2u}$ orbital is located just below the Fermi level with an indication of a cut-off at the Fermi level. The adsorbate character in this orbital is rather low. This is also reflected in the relatively strong peak from the anti-bonding orbital in the XAS spectrum in Figure 2.39. Our observations suggest that the unoccupied e^*_{2u} orbital interacts with a broad continuum of metal states and becomes broadened with a tail extending below the Fermi level. There is no distinct separation between the bonding and anti-bonding states.

In the σ-symmetry XES spectra in Figures 2.39 and 2.40 (left, solid lines), we find, at binding energies larger than 5 eV, an orbital structure identical to that of the free molecule [90]. In addition, we find for benzene on Cu, two new weak structures at binding energies below 5 eV. These structures appear at the same binding energies as the $1e_{1g}$ and e_{2u} orbitals seen in the π-symmetry spectrum. We can anticipate some symmetry-mixing due to rehybridization as discussed for ethylene in the previous section. For benzene adsorbed on Ni the σ-symmetry spectrum is different. Here we find significant σ-intensity all the way to the Fermi level, marked $\tilde{\sigma}$ in Figure 2.40. A simple admixture of σ and π states is not sufficient to explain the appearance of new σ-states in the spectrum, since then we would expect a σ-electron distribution which resembles the π states with a peak at 1.7 eV (the backbonding e_{2u} orbital). The 2*p* valence states parallel to the surface must therefore interact with the metal somewhat differently than the perpendicular 2*p* states. The band-like character of the σ states indicates a hybridization of benzene σ states with the broad Ni *sp*-band in addition to the *d*-band, which is indicative of a major distortion in the carbon skeleton going towards tetrahedral coordination. Such a distortion has in fact been proposed for the inverted boat chemisorption structure of benzene that would result from bonding of benzene in the triplet excited quinoid state; this excited state corresponds to the molecule prepared for forming two covalent bonds to the substrate, as will be discussed in the following section [84]. All in all however, the mixing of highest occupied molecular orbital (HOMO) and the LUMO-like states of the adsorbate/substrate complex is less pronounced for benzene than for ethylene, a fact that we ascribe to the strongly aromatic character of the benzene ring.

7.3. Bond energetics and rehybridization from spin-uncoupling

An alternative viewpoint to the DCD model in terms of adsorbate spin-uncoupling, or bond-preparation, focuses on the available excited states of the adsorbate [84]. The fundamental idea is that in the chemisorbed state one internal π-bond has been broken in order to form the new bond to the surface. From an electronic structure viewpoint this corresponds precisely to a $\pi \rightarrow \pi^*$ excitation, leading to an internal triplet spin-coupling on the adsorbate. The bond is then formed by spin-pairing with the appropriate open shells of the surface electronic structure. Since, in this case, a change in the electronic structure with a concomitant rehybridization cost is required for the adsorbate, this leads to the prediction of avoided crossings and possible resulting barriers in the pathway to the chemisorbed state as illustrated in Figure 2.41.

In this picture, the starting point for the interaction is an internal transfer of one full π-electron to the π* through a singlet-triplet transition. In this step, the molecule is prepared for bonding and the structure is distorted to a structure very similar to that of the adsorbed molecule. The thus formed π-electron biradical

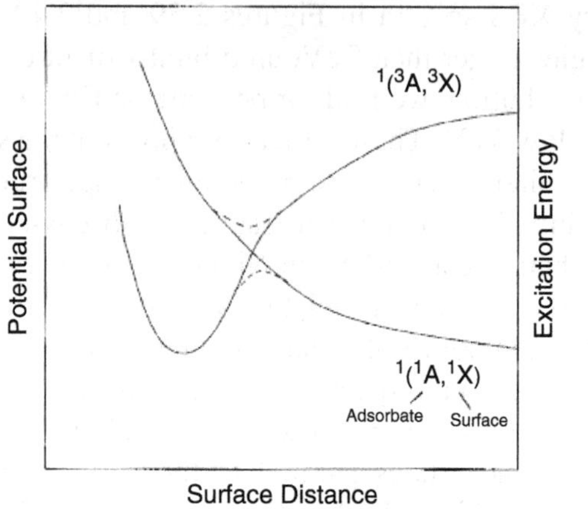

Figure 2.41. The closed-shell ground-state leads to a repulsive interaction with the substrate. An internal bond is broken to form the bond to the substrate. This bond-prepared adsorbate state corresponds to an excited gas-phase triplet state. From Ref. [84].

can form two σ-bonds to the substrate, as is the case for acetylene and ethylene in the di-σ configuration (see Chapter 1). The initial excitation rehybridizes the molecule, populating the π^*-orbital with 0.5 electrons/atom. The bonding to the surface is then formed through the loss of π^*-electron charge and gain in the π-states yielding the same final charge distribution as predicted by the DCD model. As for the DCD model, the amount of π-backdonation and π^*-donation will depend on the electronic structure of the surface. The advantage with explicitly considering the spin-uncoupling or bond-preparation is that the energetics of the bond-formation can be estimated based on bond-preparation cost and energy gain from the bond formation [84]. Here we will give a few examples showing what can be gained through this approach.

Let us begin by considering the simplest unsaturated hydrocarbon, acetylene or C_2H_2, with a triple CC-bond leading to a very short C−C distance of 1.208 Å in the gas phase [91]. When chemisorbed on various Cu surfaces DFT optimizations result in a substantially elongated C−C distance of 1.30–1.40 Å for different sites on the (100), (110) and (111) faces of the Cu surface [84] (Figure 2.42). The hydrogens are found to bend away from the surface creating a C−C−H angle between 46° and 56° smaller than the gas phase 180°. Comparison with the gas phase optimized structure of the lowest excited triplet state of acetylene reveals a structure very similar to that of the surface adsorbate with an elongation of the internal C−C bond of 0.12 Å and a reduction of the C−C−H angle by 52°; we immediately conclude that the bending up of the hydrogens away from the surface is not directly due to the repulsion from the substrate atoms, but rather due to the internal rehybridization of the molecule in its chemisorbed state.

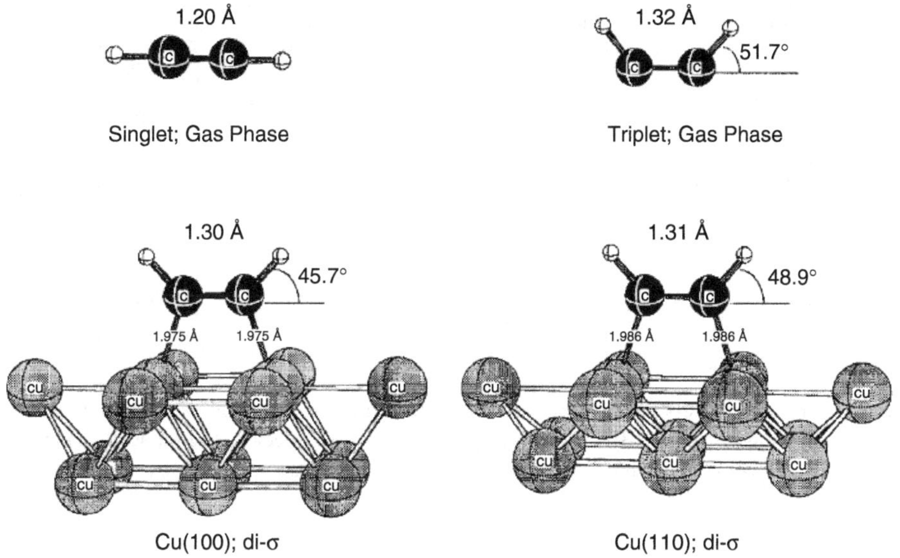

Figure 2.42. (Top) DFT optimized structure of gas phase ground-state singlet and excited triplet state acetylene. (Bottom) Optimized structures of acetylene on cluster models of Cu(100) (left) and Cu(110) (right). From Ref. [84].

Considering the energetics we find that the computed chemisorption energy varies between 0.56 and 0.87 eV when considering several different adsorption sites and structures on the Cu(111), Cu(110), and Cu(100) substrates. For each structure we take the molecule in its optimized geometry at the surface and for the isolated molecule we calculate the excitation energy required to reach the triplet state; the resulting energies vary between 3.73 and 3.86 eV depending on the structure [84]. Since the net effect is a stable bond to the surface the established two C−Cu σ-bonds must more than compensate for this excitation energy and we can establish an average individual C−Cu σ-bond strength of 2.23 eV in this case.

To further underline the relevance of the rehybridization using the triplet excited state we compare the charge density difference taken between chemisorbed acetylene and the component gas phase molecule and isolated cluster with that between singlet and triplet acetylene in the geometry of the triplet (Figure 2.43). Acetylene is found to adsorb in two different sites on the Cu(110) surface that are distinguishable through their XPS binding energies as a low-binding energy (LBE) and high-binding energy (HBE) species [92]. Both species adsorb in a di-σ configuration with their C−C axes mainly in the surface plane. The LBE species binds in a long-bridge site with its C−C axis along the (001) direction while the HBE species occupies a low-symmetry site with its C−C axis on average at an angle of 35° off the (110) direction [92]. The changes in the charge density generated already in the gas

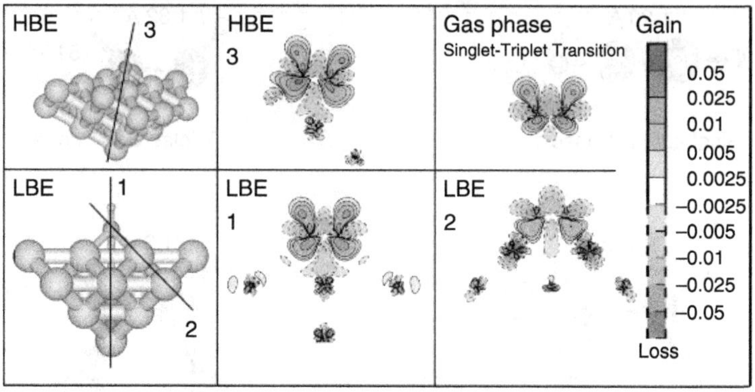

Figure 2.43. Charge density differences induced by adsorption of acetylene in the two different sites on Cu(110) showing (top) the HBE species and (bottom) the LBE species. The induced changes in the charge density are compared with those generated for the gas phase molecule by a singlet to triplet $\pi \rightarrow \pi^*$ excitation. From Ref. [92].

phase by a $\pi \rightarrow \pi^*$ excitation are compared in Figure 2.43 with the charge density differences induced by the adsorption for the LBE and HBE species, respectively. To compare we use the triplet state geometry for the gas phase comparison and it is immediately clear that the major part of the chemisorption-induced charge rearrangements can be viewed simply as due to the $\pi \rightarrow \pi^*$ excitation, i.e., as related to the internal bond-preparation of the adsorbate.

Ethylene represents the next case and a slightly different situation since the lowest excited gas-phase triplet state is the staggered or twisted conformation with a calculated excitation energy of 2.78 eV. However, in this conformation only one bond can be formed to the substrate and with a C−Cu σ-bond energy of 2.23 eV, as obtained from acetylene, the large rehybridization cost cannot be overcome in this conformation. Instead it is the triplet state in an eclipsed conformation with the hydrogens slightly bent upwards that has a structure suitable for the interaction with the substrate; this is actually a transition state in the twist around the C−C axis. Here the excitation energy is higher, 3.73 and 3.99 eV for the di-σ conformation on Cu(100) and Cu(110), respectively. However, two C−Cu bonds of around 2.23 eV each can clearly more than compensate for this rehybridization investment. From the computed chemisorption energies in the two cases we find a C−Cu σ-bond energy of 2.14 eV for ethylene [84].

To gain a deeper understanding of the bonding we can again look at the rearrangement of electronic charge density upon adsorption and compare adsorption on Cu(110) and Ni(110) as well as the charge rearrangement upon excitation from the singlet to the relevant triplet state for the free molecule. We take an increase in electronic density between two atoms to indicate bond strengthening, and a decrease to indicate bond weakening. The charge density difference plots for the different

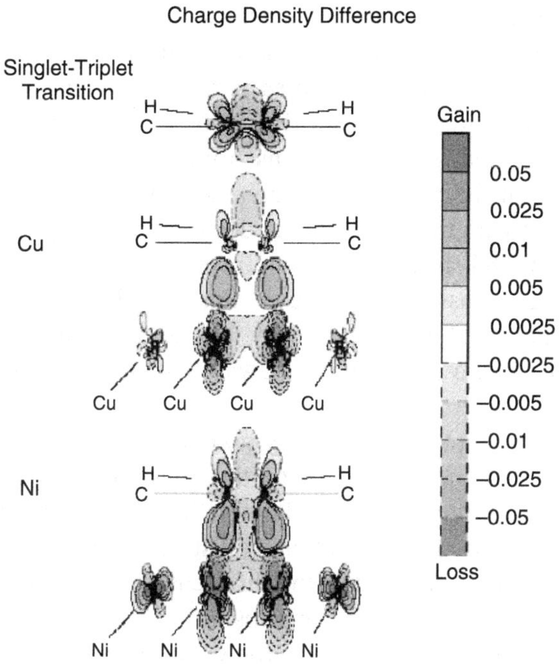

Figure 2.44. Charge density difference plotted in a plane containing the metal atoms and the carbon skeleton of the ethylene molecule. The difference is taken between interacting and non-interacting molecules and metal cluster for the adsorbed cases. For the gas phase molecule (top), the difference between the singlet and triplet state is shown. From Ref. [85].

situations are shown in Figure 2.44. The plots are cut along a plane containing the adsorbate molecular axis and the central metal row, on which the molecule is adsorbed. In the figure, increased electron density is denoted by solid lines whereas electron depletion is marked by dashed lines.

The top of Figure 2.44 shows the charge density difference between singlet and triplet electronic structures of free ethylene forced into the geometric structure of ethylene adsorbed on Ni(110). The difference in electronic structure in this case is that for the singlet molecule the π-orbital is doubly occupied, whereas for the triplet one electron is excited from the π to the π^*-orbital. This shows up in the charge density difference plot as an increase of electronic charge in the region of the π^*-orbital, above and below the molecular axis, outside of the carbon atoms, and a decrease of electronic charge in the area of the π-orbital, above and below the molecule, between the carbon atoms.

This change is very similar to what is seen upon adsorption on both the Cu and Ni surfaces, shown in the middle and lower parts of Figure 2.44. On the surfaces the charge rearrangements are smaller because only a fraction of an electron is excited into the π^*-orbital. This is in perfect agreement with the DCD model, which predicts charge depletion from the π-orbital and increase in the π^*-orbital. It can be noted that the increase in the π^* dominated area is approximately twice as large on Ni as on Cu, in complete agreement with the π^* occupation, which upon adsorption increases from unoccupied to 0.16 electrons/atom on Cu and to 0.31 electrons/atom on Ni. The decrease in the π-dominated area is only slightly larger on Ni than on Cu. This is consistent with the almost equal decrease in π-population on Ni and Cu, resulting in 0.78 and 0.84 π-electrons/atom, respectively. There is a larger increase in the electron charge density below the π^*-dominated area of the molecule. This arises as a consequence of the bonding to the surface. There is also an internal change in the metal, where charge moves away from the direction directly towards the molecule.

It can be noted that there are essentially no changes in the charge density along the direction of the C−H bonds, indicating that these bonds are not much affected by the bonding to the surface. It should be noted, however, that the charge density difference was computed with the already distorted geometry, thus leaving out electronic structure changes related to the difference in geometry, such as, e.g., the σ–π-mixing upon rotation of the CH_2 groups.

The final case to consider is benzene for which there are two near-degenerate triplet π to π^* excited states, the quinoid and anti-quinoid structures, where the former localizes the unpaired spins in para-position while the anti-quinoid has spin density on the other four carbons [84]. Here we will focus on the quinoid structure which already in the gas-phase has its carbon backbone slightly distorted in an inverted boat conformation and the hydrogens slightly bent (Figure 2.45). The excitation energy in this case, i.e., the rehybridization cost, is 4.25 eV which again is less than twice the expected C−Cu σ-bond energy where we assume the formation of two σ-bonds. The computed chemisorption energy for the quinoid form on Cu(110) of 0.78 eV together with the rehybridization cost of 4.25 eV could seem to indicate a substantially stronger Cu−C bond in this case. However, comparison with undistorted benzene on Cu(110) binding through polarization of the π-system (Figure 2.45, right) shows a significant contribution also from this effect. Correcting for the polarization contribution in the quinoid state gives a total σ-bond energy of 4.42 eV, i.e., 2.21 eV per C−Cu bond formed. We note that upon adsorption from gas phase, X-ray spectroscopic measurements clearly show that only the undistorted form of benzene is obtained on Cu(110) and it was speculated that a barrier could prevent the formation of the strongly rehybridized quinoid structure on this substrate. On the other hand, comparison of calculated and experimental XAS spectra for benzene on Ni(100) and Mo(110) gave support for the quinoid adsorption structure on these substrates [93].

In summary, analysis of the rehybridization cost in terms of the involved π to π^* excited state results in an average C−Cu σ-bond energy of 2.21 ± 0.09 eV based

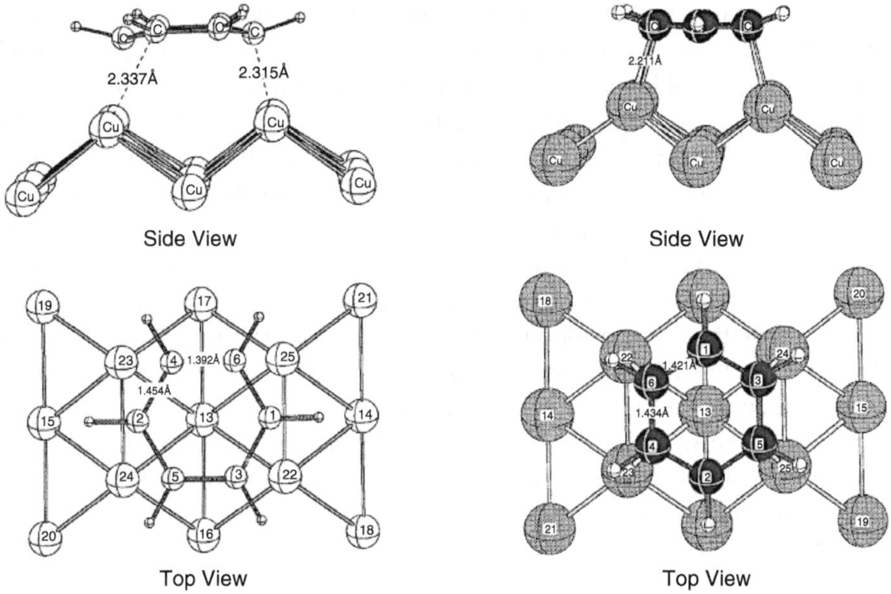

Figure 2.45. Adsorption structures of benzene on Cu(110) optimized using DFT. (Left) The quinoid structure with carbon backbone distorted in an inverted boat conformation and (Right) undistorted benzene interacting through polarization of the π-system and with minor upwards bending of the hydrogens. From Ref. [84].

on the calculated chemisorption energies for acetylene, ethylene and benzene on Cu(100), Cu(110) and Cu(111) [84]. This can be compared with the very similar value of 2.30 eV used by Carter and Koel as estimate of the Pt−C bond strength in ethylene decomposition reactions on Pt(111) [94]. All in all this provides a nice illustration of the 'economics of chemical bond-formation': to make a chemical bond, an investment (rehybridization cost) must be made. The return is the bond energy and the 'profit' is the binding energy to the surface as measured, e.g., by temperature programmed desorption (TPD). However, it is also clear from this that a low chemisorption energy does not necessarily imply a weak interaction.

8. Saturated hydrocarbons

The adsorption of saturated hydrocarbons on metallic substrates is typically considered as an example of a weak physical interaction, which is dominated by van der Waals forces. The classification of this type of interaction, denoted physisorption where no direct chemical bonds are formed between the adsorbate and substrate, has been based on the heat of adsorption. A physisorbed state is considered to be one in which the heat of adsorption is comparable to the heat of vaporization or sublimation. This is the case for the interaction of alkanes with metal surfaces and the interaction has consequently been characterized as a weak physical adsorption. However, as underlined in the previous section on unsaturated hydrocarbons, a small adsorption energy cannot by itself be used to conclude a weak interaction. Indeed in this section we will demonstrate that for *n*-octane physisorbed on Cu(110) there are still surprisingly large and important chemical bonding interactions with the surface that are beyond a physical adsorption picture [95–97]. There are relatively large internal geometry distortions in the molecule and a relatively short H−Cu bond distance due to this interaction. The C−C bond is shortened and the C−H bonds pointing towards the surface elongated due to the Cu−H interaction. This means that the molecule has taken a small step towards dehydrogenation. There is thus an important interaction of the molecular orbitals involving the CH groups that point to the surface with the *sp* and *d*-bands in the metal. It leads to a weak electron-pairing between the CH and Cu atoms. Similar effects are observed even for very weakly physisorbed methane, CH_4, on Pt(111) [98].

8.1. n-Octane adsorbed on Cu(110)

Adsorption of saturated hydrocarbons on a Cu substrate provides a good model system for investigating the electronic structure since the *d*-band interaction appears entirely in the occupied states, making the effects more clearly visible and the analysis of the electronic structure easier. There is an advantage to use the (110) surface with a two-fold symmetry if the molecule adsorbs with preferential alignment allowing projection of the electronic structure in three directions as discussed in the previous sections.

The XES and XAS spectra, symmetry-resolved in three directions, are shown in Figure 2.46 [95,96]. Note that both the XES and XAS spectra are shown on a binding energy scale relative to the Fermi level with the XAS spectra shown in the right part of the figure. The top spectrum, showing the p_x contribution is characterized by a main feature at −8.8 eV binding energy, with a shoulder on the high binding energy side (i.e., closer to the Fermi level), and in the p_y spectrum there is a main feature at −6.8 eV. Both these features can be assigned to different CCσ^* shape resonances. There are spectral features around −3.5 eV binding energy both in the p_y and p_z spectra, assigned to transitions into CH* orbitals. The p_z spectrum also shows significant intensity right at the Fermi level, previously denoted M* [99]. Since the strong CCσ^* shape resonance has its maximum intensity in the p_x spectrum we can conclude that the molecular axis is oriented parallel with the Cu rows as shown in the insert of Figure 2.46. This observation is consistent with other results using complementary techniques such as reflection absorption infrared spectroscopy (RAIRS) [100]. The XES spectra in the left panel of Figure 2.46

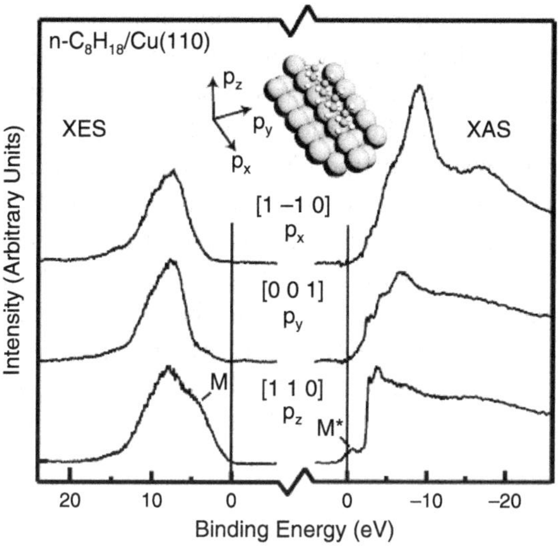

Figure 2.46. Experimental symmetry-resolved XES and XAS spectra of *n*-octane adsorbed on Cu(110). The spectra are projected along the three high-symmetry directions of the surface. The alignment of the *n*-octane molecule on the surface is shown in the inset. From Ref. [95].

show the occupied C 2*p* projected density of states. All these spectra are characterized by a broad feature around 8 eV binding energy. This large width is typical for alkanes, and arises due to band formation [101] and vibronic coupling of the C 2*p*-derived molecular orbitals. The p_z spectrum, however, shows an additional feature at 5 eV appearing as a consequence of the interaction with the metal surface; we denote this feature M. We will now turn our attention to a theoretical determination of the adsorbate geometry and electronic structure. This is at present a non-trivial task since, apart from uncertainties due to the choice of functional in the DFT treatment, the van der Waals interaction is not accounted for in present DFT approaches.

Since dispersion forces are not accounted for within DFT, it will not be possible to use a total energy criterion to determine the geometries of systems that are characterized by these weak interactions. Typically, the potential energy curves for weakly bound surface adsorbates are very shallow, leading to a large uncertainty in the adsorbate-substrate distance. The total energy in such cases may thus not be a reliable measure for the theoretical determination of the geometric structure, but the electronic structure should be. This follows from the fact that the dispersion interaction in itself does not in a direct way alter the orbital structure except for the indirect changes caused by distortions of the geometry due to the enhanced interaction with the surface. The molecular orbitals provide a more sensitive measure in this case since orbital interactions depend on the overlaps which depend near-exponentially on the distance. Having available a reliable and accurate way to compute molecular orbital overlap-dependent properties we can thus instead perform a more sensitive systematic investigation of how the XAS and XES spectra are influenced by changes of different structural parameters. Comparison with the experimental results allows us then to draw conclusions on the structure of the adsorbed molecule [95].

The positions and intensities of the CCσ* shape resonances are sensitive to the C−C bond length. The geometry of the adsorbed molecule was determined by systematically computing theoretical XAS spectra for different geometries as shown in Figure 2.47. The structure of the carbon skeleton was determined by fitting the CCσ* shape resonances in the XAS spectra to the experiment. The best agreement was obtained for a C−C bond length of 1.49 Å [95], which is shorter than for the optimized isolated molecule (1.53 Å). The occupied metal-induced states are not sensitive to changes in the C−C skeleton of the molecules (Figure 2.47, top

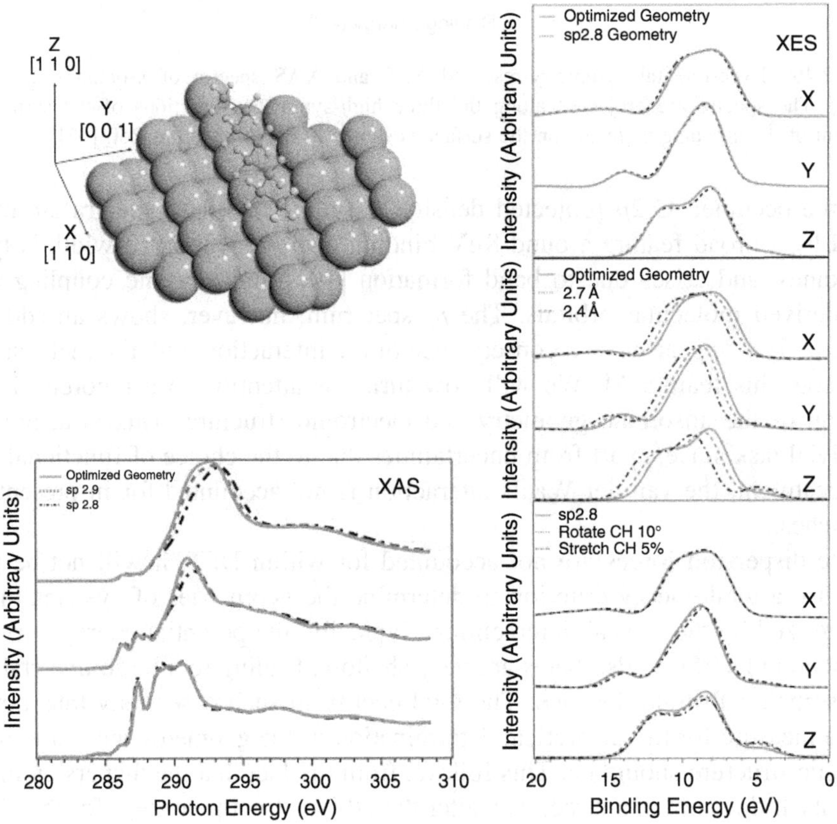

Figure 2.47. (Bottom, left) Calculated XAS spectra of adsorbed *n*-octane for different C−C bond lengths and C−C−C angles corresponding to the indicated hybridization schemes. (Right) Calculated XES spectra for change in (top) carbon hybridization, (middle) distance to surface and (bottom) CH distance and orientation. From Ref. [96].

right), but to the molecule-surface distance (Figure 2.47, center right) as well as to both bond length and direction of the C−H bonds (Figure 2.47, bottom right). To reproduce the M feature in the spectral calculations, the molecule had to be pushed in to a C−Cu distance of 2.7 Å and the C−H bonds pointing towards the surface had to be elongated from 1.10 to 1.18 Å, and rotated towards the surface [95]. In some sense the molecule has made a small step towards dehydrogenation by weakening the C−H bond and strengthening the C−C bond.

The energy cost for pushing the molecule in from the optimized carbon-metal distance of 3.0 to 2.7 Å, without changing the internal structure, is 0.75 eV according to the calculations [95]. The energy increase upon rehybridization from sp^3 to the obtained $sp^{2.8}$ is 0.21 eV at 2.7 Å. The cost of the rotation of the C−H bonds is 0.39 eV and the stretch of the bonds costs 0.57 eV. This means that the total energy cost for the change in geometry is 1.92 eV. This should be compared with the energy we can expect to gain from the dispersion interaction. If we assume that this is the same as the total adsorption energy for alkanes, the energy is approximately 2–4 kcal/mole per carbon atom [102]. This means that the dispersion energy will be between 0.7 and 1.4 eV for octane, which may not seem energetically favorable. However, considering that the structural changes were performed 'by hand' without concomitant geometry optimizations on the surface and furthermore that the experimental estimate includes also rehybridization, the energy cost for the proposed distortion is still reasonable and within the theoretical error bars considering the size of the system.

With this agreement between the experimental and theoretical XES and XAS spectra, we can go further and analyze the molecular orbitals involved in the interaction between the adsorbate and surface. Comparing the occupied molecular orbitals with the corresponding gas phase orbitals, we find that the CH bonding orbitals pointing in the out-of-plane direction mix with the copper *d*-band to form new three-center orbitals. The out-of-plane CH bonding orbitals span the energy range from 13 to 7 eV. All these orbitals are similar to the corresponding gas phase orbitals, but they also show bonding interaction with the copper *d*-orbitals. Some of these orbitals and the corresponding gas phase orbitals are plotted in Figure 2.48. In the region of the copper *d*-band there is a large number of orbitals mainly with copper *d*-character, but also some CH bonding character, showing mainly antibonding character with the metal atoms, as shown in the orbital on the right in Figure 2.48. The intensity in the XES spectra around 5 eV is assigned to these states.

We can summarize the orbital interactions with the Cu substrate in a schematic diagram shown in Figure 2.49. The new M band has mainly metal character, but also some carbon character. The octane orbitals contributing to this band are mainly the bonding CH orbitals, since these are closer in energy to the Cu *d*-band, but also to a smaller extent antibonding CH* orbitals. The band, which is completely occupied, is both bonding and antibonding with the surface, leading to a net repulsion. Also the

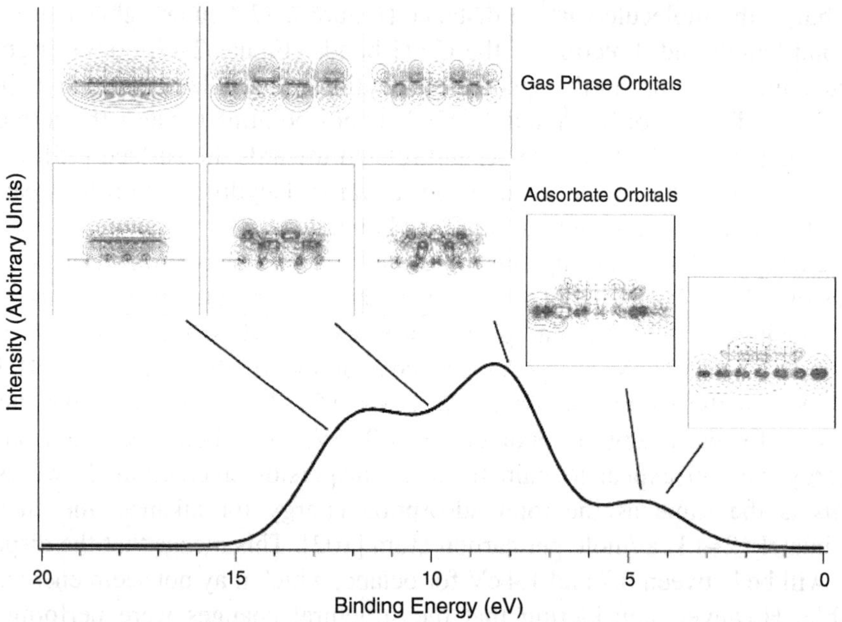

Figure 2.48. Computed XES spectrum along the (110) direction, together with plots of the xz-projection of some of the corresponding molecular orbitals. The corresponding gas phase orbitals are also shown. From Ref. [96].

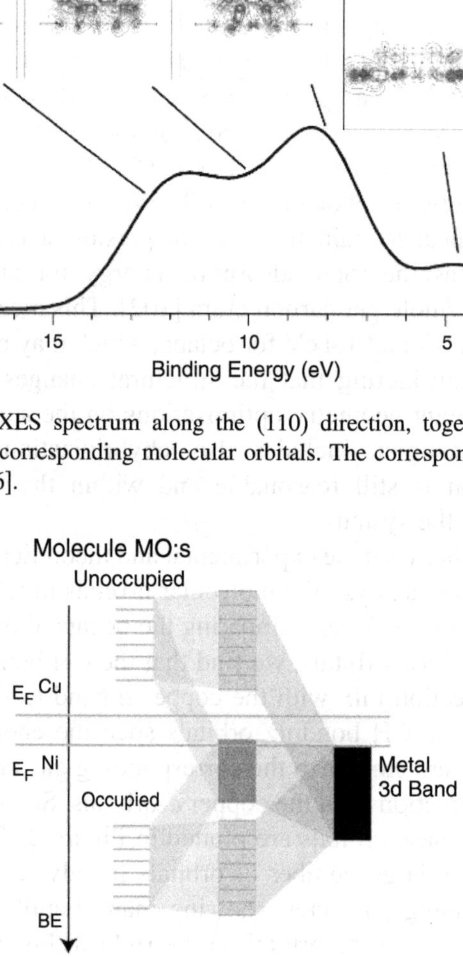

Figure 2.49. A schematic figure showing the interaction between the C−H orbitals and the metal *d*-band, which leads to a strong mixing of molecular orbitals. From Ref. [95].

much broader Cu *sp*-band interacts with the octane orbitals, leading to an extended weak band. The unoccupied part of this band can be seen in the XAS spectra as the M* spectral feature. This indicates that all bonding and antibonding states are not fully occupied. Can we anticipate a potential covalent bonding?

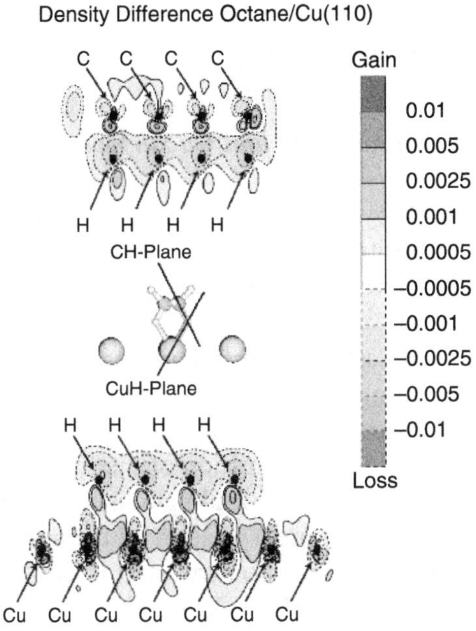

Figure 2.50. Charge density difference plotted along the C−H plane (top) and the Cu−H plane (bottom). From Ref. [95].

Figure 2.50 shows the difference in charge density compared with the free, but distorted molecule, as it is allowed to interact with the surface. We observe small but significant changes in the charge density. Along the CH-plane, there is a polarization of the C−H bonds towards the C atoms and charge depletion close to the hydrogen atoms. This is part of breaking up the C−H bonds due to mixing of C−H bonding and antibonding orbitals. This mixing leads to a weakening of the C−H bonds and results in the observed elongation of the C−H bonds pointing towards the surface. Along the CuH-plane, there is a loss of charge on the H atoms, and a build up of charge between the H and Cu atoms, which means that there is electron sharing between the adsorbate and substrate. From a closer inspection, we can derive that it is mainly the Cu p_z states that are involved in the bonding. Furthermore there are some changes in the substrate, which can be related to an *spd*-rehybridization in the Cu. The net charge does not increase on the molecule, indicating that the bonding model suggested by Wöll *et al.* [103] is not appropriate for a non-strained molecule. Their model does not allow for much interaction, and the adsorbate-surface distance is too long to allow for any interaction with the relatively small bonding CH-orbitals. Based on our charge density difference plots, we conclude that the interaction is a combination of both donation and backdonation, involving the Cu 4*sp* band. Even though the bond is weak, it is likely that this is the driving force behind the changes observed in the structure of the molecule.

8.2. Difference between octane on Ni and Cu surfaces

As shown in previous work by Hammer and Nørskov [31], the nobility of metals is directly related to the position of the metal *d*-band with respect to the Fermi level. Can we predict the differences in interaction of octane with different transition metals, based on that picture?

In Figure 2.51, computed carbon local *p*-densities of states, projected along the out-of plane (p_z) direction, are shown for octane adsorbed on both Cu and Ni. The main spectral difference between the two substrates is the position of the 3*d* metal-induced states (M), which follows the position of the metal *d*-band. In Cu, all these states are entirely below the Fermi level, leading to a net repulsion in the interaction with the 3*d* contribution. However, in the case of Ni the antibonding states continue above the Fermi level, and this interaction can start to contribute to the adsorption energy as also illustrated in Figure 2.49. In this case, we can anticipate that the molecule will move closer to the surface, due to the increased bonding, leading to further rehybridization and weakening of the C−H bonds. As the *d*-band moves towards and over the Fermi level, this increases the possibility for charge donation

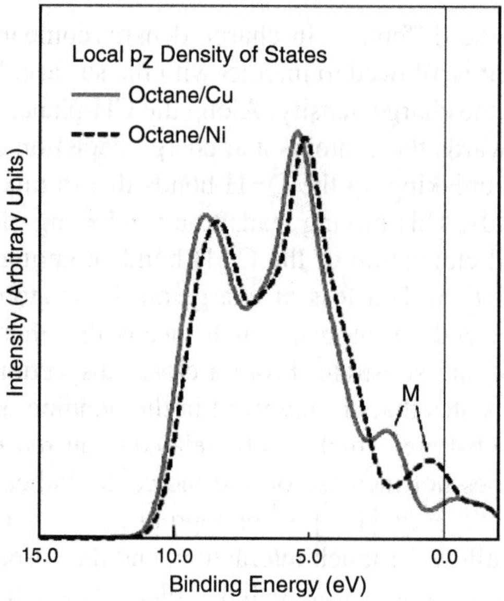

Figure 2.51. The calculated local *p*-density of states projected onto the out-of-plane (p_z) direction for *n*-octane on Cu and Ni. From Ref. [95].

to the surface, and formation of a more dative bond similar to the agostic bonding in organometallic compounds. From these results we can predict, in accordance with Ref. [31], that when we move further to the left in the periodic table the bonding will become even stronger, and all C−H bonds will be broken, depositing strongly bound carbon on the surface, thus yielding a less efficient catalyst.

9. Lone pair interactions

In the following section, we will discuss adsorbate-metal interactions in molecular systems without radical character (i.e., unpaired spins) or unsaturated π-system that can be involved in the surface bonding. The molecules in this section all have some extent of lone-pair character. This will in general lead to rather weak adsorption. We will discuss water adsorption on Pt(111) where there are two different bonding configurations to the surface: oxygen lone-pair metal interaction and agostic hydrogen-bonding with the substrate. Large molecules can often be viewed as built from different functional groups joined together by molecular bonds of strength comparable to the intramolecular bonds within the respective building units. By using the bond-prepared functional groups as a starting point, new electronic states appearing for the more complex molecules can be accounted for. Thus, to be able to address questions regarding the reactivity of biologically interesting molecules on metal surfaces, a detailed knowledge about the chemically important functional groups of the system can be very helpful. An example is the adsorption of ammonia on Cu(110), which can be taken to represent the interaction of the amino group in glycine (NH_2-CH_2-COOH) with the same metal surface.

9.1. Water adsorption on Pt and Cu surfaces

The bonding mechanism of water on different surfaces has recently inspired intense debate, from which general agreement has emerged that at low temperatures water adsorbs intact on the close-packed Pt(111) and Ru(0001) surfaces [104–106]. The intact monolayer is thought to form a hydrogen-bonded structure comprised of hexagonal rings of water, arranged in two layers to form an extended honeycomb network. The inner layer of water has the oxygen lone pair directed toward the surface, forming a weak bond to the metal with the OH groups nearly parallel to the surface, while the second layer water completes the hydrogen bonding structure, see Figure 2.52. The nature of this outer layer has been an issue of controversy. Conventionally, it has been believed that the non-hydrogen-bonded OH of the second-layer water molecule points up towards vacuum (H-up) (Figure 2.52(a)) but recently it was shown on Pt(111) that those OH groups are pointing towards the surface (H-down) (Figure 2.52(b)) [106]. This leads to two different bonds with

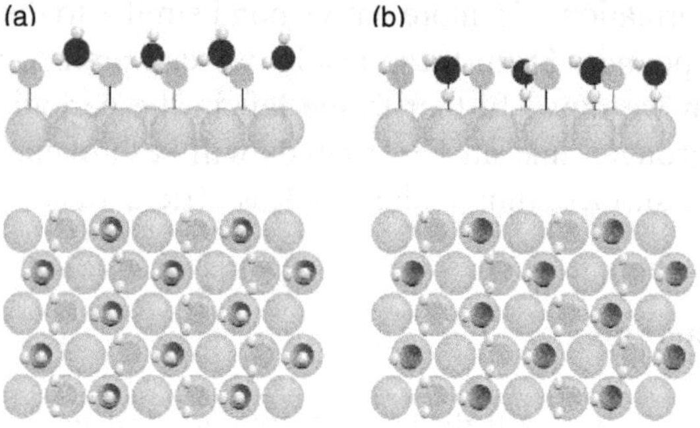

Figure 2.52. (a) The bilayer ice structure on metal surfaces. Dark and grey spheres represent O atoms. Half of the water molecules bind directly through the oxygen to the surface metal atoms. The remaining molecules are displaced toward the vacuum in the H-up configuration. (b) The 'flat ice' structure on metal surfaces with O atoms in Pt−O and Pt−HO bonding water molecules, respectively. From Ref. [106].

both O and H atoms binding to the Pt substrate at distances of 2.35 and 2.6 Å, respectively (Figure 2.52(b)). A similar molecular orientation of the outer layer has also been proposed for the monolayer molecular phase of water on Ru(0001) [104] and Pd(111) [107]. Only a small energy difference between the H-up and H-down configurations has been reported for many water-metal systems based on theoretical calculations alone [108]. The behavior of water on an open surface is quite different in comparison to the close-packed surfaces. It was recently shown that on Cu(110) water adsorbs in a mixed H-down and H-up configuration with approximately a third of the waters binding H-up in a (7 × 8) unit cell [109].

Figure 2.53 shows XAS and XES spectra of the p_z component (perpendicular to the surface) of water on Pt(111) [106]. The in-plane spectra are very similar to spectra of bulk ice. Two different theoretical model calculations of the z-component XAS spectrum are shown in the figure corresponding to two different orientations of the OH groups that are not involved in hydrogen-bonding between water molecules. The uncoordinated OH groups can either point towards vacuum (H-up) or towards the surface (H-down). The computed H-up structure gives rise to a strong feature at 536.5 eV assignable to orbitals localized at the uncoordinated OH bond pointing toward vacuum [110]. In the H-down configuration, the peaks are broadened and lose intensity due to interactions with the Pt surface. Comparison with the experimental XAS spectrum immediately shows that this feature is missing except for a weakly discernible shoulder; this provides evidence that the H-down configuration is preferred. This picture is also confirmed by infrared absorption spectroscopy measurements through the absence of the free OH vibrational band that would be present for an H-up configuration [111].

By carefully tuning the excitation photon energy, we can use XES to selectively observe the occupied electronic states projected on the oxygen atom of, respectively,

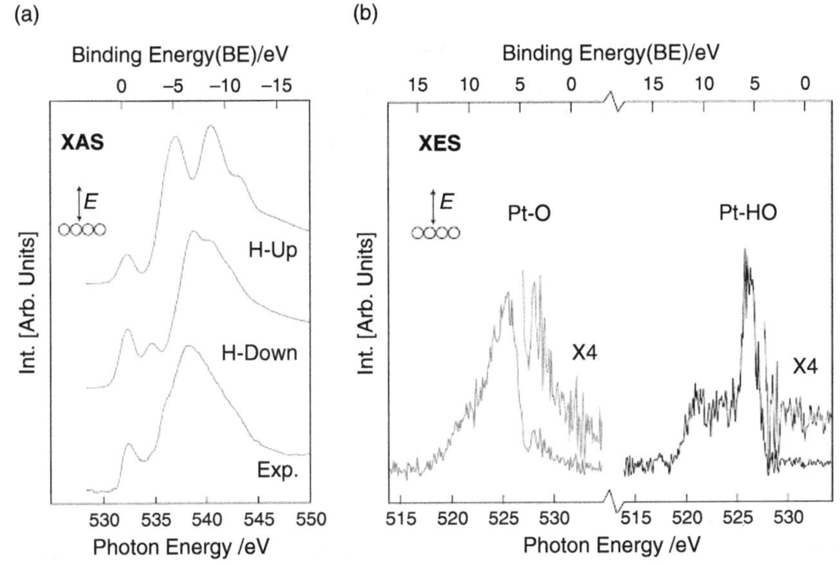

Figure 2.53. (a) Computed p_z component XAS spectra of water on Pt(111) with uncoordinated OH toward the vacuum (H-up) and toward the substrate (H-down). The H-down spectrum is in agreement with the experimental XAS spectrum (bottom). (b) XES spectra (p_z components) from Pt−O (left) and Pt−HO (right) bonding species showing the occupied orbital structure. From Ref. [106].

the Pt−O and Pt−HO bonding species as shown in Figure 2.53(b). We compare the XES spectra for the water molecules binding through oxygen and hydrogen and focus on the *p*-components projected along the (vertical) *z*-direction, which corresponds to the direction of the surface bond. The interaction of the O lone pair (*lp*) orbital of the Pt−O bonding species with the substrate *d*-band will result in bonding and anti-bonding combinations, see Figure 2.54. We can find the bonding combination as the strong asymmetric peak between 6 and 15 eV. We can recognize a weak feature between 4 eV and the Fermi level that has an anti-bonding character in the molecular orbital analysis of the computed XES spectrum. This band is not fully occupied and we can observe the unoccupied part of this band in the XAS spectrum as the peak at 532 eV (Figure 2.53a). We can see the repulsive character of the Pt−O

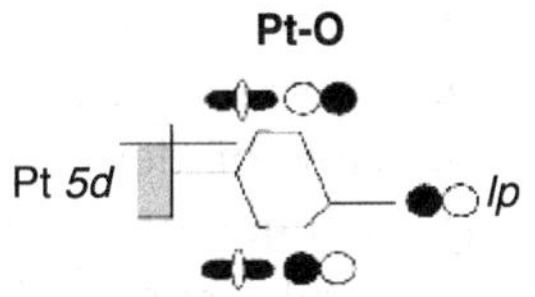

Figure 2.54. Schematic illustration of the orbitals arising from the interaction of the lone-pair level in a water molecule and the 5*d* band in Pt forming bonding and antibonding combinations.

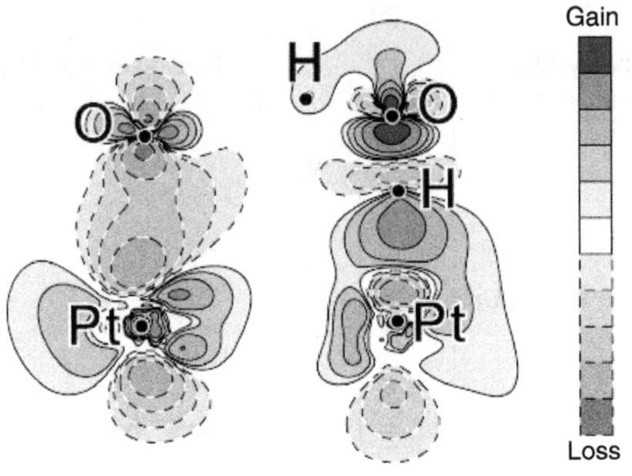

Figure 2.55. Charge density difference plots for Pt−O (left) and Pt−HO (right) bonding species. From Ref. [106].

bonding in the charge density difference plot (Figure 2.55). The charge density is decreased along the bond indicating charge polarization to the surroundings to minimize the Pauli repulsion between the lone pair orbital and the Pt *d*-orbital. The Pt−O species will be oriented on the surface with the molecular plane parallel to the surface. This means that the molecular dipole will maximize its interaction with the substrate through a metal image dipole. Since the water molecule dipole is very large, 1.8 D, this interaction will be substantial. Although the orbital interaction for the Pt−O species is repulsive we anticipate that there will be an overall attractive bonding due to electrostatic dipole-image dipole interaction and polarization of the *sp*-electrons in the metal away from the bonding Pt atoms. The latter effect allows for a dative bond to form, i.e., a stronger electrostatic interaction of the less polarizable lone pair with the positive core of the Pt atom through the removal of the easily polarized *sp*-electrons, similar to the discussion below in connection to ammonia. The overall interaction is rather weak. As we move more towards the left in the periodic table the repulsive interaction with respect to the *d*-states in the metal decreases and eventually becomes bonding in a similar way as for the σ-states in CO and N_2. However, in the case of water the repulsion with the *d*-states is much weaker since the bond length is longer.

For the Pt−HO bonding species the two OH groups in the water molecule have different surroundings: either coordinating to the surface or involved in a donating H-bond to another water molecule in the overlayer. The rearrangement of charge due to net bonding contributions can be seen in the charge density difference plot in Figure 2.55. We observe an increase of density between the H and Pt atoms (electron pairing) from the bonding combination and a decrease on the H atom along the OH bond from the non-bonding combination. We can thus say that the Pt−HO bonding results in the formation of a Pt−H bond together with a weakening of the water internal O−H bond. This bonding mechanism is similar to the interaction of the C—H groups in *n*-octane on Cu as discussed in the previous section. We also note the different response in the charge distribution in the substrate between the two configurations. There is more depletion in the surface region for the oxygen-down species and an accumulation for the Pt—HO species. This will have some implications on the surface electrochemical potential upon water interaction with electrodes as discussed in Chapter 6.

9.2. Adsorption of ammonia and the amino group in glycine on Cu(110)

The occupied orbitals of NH_3 are, using C_{3v} notation, $1a_1$ $2a_1 1e$ $3a_1$. For free NH_3, the two lowest unoccupied orbitals are the $4a_1$ and the doubly degenerate $2e$ orbital. Choosing the z-axis along the C_3-axis, the $2p_x$ and $2p_y$ orbitals correspond to the E representation and the $2p_z$ and all nitrogen s orbitals to the A_1 representation. The XES spectra for ammonia/Cu(110) are depicted in Figure 2.56 together with the corresponding calculated spectra [112]. Assuming an adsorption structure where the ammonia C_3 axis is positioned near the surface normal, states of mainly e-symmetry are expected in normal emission and the a_1-derived electronic states in the $2p_z$ spectrum. However, the saturated NH_3 monolayer is found to induce strong adsorbate–adsorbate interaction, causing the molecules to tilt, possibly in a

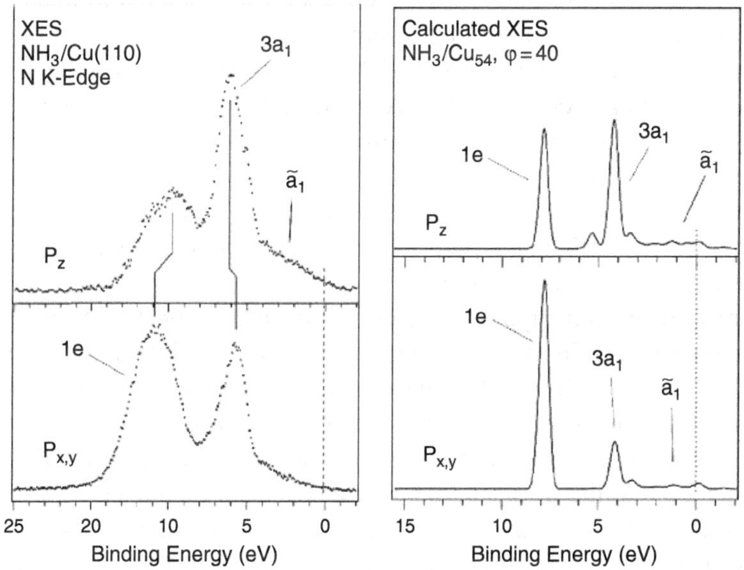

Figure 2.56. Experimental (left) and computed (right) decomposed XES spectra of NH_3/Cu(110). The upper panels contain states of mainly a_1 symmetry and the lower ones mainly e-symmetry. To find agreement with experiment the molecules must be tilted (≈40°) in the calculations. From Ref. [112].

bilayer-like structure with two NH_3 per unit cell. In the XES spectra this is revealed by the fact that both the $3a_1$ and the $1e$ transitions are visible in both spectra. Based on the angular distribution of the XES spectra, we have been able to estimate a mean tilt angle from the surface normal of 40–45° for the saturated monolayer. It is clear that hydrogen-bonding between the molecules is a significant factor in causing the observed tilt and building up of the overlayer structure.

The chemical bonding to the surface is achieved via orbitals of a_1 symmetry. The adsorbate-substrate hybrid levels exhibiting mainly metal character are represented by the $\widetilde{a}_1$ states. It has been shown that backdonation into the previously unoccupied ammonia $4a_1$ orbital, and a simultaneous $3a_1$ donation into the substrate, plays an important role in the surface chemical bond [112].

The adsorption of NH_3 on transition metals has traditionally been explained as the formation of a $3a_1$-metal bonding combination together with a strong dipole interaction with the substrate. However, while this simple line of reasoning regarding the covalent bond might hold for some metals, it is certainly not true for all. For instance, both the position and population of the metal valence bands are crucial in the bond formation. In the case of copper, the *d*-band position implies that both the bonding and antibonding $3a_1$ hybrids will be occupied. In previous studies, this has led to the conclusion that the covalent contribution to the adsorption energy instead goes via the metal 4*sp* band [113] similar to the above discussion of water on Pt.

Previous theoretical work on small Cu clusters has attempted to make a subdivision between covalent and electrostatic contributions to the total adsorption energy. The conclusion was that the major part of the adsorption energy can be associated to the electrostatic contribution [18,113]. A standard CSOV analysis does not resolve this issue since there is no straightforward way to make a subdivision between the contributions; polarization steps are also important for the formation of covalent bonds. Instead we choose to investigate how much of the adsorption energy that can be accounted for by utilizing a purely electrostatic model; this provides an upper limit to the electrostatic contribution.

There are two different types of electrostatic interaction to be considered: the ammonia dipole interaction with the surface and the ionic interaction due to charge transfer between the molecule and the substrate. The first interaction can be estimated to be 0.02 eV using the calculated dipole moment of the free NH_3 molecule of 1.8 Debye and an image plane centered within the first Cu layer. From this low value we can conclude that the direct dipole contribution can be neglected. To obtain the ionic contribution, we estimated the charge transfer by using the Mulliken charge on the adsorbate. The result was a charge of about +0.15, giving a total electrostatic contribution of ~0.1 eV, assuming the same image plane as above. Although we have to be careful with the absolute values due to the approximate character of the Mulliken charges, the obtained electrostatic contributions are significantly lower than the calculated DFT adsorption energy of ~0.78 eV. This means that the bonding only can be viewed within a covalent description, with more or less polar character.

Analyzing the covalent contribution in more detail reveals that the charge transfer goes via the A' symmetry states, i.e., through the adsorbate a_1 levels. The results indicate that the important covalent contribution to the bonding comes from the donation of the $3a_1$ molecular orbital into the substrate and a subsequent backdonation into the $4a_1$ orbital. Since the electronic charge on the adsorbate decreases upon adsorption (+0.15) the donation of the $3a_1$ into the substrate dominates. However, backdonation into the $4a_1$ orbital also contributes to the bonding. In addition, based on overlap population analysis, we find the overall interaction between the ammonia $3a_1$ and the Cu 4*sp* valence band to be slightly bonding; this is in agreement with previous work on copper surfaces. In terms of the overall contribution to the adsorption energy, this is, however, not found to be the dominating interaction of the Cu 4*sp* band. In the initial interaction with the surface, the doubly occupied $3a_1$ orbital will experience repulsion by the metal states of σ-symmetry. This repulsion must be reduced in order for the molecule to approach close enough to form a bond to the surface. As indicated by substrate population analysis, this is accomplished by polarizing the central 4*sp* density out towards the surrounding metal centers. The result is a dative-type, stabilizing [19] interaction between the $3a_1$ orbital and the substrate, allowing for the surface chemical bond to be formed.

Let us now inspect the bonding of the amino group in glycine interacting with the Cu(110) surface. The most stable adsorption structure is shown in Figure 2.57 and was deduced from a combination of core-level spectroscopy and theory [114] and confirmed using photoelectron diffraction (PhD) as discussed in Chapter 1. In this system, the oxygen atoms are aligned along the $(1\bar{1}0)$ direction of the substrate and a second chemisorption bond is formed at the nitrogen end of the molecule involving Cu atoms in the neighboring $(1\bar{1}0)$ row. This enables a complete partition into atomic $2p_x$, $2p_y$, and $2p_z$ contributions providing a truly detailed picture of the electronic structure of the adsorbed glycine [115,116]. This indeed opens fascinating perspectives as to the applications of XES to the study of larger, biologically interesting molecules in interaction with substrates.

In the chemisorbed geometry, the nitrogen binds with a lone-pair sticking into the surface similar to the case of ammonia. The X-ray emission data of nitrogen show the same hybridization of p_z levels with the surface forming bonding and antibonding states. We can directly see that the antibonding states of the p_z-level show a stronger feature than in ammonia indicating a stronger interaction with the surface. However, the bond length to the surface is the same, 2.04 Å [117]. The increased interaction is instead a consequence of the more upright geometry found for the amino group in glycine as compared to adsorbed ammonia [115].

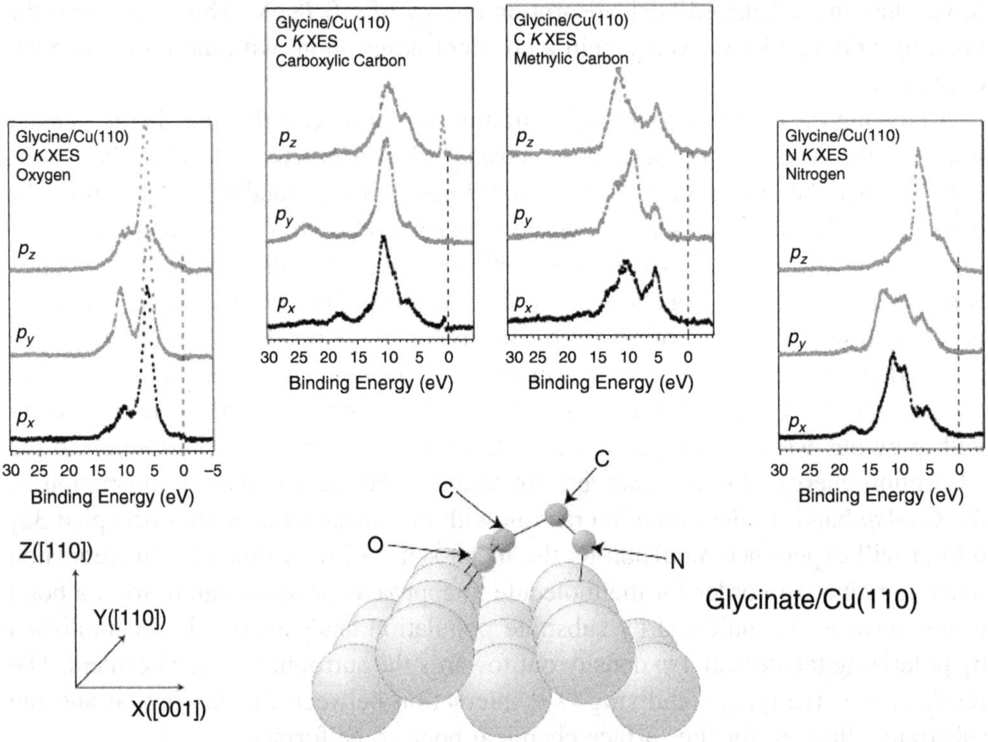

Figure 2.57. Symmetry- and atom-resolved XES spectra of deprotonated glycine on Cu(110) displaying clockwise from the left: oxygen, carboxylic carbon, methyl carbon, and nitrogen K-shell XES spectra with from top to bottom the p_z, p_y, and p_x spectrum contributions. From Ref. [3].

10. Summary

Most adsorbates presented in this chapter interact with the metal substrate via covalent bonding, i.e., electron-pair sharing, see Figure 2.58(a–c). In the formation of electron-pair sharing we need to create a radical state in the interacting atom or molecule to form the bond to the substrate. This is the essence of the spin-uncoupling concept, which was applied specifically to the adsorption of unsaturated hydrocarbons, but the principle has greater validity. For a molecular adsorbate the bond-prepared radical state can be obtained upon internal (partial) bond-breaking where the fragments will have unpaired electrons that can interact with unpaired electrons in the metal surface. We showed examples of the most simple case where the fragment is an atom, such as N or O. The unpaired radical interacts with the metal *d*-states in the metal to form bonding and antibonding states. If only the bonding states are occupied a strong bond is created, as in adsorption on many transition metals (Figure 2.58(a)), whereas if both bonding and antibonding states

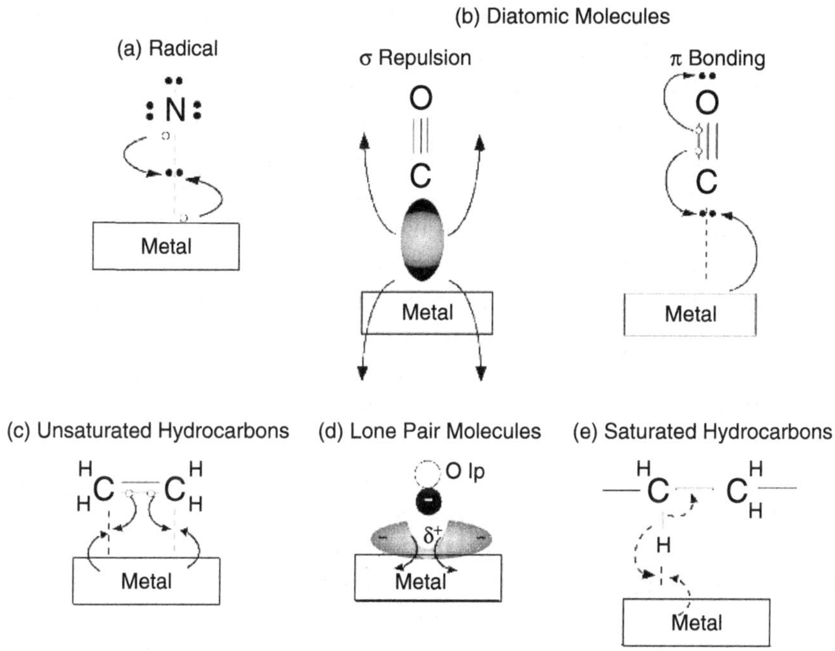

Figure 2.58. Schematic illustrations of the five different types of chemical bond formation on metal surfaces.

are occupied Pauli repulsion will make the resulting bond rather weak, as on noble metals. As we go towards the left in the periodic table we depopulate more and more of the antibonding state making the overall bond stronger and stronger. As with simple atomic adsorbates this picture would hold for any radical state in molecular fragments such as –OH, $-CH_3$, $-NH_2$ etc. In the case where the interaction with the *d*-states leads to weak or repulsive interaction, there could still be some bonding involving *sp*-electrons. Ionic contributions can also be important, in particular for cases where the adsorbate has a high- or low-electronegativity with respect to the metal, as is the case for oxygen, halogen, and alkali atom adsorption.

In molecular adsorbate systems, with no unpaired electrons, but with an unsaturated π-electron system it can be possible to obtain a bond-prepared radical state by mixing π and π^* orbitals (Figure 2.58(b)). The virtual radical state can now interact with the metal *d*-states and again form bonding and antibonding states. The total adsorption bond strength will depend on the filling of these states. The mechanism of the bond-preparation is different depending on the orientation of the π-electron system. Diatomic molecules such as CO and N_2 typically adsorb with the molecular axis perpendicular to the surface whereas the unsaturated hydrocarbons adsorb parallel to the surface. The π-electron system assumes the allylic configuration for the perpendicular geometry. The mixing of the 1π and $2\pi^*$ orbitals will partly break the internal π-bond and restore one part of the electron pair into a radical state on the inner atom to become available for bonding to the surface and the other part to become a lone pair on the outer atom (Figure 2.58(b)). For a lying-down species, such as adsorbed ethylene, the symmetry due to the interaction with the surface is different and the 1π and $2\pi^*$ levels cannot mix and will instead interact independently with different *d*-orbitals of appropriate symmetry in the metal. This bonding mechanism can be described in the discussed spin-uncoupling model. We prepare the interaction by exciting an electron from the 1π to the $2\pi^*$ level generating two unpaired radical orbitals that can interact with the metal. In the case of CO and N_2 it is also important to consider the σ system which is repulsive and there is a balance between σ repulsion and π bonding. This balance is affected by the metal coordination which leads to that N_2 only adsorbs in on-top sites whereas CO can populate a large range of different sites with nearly equal adsorption energy. We could ask the question why CO and N_2 do not adsorb with the molecular axis parallel to the surface similar to adsorbed ethylene?

If we look at the excitation energy of the molecule and analyze the interaction energy of the bond-prepared orbitals based on a π to π* excitation, the adsorption energetics and geometries have successfully been predicted for a large number of unsaturated hydrocarbons adsorbed on surfaces (Figure 2.58c) [84]. In the case of CO and N_2 the excitation energy is of the order of 6 eV, which is rather high. It means that the bond-prepared orbital interaction will not be strong enough on most surfaces to compensate this large excitation energy. It is therefore more efficient to slightly mix the π-system in the allylic configuration to open for bonding to the surface. In a simple perturbative treatment this involves only 10–20% of the excitation energy. This geometry is more stable in spite of the σ-repulsion that can be avoided in the lying-down geometry. The latter also compensates the internal bond length, where σ interaction shortens and π interaction elongates the bond resulting in essentially no change from the free molecule; as discussed previously the 5σ is an antibonding orbital which loses antibonding character through the interaction leading to a strengthening of the σ-bond. In the case of a lying-down species there is no compensation from the σ-system resulting in a strong elongation of the internal bond leading to a precursor state to dissociation. On specific surfaces such as Fe(111) and stepped surfaces, where both atoms can more efficiently bond to active sites with energetics overcoming the excitation energy, the lying-down species becomes more favorable [118]. A similar balance should be expected for O_2 where the 1π to $2\pi^*$ excitation energy is much smaller, 4.1 eV [65], and where unpaired electrons are already available in the π-system for bonding. O_2 usually adsorbs on most metal surfaces with the molecular axis parallel to the surface prior to dissociation [13,119]. It becomes energetically favorable for the molecule to avoid the perpendicular geometry and instead adsorb with the molecular axis parallel to the surface. It has been estimated experimentally that for O_2 on Ni the lying-down geometry is favored over the perpendicular by a difference in bond energy of 2 eV [120].

The $\pi \rightarrow \pi^*$ excitation energies of the unsaturated hydrocarbons, ethylene, and benzene, discussed in the present chapter are of a similar magnitude, 3.5–4 eV, as for O_2 [84]. They thus have easy access to the unoccupied π^* in the bond-formation with the substrate and can be expected, as is also found, to bind with the molecular axis or plane parallel to the surface. The resulting bond-strength per σ-bond formed is surprisingly constant which means that the binding energy *per carbon atom* for benzene becomes substantially smaller than for acetylene and ethylene. This is due to the aromatic stabilization in the benzene ring: in order to have the same *per carbon* binding energy six σ-bonds would have to be formed. This would require a prohibitive excitation energy since the complete aromatic structure of the ring system would need to be broken.

Molecules that lack radical bonding either through broken bonds or virtual via the π-electron system usually have much lower adsorption strength. The interaction of molecules, such as water and ammonia, that bond through their lone-pairs is stronger than physisorption. The lone-pair interaction is similar to the σ-system interaction in CO and N_2 but occurs at much longer bond distances where it becomes attractive. The bond is mainly of electrostatic character, both through the permanent dipole of the adsorbate but also through a contribution induced via strong polarization of the *sp*-electrons away from the bonded metal atom to increase the local positive charge on the interacting metal atom (Figure 2.58d). It becomes similar to the solvation bond of water to a metal ion in aqueous solutions. The interaction with the *d*-band leads to the formation of bonding and antibonding states. The bonding state will mainly have molecular character and the antibonding state metal *d*-character. Since, for the late transition metals, the *d*-band is nearly fully occupied the antibonding state will mostly be filled resulting in Pauli repulsion. In comparison with the previously discussed radical adsorption the much stronger interaction involving unpaired electrons leads to larger splitting of the bonding and antibonding states where the latter will be above the *d*-band in its energetic position and therefore not populated for transition metals.

Finally, let us consider the bonding of X−H groups to metal surfaces. This constitutes the weakest bonding covered in the present chapter. We have discussed two types of X−H interactions on surfaces, the Pt−HO species in adsorbed water on Pt(111) and *n*-octane adsorbed on Cu(110). In the case of *n*-octane on Cu there is a weak electron-pairing between the hydrogen atom in the C−H group and the metal surface. This arises through interaction of both C−H bonding and antibonding orbitals with the *d*- and *sp*-bands. In order to create the bond the internal C−H bond is weakened to reduce the bond-order and simultaneously the C−C bond is strengthened (Figure 2.58(e)). We anticipate that the O−H group in water on Pt will behave in a similar manner where the internal O–H bond is weakened due to bonding to the metal surface.

Acknowledgement

We gratefully acknowledge all the people involved in the various projects upon which this chapter is based. This work was supported by the National Science Foundation under Contract No. CHE-0431425 (Stanford Environmental Molecular Science Institute); the U.S. Department of Energy, Office of Basic Energy Sciences, through the Stanford Synchrotron Radiation Laboratory, Contract No. DE-AC02-05CH11231, and under the auspices of the President's Hydrogen Fuel Initiative; the Swedish Foundation for Strategic Research and the Swedish Natural Science Research Council.

References

[1] D. P. Woodruff and T. A. Delchar, *Modern Techniques of Surface Science*. (Cambridge University Press, New York, 1986).
[2] A. Nilsson, J. Hasselström, A. Föhlisch, O. Karis, L. G. M. Pettersson, M. Nyberg, and L. Triguero, J. El. Spec. Rel. Phenom. **110/111**, 15 (2000).
[3] A. Nilsson and L. G. M. Pettersson, Surf. Sci. Reps. **55** (2-5), 49 (2004).
[4] A. Nilsson, J. El. Spec. Rel. Phenom. **126**, 3 (2002).
[5] A. Nilsson, H. Tillborg, and N. Mårtensson, Phys. Rev. Lett. **67**, 1015 (1991).
[6] A. Nilsson, M. Weinelt, T. Wiell, P. Bennich, O. Karis, N. Wassdahl, J. Stöhr, and M. Samant, Phys. Rev. Lett. **87**, 2847 (1997).
[7] A. Nilsson, P. Bennich, T. Wiell, N. Wassdahl, N. Mårtensson, J. Nordgren, O. Björneholm, and J. Stöhr, Phys. Rev. B **51**, 10244 (1995).
[8] S. D. Kevan, *Angle-resolved photoemission*. (Elsevier, Amsterdam, 1992).
[9] K. Horn, J. Dinardo, W. Eberhardt, and H. J. Freund, Surf. Sci. **118**, 465 (1982).
[10] P. Bennich, T. Wiell, O. Karis, M. Weinelt, N. Wassdahl, A. Nilsson, M. Nyberg, L. G. M. Pettersson, J. Stöhr, and M. Samant, Phys. Rev. B **57**, 9274 (1998).
[11] H. Kuhlenbeck, H. B. Saalfeld, U. Buskotte, M. Neumann, H. J. Freund, and E. W. Plummer, Phys. Rev. B **39**, 3475 (1989).
[12] H. J. Freund and H. Kuhlenbeck, in *Applications of Synchrotron Radiation; High resolution studies of molecules and molecular adsorbates*, edited by W. Eberhardt (Springer-Verlag, Berlin-Heidelberg, 1995), Vol. 35.
[13] J. Stöhr, *NEXAFS Spectroscopy*. (Springer-Verlag, Berlin-Heidelberg, 1992).
[14] A. Nilsson and N. Mårtensson, Physica B **208&209**, 19 (1995).
[15] P. D. Johnson and S. L. Hulbert, Phys. Rev. B **35**, 9427 (1987).
[16] P. Hohenberg and W. Kohn, Phys. Rev. **136**, 864 (1964).
[17] C. Kolczewski, R. Puttner, O. Plashkevych, H. Ågren, V. Staemmler, M. Martins, G. Snell, M. Sant'anna, G. Kaindl, and L. G. M. Pettersson, J. Chem. Phys. **115**, 6426 (2001).
[18] P. S. Bagus, K. Hermann, and C. W. Bauschlicher, J. Chem. Phys. **81**, 1966 (1984).

[19] P. S. Bagus and K. Hermann, Phys. Rev. B **33**, 2987 (1986).
[20] C. W. Bauschlicher, Chem. Phys. **106**, 391 (1986).
[21] A. Föhlisch, P. Bennich, J. Hasselström, O. Karis, A. Nilsson, L. Triguero, M. Nyberg, and L. G. M. Pettersson, Phys. Rev. B **16**, 16229 (2000).
[22] J. C. Slater, Adv. Quant. Chem. **6**, 1 (1972); J. C. Slater and K. H. Johnsson, Phys. Rev. B **5**, 844 (1972).
[23] L. Triguero and L. G. M. Pettersson, Surf. Sci. **398**, 70 (1998).
[24] M. Cavalleri, D. Nordlund, M. Odelius, A. Nilsson, and L. G. M. Pettersson, Phys. Chem. Chem. Phys. **7**, 2854 (2005).
[25] P. W. Anderson, Phys. Rev. **124**, 41 (1961).
[26] D. M. News, Phys. Rev. **178**, 1123 (1969).
[27] T. Wiell, J. E. Klepais, P. Bennich, O. Björneholm, N. Wassdahl, and A. Nilsson, Phys. Rev. B **58**, 1655 (1998).
[28] A. Sandell, O. Hjortstam, A. Nilsson, P. A. Bruhwiler, O. Eriksson, P. Bennich, P. Rudolf, J. M. Wills, B. Johansson, and N. Martensson, Phys. Rev. Lett. **78** (26), 4994 (1997).
[29] H. Öström, PhD thesis, Stockholm University, 2004.
[30] N. D. Lang and A. R. Williams, Phys. Rev. B **18**, 616 (1978).
[31] B. Hammer and J. K. Nørskov, Nature **376**, 238 (1995).
[32] A. Nilsson, L. G. M. Pettersson, B. Hammer, T. Bligaard, C. H. Christensen, and J. K. Nørskov, Catal Lett **100** (3–4), 111 (2005).
[33] H. C. Zeng, R. N. S. Sodhi, and K. A. R. Mitchell, Surf. Sci. **188**, 599 (1987).
[34] T. Lederer, D. Arvanitis, M. Tischer, G. Comelli, L. Troeger, and K. Baberschke, Phys. Rev. B **40**, 11277 (1993).
[35] T. Wiell, H. Tillborg, A. Nilsson, N. Wassdahl, N. Mårtensson, and J. Nordgren, Surf. Sci. **304**, L451 (1994).
[36] L. Triguero and L. G. M. Pettersson, Surf. Sci. **398**, 70 (1998).
[37] R. Hoffmann, *Solids and Surfaces: A Chemists View of Bonding in Extended Structures*. (VCH Publishers, Inc., New York, 1988).
[38] M. Wuttig, R. Franchy, and H. Ibach, Surf. Sci. **213**, 103 (1989).
[39] T. Lederer, D. Arvanitis, M. Tischer, G. Comelli, L. Tröger, and K. Baberschke, Phys. Rev. B **48**, 11277 (1993).
[40] W. Oed, H. Lindner, U. Starke, K. Heinz, K. Muller, and J. B. Pendry, Surf. Sci. **224**, 179 (1989).
[41] D. Westphal and A. Goldmann, Surf. Sci. **131** (1), 113 (1983).
[42] L. G. M. Pettersson and P. S. Bagus, Phys. Rev. Lett. **56**, 500 (1986).
[43] W. Eberhardt, F. Greuter, and E. W. Plummer, Phys. Rev. Lett **46**, 1085 (1981); W. Eberhardt, S. G. Louie, and E. W. Plummer, Phys. Rev. B **28**, 465 (1983); S. G. Louie, Phys. Rev. Lett. **42**, 476 (1979).
[44] P. J. Feibelman, D. R. Hamann, and F. J. Himpsel, Phys. Rev. B **22**, 1734 (1980).
[45] C. L. Allyn, T. Gustafsson, and E. W. Plummer, Solid. State. Comm. **24**, 531 (1977).
[46] G. Blyholder, J. Phys. Chem. **68**, 2772 (1964).
[47] F. Delbecq and P. Sautet, Phys. Rev. B **59**, 5142 (1999); D. E. Eastman and K. Cashion, Phys. Rev. Lett. **27**, 1520 (1971); B. Gumhalter, K. Wandelt, and P. Avouris, Phys. Rev. B **37**, 8048 (1988); S. S. Sung and R. Hoffman, J. Am. Chem. Soc. **107**, 578 (1985).
[48] T. Gustafsson and E. W. Plummer, in *Photoemission from surfaces*, edited by B. Feuerbacher, B. Fitton, and R. Willis (Wiley, London, 1977).
[49] B. Hammer, Y. Morikawa, and J. K. Nørskov, Phys. Rev. Lett **76**, 2141 (1996).
[50] B. Hammer and J. K. Nørskov, Advances in Catalysis, Vol **45**, 71 (2000).

[51] M. Nyberg, A. Föhlisch, L. Triguero, A. Bassan, A. Nilsson, and L. G. M. Pettersson, J. Mol. Struct. (THEOCHEM) **762**, 123 (2006).
[52] M. Grunze, P. A. Dowben, and R. G. Jones, Surf. Sci. **141**, 455 (1984).
[53] J. Stöhr and R. Jaeger, Phys. Rev. B **26**, 4111 (1982).
[54] D. I. Sayago, J. T. Hoeft, M. Polcik, M. Kittel, R. L. Toomes, J. Robinson, D. P. Woodruff, M. Pascal, C. L. A. Lamont, and G. Nisbet, Phys. Rev. Lett. **90**, 116104 (2003).
[55] A. Sandell, O. Björneholm, A. Nilsson, E. Zdansky, H. Tillborg, J. N. Andersen, and N. Mårtensson, Phys. Rev. Lett **70**, 2000 (1993).
[56] D. Dubois and R. Hoffman, Nouv. J. Chim. **1**, 479 (1977); R. Hoffmann, M. M. L. Chen, and D. Thorn, Inorg. Chem. **16**, 503 (1977).
[57] T. A. Albright, J. K. Burdett, and M. H. Whengbo, *Orbital Interactions in Chemistry*. (Wiley, New York, 1985); M. Dewar, *The Molecular Orbital Theory of Organic Chemistry*. (McGraw-Hill, New York, 1969).
[58] M. Gajdos, A. Eichler, and J. Hafner, J. Phys: Condens. Matter **16**, 1141 (2004).
[59] P. Hu, D. A. King, M. H. Lee, and M. C. Payne, Chem. Phys. Lett. **246**, 73 (1995); G. Kresse, A. Gil, and P. Sautet, Phys. Rev. B **68**, 073401 (2003).
[60] O. Björneholm, A. Nilsson, E. Zdansky, A. Sandell, H. Tillborg, J. N. Andersen, and N. Mårtensson, Phys. Rev. B **47**, 2308 (1993).
[61] P. S. Bagus, C. J. Nelin, and C. W. Bauschlicher, Phys. Rev. B **28**, 5423 (1983).
[62] A. Föhlisch, M. Nyberg, P. Bennich, L. Triguero, J. Hasselström, O. Karis, L. G. M. Pettersson, and A. Nilsson, J. Chem. Phys. **112**, 1946 (2000).
[63] J. T. Stuckless, N. Al-Sarraf, C. Wartnaby, and D. A. King, J. Chem. Phys. **99**, 2202 (1993).
[64] H. Tillborg, A. Nilsson, and N. Mårtensson, Surf. Sci. **273**, 47 (1992).
[65] K. P. Huber and G. Hertzberg, *Molecular Spectra and Molecular Structure*. (Van Nostrand Reinhold, New York, 1979).
[66] E. Wimmer, C. Fu, and A. Freeman, Phys. Rev. Lett. **55**, 2618 (1985).
[67] A. Eichler and J. Hafner, Phys. Rev. B **57**, 10110 (1998); J. T. Hoeft, M. Polcik, D. I. Sayago, M. Kittel, R. Terborg, R. L. Toomes, J. Robinson, D. P. Woodruff, M. Pascal, G. Nisbet, and C. L. A. Lamont, Surf. Sci. **540**, 441 (2003).
[68] A. Föhlisch, M. Nyberg, J. Hasselström, O. Karis, L. G. M. Pettersson, and A. Nilsson, Phys. Rev. Lett **85**, 3309 (2000).
[69] M. Mavrikakis, B. Hammer, and J. K. Nørskov, Phys. Rev. Lett. **81**, 2819 (1998).
[70] S. Andersson and J. B. Pendry, Phys. Rev. Lett **43**, 363 (1979).
[71] J. C. Tracy, J. Chem. Phys. **56**, 2748 (1972).
[72] S. Y. Tong, A. Maldonado, C. H. Li, and M. A. van Hove, Surf. Sci. **94**, 73 (1980).
[73] A. Föhlisch, M. Stichler, C. Keller, W. Wurth, and A. Nilsson, J. Chem. Phys. **121**, 4848 (2004).
[74] H. P. Bonzel, Surf. Sci. Reps. **8**, 43 (1987).
[75] H. P. Bonzel and G. Pirug, in *The chemical physics of solid surfaces and heterogeneous catalysis*, edited by D. A. King and D. P. Woodruff (Elsevier, Amsterdam, 1992), Vol. 6.
[76] N. Al-Sarraf, J. T. Stuckless, and D. A. King, Nature **360**, 243 (1992).
[77] J. Hasselström, A. Föhlisch, R. Denecke, A. Nilsson, and F. de Groot, Phys. Rev. B **62**, 11192 (2000).
[78] F. Sette, J. Stöhr, E. B. Kollin, D. J. Dwyer, J. L. Gland, J. L. Robbins, and A. L. Johnson, Phys. Rev. Lett. **54**, 935 (1985); W. Wurth, C. Schneider, E. Umbach, and D. Menzel, Phys. Rev. B. **34**, 1336 (1986).
[79] R. Davis, D. P. Woodruff, O. Schaff, V. Fernandez, K. M. Schindler, P. Hofmann, K. U. Weiss, R. Dippel, V. Fritzche, and A. M. Bradshaw, Phys. Rev. Lett. **74**, 1621 (1995).
[80] S. Jenkins and K. D. A., J. Am. Chem. Soc. **122**, 10610 (2000).

[81] J. Hasselström, A. Föhlisch, P. Väterlein, M. Nyberg, L. G. M. Pettersson, C. Heske, and A. Nilsson, unpublished.
[82] M. J. S. Dewar, Bull. Soc. Chim. France **18**, C79 (1951); J. Chatt and L. A. Duncanson, J. Chem. Soc., 2939 (1953).
[83] L. Triguero, A. Föhlisch, P. Väterlein, J. Hasselström, M. Weinelt, L. G. M. Pettersson, Y. Luo, H. Ågren, and A. Nilsson, J. Am. Chem. Soc. **122**, 12310 (2000).
[84] L. Triguero, L. G. M. Pettersson, B. Minaev, and H. Ågren, J. Chem. Phys. **108**, 1193 (1998).
[85] H. Öström, A. Föhlisch, M. Nyberg, M. Weinelt, C. Heske, L. G. M. Pettersson, and A. Nilsson, Surf. Sci. **559**, 85 (2004).
[86] C. J. Jenks, B. E. Bent, N. Bernstein, and F. Zaera, Surf. Sci. Letters **277**, L89 (1992).
[87] J. A. Stroscio, S. R. Bare, and W. Ho, Surf. Sci. **148**, 499 (1984).
[88] W. A. Brown, R. Kose, and D. A. King, J. Mol. Catal. A: Chemical **141**, 21 (1999).
[89] D. Arvanitis, U. Döbler, L. Wenzel, and K. Baberschke, Surf. Sci. **178**, 686 (1986).
[90] M. Weinelt, N. Wassdahl, T. Wiell, O. Karis, J. Hasselström, P. Bennich, A. Nilsson, J. Stöhr, and M. Samant, Phys. Rev. B **58**, 7351 (1998).
[91] G. Herzberg, *Molecular Spectra and Molecular Structure III. Electronic Spectra and Electronic Structure of Polyatomic Molecules.* (Krieger, Malabar, 1991).
[92] H. Öström, D. Nordlund, H. Ogasawara, K. Weiss, L. Triguero, L. G. M. Pettersson, and A. Nilsson, Surf. Sci. **565**, 206 (2004).
[93] L. G. M. Pettersson, H. Ågren, Y. Luo, and L. Triguero, Surf. Sci. **408**, 1 (1998).
[94] E. A. Carter and B. E. Koel, Surf. Sci. **226**, 339 (1990).
[95] H. Öström, L. Triguero, M. Nyberg, H. Ogasawara, L. G. M. Pettersson, and A. Nilsson, Phys. Rev. Lett. **91**, 046102 (2003).
[96] H. Öström, L. Triguero, K. Weiss, H. Ogasawara, M. G. Garnier, D. Nordlund, M. Nyberg, L. G. M. Pettersson, and A. Nilsson, J. Chem. Phys. **118**, 3782 (2003).
[97] K. Weiss, H. Öström, L. Triguero, H. Ogasawara, M. G. Garnier, L. G. M. Pettersson, and A. Nilsson, J. El. Spec. Rel. Phen. **128**, 179 (2003).
[98] H. Öström, H. Ogasawara, L.-Å. Näslund, L. G. M. Pettersson, and A. Nilsson, Phys. Rev. Lett. **96**, 146104 (2006).
[99] G. Witte, K. Weiss, P. Jakob, J. Braun, K. L. Kostov, and C. Wöll, Phys. Rev. Lett. **80**, 121 (1998).
[100] W. L. Manner, G. S. Girolami, and R. G. Nuzzo, Langmuir **14**, 1716 (1998); M. A. Chesters, P. Gardner, and E. M. MacCash, Surf. Sci. **209**, 89 (1989).
[101] J. T. Pireaux, S. Svensson, E. Basilier, P. Å. Malmquist, U. Gelius, R. Gaudario, and K. Siegbahn, Phys. Rev. A **14**, 2133 (1976).
[102] T. E. Madey and J. T. Yates, Surf. Sci. **76**, 397 (1978).
[103] C. Wöll, K. Weiss, and P. S. Bagus, Chem. Phys. Lett. **332**, 553 (2000).
[104] K. Andersson, A. Nikitin, L. G. M. Pettersson, A. Nilsson, and H. Ogasawara, Phys. Rev. Lett. **93**, 196101 (2004); C. Clay, S. Haq, and A. Hodgson, Chem. Phys. Lett. **388**, 89 (2004); D. N. Denzler, C. Hess, R. Dudek, S. Wagner, C. Frischkorn, M. Wolf, and G. Ertl, Chem. Phys. Lett. **376**, 618 (2003).
[105] N. S. Faradzhev, K. L. Kostov, P. Feulner, T. E. Madey, and D. Menzel; G. Materzanini, G. F. Tantardini, P. J. D. Lindan, and P. Saalfrank, Phys. Rev. B **71**, 155414 (2005); A. Michaelides, A. Alavi, D. A. King, J. Am. Chem. Soc. 125, 2746 (2003).
[106] H. Ogasawara, B. Brena, D. Nordlund, M. Nyberg, A. Pelmenschikov, L. G. M. Pettersson, and A. Nilsson, Phys. Rev. Lett. **89** (27), 276102 (2002).
[107] J. Cerda, A. Michaelides, M.-L. Bocquet, P. J. Feibelman, T. Mitsui, M. Rose, E. Fomin, and M. Salmeron, Phys. Rev. Lett. **93** (11), 116101 (2004).

[108] S. Meng, E. G. Wang, and S. Gao, Phys. Rev. B **89**, 195404 (2004); A. Michaelides, A. Alavi, and D. A. King, Phys. Rev. B **69** (113404) (2004).

[109] T. Schiros, S. Haq, H. Ogasawara, O. Takahashi, H. Ostrom, K. Andersson, L. G. M. Pettersson, A. Hodgson, and A. Nilsson, Chem. Phys. Lett. **429** (4-6), 415 (2006).

[110] Ph. Wernet, D. Nordlund, U. Bergmann, H. Ogasawara, M. Cavalleri, L. Å. Näslund, T. K. Hirsch, L. Ojamäe, P. Glatzel, M. Odelius, L. G. M. Pettersson, and A. Nilsson, Science **304**, 995 (2004).

[111] S. Haq, J. Harnett, and A. Hodgson, Surf. Sci. **505**, 171 (2002).

[112] J. Hasselström, A. Föhlisch, O. Karis, M. Weinelt, A. Nilsson, M. Nyberg, L. G. M. Pettersson, and J. Stöhr, J. Chem. Phys. **110**, 4880 (1999).

[113] G. J. C. S. Van de Kerkhof, W. Biemolt, A. P. J. Jansen, and R. A. Van Santen, Surf. Sci. **284**, 361 (1993).

[114] J. Hasselström, O. Karis, M. Weinelt, N. Wassdahl, A. Nilsson, M. Nyberg, L. G. M. Pettersson, M. Samant, and J. Stöhr, Surf. Sci. **407**, 221 (1998).

[115] J. Hasselström, O. Karis, M. Nyberg, L. G. M. Pettersson, M. Weinelt, N. Wassdahl, and A. Nilsson, J. Phys. Chem. B **104**, 11480 (2000).

[116] M. Nyberg, J. Hasselström, O. Karis, N. Wassdahl, M. Weinelt, A. Nilsson, and L. G. M. Pettersson, J. Chem. Phys. **112**, 5420 (2000).

[117] N. A. Booth, D. P. Woodruff, O. Schaff, T. Giessel, R. Lindsay, P. Baumgärtel, and A. M. Bradshaw, Surf. Sci. **397**, 258 (1998).

[118] S. Dahl, A. Logadottir, C. Egeberg, J. H. Larsen, I. Chorkendorff, E. Tornqvist, and J. K. Nørskov, Phys. Rev. Lett **83** (Issue 9), 1814 (1999); S. Dahl, A. Logadottir, C. H. Jacobsen, and J. K. Nørskov, Appl. Catal. **A222** (19) (2001); M. Grunze, M. Golze, W. Hirschwald, H. J. Freund, H. Plum, U. Selp, M. Tsai, G. Ertl, and J. Kuppers, Phys. Rev. Lett. **53** (8), 850 (1984).

[119] R. J. Guest, B. Hernnäs, P. Bennich, O. Björneholm, A. Nilsson, R. E. Palmer, and N. Mårtensson, Surf. Sci. **1992** (278), 239 (1992); C. Puglia, A. Nilsson, B. Hernnäs, O. Karis, P. Bennich, and N. Mårtensson, Surf. Sci. **342** (119) (1995).

[120] A. Sandell, A. Nilsson, and N. Mårtensson, Surf. Sci. **241**, L1 (1991).

Chapter 3

The Dynamics of Making and Breaking Bonds at Surfaces

A. C. Luntz

Physics and Chemistry Institute, University of Southern Denmark, Campusvej 55, 5230 Odense M, Denmark

1. Introduction

The dynamics of making and breaking bonds at surfaces is concerned with understanding the elementary steps occurring at surfaces in terms of the individual atomic and molecular forces, and the motions that these induce during the process of chemical change. Many surface chemical reactions represent the sum of several elementary dynamical steps occurring on different time scales. For example, even the simple dissociation of a diatomic molecule such as O_2 from the gas phase on a single crystal metal surface such as Pt(111) involves energy transfer to the surface upon impact, molecular adsorption on the surface, diffusion to a transition state, dissociation, and then diffusion of the atoms away from the site of dissociation. The elementary steps of energy transfer, adsorption, diffusion, reaction, and desorption are linked via underlying dynamical principles. Some surface chemistry occurs between atoms/molecules adsorbed on surfaces and the products remain adsorbed on the surface. Usually this chemistry is only measured as a kinetic process at thermal equilibrium so that details of the dynamics are unobservable. However, when one of the reactants or products is in the gas phase, i.e., in gas-surface chemistry, experiments/theory probing the detailed dynamics are possible.

This chapter focuses on the dynamics of gas-surface chemistry as defined above. Both the theoretical and experimental methodology inherent in such an approach borrow much from an older sibling, i.e., the study of the dynamics of atom–molecule chemical reactions in the gas phase [1]. However, gas-surface reactions are more complicated than the gas phase analogy, which can usually be described homogeneously in terms of a limited number of degrees of freedom. Surfaces introduce infinite dimensionality into the dynamics and this provides a heat bath coupled to the reactive coordinates. In addition, real surfaces are not perfect and contain steps, other defects, and impurities. These are often poorly known or controlled in experiments. Especially for activated dissociation of molecules at surfaces, e.g., when there is an energy barrier to dissociation, these minority sites may have reactivities many orders of magnitude larger than majority terrace sites. Thus, sample heterogeneity is a significant issue in real world dynamics, especially when trying to compare theory with experiment.

There are several reasons why the dynamics of making and breaking bonds at the gas-surface interface is now topical and appropriate for a general discussion of bonding at surfaces. Over the past roughly 25 years, great experimental advances have produced ever-refined measurements of reaction probabilities for a variety of gas-surface chemical systems; e.g., molecular beam and laser state-resolved studies of adsorption and scattering. Coupled with this, is the recent emergence of 'first principles' theoretical understanding/predictions of reaction probabilities based on the Born–Oppenheimer approximation. Thus, detailed comparison of first principles theory with experiment is now meaningful. There are, of course, significant limitations in the current state of both experiments and theory. In addition to fundamental interest in the dynamics, there are practical reasons as well for trying to develop a detailed dynamic understanding. As outlined in other chapters in this book, the kinetics of chemical change at the gas-surface interface underlies many technological fields as well; e.g., catalysis, semiconductor processing, etc. Kinetics is simply a thermal averaging of individual dynamic processes. Understanding the individual dynamic processes and how these define the kinetic rates and selectivity of the chemistry can in principle lead to better design of technology.

This chapter first gives a theoretical (Section 2) and experimental (Section 3) overview of the dynamics of making/breaking bonds at the gas-surface interface. It then focuses on several different types of chemical processes and discusses these in terms of the elementary dynamic principles (Section 4). These include atomic adsorption/ desorption/scattering (Section 4.1), molecular adsorption/desorption/scattering (Section 4.2), direct dissociation/associative desorption (Section 4.3), precursor-mediated and indirect dissociation/associative desorption (Section 4.4), direct and precursor-mediated dissociation (Section 4.5), Langmuir–Hinschelwood chemistry (Section 4.6), Eley-Rideal/hot atom chemistry (Section 4.7), and hot electron-induced chemistry (Section 4.8). The various bond making/breaking processes considered here are outlined schematically in Figure 3.1. A few examples of each are treated in some detail in Section 4. The choice is based principally on those systems, which have been well studied *both* experimentally and theoretically, with a

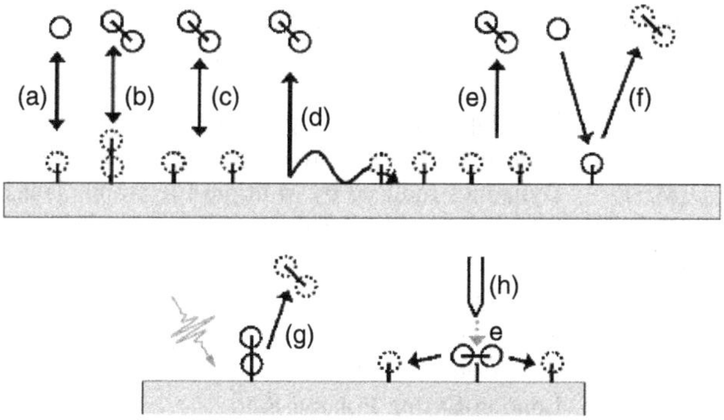

Figure 3.1. Schematic of bond making/breaking process considered in this chapter: (a) atomic adsorption/desorption/scattering, (b) molecular adsorption/desorption/scattering, (c) direct dissociation/associative desorption, (d) precursor-mediated dissociation/associative desorption, (e) Langmuir-Hinschelwood chemistry, (f) Eley-Rideal chemistry, (g) photochemistry/femtochemistry, and (h) single molecule chemistry. Solid figures generally represent typical intial states of chemistry and dashed figures the final states of the chemistry.

strong emphasis on 'first principles' theory. The discussion of these examples tries to emphasize the dynamic principals outlined in Section 2.

There are of course limitations in the coverage of this chapter. Most importantly, only the making/breaking of bonds occurring at the gas-surface interface of single crystal metal surfaces is treated in detail. Gas-surface chemistry on semiconductor surfaces is covered extensively in another chapter in this book. Except in Sections 4.6 and 4.7, only chemistry occurring at a bare surface is treated in detail, i.e., in the limit when there are no adsorbates. While covered or partly covered surfaces are important in technology, nearly all dynamics experiments (and most theory) have been performed in the limit of low surface coverage. When detailed dynamic information exists for the role of co-adsorbates or coverage in the examples presented, it is included in the discussion. In addition, dynamics and kinetics can be strongly influenced by steps and defects. Where appropriate, these issues are also discussed briefly. Femtosecond laser pump-probe techniques that follow dynamic processes occurring on the surface in real time and scanning tunneling microscopy (STM) techniques that follow the motion of individual atoms/molecules adsorbed on the surface are beginning to unravel many of the mysteries of dynamics that occurs entirely on the surface. However, to limit the scope of this chapter, these contributions to the dynamics of bond making/breaking at surfaces are not treated separately in any detail. However, they are discussed in terms of their contributions to the examples of gas-surface chemistry discussed in this chapter. Abbreviations that are used repeatedly in this chapter are summarized in Table 3.1.

Table 3.1
Principal abbreviations used in this chapter.

BOA	Born-Oppenheimer Approximation
DIET	Desorption Induced by Electronic Transitions
DIMET	Dynamics Induced by Multiple Electronic Transitions
DFT	Density Functional Theory
e-h (pair)	electron-hole (pair)
ER	Eley-Rideal
fs	Femtosecond
HA	Hot-Atom
LEPS	London-Eyring-Polanyi-Sato
LH	Langmuir-Hinschelwood
LIF	Laser Induced Fluorescenc
MDEF	Molecular Dynamics with Electronic Frictions
nD	n-Dimensional
PES	Potential Energy Surface
ps	Picoseconds
REMPI	Resonantly Enhanced Multi-Photon Ionization
SFG	Sum Frequency Generation
STM	Scanning Tunneling Microscopy
TOF	Time Of Flight
TPD	Thermal Programmed Desorption
TST	Transition State Theory

2. Theoretical background

This section introduces some of the main theoretical concepts necessary to describe the dynamics of making and breaking bonds at surfaces. Several simple models that often capture the essence of the dynamics are also introduced. For more details, a recent book that emphasizes many aspects of theoretical dynamics is Ref. [2].

2.1. Adiabatic dynamics (Born-Oppenheimer approximation)

Formally, the Hamiltonian describing a collection of atoms with electrons of coordinates **r** and nuclei of coordinates **R** is

$$\hat{H} = \hat{T}_n(\mathbf{R}) + \hat{H}_e(\mathbf{r}, \mathbf{R}) \tag{2.1}$$

with $\hat{T}_n(\mathbf{R})$ the nuclear kinetic energy and $\hat{H}_e(\mathbf{r}, \mathbf{R})$ is the usual electronic Hamiltonian and includes the internuclear repulsion term. The total wavefunction $\Psi(\mathbf{R}, \mathbf{r}, t)$ is non-separable in the electronic and nuclear coordinates because $\hat{H}_e$ involves coupling terms [3,4]. However, because electronic velocities are much faster than nuclear velocities, it is generally possible to approximately separate the wavefunction by treating the nuclei as static relative to the electronic motion, i.e.,

$$\Psi(\mathbf{R},\mathbf{r},t)=\sum_{\alpha}\Phi_{\alpha}(\mathbf{r};\mathbf{R})\,\chi_{\alpha}(\mathbf{R},t), \tag{2.2}$$

where the Φ_{α} is the electronic wave functions for the nuclear coordinates treated only parametrically and χ_{α} is the nuclear wavefunction for that electronic state. The purely electronic energies of α for a given $\mathbf{R}$ and labeled $\varepsilon_{\alpha}(\mathbf{R})$ define the adiabatic potential energy surfaces (PES). Insertion of this coupled wavefunction into the time-dependent Schrödinger equation leads to a set of coupled equations for the nuclei on each of the electronic PES, i.e., that nuclear motion on the β PES is governed by

$$i\hbar\frac{\partial\chi_{\beta}(\mathbf{R},t)}{\partial t}=\left[\hat{T}_{n}(\mathbf{R})+\varepsilon_{\beta}(\mathbf{R})\right]\chi_{\beta}(\mathbf{R},t)+\sum_{\alpha}\hat{K}_{\beta\alpha}\chi_{\alpha}(\mathbf{R},t). \tag{2.3}$$

The coupling terms that are off diagonal in the PES are of the form

$$\hat{K}_{\beta\alpha}=-\frac{\hbar^{2}}{2M_{j}}\sum_{j}\left[2\left\langle\Phi_{\beta}\middle|\nabla_{R_{j}}\middle|\Phi_{\alpha}\right\rangle\nabla_{R_{j}}+\left\langle\Phi_{\beta}\middle|\nabla_{R_{j}}^{2}\middle|\Phi_{\alpha}\right\rangle\right] \tag{2.4}$$

with the sum over all j nuclei of coordinates R_j and mass M_j.

The well-known Born-Oppenheimer approximation (BOA) assumes all couplings $\hat{K}_{\beta\alpha}$ between the PES are identically zero. In this case, the dynamics is described simply as nuclear motion on a single adiabatic PES and is the fundamental basis for most traditional descriptions of chemistry, e.g., transition state theory (TST). Because the nuclear system remains on a single adiabatic PES, this is also often referred to as the adiabatic approximation.

For much of the discussion in this chapter, the BOA is assumed valid so that the bond making/breaking is simply described by motion of nuclei on a multi-dimensional ground state PES. For example, dissociation of a molecule from the gas phase is described as motion on the PES from a region of phase space where the molecule is far from the surface to one with the adsorbed atoms on the surface. Conversely, the time-reversed process of associative desorption is described as motion on the PES from a region of phase space with the adsorbed atoms on the surface to one where the intact molecule is far from the surface. For diatomic dissociation/associative desorption, this PES is given as $V(Z, R, X, Y, \vartheta, \varphi, \{\vec{q}_i\})$, where Z is the distance of the diatomic to the surface, R is the distance between atoms in the molecule, X and Y are the location of the center of mass of the molecule within the surface unit cell, ϑ and φ are the orientation of the diatomic relative to the surface normal and $\{\vec{q}_i\}$ represent the thermal distortions of the ith metal lattice atom from its equilibrium position. For dissociative adsorption/associative desorption of more complex molecules, R is replaced by a set of coordinates $\left\{\vec{R}_j\right\}$ describing all the internal co-ordinates of the molecule. However, for the purposes of discussion, diatomic dissociation/associative desorption will be used as the preeminent example. It is shown in Section 2.4 that adsorption and desorption generally contain identical dynamic information. Therefore, Section 2 discusses only adsorption phenomena.

Within the adiabatic approximation, there are two major separate parts to a theoretical description of the chemical dynamics; first, knowing the instantaneous forces on all atoms at all nuclear configurations, i.e., knowing the multi-dimensional PES, and second, solving the multi-dimensional nuclear dynamics or scattering on this PES to obtain cross sections or probabilities for the reactive chemistry in various initial states. Finally, appropriate statistical averages of the dynamical probabilities can then be compared with experiments. Thermal averaging over all probabilities yields kinetic rates. It is also possible to simultaneously solve the electron dynamics quantum mechanically and the nuclear dynamics classically in the so-called ab initio molecular dynamics method [5]. This has had only limited application so far to bond making/breaking at surfaces because computational demands make it impossible to do enough trajectories via this method to obtain reasonable statistics [6].

Accurate quantum chemistry calculations of the PES describing surface chemistry are not currently possible for extended systems, and cluster models of surfaces have some artifacts. However, practical computational techniques in density functional theory (DFT) for slab type representations of extended systems have evolved significantly during the past two decades [5] so that calculations of the ground state PES are now quite feasible and have been performed for several interesting reaction systems, e.g., H_2, O_2, and N_2 interacting with various metals. Although it is not yet possible to obtain absolute chemical accuracy in DFT because of uncertainties in the exact exchange correlation functional, the general topology of the PES is very well described semi-quantitatively and relative energies within a given system (comparing two similar parts of a PES) appear almost quantitative in good calculations when comparing with experiments. It is, however, still a daunting computational task to map out large regions of a multi-dimensional PES necessary for full dynamical calculations of bond making/breaking, and the list of systems for which detailed multi-dimensional PES is available is still relatively short. For obvious practical reasons, these are mostly for H_2 interacting with metals. In addition, essentially all DFT PES to date have neglected the dependence on lattice distortion $\{\vec{q}_i\}$, e.g., are calculated only for frozen substrate atoms. At present, this limitation is only removed in a model dependent way (see Section 2.5). However, even with this limitation, the PES is six-dimensional (6D) for a diatomic interacting with a surface. In addition, dynamical calculations require interpolation between the DFT calculated points on the PES, so that an accurate analytical representation to the entire DFT PES is necessary. This is no easy task as there is no theoretically based fitting function for PES. Recently, several novel schemes have shown great promise as general interpolation procedures between the DFT calculated points, e.g., the corrugation reducing procedure [7], modified Shepard interpolation [8] and the use of neural networks [9].

There are now very many DFT calculations for chemical systems in which only regions of the PES around the transition states are explored, especially the barriers V^* for activated processes, since only this region of the PES is necessary to describe the overall chemical rate within transition state theory (TST) (see the chapter by Bligaard and Nørskov in this book). TST requires zero point corrected adiabatic barriers $V^*(0)$, but the zero point corrections are generally <0.1 eV for most chemistry at surfaces. Therefore, the distinction between V^* and $V^*(0)$ will usually be ignored in this chapter.

Formally, the nuclear dynamics describing the bond making/breaking is a quantum mechanical phenomenon. In many low-dimensional dynamic models, quantum effects (tunneling, resonances, zero point effects, etc.) are very important. As a result, there has been a huge investment in developing theoretical quantum dynamics for higher dimensional systems. Several examples of 6D quantum dynamics on 6D DFT PES for H_2 dissociation on metals now exist and give fully state-resolved dissociation probabilities [10,11]. However, most experiments seem to be sufficiently averaged and involve high enough energies that observable quantum effects disappear, even for H_2 dissociation on metals [12]. This means that much simpler classical nuclear dynamics is a reasonable approximation and generally sufficient to describe the chemistry. This introduces, however, some ambiguity as to how to treat vibrational zero point. For activated processes, quasi-classical dynamics (where vibrational zero point energy is added artificially in the initial scattering states) seems to be the best approximation, while for non-activated processes classical dynamics neglecting initial vibrational zero point energy entirely (or perhaps correcting the PES by the adiabatic softening of the vibrational zero point energy) seems to generally be a better approximation. Classical and quasi-classical dynamics in multi-dimensions is a straightforward molecular dynamics calculation, averaging many 'mindless' trajectories for random different starting conditions (e.g., X, Y, ϑ, φ, initial vibrational phase, etc.). Thus, knowledge of the multi-dimensional PES (and its analytical representation) is really the limiting factor in the theoretical description of bond making/breaking and ones ability to relate adiabatic theory to experiments quantitatively.

2.2. Generic PES topologies

Before discussing the dynamics of bond making/breaking at surfaces, it is helpful to consider generic PES topologies encountered in the various experimental systems and neglect the coupling to the lattice coordinates.

For atomic adsorption/desorption of physisorption systems (He, Ne, Ar, Xe on metals), $V(Z, X, Y)$ is reasonably well represented as a one-dimensional (1D) potential $V(Z)$. The laterally averaged PES is a balance of Pauli repulsion and van der Waals attraction and is given as $V(Z) = V_0 \exp(-\alpha Z) - C_6 \left(1/(Z - z_0)^3\right)$, where V_0, α, the van der Waals coefficient C_6 and the image plane z_0 are all constants [13]. The well depth W is small (typically less than 0.25 eV) and increases with the polarizability of the atom. Such PES are only weakly corrugated (i.e., variation with X, Y), with the corrugation increasing somewhat with E as the atom probes closer to the metal ion cores [14]. When the atom interacts chemically with the surface, the PES $V(Z, X, Y)$ is far more complicated. There is generally very strong corrugation throughout the PES that reflects the specific chemical nature of the interaction. For fixed impact sites X, Y, a 1D slice through the PES $V(Z)$ is roughly a Morse potential. Typical minimum well depths W for chemisorption at the optimum adsorption site X, Y are ~1–3 eV, but variations in W with X, Y can be typically ~20 % of W so there is no simple way to represent the PES in 1D.

Molecular potentials $V(Z, R, X, Y, \vartheta, \varphi)$ are even more complicated. In molecular adsorption/desorption (without dissociation), even in the physisorption limit, Z, ϑ coupling in the PES (anisotropy) is generally large in the repulsive part of the PES. For chemisorption systems, there is generally very strong coupling in all coordinates due to the specific chemical nature of the interactions, i.e., in both corrugation and anisotropy. Schematic 1D potentials for the dissociative adsorption/associative desorption of a diatomic molecule on a surface are given in Figure 3.2. Because they are only 1D, the richness of the dynamics/couplings is ignored, but they do introduce different classes of molecular dissociation behavior. Figure 3.2(a) results when the molecular interaction with the surface is quite weak, e.g., physisorption, while the atomic/fragment states are often more strongly bound so that the net energy change of the surface dissociation is ΔH. This PES topology is typical for H_2 and CH_4 dissociation on metals, i.e., molecules that contain only σ-bonding. This PES topology was originally discussed in the pioneering paper of Lennard-Jones [15] that first introduced a dynamic discussion to dissociation of molecules at surfaces. In terms of diabatic potentials, the A-B molecule–surface interaction is given as potential $V_m(Z)$, while the interaction of the dissociated fragments is given as $V_a(Z)$. For V_m, the A–B internuclear coordinate R is the asymptotic equilibrium molecular distance while for V_a, $R = \infty$. There is a diabatic crossing between V_m and V_a at some intermediate distance to the surface and this curve crossing generates a lower energy route to dissociation at the surface than in the gas phase. This is of course just the basis of heterogeneous catalysis. In terms of adiabatic potentials, the interaction of V_m and V_a lead to an avoided crossing with barrier V^*. If $V^* > 0$, then the dissociation is 'activated', i.e., requires some excitation relative to the molecule + surface asymptote in order to dissociate. If $V^* < 0$, the dissociation is non-activated. The equivalent 1D adiabatic potential gives the electronic ground

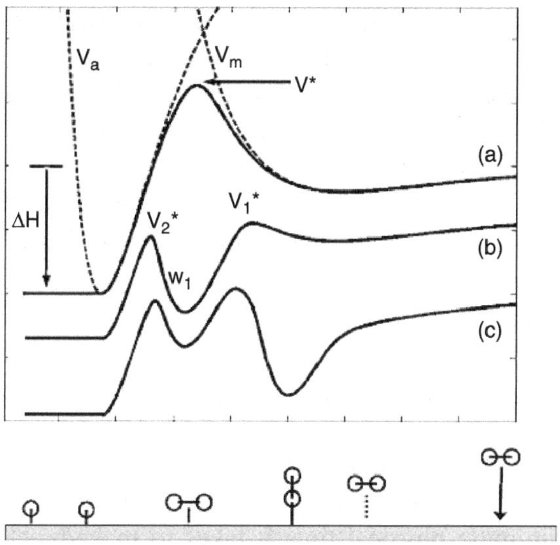

Figure 3.2. Schematic PES topologies typically encountered in diatomic molecular dissociation at surfaces. Idealized molecular geometries are given below the PES. Solid lines are adiabatic PES and dashed curves are diabatic PES. ΔH is exothermicity in dissociation, V_i^* are potential barriers and W_i are molecular adsorption wells, all relative to the molecule + surface asymptote. (a) PES is example of activated adsorption as described by Lennard-Jones, (b) PES with intermediate molecular adsorption well and (c) PES with multiple intermediate molecular adsorption wells.

state along the reaction path, i.e., the electronic energy including relaxation of all molecular coordinates to minimize the energy. It was well appreciated even by Lennard-Jones that such a 1D picture of dissociation is an oversimplified description of the dynamics, and he already suggested in his early paper (as a note added in proof) that the barrier would depend on impact site of the molecule on the surface. Thus, there is a distribution of barrier heights, which depend upon the impact site and orientation of the molecule to the surface.

Another PES topology for molecular dissociation occurs when an intermediate molecularly chemisorbed state lies parallel to the surface between the physisorption well and the dissociated species as shown in Figure 3.2(b). This molecular state is usually described in terms of a diabatic correlation to a state formed by some charge transfer from the surface to the molecule [16]. In this case, there can be two activation barriers, V_1^* for entry into the molecular chemisorption state of depth W_1 and barrier V_2^* for dissociation of the molecularly chemisorbed state. This PES topology is relevant to the dissociation of some π bonded molecules such as O_2 on metals, although this is often an oversimplification since distinct molecularly adsorbed states may exist at different sites on the surface [17]. In some cases, $V_1^* < 0$ so that no separate physisorbed state exists [18]. If multiple molecular chemisorption states exist for adsorbates, bonded both vertical and parallel to the surface and possibly at different sites on the surface, complicated PES topologies similar to Figure 3.2(c) are possible. There can be many different barriers and molecular wells and often the parallel molecular states are only metastable. Since these parallel states are the only ones that can readily dissociate on the surface, the dissociation dynamics is indirect and complicated. This PES topology is appropriate for N_2 and CO dissociation on many transition metals, e.g., Fe(111), W(100), etc.

2.3. Dynamics vs. kinetics

There are three different generic time scales defining chemical processes of molecules on surfaces; τ_v, a typical vibrational period, τ_E, the typical energy relaxation time, and τ_R, the dissociation/ residence time for a molecule on the surface. Typical times are $\tau_v \approx 0.2$ picoseconds (ps), $\tau_E \approx 1-10$ ps and $\tau_R \approx 0.2$ ps–∞. τ_v is essentially the minimum physical time it takes to make/break a chemical bond. When $\tau_R >> \tau_E$, the molecule becomes thermalized with the surface and one can only describe the chemistry as a thermal rate process, albeit one subject to the constraints of unitarity, thermal equilibrium, and reciprocity [19] (see Section 2.4). For example, molecular adsorption/desorption can only be described in terms of kinetic rates; i.e., as an adsorption rate defined by a trapping coefficient α and a desorption rate defined by desorption flux D_f. Of course α and D_f still depend on the energy and quantum states of the gas phase species. When $\tau_R << \tau_E$, the detailed dynamics occurring over the few ps time scale is important and can be followed theoretically. Thus, although atom/molecules can still exchange energy with the surface in the initial collision, subsequent equilibration with the surface is minimal. This is the scenario for direct inelastic scattering or direct dissociation. Because different modes of adsorbate energy relax at different rates, the intermediate case of $\tau_R \sim \tau_E$, requires a good description of both the dynamics occurring on initial impact with the surface and subsequent energy transfer processes to the surface. This is the general scenario for indirect scattering or indirect dissociation.

The distinction between dynamics vs. kinetics has led to a great divide in categorizing molecular dissociation at surfaces; i.e., as direct dissociation or as precursor-mediated dissociation. These two different limiting types of dissociation are indicated in Figure 3.3 on a schematic 1D PES. The direct dissociation occurs on initial impact with the surface. PES topologies, described by Figure 3.2(a), generally favor direct dissociation. On the other hand, if an incident molecule first adsorbs into a molecular chemisorption state and thermalizes with the surface prior to dissociation, then its progress to the dissociated state can only be described in terms of chemical rates. Since the molecularly chemisorbed state is a precursor to the dissociation step, this is referred to as precursor-mediated dissociation. Generally, PES topologies of Figure 3.2(b) and (c) give precursor-mediated dissociation. In some cases, the

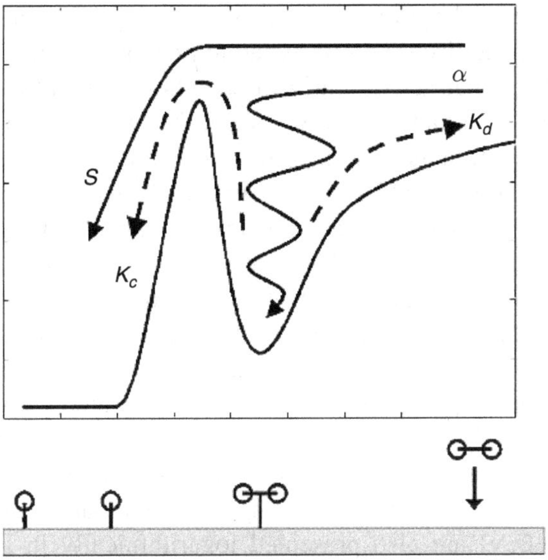

Figure 3.3. Schematic of direct and precursor-mediated dissociation processes on a typical adiabatic PES (given by the solid line). Solid arrow labeled S represents direct dissociation and that labeled α represents trapping into a molecular adsorption well. Dashed arrows represent competing thermal (Arrhenius) rates for desorption (k_d) and dissociation (k_c) from the molecular well.

dissociation dynamics is more complicated because $\tau_R \sim \tau_E$ and only some molecular modes may be equilibrated with the surface prior to reaction. Such a scenario may be especially relevant when quasi-trapping and dynamic trapping are important, especially when lattice equilibration times are long, e.g., for H_2 non-activated dissociation on surfaces (see Section 4.3.3.1).

2.3.1. Direct dissociation

As outlined previously, 6D dynamical calculations for direct dissociation are relatively straightforward once an analytical representation of the PES is known, even 6D quantum dynamical calculations. Generically, depending upon the distribution of barriers in the PES, the direct dissociation can be strongly activated, only weakly activated or non-activated. A wide variety of dynamical phenomena emerge from 6D calculations of these dissociations and are used to interpret experiments. The theoretical dynamics relevant to the examples presented in this chapter will be discussed in more detail in Section 4.3. Since the availability of reasonable 6D PES is quite recent, reduced dimensionality dynamical models have played a significant role in initially building understanding of the dynamics [4,20]. These have been based both on model PES and on using low-dimensional slices through DFT PES. A 1D model along the minimum energy path for dissociation on the PES only allows a kinetic description of the processes, e.g., in TST. The 2D models allow the beginning of a discussion of dynamic phenomena, i.e., the importance of translational vs. vibrational activation in dissociation and isotope effects. In direct activated dissociation of a diatomic at a surface, the two essential coordinates are Z and R since the molecule must approach the surface and the molecular bond must break. Figure 3.4(a) shows a 2D PES for activated H_2 dissociation on Cu(111) [21]. It is obtained as the 2D slice through the lowest barrier configuration in a DFT calculation keeping all other molecular coordinates (X, Y, θ, φ) frozen at those that give the minimum barrier. Figure 3.4(b) gives an equivalent 2D PES for activated N_2 dissociation on Ru(0001) [22]. Dissociation dynamics on these 2D PES give $S(E, v)$, the dissociation probability as a function of translational energy E and initial vibrational state v. An example of $S(E, v)$ calculated both by quasi-classical and quantum dynamics for a PES similar to Figure 3.4(a) is given in Figure 3.5(b). The sharp 'S'-shaped classical translational excitation functions $S(E, v)$ are simply broadened in quantum dynamics by tunneling below the barrier and reflection above the barrier. The $S(E, v)$ are also presented logarithmically in Figure 3.5(a) since experiments are often reported in this manner to present the large dynamic range measured. Both translational energy and vibrational excitation promote dissociation. The relative importance of each in dissociation is often discussed in terms of barrier location in the PES in analogy with the well-known Polanyi rules for atom/molecule reactions in the gas phase [1]. When the barrier is principally in the entrance channel

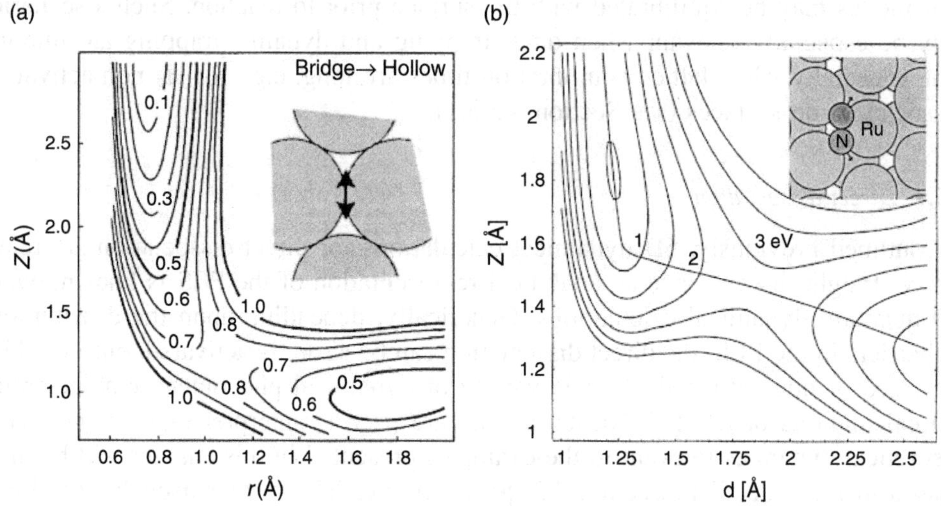

Figure 3.4. 2D DFT PES for dissociative adsorption through approximately the minimum barrier configuration (transition state). Contours in eV are as indicated on the figures. (a) H_2 dissociation on Cu(111), $r \equiv R$ in the notation of this chapter. From Ref. [21]. (b) N_2 dissociation on Ru(0001), $d \equiv R$ in the notation of this chapter. From Ref. [22].

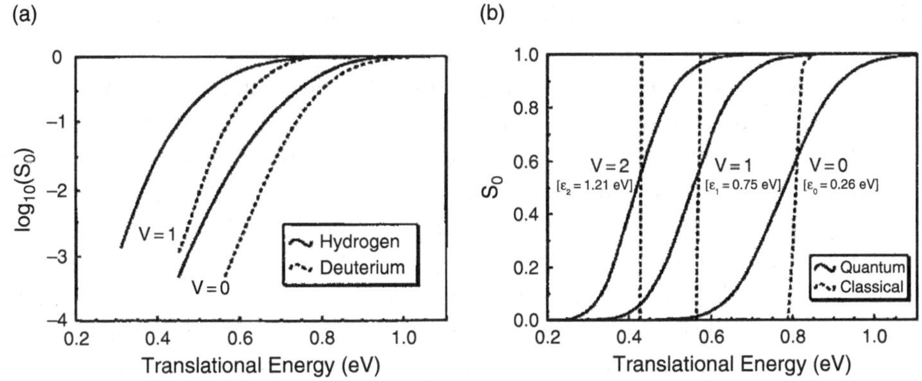

Figure 3.5. 2D dissociation probability S_0 ($\equiv S$) as a function of translational energy and vibrational state v for H_2 (D_2) dissociation on a PES similar to (but not identical) to that of Figure 3.4(a). (a) Quantum dissociation probabilities plotted logarithmically. (b) Dotted lines are results of quasi-classical dynamics and solid lines are from quantum dynamics. From Ref. [222].

to dissociation (along the translational coordinate), initial vibrational excitation is of little help in overcoming the barrier, while if it is principally in the exit channel to dissociation (along the vibrational coordinate), initial vibrational energy can be even more important than translational energy in surmounting the barrier. In terms of the $S(E, v)$, this is essentially a question of how much the 'S'-shaped translational excitations functions shift with v relative to the magnitude of the vibrational energy. One defines a vibrational efficacy η_v

$$\eta_v = \frac{E_0(v-1) - E_0(v)}{\varepsilon_v - \varepsilon_{v-1}} = \frac{\Delta E_0}{\Delta \varepsilon_v}, \tag{2.5}$$

where $E_0(v)$ is the center E of the $S(E, v)$ 'S'-shaped curve and ε_v is the vibrational energy for the v state. η_v provides a quantitative measure of how effective vibration is in promoting dissociation on the 2D PES. For barriers completely in the entrance channel, $\eta_v \approx 0$ while $\eta_v >> 0$ for barriers in the exit channel. $\eta_v \approx 0.5$ for the PES of Figure 3.4(a), while $\eta_v > 1$ for the PES of Figure 3.4(b). Thus, initial vibrational excitation is anticipated to be significantly more important in N_2 dissociation on Ru(0001) than in H_2 dissociation on Cu(111).

Multi-dimensional DFT calculations show that the PES actually varies considerably with molecular orientation and impact point on the surface [21,23]. Universally, barriers for dissociation are substantially lower with molecular bonds nearly parallel to the surface. V^* is also laterally corrugated in the unit cell, both in magnitude and in location relative to the surface plane. The variation in the magnitude of V^* with X,Y has been labeled energetic corrugation, while the variation in its position relative to the surface plane with *X,Y* labeled geometric corrugation [24]. Since all dissociation experiments average over impact site and orientation, even neglecting all dynamic couplings, they must sample a distribution of activation barriers. One consequence of this is that $S(E, v)$ measured in experiments is broader than that predicted by the 2D model with a single barrier V^*. Full 6D dynamic calculations show that S depends on all initial quantum states, i.e., $S(E_i, \theta_i, v, J, M)$ where E_i is the incident energy, θ_i is the incident angle to the surface, v is the vibrational state, J the rotational state and M is the projection of J on a space fixed axis (i.e., surface normal). Typically, each independent quantum state gives a different 'S'-shaped curve and the dependence on E_i, θ_i depends upon the corrugation [24]. When energetic corrugation dominates over geometric corrugation in the dynamics, the component of energy parallel to the surface $E_{||} = E_i \sin^2\theta_i$ inhibits dissociation relative to that from just the normal component $E_n = E_i \cos^2\theta_i$. On the other hand, when geometric corrugation dominates the dynamics, $E_{||}$ aids dissociation relative to that produced by the E_n component.

One attempt to remedy the limitations of the 2D model, and yet retain its simplicity, is the so-called 'hole' model [25] which represents a simple static way to average over this distribution of barriers. If the variation of barrier height with these 'spectator' variables is given as $\Delta V^*(X, Y, \vartheta, \phi)$, then the 6D dissociation probability S_{6D} is approximated in terms of the 2D S_{2D} as

$$S_{6D}(E, v) \approx \frac{1}{2\pi A_s} \int S_{2D}\left([E - \Delta V^*(X, Y, \vartheta, \phi)], v\right) \sin\theta d\theta d\phi dX dY, \qquad (2.6)$$

where A_s is the surface unit cell area. This equation is based on two simple approximations; a classical sudden approximation for the dynamics of the spectator coordinates and a static approximation to account for the variation in S_{2D} with changes in barrier height. $\Delta V^*(X, Y, \vartheta, \phi)$ can be approximated from the vibrational frequencies of modes perpendicular to the reaction coordinate at the top of the barrier, so the full 6D PES is not required. The hole model is in reasonable agreement with 6D dynamical calculations for activated adsorptions [26,27]. However, since all dynamical coupling beyond that of *Z,R* is neglected in the 'hole' model, dynamical effects such as steering, rotational hindering, dynamical trapping, etc. that are key aspects of non-activated dissociation are not included (see Section 4.3.3.1). Therefore, the 'hole' model is only a first approximation for activated dissociation at incident energies only slightly above V^*. In non-activated dissociation, the dynamical couplings induced by the 6D PES must be included at the outset.

2.3.2. Precursor-mediated dissociation

After a molecule traps and thermally equilibrates in a molecularly adsorbed state, the net dissociation probability S is given by competition between the thermal dissociation rate, $k_c = \nu_c \exp(-E_c/k_B T_s)$, and the thermal desorption rate, $k_d = \nu_d \exp(-E_d/k_B T_s)$, from this adsorbed state. T_s is the surface temperature and E_i and ν_i are the activation energy and pre-exponential in Arrhenius rate expressions. These rates are sketched on Figure 3.3. In this case, S is given at steady state as

$$S = \alpha(E_i, \theta_i, T_s)\left(\frac{k_c}{k_c + k_d}\right) = \frac{\alpha(E_i, \theta_i, T_s)}{1 + \frac{\nu_d}{\nu_c}\exp\left[-(E_d - E_c)/k_B T_s\right]}, \quad (2.7)$$

where α is the trapping probability into the molecular precursor state and is a function of E_i, θ_i and T_s. Trapping generally decreases with incident energy (see Section 2.5.1). In the terminology of Figure 3.2(b), $E_d = W_1$ (assuming no barrier to molecular adsorption) and $E_c = W_1 + V^*$. Rearranging eq. (2.7) shows that a plot of ln $[(\alpha/S) - 1]$ vs. $1/T_s$ is linear with a slope $-(E_d - E_c)/k_B$ and intercept ln $[\nu_d/\nu_c]$. When $E_d > E_c$, S decreases roughly exponentially with T_s and conversely when $E_d < E_c$, S increases roughly exponentially with $T_s \cdot \frac{\nu_d}{\nu_c} >> 1$, reflecting the larger phase space available for desorption relative to that at the barrier for dissociation [28]. Similar, but more complicated kinetic equations can also be derived for the topologies of Figure 3.2(c) when sequential precursors are involved [29,30]. In some cases, if the barrier into the second precursor state $V_2^* > 0$, then S increases with incident energy as in a direct activated process, but there will still be a strong dependence on T_s. This has been called an activated precursor [31].

2.4. Detailed balance

Detailed balance, or as more properly stated reciprocity, relating desorption from a surface to adsorption is a rigorous relation under conditions of equilibrium. This relation is easily derived by considering a dividing plane parallel to but far from the surface, with the rate of passage of a given internal state β with translational energy E at angle θ towards the surface balanced by the equivalent rates away from the surface. This leads quite simply to the well known equation relating the desorption flux D_f through a given solid angle $d\Omega = \sin\theta d\theta d\phi$ to the adsorption coefficient S at the same incident angle

$$D_f(E, \theta, \beta, T_s) \propto E \exp\left(-E/k_B T_s\right) \exp\left(-E_\beta/k_B T_s\right) \cos\theta\, S(E, \theta, \beta, T_s), \quad (2.8)$$

where E_β is the internal energy of the β state. Thus, the dynamics of desorption is equivalent to that in adsorption and is fully contained in S. There will be many examples throughout the remainder of this chapter where this relationship is used to compare desorption with adsorption, e.g., in comparing atomic desorption to trapping [32] or state-resolved associative desorption fluxes with state-resolved dissociation experiments [33,34].

Because all adsorption/desorption experiments are performed under non-equilibrium conditions, there has been much discussion as to whether experiments 'fulfill' detailed balance [35]. However, as pointed out by several authors [36,37], the deviations due to departure from equilibrium in the experiments are insignificant if the adsorbate–substrate subsystem attains a Boltzmann distribution. This does not mean, however, that detailed balance is necessarily a good approximation in comparing any given pair of desorption/adsorption experiments. The two experiments must effectively sample the same phase space. Often adsorption experiments are done in the limit of low atomic coverage Θ and at low T_s, while desorption experiments are usually done at high Θ and higher T_s. Therefore, if there is a strong dependence of the PES or dynamics on Θ or T_s, detailed balance cannot be used to relate the two experiments. For example, activated adsorption in the limit of low adsorbate coverage is often dominated by dissociation at steps or defects. On the other hand, associative desorption experiments at high adsorbate coverage may measure desorption preferentially from the terraces, and consequently detailed balance does not relate these two experiments [38].

2.5. *Lattice coupling*

In the PES and dynamics discussed so far, only the molecular internal and external coordinates have been considered (e.g., 6D for diatomic dissociation). However, the PES also depends on the lattice coordinates $\{\vec{q}_i\}$ as well and this leads to energy exchange between the molecule and the lattice. There is now considerable theoretical and experimental evidence that coupling of the lattice to the translational coordinates $Z, X, Y \leftrightarrow \{\vec{q}_i\}$ in the repulsive part of the PES causes large energy transfer between the translational energy E_i and the lattice. On the other hand, coupling to the intramolecular coordinate $R \leftrightarrow \{\vec{q}_i\}$ gives minimal energy transfer to the lattice because of the large frequency mismatch between the high frequency intramolecular vibration and phonon frequencies [39]. This requires the coupling to be high order in $\{\vec{q}_i\}$ and the energy transfer correspondingly weak. In fact, this coordinate generally couples more strongly to the metal electrons than the lattice (see Section 2.6.1). The strength of rotation-lattice energy transfer from $(\vartheta, \varphi) \leftrightarrow \{\vec{q}_i\}$ coupling in the PES is generally in between these two extremes [40].

Translation to lattice energy transfer is *the* dominant aspect of atomic and molecular adsorption, scattering and desorption from surfaces. Dissipation of incident translational energy (principally into the lattice) allows adsorption, i.e., bond formation with the surface, and thermal excitation from the lattice to the translational coordiantes causes desorption and diffusion i.e., bond breaking with the surface. This is also the key ingredient in trapping, the first step in precursor-mediated dissociation of molecules at surfaces. For direct molecular dissociation processes, the implications of $Z, X, Y \leftrightarrow \{\vec{q}_i\}$ coupling in the

PES is far less well understood (and generally ignored). It may, however, be quite important for direct dissociations of molecules heavier than H_2 or D_2.

The classical molecular dynamics of gas-surface interactions including the coupling to the lattice can be recast as a generalized Langevin equation where one focuses on a small primary lattice region which is coupled to the rest of the lattice via a friction and fluctuating forces [41,42]. The Langevin equation can be solved in a straightforward manner via stochastic trajectories if the potential $V\left(Z, R, X, Y, \vartheta, \varphi, \{\vec{q}_i\}\right)$ is known. Although the dependence of V on $\{\vec{q}_i\}$ can in principal be obtained from DFT calculations, this is an overwhelming computational task because of the high dimensionality of the problem. Therefore, lattice couplings have in general only been included semi-empirically, by V, which is approximated as a sum of pair-wise atomic or at least pair-wise functional additive interactions to make the molecular dynamics efficient. Some examples of such calculations will be discussed in Section 4.1.1. Recently, a good approximation to include lattice coupling in an atom-surface DFT PES has been proposed that is based on the corrugation reduction procedure for fitting the DFT PES [43,44]. However, there is at present no way to generalize this procedure to include the lattice coupling to the molecular region of a PES, e.g., in the entrance channel to molecular dissociation.

Because first principles knowledge of the $\{\vec{q}_i\}$ dependence of V is generally poor, simple models of the lattice coupling and its consequences for dynamics have played an important role historically. A particularly good discussion of energy transfer to surfaces based on a simple model of atom/molecules interacting with a 1D lattice chain is in Ref. [45]. Results from a few simple models are presented below.

2.5.1. Energy transfer in adsorption/scattering

It is no surprise that when an atom/molecule collides with a surface that it transfers some of its translational energy to the lattice through momentum transfer. The simplest example is the interaction of a rare gas with a metal surface since the PES is weakly corrugated, and these have been studied extensively to learn about the dynamics of energy transfer to surfaces. Because of the weak corrugation, the dominant PES coupling to the lattice is $Z \leftrightarrow \{\vec{q}_i\}$ and energy transfer to the surface occurs principally through the normal component of incident translational energy E_n. In the impulsive collision limit and assuming conservation of parallel momentum, simple kinematics gives the energy transfer of an atom of mass m to a stationary surface 'cube' of mass M_s as

$$\Delta E_q = \frac{4\mu}{(1+\mu)^2}\left[E_n + W\right], \tag{2.9}$$

where $\mu = m/M_s$ and W is the well depth in the atom-surface potential. This is illustrated schematically in Figure 3.6. The ΔE_q is commonly known as the Baule

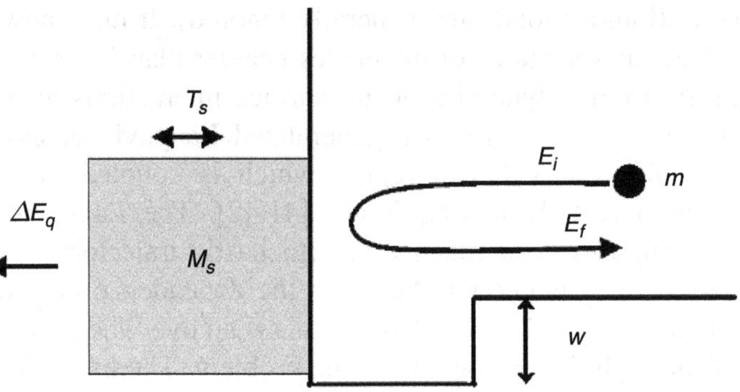

Figure 3.6. Schematic of the 'cube' model for energy transfer (ΔE_q) of an atom/molecule of mass m incident with energy E_i to the lattice represented by a cube of mass M_s. The atom/molecule adsorption well depth is W. The double arrow labeled T_s emphasizes that the cube also has initial thermal motion in the scattering.

limit of energy transfer. The classical energy transfer is therefore large when $\mu \sim 1$ and for large E_n and/or W, and can often exceed 50 % of the incident energy. Including initial surface atom motion from T_s in the kinematics of energy transfer in the cube model gives an energy transfer [46]

$$\Delta E_q \approx \frac{4\mu}{(1+\mu)^2}[E_n + W] - \frac{\mu(2-\mu)}{(1+\mu)^2}(2k_B T_s) \tag{2.10}$$

The second term allows for energy transfer from the surface at finite T_s to the atom in addition to the energy loss from the atom to the surface of the first term. If the atom-surface PES is somewhat corrugated, then the cube of Figure 3.6 is replaced by overlapping hard spheres [47] or a 'washboard' corrugation [48]. In this case, an impulsive model for the scattered energy E_f gives the more general relation

$$\langle E_f \rangle \approx a \langle E_e \rangle + b(2k_B T_s), \tag{2.11}$$

where $E_e = E_i \cos^n\theta_i$, with $n = 0{-}2$ depending upon the corrugation. However, because the corrugation varies with E_i, the coefficient n in E_e may not be the same constant for all incident energies. There is no fundamental basis for the functional form of E_e. It is simply a convenient way to interpolate between normal energy scaling ($n = 2$) and total energy scaling ($n = 0$).

For the cube model, when $\Delta E_q > E_n$, trapping into the adsorption well occurs. Therefore, there is a critical energy E_c such that the trapping coefficient $\alpha = 1$ for $E_n < E_c$, while $\alpha = 0$ and only direct inelastic scattering occurs for $E_n > E_c$. For the stationary cube,

$$E_c = \frac{4\mu}{(1-\mu)^2} W. \tag{2.12}$$

From eq. (2.10) for ΔE_q, E_c and hence trapping also decreases with T_s. Real atom trapping $\alpha(E_i, \theta_i, T_s)$ involves a gentle almost exponential fall off with E_e and T_s because of averaging the collision impact over the unit cell, thermal smoothing, etc. However, E_c still gives a reasonable estimate for the range of E_e for which trapping occurs in these weakly adsorbed systems.

Trapping on a corrugated surface requires that the total translational energy $E < 0$ and not just $E_n < 0$ as in the simple cube model for trapping. Since the energy parallel to the surface $E_{||}$ generally couples to the lattice more weakly than E_n, this is not always true when $\theta_i >> 0$. Some authors use the terminology trapping to refer to $E_n < 0$ only and sticking to refer to $E < 0$. However, this chapter uses the terminology quasi-trapping for $E_n < 0$ only and trapping for $E < 0$. The term sticking and symbol S in this chapter is reserved for the case of dissociative adsorption.

It is important to note that even if there is approximate conservation of parallel momentum or velocity $\langle \Delta v_{||} \rangle \approx 0$ in scattering, there can still be energy transfer to the lattice from $E_{||}$ due to a broadening in the scattered $v_{||}$ distribution. There is now plenty of evidence that even in the simplest examples of energy transfer to surfaces, e.g., Ar/Pt(111), that $\Delta E_{||}$ is significant, especially at high E_i and θ_i (see Section 4.1.1). Nevertheless, the Baule limit, eq. (2.9), is still a reasonable first estimate for the energy transfer, and especially for near normal incidence conditions. Often inelastic scattering plotted as E_f/E_i as a function of the final scattering angle θ_f is compared to two limits; conservation of parallel momentum and an impulsive model-treating scattering as the binary collision of two hard spheres. The former limit is more appropriate for smooth surfaces (low E_i, θ_i, physisorption systems), while the latter is more appropriate for highly corrugated surface and/or large $E_{||}$ (high E_i, θ_i).

Because of the large W, $\alpha \approx 1$ for reactive atoms. However, the dynamics within the well before thermalization is anything but simple. Because of the strong corrugation, there is considerable scrambling of E_n and $E_{||}$. When μ is very small (H and D), energy dissipation to the lattice in the well is slow and 'hot atoms' are formed which have considerable reactivity with other species adsorbed on the surface(see Section 4.1.2).

Energy transfer from molecules to the lattice are significantly more complicated because translational energy may simultaneously be transferred to molecular coordinates, e.g. rotation J, as well as the lattice. Even for weak molecule-surface interactions, there is still strong anisotropy in the repulsive part of the PES, which causes strong translational-rotational conversion, $E_n \leftrightarrow J$. Because of the competition for E_n by both ΔE_q and J, an anti-correlation is observed between energy transfer to the lattice and rotation [49]. Nevertheless, energy transfer to the lattice still dominates at even modest E_i. This energy transfer is discussed more fully in Section 4.2.1. In the molecular chemisorption regime, there is generally both very strong anisotropy and corrugation in the attractive part of the PES as well because of the detailed chemical nature of the interactions. For example, CO on Pt(111) preferentially binds with the C end down and at on-top sites. This induces significant conversion of $E_n \leftrightarrow J, E_{||}$. Since $E_{||}$ accommodates to the lattice more slowly than E_n, there can be a significant difference between quasi-trapping ($E_n < 0$) and trapping ($E < 0$). Molecules may quasi-trap at the surface, making several round trips in the well and finally scatter back into the gas phase via conversion of J, $E_{||} \rightarrow E_n$, retaining partial memory of the initial $E_{||}$ in scattering [50]. The effect of the strong anisotropy and corrugation on trapping is also dramatic [50], decreasing gradually with E_i over several eV and almost independent of θ_i. Again, despite the complexity of the adsorption/scattering, energy transfer to the lattice is still the dominant process. A well-studied example of adsorption/scattering in the molecular chemisorption regime is discussed more fully in Section 4.2.2.

For light particles, i.e., He, Ne and H_2, and low E and W, where energy transfer to the lattice is less than the phonon bandwidth, the classical approximation for energy transfer breaks down and quantum dynamics is necessary [45]. Of principal importance, especially at low T_s, is the quantum nature of the lattice, which causes the emergence of a no-loss line, i.e., purely elastic scattering. A good description of this phenomenon is given by the forced oscillator model which treats the particle trajectory classically so that its motion induces a time dependent force $F(t)$ on a set of quantum harmonic oscillators representing the lattice [45,51]. The intensity of the elastic scattering $P_{el}(E)$ is defined by a Debye-Waller factor W_F for the collision at surface temperature T_s, i.e., $P_{el}(E) = e^{-2W_F(T_s)}\,\delta(E)$. Within the forced oscillator model, $2W_F$ is defined

$$2W_F(T_s) = \int d\varepsilon\,[1+2\bar{n}(\varepsilon)]\,\frac{\left|\int_{-\infty}^{\infty} dtF(t)\,e^{i\varepsilon t}\right|^2}{2M_s\varepsilon}\,\rho(\varepsilon) \approx \Delta P_0^2\left\langle z_0^2\right\rangle, \qquad (2.13)$$

where $F(t)$ is the force induced on the lattice by the approaching atom of mass m, $\rho(\varepsilon)$ is the phonon density of states, $\bar{n}(\varepsilon) = \left[\exp\left(\varepsilon/k_B T_s\right) - 1\right]^{-1}$ is the Bose–Einstein factor for finite T_s. The right-hand part of eq. (2.13) results from an impulsive collision approximation where ΔP_0 is the momentum transferred to the particle in the collision and z_0 is the mean displacement of the lattice atom. W_F increases approximately linearly and hence $P_{el}(E)$ decreases exponentially with E, W, T_s and projectile mass m. Thus, only for low E, W, T_s and m is elastic scattering observed, i.e., typically for He and H_2 physisorption systems. Elastic scattering makes possible the observation of diffractive scattering due to lateral or orientational corrugation in the PES [52] and the detection of selective adsorption resonances due to interference of the bound states in the laterally averaged particle surface PES with the continuum of elastic scattering states [53]. Elastic scattering also affects trapping since the limit of α at $E = 0$ is no longer unity but $1 - P_{el}(0)$. This quantum effect has been observed for H_2 trapping on Cu(100) [54] and rare gas trapping on Ru(0001) [55,56].

2.5.2. Lattice coupling in direct molecular dissociation

The importance of lattice coupling in direct molecular dissociation is at present poorly understood. However, there are at least two ways in which inclusion of the lattice can affect direct dissociative adsorption. First, conversion of E_i to ΔE_q competes with translational activation in dissociation. Second, thermal distortion of lattice atoms from their equilibrium positions may affect the PES, e.g., the barriers to dissociation $V^*\left(\{\vec{q}_i\}\right)$. These two effects can be most simply thought of as a phonon induced modulation of the barrier along the translational coordinate and in amplitude, respectively.

A simple model has been proposed to account for energy loss to the lattice during the act of activated dissociative adsorption, so-called 'dynamic recoil' [57]. In this model, the lattice is replaced by a single Einstein oscillator q in the Z direction that remains in thermal equilibrium with the rest of the lattice. The coupling of the incident molecule to the lattice is assumed to be simply a modulation of the 'stiff' lattice PES in and out with the lattice coordinate, i.e., as a coupling to the translational coordinate only so that $V(Z, R, X, Y, \vartheta, \varphi, q) = V(Z - q, R, X, Y, \vartheta, \varphi) + \frac{1}{2}kq^2$, where k is the force constant of the lattice Einstein oscillator. Classical dynamics using this simple model PES describes reasonably well the average energy transfer to the lattice in molecular scattering and its dependence on E_n [58]. For dissociative adsorption, low-dimensional dynamic simulations show that the key effect compared to a stationary lattice is a shift of the 'S' shaped translational excitation functions to *higher* energies by $\delta E_0^q(v) \approx \mu E_0(v)$. Thus, the effect of the lattice is approximately described as a larger effective V^*, i.e., as 'dynamic recoil' of the barrier $V^* \to (1 + \mu)V^*$. These effects are especially significant for large μ (large m) and should be minimal for activated H_2 dissociation. When the lattice is not stationary due to thermal excitation, low-dimensional model dynamics [57,59] also predict a moderate broadening of the 'S' shaped $S(E, v)$ with T_s so that they should be more properly described as $S(E, v, T_s)$. Dynamic recoil and its T_s dependence has been extensively discussed for CH_4 dissociation on metals [59] (see Section 4.3.1.3). This was originally described in terms of thermally assisted tunneling, but the effects of the lattice coupling are equivalent in purely classical dynamics as well.

The other lattice coupling $V^*\{\vec{q}_i\}$ is also likely to be a general phenomenon in activated direct dissociation since barriers are generally lower for less coordinated sites, i.e., distortion of the lattice can give lower/higher barriers. However, this barrier modulation is not generally anticipated to be too large since the usual transition states at bridge and hollow sites are distributed over several lattice atoms, and the thermal modulations from the different lattice atoms tend to cancel. Nevertheless, significant variation of V^* with a single q_i distortion has been calculated for H_2 dissociation on Pt(100) by DFT and used as a rationalization to explain the observed modest T_s dependence of S [60].

For CH_4 dissociation on metals these effects can be much larger since there is an atop transition state which should be strongly affected by only a single lattice atom motion. DFT calculations for CH_4 dissociation on Ir(111) [61] and Ni(111) [62,63] show that V^* is lowered by ~0.3 eV relative to that of the fixed lattice when the single lattice atom below the atop transition state is pulled out of the surface by ~0.3 Å. The strain induced by this local lattice distortion is minimized by long-range relaxation of the rest of the lattice. Low-dimensional dynamics calculations suggest that the dominant dynamical effect of this is to give a shift $\delta E_0^q(v)$ of the translational excitation functions to *lower* energies because of the lower barrier with the lattice distortion [63]. The magnitude of the shift should depend on the extent of lattice response during the approach of CH_4 to the transition state, and this will depend on details of the PES lattice coupling and the mass and force constants of the surface atoms. In addition, this type of lattice coupling also predicts a T_s dependent broadening in $S(E, v, T_s)$ similar to that of dynamical recoil. In essence, thermal fluctuations cause a wider distribution of barriers, some lower and some higher than that for the frozen surface.

The presence of lattice coupling in direct dissociation complicates somewhat the definition of V^* and how V^* extracted from measurement should be compared to DFT calculations. There are two well defined limits for DFT calculations of V^*, for a frozen lattice ($T_s = 0$) and for the case where the lattice is fully relaxed at the transition state ($T_s = \infty$). Since experiments are performed at finite T_s, the best V^* description is probably intermediate between these two extremes.

Unfortunately, the only experimental signature of lattice coupling in direct dissociation is a T_s dependent broadening of S. This broadening causes only a modest change in the dissociation dynamics, although it significantly affects activation energies extracted in thermal kinetics experiments (see Sec 4.3.1.3). However, lattice coupling also causes significant shifts δE_0^q in S. There is unfortunately no way to obtain these shifts independently in measurements. Note that this shift is anticipated to be in opposite directions for the two different mechanisms of lattice couplings, and there is at present little definitive guide as to which dominates for any given system.

2.6. Non-adiabatic dynamics

The basis of the BOA for bond making/breaking at metal surfaces is the neglect of the coupling terms in eq. (2.3). These terms are small if (1) nuclear velocities are small, (2) the change in the adiabatic wave functions with nuclear coordinate are gradual and (3) if the adiabatic states are well separated in energy. Because there is a continuum of electron–hole (e–h) pair excitations in the metal about the Fermi level (i.e., no energy gap), condition (3) is formally violated for all chemical dynamics on metal surfaces and it is a-priori difficult to predict the validity of the BOA. This has two aspects; first that chemistry occurring on surfaces may dissipate energy by creating e-h pairs in the metal and second, that e-h pairs created thermally or by other means in the metal (e.g., absorption of photons) can induce surface chemistry. These two scenarios for non-adiabatic dynamics are indicated in Figure 3.7. This figure shows schematically that chemistry induced by coupling to the lattice q, i.e., thermal associative desorption, can create e-h pairs by a breakdown of the BOA (via frictional damping in the figure). This also shows that hot electrons (created by laser excitation of the metal) can exert fluctuating forces on the nuclei and hence induce chemistry, i.e., hot electron-induced associative desorption. These two scenarios, hot electrons from chemistry and chemistry from hot electrons are discussed separately below.

2.6.1. *Hot electrons from chemistry*

The conditions (1)–(3) above for the validity of the BOA give some indication as to when its breakdown is most likely to be important in chemical dynamics. For example, very high velocities, e.g., hyperthermal atoms or high vibrational states interacting with surfaces, should promote non-adiabatic contributions to the

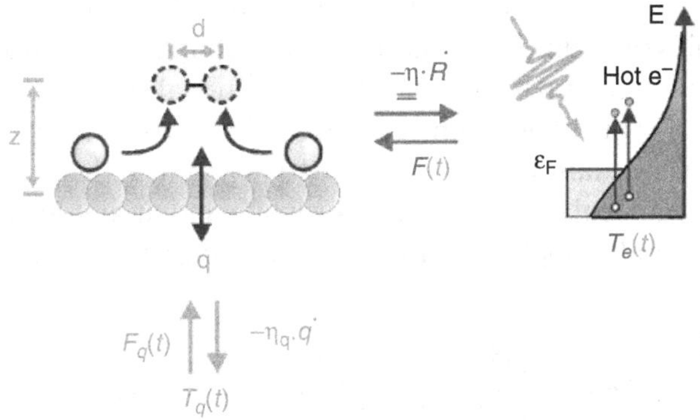

Figure 3.7. Schematic showing that associative desorption induced by thermal excitation of the lattice can create e-h pairs (hot electrons) via non-adiabatic damping of nuclear coordinates (described here as an electronic frictional damping $\underline{\underline{\eta}} \cdot \dot{\vec{R}}$). The lattice coordinate q coupled to the dissociation is kept in equilibrium with the rest of the lattice at T_q via Langevin dynamics. This also shows that thermally or photochemically produced e-h pairs (described here as a hot electron distribution $T_e(t)$) can also induce associative desorption via the fluctuating forces $F(t)$ that hot electrons exert on nuclei. From Ref. [101].

dynamics [64,65]. Abrupt or strong changes in the electronic structure of the adiabatic states should also promote non-adiabatic behavior, e.g., when a narrow adsorbate resonance crosses the Fermi level of the metal. This occurs when electronegative adsorbates (NO, O_2, Cl_2) adsorb on metals, especially low work function metals [66,67]. Finally, more gradual but very strong changes in the electronic nature of the adiabatic states also occur at avoided crossings, e.g., at the transition states in activated dissociative adsorption/associative desorption [68].

When the non-adiabatic coupling is weak and the frequencies associated with the nuclear motion are small relative to structure in the electronic density of states, the breakdown in the BOA can be described in classical dynamics as a frictional damping of the nuclear coordinate X, with friction η_{XX} [69,70]. The basic assumption is that the energy loss to e-h pairs occurs as a series of random uncorrelated small energy losses. The rate of non-adiabatic energy transfer between the nuclear coordinate and the electron hole pairs is then given as $\dot{E}_{NA} = \eta_{XX}\dot{X}^2$, and the magnitude of the breakdown in the BOA is then simply a question of how large are η_{XX} and $\dot{X}$. Because there is damping of nuclear motion into e-h pairs, excited e-h pairs at electron temperature T_e must also be able to excite nuclear coordinates by fluctuating forces F_X that satisfy the second fluctuation dissipation theorem given as

$$\langle F_X(t) F_X(t') \rangle = 2k_B T_e \eta_{XX} \delta(t-t'). \tag{2.14}$$

Including η_{XX} and F_X in Langevin-type classical dynamics has been termed molecular dynamics with electronic frictions (MDEF) [70], and has now been used in several simulations of non-adiabatic dynamics. Of course, the key unknown is the magnitude of the electronic frictions (since they also determine F_X).

Fortunately, the same limiting conditions that validate the friction approximation can also be used with time-dependent density functional theory to give a theoretical description of η_{XX}. This expression was originally derived to describe vibrational damping of molecules adsorbed on surfaces [71]. It was later shown to also be applicable to any molecular or external coordinate and at any location on the PES, and thus more generally applicable to non-adiabatic dynamics at surfaces [68,72]. The expression is

$$\eta_{XX} = 2\pi\hbar \sum_{\alpha,\beta} \left| \left\langle \psi_\alpha \left| \frac{\partial v}{\partial X} \right| \psi_\beta \right\rangle \right|^2 \delta(\varepsilon_\alpha - \varepsilon_F)\, \delta(\varepsilon_\beta - \varepsilon_F), \tag{2.15}$$

where ψ_i, ε_i are Kohn–Sham orbitals and orbital energies, respectively, $\partial v/\partial X$ is the derivative of the Kohn–Sham potential v with respect to nuclear coordinate X and ε_F is the Fermi energy. Conventional DFT calculations give ψ_i, ε_i and finite differences can be used to evaluate the matrix element [73]. Thus, DFT based calculations of η_{XX} are possible and these can be used as 'first principles' estimates of the importance of the breakdown of the BOA in surface chemical dynamics.

To date, there are only a few 'first principles' estimates of η_{XX} and their effects on surface dynamics. One well-known example is the vibrational damping of molecules adsorbed on surfaces, where adiabatic theory cannot account for the short lifetimes/ large linewidths of adsorbed species [74]. Calculations of η_{XX} for CO/Cu(100) [73,75,76] and on other surfaces [77] are in good agreement with lifetimes/ linewidths observed in experiments. Low-dimensional MDEF with ab initio η_{XX} [68] suggest that the associative desorption of H_2 from Cu(111) is essentially adiabatic (see Section 4.3.1.1.), while non-adiabatic damping of vibration is significant in associative desorption of N_2 from Ru(0001) (see Section 4.3.1.2). Other examples are comparisons of MDEF with ab initio η_{XX} for vibrational de-excitation of H_2 in scattering from Cu [78] (see Section 4.3.1.1.), CO scattering and sticking on Cu(100) [79] and ab initio η_{XX} combined with the forced oscillator model to estimate the chemicurrent for H adsorption on Cu [72,80] (see Section 4.1.2.).

Recently, equations for non-adiabatic quantum vibrational dynamics on metal surfaces have been derived by Krishna [81]. This derivation clarifies approximations necessary for a quantum version of the generalized Langevin equation and justifies MDEF, which was originally derived by mixed classical-quantum Ehrenfest dynamics. The quantum dynamics also shows how the fluctuation-dissipation theorem, originally postulated aposteriori, arises naturally under appropriate limitations. Most importantly, however, is that the derivation suggests a criterion as to when the friction (or mean-field) approximation is valid to describe the non-adiabatic coupling. Krishna suggests that the mean-field approach is no longer valid whenever the time-scale of the variation of non-adiabatic coupling is comparable to the time-scale of e-h pair excitations. For vibrational damping in a state with quantum number v, the mean field criterion can be expressed as

$$\tfrac{1}{2}\pi\nu X_v \cdot \nabla_X \ln(\eta_{XX}) << 1, \tag{2.16}$$

where X_v is the amplitude of vibrational motion for the v state. When the variation of non-adiabatic coupling over the vibrational motion is not negligible, then the corrections to the frictional non-adiabatic coupling can be formally expressed as Lorentz forces arising from the created e-h pairs.

When the limiting conditions of the friction approximation are not valid, e.g., there is strong non-adiabatic coupling or rapid temporal variation of the coupling, there is at present no well-defined 'first principles' method to calculate the breakdown in the BOA. The fundamental problem is that DFT cannot calculate excited states of adsorbates and quantum chemistry techniques, that can in principle calculate excited states, are not possible for extended systems.

In some cases, the electronic structure for the surface chemistry can be approximated as a narrow (quasi-localized) single affinity resonance [with center position ε_a and width Δ_a] crossing the Fermi level of the metal. When the rate of electron transfer between the metal and the resonance lags adiabatic or instantaneous filling as it crosses ε_F, $\frac{\hbar\dot{\varepsilon}_a}{2\Delta_a} = \hbar\dot{Z}\frac{d\varepsilon_a}{dZ}\left(\frac{1}{2\Delta_a}\right) >> 0$, significant breakdown of the BOA is anticipated [82]. In this case, a diabatic representation may be a more appropriate way to discuss this breakdown. Instead of using adiabatic electronic states to define $\Psi(\mathbf{R}, \mathbf{r}, t)$, diabatic electronic states are defined that are independent of nuclear coordinates

$$\Psi(\mathbf{R}, \mathbf{r}, t) = \sum_{\alpha} \Phi_{\alpha}(\mathbf{r})\chi_{\alpha}(\mathbf{R}, t). \tag{2.17}$$

The diabatic PES represent essentially frozen electronic configurations between the molecule and the surface, e.g., do not allow for charge transfer between the surface and the molecule during the chemistry. In this case, the coupling terms of eq. (2.4) are identically zero but there is now coupling between the diabatic electronic states in terms of off-diagonal potential terms. Although there is no rigorous or unique way to define and obtain the diabatic PES, several methods have been proposed based on DFT [83,84]. They do require, however, that integer charge and spin states be pre-defined to describe the ground and excited diabatic states, and it is unclear how closely the diabatic state resembles the actual excited state of the adsorbates (since they cannot be calculated). Unfortunately, there is also no 'first principles' way to obtain the off-diagonal potential terms so that only models with adjustable parameters can describe the breakdown in the BOA in the diabatic representation. However, such models have been used to successfully describe exoelectron and photon emission in halogen adsorption on alkali metals [66] and O_2 dissociation on Al(111) [67].

At present, there is considerable controversy over the importance of the breakdown in the BOA in surface chemistry on metals. There are only a few examples where direct experimental evidence is available for the creation of hot electrons via chemistry, e.g., exoelectron and photon emission [66,85] or the observation of chemicurrents following atomic/molecular adsorption [86] (see Section 4.1.2.). There is, however, indirect evidence in several experiments. This means that DFT based adiabatic dynamical theory is incapable of qualitatively explaining the experiments, while the inclusion of non-adiabatic effects gives good/better agreement. Examples are the vibrational damping of molecules adsorbed on surfaces [74,87], vibrational de-excitation in scattering from surfaces [65,78,88], vibrational damping in associative desorption [68] and the failure to predict activated adsorption where it exists in experiment [67,89,90]. Some of these aspects are discussed in more detail for the specific systems discussed in Section 4. However, despite interest in this area at the present time, it is important to keep in mind that the BOA seems valid to describe most aspects of chemistry at surfaces, and only when special situations exist is the breakdown of the BOA important.

2.6.2. Chemistry from hot electrons

While the importance of the breakdown of the BOA in thermal chemistry is still controversial, the time-reversed process of creating chemistry from hot electrons is well established. Because experiments are generally performed under conditions where there is *no* adiabatic chemistry, hot electron induced chemistry is easily identified and studied, even when the cross-section for the chemistry is very small. Typical scenarios involve photochemistry, femtochemistry and single molecule chemistry on surfaces. A few well-studied examples are discussed briefly in Section 4.8. Because a detailed discussion of these active fields would take this chapter far from its original purpose, they are only treated briefly to illustrate the relationship to other aspects of bond making/breaking at surfaces.

The theoretical description of photochemistry is historically based on the diabatic representation, where the diabatic models have been given the generic label desorption induced by electronic transitions (DIET) [91]. Such theories were originally developed by Menzel, Gomer and Redhead (MGR) [92,93] for repulsive excited states and later generalized to attractive excited states by Antoniewicz [94]. There are many mechanisms by which photons can induce photochemistry/desorption; direct optical excitation of the adsorbate, direct optical excitation of the metal-adsorbate complex (i.e., via a charge-transfer band) or indirectly via substrate mediated excitation (e–h pairs). The differences in these mechanisms lie principally in how localized the relevant electron and hole created by the light are on the adsorbate.

Figure 3.8 illustrates the photon-induced excitation of NO chemisorbed on Pt(111) based on the scenario of substrate mediated excitation by hot electrons (see Section 4.8.1.1). Hot electrons (and hot holes) are created in the optical skin depth

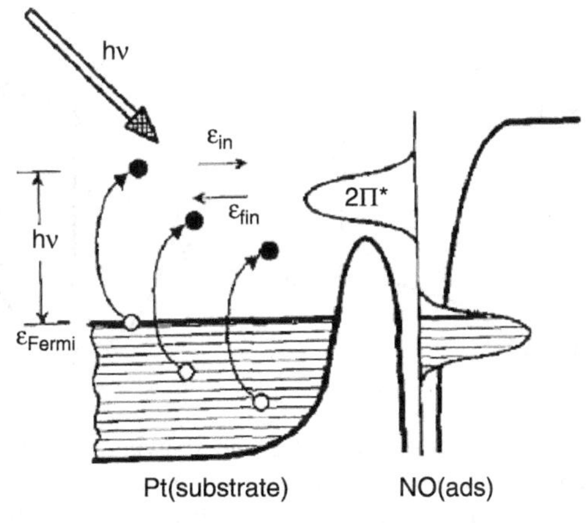

Figure 3.8. Schematic energy level diagram for NO adsorbed on a Pt metal substrate. This also shows inelastic hot electron scattering through the $2\pi^*$ resonance associated with the NO, with the hot electrons produced by Pt absorption of photons of energy *hv*. From Ref. [95].

of the metal by optical absorption. Given typical mean free paths of low-energy electrons in metals, most of the hot electrons created can reach the surface. Thus, absorption of photons of frequency $\hbar\omega$ by the metal creates a wide distribution of hot electrons at the surface of $0 - \hbar\omega$, and these can scatter inelastically from the $2\pi^*$ resonance of the adsorbate to induce chemistry. This is a common mechanism in the photochemistry of chemisorbed molecules on metals since absorption of light in the thin adsorbate layer is weak compared to absorption in the underlying metal. In addition, hot hole mobilities are generally less than that of hot electrons so that there is less likelihood of holes reaching the surface. In the following, only the indirect hot electron mechanism is discussed.

In terms of diabatic PES, this resonant scattering process is represented schematically in Figure 3.9. Inelastic scattering by the hot electron creates a transient negative ion state with a lifetime t_R defined by the exponential probability distribution $f(t_R, \tau^*) = \exp(-t_R/\tau^*)/\tau^*$ with $\tau^* = \hbar/2\Delta_a$. Because the adsorbate structure is generally different in the negative ion state than the ground state, forces exist on the nuclei after excitation and induce motion for t_R until electron transfer from the adsorbate returns the electron to the metal. If t_R is longer than some critical time t_c, the projection back onto the ground state PES can contain adsorbates with a contribution of nuclear wave functions in the molecule-surface bond continuum, i.e., for desorption. Although τ^* is still largely unknown for adsorbate excited states, it is estimated to be typically ~1–10 fs in typical chemisorption systems. Treating hot electron production at the surface, inelastic electron scattering and nuclear dynamics as separable processes, the photoprocess can be modeled via Gaussian wavepacket

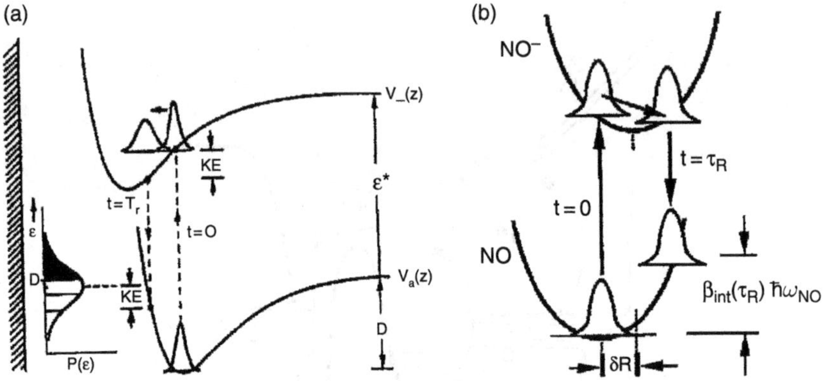

Figure 3.9. (a) Diabatic PES for NO–Pt coordinate and showing wavepacket propagation through the time sequence involving excitation/de-excitation of the negative ion $2\pi^*$ resonance of NO. The distribution of final NO–Pt states between vibrationally excited bound and desorptive continuum states is shown as P(ε) vs. energy ε. (b) Diabatic PES for the intra-molecular N–O bond and wavepacket propagation through the same time sequence as in (a). This illustrates the mechanism for internal vibrational excitation in desorbed NO. From Ref. [95].

dynamics coupled with Franck–Condon projections for the electronic transitions to give desorption yields and E_f distributions [95]. Multi-dimensional PES allows dynamical evolution in several coordinates simultaneously to give v, J distributions of desorption products as well [96,97]. This is schematically indicated in Figure 3.9 by the second set of oscillators representing the N−O bond. In this case, resonant scattering of hot electrons causes vibrational excitation in the N−O bond. Unfortunately, excited state PES and τ^* are completely unknown so that these treatments are at best only phenomenological models with adjustable parameters. They do, however, rationalize a wide range of phenomena in single-photon photochemistry experiments (see Section 4.8.1.1).

Absorption of light in a metal initially creates a nascent hot electron distribution governed by the optical absorption of the metal. However, this distribution temporally evolves by electron–electron scattering and electron-phonon scattering. The former scattering produces a hot electron cascade and ultimately a Fermi-Dirac distribution at the surface defined by an electron temperature T_e, while the latter cools the hot electron distribution by heating the lattice. The electron thermalization time t_e depends both on the material (electron density of states) and the level of optical excitation, with $t_e \sim 0.1 - 1$ ps. High levels of optical excitation cause significantly more rapid thermalization [98]. DIET processes are generally believed to result from non-thermalized electrons, although it is not entirely clear whether this is strictly the nascent distribution of hot electrons produced in photoabsorption. When photons arrive at the surface as intense ~100 femtosecond (fs) laser pulses, the high intensity causes a rapid thermalization ($t_e \sim 100$ fs) to produce a hot Fermi-Dirac distribution at the surface with temperature T_e. Because of the small heat capacity for electrons in metals, T_e up to 5000 K can readily be achieved. Over periods of ~1 picoseconds (ps), the hot electrons thermalize with the lattice via electron–phonon scattering to produce lattice temperatures (T_q) such that ultimately $T_q = T_e \sim 1000$ K for typical fs laser intensities. The surface then cools slowly (20–40 ps) by thermal conduction into the bulk. Thus, for periods of ~1 ps, there is an exceedingly hot electron distribution T_e that can cause chemistry before the lattice T_q heats up. Neglecting the initial electron thermalization, the temporal evolution of T_e and T_q are well described as two coupled heat baths via the so-called two temperature model [99]. Using intense fs lasers to induce (and probe) hot electron chemistry is generically called femtochemistry. It appears that most femtochemistry experiments are dominated by a fully thermalized distribution of hot electrons so that in many ways this is a more defined experiment than that of the low intensity DIET regime.

Initial femtochemistry theoretical models were simply generalizations of those for the single-photon DIET processes, i.e., as dynamics induced by multiple electronic transitions (DIMET) [100]. The idea is simply that even if the excited state residence is too short to cause excitation to a ground state continuum after resonant scattering ($t_R < t_c$), it can still cause some vibrational excitation in the ground state. If resonant scattering occurs again prior to vibrational de-excitation in the ground state, ladder climbing on the ground state PES can occur by multiple excitations until dynamics is induced. It is the high density of hot electrons created by the fs laser that allow the multiple excitations to compete with vibrational de-excitation. Again, phenomenological models with adjustable parameters can fit experiments.

The ladder climbing in the DIMET model is essentially the time-reversed process to non-adiabatic vibrational damping, which is well described in the adiabatic representation by an electronic friction (Section 2.6.1). Therefore, the femtochemistry can in principle also be described as the result of fluctuating forces caused by the thermalized hot electrons created by the fs laser as shown schematically in Figure 3.7. Multi-dimensional dynamics based on MDEF as described earlier is then a way to also describe the femtochemistry. In fact, if the PES and electronic frictions are obtained by DFT, this becomes a 'first principles' theoretical treatment of the femtochemistry. At present, only two examples of first principles theory exist [70,101] and one will be discussed in more detail in Section 4.8.1.3. Several femtochemistry experiments have also been fit to a phenomenological model representing ladder climbing induced by electronic friction η_e in a 1D PES defined as a truncated harmonic oscillator of well depth V_0 [102,103]. This model was originally developed to describe fs laser induced desorption of intact molecules from surfaces, in principle a 1D dynamical problem. In typical applications, V_0 and η_e are treated as adjustable parameters.

Finally, hot electrons (or hot holes) of specific energy and at a given flux can be injected to scatter inelastically from a single molecular resonance on the surface by placing a scanning tunneling microscope (STM) tip at a given bias/ current just above the molecule, and measuring the chemistry this induces with the STM. One example is discussed in Section 4.8.2.

3. Experimental background

This section introduces the principal experimental methods used to study the dynamics of bond making/breaking at surfaces. The aim is to measure atomic/molecular adsorption, dissociation, scattering or desorption probabilities with as much experimental resolution as possible. For example, the most detailed description of dissociation of a diatomic molecule at a surface would involve measurements of the dependence of the dissociation probability (sticking coefficient) S on various experimentally controllable variables, e.g., $S(E_i, \theta_i, v, J, M, T_s)$. In a similar manner, detailed measurements of the associative desorption flux D_f may yield $D_f(E_f, \theta_f, v, J, M, T_s)$ where E_f is the produced molecular translational energy, θ_f is the angle of desorption from the surface and v, J and M are the quantum numbers for the associatively desorbed molecule. Since dissociative adsorption and associative desorption are time-reversed processes of each other, they are in principle related by detailed balance (see Section 2.4). Of course, most experiments only have partial resolution of the molecular internal states. Note also that the only characterization of the surface that is possible is specification of T_s.

3.1. Experimental techniques

Two techniques are principally responsible for the experimental development of dynamics in surface chemistry. These are the application of molecular beams and laser 'state-to-state' techniques to gas-surface interactions. This roughly parallels their application to gas phase chemistry, although there are certainly some different technical requirements. More detailed discussion of some of these experimental techniques are in Refs. [104] and [105].

A schematic diagram of a state-of-the-art molecular beam-surface machine is shown in Figure 3.10. In this machine, atomic/molecular beams impinge on a clean well-ordered single-crystal surface. Typically, seeded supersonic molecular

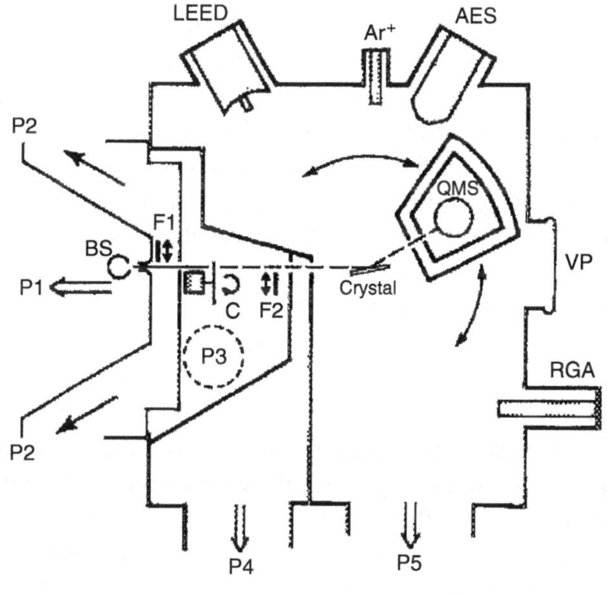

Figure 3.10. Molecular beam apparatus for determining scattering distributions and dissociative chemisorption probabilities. Key: P1–P5, pumping stages; BS, beam source; C, chopper; F1 and F2 beam flags; QMS, doubly differentially pumped quadrupole mass spectrometer; LEED, low energy electron diffraction, Ar+, argon ion sputtering gun; AES, Auger electron spectrometer; VP, viewport; RGA, residual gas analyzer. The molecular beam source detaches at the P3 region and is nonbakeable. The crystal is fixed to a manipulator that allows for heating, cooling, and rotation about the same axis as the QMS. From Ref. [420].

beams are employed which have a wide range of translational energies E_i and a narrow translational energy distribution ΔE_i. The number of molecules adsorbing irreversibly on the surface can be followed by post surface analysis, e.g., temperature programmed desorption (TPD), Auger, etc. Those scattering/desorbing from the surface can be measured via a small aperture rotating differentially pumped mass spectrometer. A chopper placed in the molecular beam path prior to the surface allows time-of-flight (TOF) measurements of the initial beam energy and either surface residence time or final translational energy for the scattered/desorbed particles, depending upon the scattering conditions.

Laser 'state-to-state' techniques include both the application of highly sensitive laser spectroscopy for internal state-resolved detection of molecules in the gas phase, e.g., desorbing or scattering from a surface, and second, for laser pumping an initial state prior to interaction with a surface. To date, laser detection of internal states has been widely applied in gas-surface dynamics experiments, while those involving optical state preparation techniques have only been applied in a limited fashion.

Tunable laser spectroscopic techniques such as laser-induced fluorescence (LIF) or resonantly enhanced multi-photon ionization (REMPI) are well-established mature fields in gas-phase spectroscopy and dynamics, and their application to gas-surface dynamics parallels their use elsewhere. The advantage of these techniques is that they can provide exceedingly sensitive detection, perhaps more so than mass spectrometers. In addition, they are detectors of individual quantum states and hence can measure nascent internal state population distributions produced via the gas-surface dynamics. The disadvantage of these techniques is that they are not completely general. Only some interesting molecules have spectroscopy amenable to be detected sensitively in this fashion, e.g., H_2, N_2, NO, CO, etc. Other interesting molecules, e.g. O_2, CH_4, etc., do not have suitable spectroscopy. However, when applicable, the laser spectroscopic techniques are very powerful.

A typical application is the use of the $(2+1)$ REMPI scheme for measuring the (v, J) distribution of H_2 produced in associative desorption from a surface. When the laser is tuned to a spectroscopic transition between individual quantum states in the $X \rightarrow E$ electronic band, resonant two-photon absorption populates the E state and this is subsequently ionized by absorption of another photon. The ion current is proportional to the number in the specific (v, J) quantum state in the ground electronic state that is involved in the spectroscopic transition. Tuning the laser to another spectroscopic feature probes another (v, J) state. Therefore, recording the ion current as the laser is scanned over the electronic band maps out the population distribution of $H_2(v, J)$ produced in the associative desorption. E_f of the (v, J) state can also often be simultaneously measured using field – free ion TOF or laser pump – probe TOF detection techniques. The $(2+1)$ REMPI scheme for detecting H_2 is almost independent of the rotational alignment and orientation $f(M)$ of molecules so that only relative populations of the internal states are measured. However, other REMPI schemes, e.g., $(1+1)$ REMPI, can be made to be very sensitive to $f(M)$ so that measuring the REMPI intensity as a function of the incident laser polarization relative to the surface normal can measure moments of $f(M)$. Usually only the first moment of $f|M|$, the quadrupolar alignment $A_0^{(2)}(E_f, \theta_f, v, J, T_s)$, is measured. Treating the rotational angular momentum as a classical vector, $A_0^{(2)}(J) = \langle 3\cos^2\vartheta_J - 1\rangle$, where ϑ_J is the angle between the rotational angular momentum vector and the surface normal. The two extremes are when $A_0^{(2)} = 2$, called colloquially 'helicoptering', when the molecule rotates in a plane parallel to the surface and $A_0^{(2)} = -1$, labeled 'cart-wheeling', when the molecule rotates in a plane normal to the surface.

Various laser techniques are available for optical pumping molecules to excited (v, J) states; direct IR absorption, Raman pumping and stimulated emission pumping. The useful technique depends on the spectroscopy of the system. Both direct IR and Raman pumping are probably familiar to all readers. Stimulated emission pumping occurs when one laser excites a molecule to an individual quantum level in an excited electronic state and a second laser dumps a large fraction of the excited state back down to vibrational levels of the ground electronic state. When Franck-Condon factors in the electronic spectroscopy are distributed over a wide range of vibrational states, very high vibrational states can be populated via stimulated emission pumping, albeit fairly weakly [106]. Gas-surface dynamics experiments involving all schemes for optical pumping of molecules to specific initial (v, J) states suffer from the fact that only a small fraction of the total molecules incident on a surface are pumped into an excited state and their effect must be isolated in the dynamics.

3.2. Typical measurements

3.2.1. Rate measurements

The least resolved measurement is determination of the isothermal rate constant $k(T)$, where T is the isothermal temperature. Although conceptually simple, such measurements are often exceedingly difficult to perform for activated process without experimental artifact (contamination) because they require high pressures to achieve isothermal conditions. For dissociative adsorption, $k(T) = k_{col}(T)\langle S(T_g = T_s = T)\rangle$, where $k_{col}(T)$ is simply the collision rate with the surface and is readily obtainable from kinetic theory and T_g and T_s are the gas and surface temperatures, respectively [107]. $\langle S\rangle$ refers to thermal averaging. A simple Arrhenius treatment gives the effective activation energy E_a for the kinetic rate as

$$E_a = -\frac{d\ln\langle S(T)\rangle}{d(1/k_B T)} + \frac{1}{2}k_B T \tag{3.1}$$

and this is nearly identical to the zero point corrected activation energy $V^*(0)$ via transition state theory. Although such rate measurements do not readily elucidate the dynamics, they do represent the key chemical rates for a variety of technological processes, e.g., heterogeneous catalysis (see the chapter by Bligaard and Nørskov), and so are very important measurements in their own right. At low gas pressure $T_g \neq T_s$, and measurements $k\left(T_g, T_s\right) = k_{col}\left(T_g\right)\langle S\left(T_g, T_s\right)\rangle$ with $T_g = 300$ K and varying T_s are relatively straightforward. However, it is difficult to interpret these experiments dynamically since $\langle S(T_g, T_s)\rangle$ is a highly averaged quantity relative to those obtained from molecular beam experiments. For example, such measurements have led to some controversy as to whether CH_4 dissociation on transition metals is direct or precursor-mediated [59].

3.2.2. Adsorption-trapping and sticking

The preeminent technique for measuring trapping and sticking probabilities (α and S) is via molecular beam techniques [104], with trapping referring to non-dissociative adsorption and sticking referring to dissociative adsorption. Note that many authors use the symbol S_0 to represent sticking on a bare metal surface, although this chapter uses simply the symbols S. Seeded supersonic molecular beams allow a large range of translational energies to be explored with a rather narrow translational energy distribution. The collimated and directed molecular beams allow straightforward measurements of $\alpha(E_i, \theta_i, T_s)$ and $S(E_i, \theta_i, T_s)$.

Measurements of $\alpha(E_i, \theta_i, T_s)$ for atomic physisorption systems have clarified the mechanism of energy transfer to the lattice (see Section 2.5.1). These experiments are discussed in detail in Section 4.1.1 for the Ar/Pt(111) example. Reactive atoms, e.g., H, generally have $\alpha \approx 1$, so detailed experiments on $\alpha(E_i, \theta_i, T_s)$ are not interesting. In molecular adsorption, energy transfer occurs from E_i to molecular coordinates as well as the lattice upon impact with the surface. These energy transfer processes change the behavior of $\alpha(E_i, \theta_i, T_s)$ from that of atoms and are discussed more fully in Sections 4.2.1 and 4.2.2. It is usual in adsorption studies to find an energy scaling relation for the adsorption, i.e., to define the dependence on E_i and θ_i so that it can be expressed in terms of an effective energy $E_e = E_i\cos^n\theta_i$, where $0 < n < 2. n = 2$ is normal energy scaling and $n = 0$ is total energy scaling.

For molecular dissociation, the qualitative behavior of $S(E_i, \theta_i, T_s)$ depends greatly on the mechanism of dissociation, i.e., whether it is direct or precursor-mediated. When the dissociation is direct, absolute values of S are generally only weakly dependent on T_s. On the other hand, S is a very strong function of T_s for precursor-mediated dissociation (eq. (2.7)), decreasing nearly exponentially with T_s when $E_c < E_d$ and increasing nearly exponentially when $E_c > E_d$. Generally, the entrance to the precursor state is non-activated and S decreases with E_i due to a fall-off of trapping in the precursor state with E_i. In a few cases, entrance to the precursor state is activated so that S increases with E_i. It is largely the vastly different dependences of S on T_s and E_i that allows the assignment of a mechanism to dissociation. Typical dependencies showing these differences are shown in Figure 3.11. Again, the dependence of S on E_i and θ_i can usually be expressed in terms of an effective energy $E_e = E_i \cos^n \theta_i$.

In molecular beam studies of direct dissociation at surfaces, three generic 'sticking' behaviors $S(E_n)$ at $\theta_i = 0$ are often encountered, corresponding roughly to non-activated, weakly activated and activated dissociative adsorption. Examples of each are presented in Section 4.3. For non-activated dissociation, most of the phase space (i.e. impact parameters X, Y and molecular orientations θ, φ lead to dissociation for all $E_n \rangle 0$. At very low energies, often steering to the optimum dissociation site or dynamic trapping causes an increase in S at very low E_n [108–111]. In weakly activated dissociation, although there is a small region of phase space with minimum barrier $V^* \approx 0$, most impact sites and orientations have finite barriers to dissociation which are overcome as E_n increases. Finally, in activated dissociation even the minimum barrier $V^* > 0$ so that high E_n is necessary to obtain any significant S. Although the classical threshold in the increase of S with E_n is related to the adiabatic barrier V^*, no sharp onset in S is measurable in experiments, probably due to thermal smearing and heterogeneity (defects).

High translational energies are available by seeding a small mole fraction of a heavy gas in a light carrier gas (He or H_2) and by raising the nozzle temperature

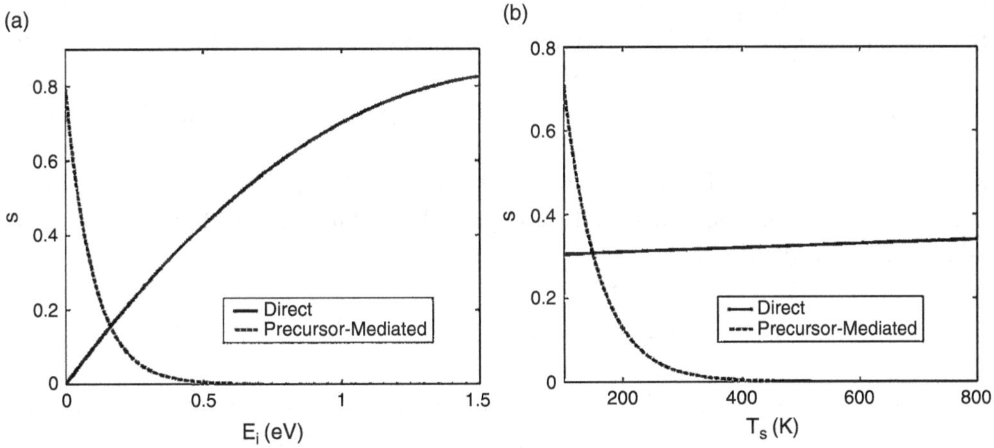

Figure 3.11. Typical experimental behaviors for dissociative adsorption probabilities S with respect to incident energy E_i in (a) and with respect to T_s in (b) for limiting dissociation behaviors. The solid lines are for direct (weakly activated) dissociative adsorption and the dashed lines are for a precursor-mediated dissociation.

T_n and this makes possible detailed dynamical studies of highly activated dissociations. Since the vibrational degree of freedom cools little in the supersonic expansion, the vibrational temperature $T_v \approx T_n$. On the other hand, rotations are strongly cooled in such expansions so that the rotational temperature $T_J \approx 0$ (except for H_2 which experiences modest rotational cooling). Thus, typical molecular beam experiments of dissociative adsorption measure $S(E_i, \theta_i, T_v, T_J \approx 0, T_s)$ or $S(E_i, \theta_i, T_v, T_J \approx 0.8T_n, T_s)$ for H_2. By varying the mole fraction of seed gas in the carrier and the mass of the carrier gas, it is possible to vary the translational energy of the beam at a fixed T_n and hence at constant T_v. Experiments of this type separate the increase in activated adsorption due to translational excitation from that of vibrational excitation. Neglecting rotation, the observed sticking can be expressed as

$$S(E_i, \theta_i, T_n, T_s) = \sum_v P_v(T_n)\, S(E_i, \theta_i, v, T_s), \tag{3.2}$$

where $P_v(T_n)$ is the Boltzmann population of vibrational state v in the beam at nozzle temperature T_n. In analyzing activated direct dissociative adsorption, it is often assumed that

$$S(E_i, \theta_i, v, T_s) = \frac{A(v)}{2}\left[1 + erf\left(\frac{E_e - E_0(v)}{W(v, T_s)}\right)\right]. \tag{3.3}$$

$A(v)$ is the high-energy sticking limit of the S shaped curve, $E_0(v)$ is the center point of the S curve and $W(v, T_s)$ is the width of the S shaped curve. E_e is the translational energy which is effective in promoting dissociation and is again usually characterized as $E_e = E_i \cos^{n_e}\theta_i$. Although the functional form for eq. (3.3) was originally proposed purely as an empirical ansatz to fit experimental data [33], a simple model shows that this form results when there is a Gaussian distribution of barriers [112]. This identification allows better understanding of the parameters used to fit experiments, and how to compare these with theory. For example, the adiabatic barrier is given as

$$V^* \approx E_0(v=0, T_s=0) - W(v=0, T_s=0) - \delta E_0^q(v=0), \tag{3.4}$$

where $\delta E_0^q(v=0)$ is the shift of $E_0(v=0)$ due to lattice coupling. Unfortunately $\delta E_0^q(v=0)$ can only be estimated from models and this introduces some uncertainty in obtaining V^* directly from molecular beam experiments. Typical shifts of $\delta E_0^q(v=0)$ are estimated as $\sim$0.2 eV from dynamic recoil for $\mu = 0.2$ and a static $V^* \sim 1.0$ eV. However, as discussed earlier, the only experimental evidence for the presence of lattice coupling in direct dissociation is a

T_s dependence of S. This seems well-described as a linear broadening of the S with T_s [113,114]

$$W(v, T_s) = W(v, T_s = 0) + a_v T_s, \tag{3.5}$$

where a_v is a small constant.

One issue that is particularly interesting for activated dissociation is the importance of translational vs. vibrational activation since this relates to the topology of the barrier location on the PES (see Section 2.3.1). In analyzing experiments, it has been traditional to define the vibrational efficacy η_v as in eq. (2.5). This analysis, however, assumes that $A(v)$ is the same for all v and this may not be universally true. In this case, describing vibrational efficacy is more complicated. Very recently, experiments for CH_4 dissociation on transition metals even combine supersonic nozzle molecular beams with laser state preparation techniques to probe the reactivity of specifically prepared vibration rotation states [115–118] (see Section 4.3.1.3).

3.2.3. Desorption

Since desorption is the time-reversed process to adsorption, experiments measuring the desorbing flux D_f from a surface should give the same information as experiments measuring the adsorption α or S, provided that detailed balance is appropriate in the comparison of the two experiments (see Section 2.4).

All readers are familiar with conventional temperature programmed desorption (TPD) that gives the total integrated desorption flux as a function of surface temperature, $D_f(T_s)$. Analysis of the TPD gives reasonable estimates for adsorption energies W for non-activated systems or desorption barriers for activated systems [119]. The TPD can also be angularly resolved [120], $D_f(\theta_f, T_s)$, and even the translational energy E_f measured as well, $D_f(E_f, \theta_f, T_s)$, both in TPD [121] and in associative desorption [122,123]. In fact, it was the measurement of peaking in $D_f(\theta_f, T_s)$ and the measurement of $E_f >> 2k_B T_s$ in associative desorption that was the first experimental evidence for the potential barrier suggested by Lennard-Jones in 1932 to account for the small dissociation probabilities in activated adsorption [15].

For molecular desorption, laser spectroscopic studies of the desorbing molecule can give full internal state distributions, $D_f(E_f, \theta_f, v, J, f(|M|), T_s)$, where $f(|M|)$ is some distribution function describing the rotational orientation/alignment relative to the surface normal. For thermal desorption in non-activated systems, most atoms/molecules have only modest (but important) deviations from a thermal distribution at T_s. However, in associative desorption of systems with a barrier, the internal state distributions reveal intimate details of the dynamics. Associative desorption results from the slow thermal creation of a transition state, with a final thermal fluctuation causing desorption. Partitioning of the energy stored in V^* into the various molecular degrees of freedom probes the same region of the PES as that for activated adsorption, and in fact $D_f(E, \theta_f, v, J, f|M|, T_s)$ is (in principle) related to $S(E, \theta_f, v, J, f|M|, T_s)$ via detailed balance. In fact, it is just these laser studies of associative desorption that have provided the most detailed information on direct activated adsorption. The systems studied in most detail via associative desorption are H_2/Cu(111), H_2/Pd(100) and N_2/Ru(0001), and they will all be discussed in Section 4.3.1. Several experimental techniques have been developed to induce associative desorption for these studies; gas permeation through a single crystal to the surface at high T_s [33,122,124], using a pulsed molecular beam to drive the atomic coverage beyond equilibrium at a given T_s [125] and using a laser T-jump to rapidly increase T_s and induce desorption of a previously covered surface [38,126].

3.2.4. Scattering

Atomic/molecular scattering is the fraction of incident particles that do not trap or stick irreversibly on impact with the surface. However, in adiabatic dynamic theory, the scattering results from the same PES as that which causes bond making/breaking, so that a study of those that 'got away' also gives complimentary information about the bond making/ breaking dynamics.

Modern molecular beam techniques of the kind represented by the apparatus of Figure 3.10 allow measurement of detailed scattering probabilities $P(E_i, \theta_i, E_f, \theta_f, T_s)$, where the subscript i refers to the incident beam and the subscript f to the scattered beam. Depending upon the initial and final scattering conditions, the scattering can often be resolved into two components; direct inelastic scattering and trapping-desorption [127]. The former is from roughly single-bounce scattering from the surface. The second occurs when an atom/molecule first traps into an adsorption well and thermalizes, but then thermally desorbs because the T_s is greater than the thermal desorption temperature. These two fractions can generally be separated since trapping-desorption 'scatters' with a distribution thermalized with the surface, i.e., with $E_f \sim 2k_BT_s$ and $\sim\cos\theta_f$ angular distribution. The direct scattering distribution retains memory of initial conditions, i.e., scatters with E_f and θ_f reflecting E_i and θ_i. Depending upon the system and upon experimental conditions, these studies can also be used to measure how the trapping coefficient α varies with incident atom/molecule conditions $\alpha(E_i, \theta_i, T_s)$ or to measure the desorption $D_f(E_f, \theta_f, T_s)$. In some cases, even the residence time on the surface τ_R can be measured. Such studies will be discussed in more detail in Sections 4.1.1 and 4.2.2. There are also of course examples where this clean separation is not possible, e.g., quasi-trapping, dynamic trapping, etc. and these will be discussed later.

Combining molecular beam techniques with laser state-resolved detection techniques has allowed state-resolved scattering measurements.

$P(E_i, \theta_i, v_i = 0, J_i \approx 0, E_f, \theta_f, v_f, J_f, f|M_f|, T_s)$ for several systems. This internal state-resolved detection in scattering is essential to understand how molecules make/break bonds at surfaces because of the strong scrambling of molecular modes in impact with the surface. Some of these experiments are discussed in Section 4.2.

3.2.5. *Initial state preparation*

Gas-surface dynamics experiments using initial state preparation techniques are still relatively uncommon. Molecules with permanent dipole moments can be oriented in hexapole electric fields. For example, NO from a supersonic nozzle can be fully quantum state selected in such fields and this allows studies of the dependence of S or scattering P on molecular orientation to the surface, i.e., N end down or O end down [128]. Some of these experiments are described in Section 4.2.

Direct vibrational excitation of CH_4 by both cw and pulsed IR lasers have recently allowed very interesting studies of the dependence of different initial vibrational states v_i on dissociative sticking on transition metal surfaces, $S(E_i, \theta_i, v_i, J, T_s)$ (see Section 4.3.1.3). In this case, the adsorption is highly activated so that the excited vibrational state preferentially reacts at the surface and the effect of the (v, J) excitation is observable over the dominant unexcited molecules. These studies have probed interesting issues relating to mode selectivity in vibrational activation and whether statistical theories are appropriate to describe the dissociation of polyatomic molecules at surfaces. These will be discussed in more detail in Section 4.3.1.3. Raman pumping of H_2 to its vibrationally excited state has made possible true 'state-to-state' scattering studies of H_2 from metal surfaces [105], $P(E_i, \theta_i, v_i = 1, J_i, E_f, \langle\theta_f\rangle, v_f, J_f, T_s)$. These will be discussed in more detail in Sections 4.3.1.1. and 4.3.1.2. Finally, stimulated emission pumping of NO to very high vibrational states, e.g., $v = 15$, allows a variety of new interesting dynamics to be probed, especially strong vibrational de-excitation in scattering which is suggestive of non-adiabatic dynamics [65] (Section 2.6.1).

3.2.6. *Photochemistry/femtochemistry*

The fundamental measurement in photochemistry/femtochemistry is the measurement of the yield Y per absorbed photon for the photochemical reaction (or equivalently cross section σ_ω), often as a function of optical wavelength $\hbar\omega$, absorbed optical fluence F_ω, angle of incidence of the light to the surface θ_i and polarization of the light relative to the surface normal P_ω, i.e., as $Y(\hbar\omega, F_\omega, \theta_i, P_\omega)$. A good indicator for hot electron (or more generally hot carrier) induced photochemistry is when the variation of $Y(\hbar\omega, \theta_i, P_\omega)$ parallels that for absorption of the light in the metal substrate. Direct adsorbate photochemistry behaves quite differently when the transition dipole is perpendicular to the surface [129]. There is, however, some ambiguity in the mechanism when the transition dipole is in the plane of the surface [130]. When pulsed lasers are used for the irradiation, state-resolved detection of the products is also possible in the manner described earlier and yields in principle $Y(\hbar\omega, F_\omega, \theta_i, P_\omega, E_f, \theta_f, v, J)$.

Depending upon F_ω (and $\hbar\omega$), two generic behaviors are generally observed corresponding to single vs. multiple excitations driving the photochemistry. The single excitation or DIET regime is generally found with excitation by cw sources and ns pulsed lasers. It is characterized by very small Y, a linear dependence of magnitude of Y on F_ω, a dependence of $Y(E_f, v, J)$, both the magnitude and final state distributions on $\hbar\omega$, and an independence of $Y(E_f, v, J)$ state distributionon F_ω. The high-fluence regime is achieved by excitation with fs lasers. In this DIMET or friction regime, Y is generally orders of magnitude larger than in the DIET regime and depends nonlinearly on $F_\omega(\sim F^n{}_\omega$ with $n \approx 3-8)$. The magnitude of Y is generally independent of $\hbar\omega$ and $Y(E_f, v, J)$ state distributions that depend on F_ω but are generally independent of $\hbar\omega$. This DIMET regime also gives different branching-rations for competing reaction channels than the DIET regime.

A key new aspect to femtochemistry is that the fs lasers can be used in pump-probe real time measurements of chemistry occurring on the surface. One key technique is to measure the two-pulse correlation in the yield of the femtochemistry (2PC). Because Y depends nonlinearly on F_ω, a measurement of Y as a function of the delay between two laser pulses (t_d) irradiating the surface essentially measures the temporal correlation of the excitation causing femtochemistry. When chemistry is due to hot electrons $Y(t_d)$ contains a significant spike of ~1 ps. On the other hand, when the chemistry is phonon-induced $Y(t_d)$ decays over some 20–30 ps. An example is presented in Section 4.8.1.3. In addition, pump-probe measurements of second harmonic generation or sum frequency generation (SFG) can be used to measure chemical process occurring on the surface in real time. For example, the time for desorption of CO from Cu(111) is measured by time resolved second harmonic generation [131] and the time for diffusion of CO from steps on Pt(533) is measured by time resolved SFG [132].

3.2.7. Single molecule chemistry (STM)

It is beyond the scope of this chapter to discuss techniques involved in STM as this important field deserves a chapter of its own. The reader should consult any of the many excellent texts devoted to this topic.

4. Processes

Before discussing the examples, the notation used in Sections 2 and 3 is briefly summarized. Principal observables are the rate constant k, trapping coefficient α, dissociative adsorption probability S, scattering probability $P(i \rightarrow f)$, desorption flux D_f, rotational alignment $A_0^{(2)}$ and yield of (hot electron) reaction Y. The only controllable variable in the k measurement is the isothermal temperature T (or separately gas and surface temperature T_g and T_s). For α and S measurements, controllable variables are surface temperature T_s, incident energy E_i, angle of incidence θ_i (which together define the normal component of energy E_n), vibrational and rotational quantum numbers v, J (or alternatively vibrational and rotational temperatures T_v, T_J) and initial orientation or alignment $f|M|$. Identical variables are in principle measurable in D_f experiments, with simply the i subscript changed to an f subscript. In scattering experiments $P(i \rightarrow f)$, both i and f variables can in principle be controlled or measured. Hot electron Y can be studied as a function of optical frequency $\hbar\omega$, laser fluence F_ω, laser polarization P_ω, laser angle of incidence θ_i, final (desorption) energy E_f, desorption angle θ_f and final quantum states v,J, $f|M|$.

4.1. Atomic adsorption/desorption/scattering

4.1.1. Ar/Pt(111)

Ar/Pt(111) is a prototypical physisorption interaction, with $W \approx 80\,\text{meV}$. The PES depends only weakly on X, Y (i.e., is weakly corrugated). However, there is increasing corrugation in the PES at small Z and this is probed at high E_i. Because of the simplicity of the PES, rare gas–metal systems such as Ar/Pt(111) have been used to study fundamental aspects of the dynamics of energy transfer to the lattice (see Section 2.5.1). Comparison of theory with experiment, however, is based only on empirical PES since current exchange-correlation functionals in common use in DFT cannot describe van der Waals interactions [133]. These empirical PES are based on sums of pair-wise additive Ar–Pt potentials [134], with often non-central Z dependent terms added to correct for the overestimation of corrugation in the purely pair-wise additive PES [135,136]. The parameters in the empirical PES are obtained by fitting experimental data and DFT calculations of Pauli repulsion [135].

At low E_i, Ar/Pt(111) scattering measurements $P(E_i, \theta_i, E_f, \theta_f, T_s)$ are resolved into two distinct channels; trapping-desorption and direct inelastic scattering [134,135,137–139]. Some scattering conditions (low E_i, large θ_i, low T_s, $E_f \approx 2k_BT_s$, $\theta_f \approx 0$) favor detection of trapping-desorption and others (high E_i, small θ_i, high T_s, $E_f >> 2k_BT_s$, $\theta_f \approx \theta_i$) favor detection of direct inelastic scattering. Under some intermediate conditions, even both processes can be observed simultaneously in the TOF, as in the similar scattering of Xe from Pt(111) [127]. Although there is no a priori reason that such a clean division into the two different channels should occur in the scattering, this does seem to be modestly general for rare gas-metal scattering (but see the qualification below).

Extensive early measurements of Ar/Pt(111) direct-inelastic scattering for 1.7 meV < E_i < 1.7 eV and 90 K < T_s < 900 K did much to clarify energy transfer processes to the lattice [137]. The scattering could be resolved into three independent directions; normal to the surface (*Z*), parallel to the surface and perpendicular to the scattering plane (*x*) and parallel to the surface and in the scattering plane (*y*). Both the *Z* and *y* directions independently satisfy relationships of the form given by eq. (2.11). Although both energy transfer coefficients into and out of the surface are approximately four times larger for the *Z* direction relative to the *y* direction, there is still significant energy transfer of E_y to the lattice. In addition, there is not complete conservation of average parallel momentum, with $\delta\langle v_y \rangle / \langle v_y \rangle \approx -0.04$, where v_y is the velocity in the *y* direction. This means that a cube model for energy transfer is a reasonable approximation for Ar/Pt(111) scattering at low E_i, but is by no means rigorous. A comparison of some experimental scattering at low E_i with molecular dynamics simulations based on an empirical PES is given in Figure 3.12 [135]. Average energies predicted by conservation of parallel momentum are also given in this figure and agree rather well with the measurements/ simulations at low E_i.

For hyperthermal incident energies, $E_i >> 1$ eV, the energy transfer results show a gradual transition from the 'cube' scattering regime to the binary sphere scattering regime [135]. This is because at higher E_i, the PES has larger corrugation since the turning point in the scattering is closer to the ion cores. This results in a strong breakdown in parallel momentum conservation in the scattering and is also demonstrated in Figure 3.12. At very high E_i, the principal disagreement between experiment and the molecular dynamics fit is that there is additional energy loss (with small momentum transfer) in the experiments relative to that predicted. The authors suggest that this is due to an additional energy loss channel not included in the molecular dynamics, i.e., energy loss to e-h pairs. There is direct experimental evidence (i.e., a current) that hyperthermal Xe atoms create e-h pairs when colliding with semiconductor surfaces [64].

At low E_i and low T_s, trapping is dominant as shown in Figure 3.13 [139]. The fall off in trapping with E_i occurs over an energy scale of $\sim E_c$ of eq. (2.12). The scaling of trapping with E_i and θ_i is approximately given as $E_e = E_i \cos^{1.5}\theta_i$. At higher T_s, trapping-desorption [139] (obtained by measuring the desorption flux at $\theta_f \approx 0$) suggested that *n* decreased with T_s, perhaps due to thermally induced corrugation. However, molecular dynamics simulations and further experiments showed that at high T_s, the experiments only measure quasi-trapping rather than true trapping [134]. Because the residence time of Ar on the surface is sufficiently short at the higher T_s, the trapped atoms desorb before their incident parallel momentum is fully thermalized with the surface. Therefore, the decrease in *n* with T_s is due to incomplete equilibration of parallel momentum prior to desorption rather than thermal roughening of the surface. The slow accommodation of parallel momentum

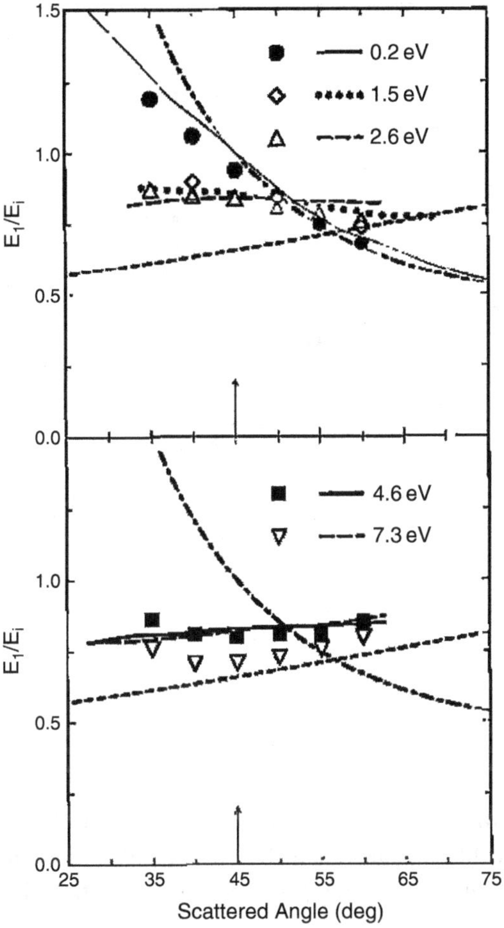

Figure 3.12. Inelastic scattering of Ar from Pt(111) at the various input energies listed in the figure and for an initial angle of incidence $\theta_i = 45°$ and $T_s = 800$ K. Results are plotted as E_f/E_i vs. the final scattered angle θ_f. Points are the experimental results and the lines marked adjacently in the label are results of molecular dynamics simulations on an empirical PES. The long dot-dashed curve is the prediction of a cube model of energy transfer, while the dashed curve is the prediction from hard sphere scattering. From Ref. [135].

for these physisorption systems means that quasi-trapped particles can explore a large spatial range on the surface, and this can make defects quite important in trapping.

At sufficiently low $T_s \approx 100$ K, Ar residence times on the surface are sufficiently long that trapping-desorption measures desorption from a fully equilibrated Ar on the surface. In this case, measurements of $D_f(E_f, \theta_f, T_s)$ showed that $\langle E_f \rangle < 2k_B T_s$ and $D_f(\theta_f)$ is broader than $\cos\theta_f$, i.e., somewhat different than the usual equilibrium assumptions [138]. However, these results are fully consistent with detailed balance and the E_i and θ_i dependence observed for $\alpha(E_i, \theta_i, T_s)$ [32,138].

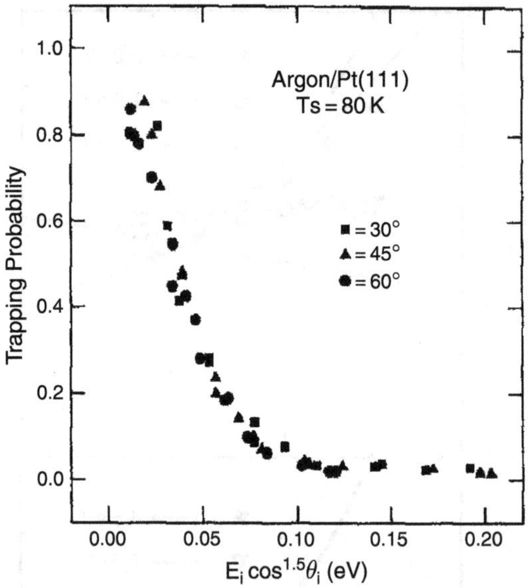

Figure 3.13. Trapping probability α for Ar on Pt(111) at $T_s = 80$ K as a function of E_i for various angles of incidence θ_i as labeled in the figure. Results are plotted vs. $E_e = E_i \cos^{1.5} \theta_i$. From Ref. [139].

4.1.2. H/Cu(111)

The strong chemical interaction of H with Cu(111) illustrates a wide variety of new chemical phenomena that are absent in the weak physisorption systems. The small mass and size of H even make some dynamic processes possible that are not present with other atom chemisorptions.

The lowest energy atomic site for H chemisorbed on Cu(111) is the fcc hollow site with $W = 2.3$ eV. Smaller and different W exists at the other high symmetry sites. Thus, the PES is very laterally corrugated, both in energy and geometry. In addition, there are metastable subsurface sites inside the surface plane, e.g., one site exists below the fcc hollow with $W \sim 0.9$ eV above that of the most stable surface adsorption site [140]. This is made metastable by a barrier of ~0.4 eV relative to the bottom of the subsurface well. Bulk octahedral absorption sites have essentially the same stability as the subsurface sites, with presumably similar barriers to migration into the bulk. Thus, populating the subsurface site represents the initial step in bulk absorption of H.

As anticipated with such a large W, trapping α is large, $0.98 < \alpha < 0.9$ by atom beam reflection techniques [141]. The α is estimated as only ~0.2 by measuring the build up of chemisorbed H on the surface [142]. The latter measurement, however, requires an uncertain calibration of the atom flux on the surface. Therefore, this author believes the beam reflection technique is more accurate. Once some H atoms are on the surface, additional incoming H atoms abstract adsorbed atoms via Eley-Rideal/hot atom process. This is covered in Section 4.7.1.

Dosing a Cu(111) surface with H atoms (or dissociative adsorption of high energy H_2 molecules) preferentially populates the fcc surface site. But there is now extensive experimental evidence that dosing with atoms also creates subsurface/ bulk H [141–144], while dissociative adsorption of high energy H_2 does not. Analysis of the leading edge of the zeroth order TPD peak of the absorbed species yields an activation energy of ~0.35 eV for conversion of subsurface to surface species [142], in very good agreement with the DFT calculations of the adiabatic barrier [140]. The probability of forming the subsurface state by the incoming H is $S_{ab} < 5\%$ [141,142]. It is not certain whether the subsurface site is only populated after the surface site is heavily H covered ($\Theta_H > 0.3$) or not. The difficulty in an experimental determination of S_{ab} and whether it starts in the limit $\Theta_H \to 0$ is that S_{ab} is more than an order of magnitude smaller than the probability of abstraction via Eley-Rideal/hot atom processes (Section 4.7.1). H atom absorption is observed on many other metals with atomic dosing as well; e.g., Ni(111) [145], Ni(100) [146], Pt(111) [147], etc. so that it appears to be a general phenomenon.

There are two classical molecular dynamics studies of H adsorption/absorption on Cu(111). One is based on an effective medium fit to DFT calculations which gives a pair-functional additive potential for traditional Langevin dynamics [140]. The other is based on a pair-wise additive modified Morse PES fit to the DFT calculations and with an extensive expansion in reciprocal lattice vectors to account for coupling to the lattice [148]. The two simulations give very similar descriptions of the hot H atoms produced by atom dosing. Because of the very small $\mu = m/M_s$, energy transfer to the lattice is very slow, requiring many impacts with the hard repulsive wall of the H/Cu PES to thermalize. Also, the strong corrugation of the PES causes much of the normal energy caused by acceleration in the well to be converted into energy parallel to the surface. These two facts combine to produce 'hot' H atoms which sample 10–100 Å before thermalizing with the surface, and this is a very important aspect of the reaction of H with H/Cu(111) discussed in Section 4.7.1. These simulations give a theoretical basis to the phenomenology proposed years ago as 'hot precursors' [149]. Both simulations predict a large α at experimentally accessible $E_i < 0.2\,\mathrm{eV}$, i.e., 0.85 [148] or 0.95 [140]. Only one of the simulations calculated absorption [140]. They find $S_{ab} \approx 0.05$ for low T_s and $E_i < 0.2\,\mathrm{eV}$, in approximate agreement with experiment.

Both simulations stress that the relaxation rate for the adsorption energy into the lattice is slow, ~1–4 ps, and that energy relaxation into e-h pairs, omitted in these molecular dynamics simulations, is likely to be of the same order of magnitude or perhaps even larger. The non-adiabatic relaxation rate is estimated to also be ~1 ps from the vibrational damping rate of the parallel mode for H adsorbed on Cu(111) [150]. The excitation of e-h pairs accompanying H adsorption on Cu has

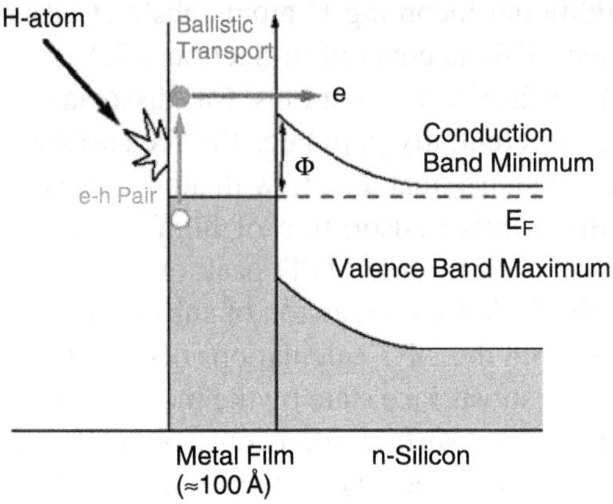

Figure 3.14. Schematic representation of experiment detecting hot electrons created by atomic/ molecular adsorption on a thin metal film. Φ is the Schottky barrier created by the metal/ Si interface. From Ref. [86].

now been observed by measuring a chemicurrent induced by the H adsorption on a Cu thin film [151,152]. This unique experiment is shown schematically in Figure 3.14. A thin Cu film on a Si substrate creates a Schottky barrier Φ. Electrons from the e-h pairs created by adsorption on the Cu surface travel ballistically to the Cu/Si interface. When the electron energies $E_e > \Phi$, they induce a current I in the Si which can be detected. Strong evidence for the non-adiabatic origin of I was the observation of an isotope effect, with I for H adsorption >> I for D adsorption. The magnitude of the current observed and the isotope effect are in good agreement with a theoretical model for the non-adiabatic process that uses electronic frictions from DFT, eq. (2.15) and the forced oscillator model (on e-h pairs) to calculate the number of electrons created with $E_e > \Phi$ [72,80].

Certainly H adsorption/absorption on other metals should be similar to that on Cu(111). The interaction of other reactive atoms (O, N, etc.) with metal surfaces should have PES topologies similar to that of H/Cu(111), i.e., deep strongly corrugated adsorption wells. Thus, although $\alpha \approx 1$, there will be strong mixing of E_n and $E_{||}$ in adsorption. However, in these cases significantly faster energy transfer occurs to the lattice because of the larger mass. Therefore, the range of hot atoms created by atomic adsorption or other means is very restricted, typically only a few lattice constants [153,154].

4.2. Molecular adsorption/desorption/scattering

4.2.1. NO/Ag(111)

Direct NO scattering from Ag(111) is perhaps the most extensively experimentally studied molecule-surface scattering system. This principally reflects the ease and sensitivity of state-resolved laser detection techniques to study NO, combined with the relatively weak interaction of NO with Ag(111). On the other hand, the emphasis on state-resolved techniques has led to only modest emphasis on trapping and desorption. Nevertheless, these studies do illustrate the full complexity inherent in molecule-surface dynamics and how this may affect trapping/desorption. Unfortunately, there is no DFT PES for a first principles theoretical comparison with experiment. However, many (qualitatively different) model PES have been used to interpret experiments.

NO is only weakly chemisorbed to the Ag(111) surface in an upright geometry with the N end down and with a well depth $W \approx 0.2\,\text{eV}$ estimated from thermal desorption measurements [155,156]. Several different nearly energetically equivalent binding sites occur on the surface. A DFT calculation [157] suggests a much higher well depth $W \approx 0.7\,\text{eV}$, but this is inconsistent with experiment. At all incident E_i (and high T_s), NO adsorption is non-dissociative as anticipated for a weakly interacting system. However, the chemistry of NO adsorbed on the surface at low T_s is quite complicated since dimers form that readily decompose into N_2O+O [155,156]. At the T_s of all scattering experiments, the NO coverage is extremely low so that no dimers form. Thus scattering/trapping can be viewed as that occurring for an isolated NO/Ag(111) interaction.

Early measurements of $P(E_i, \theta_i, \theta_f, T_s)$ and $P(E_i, \theta_i, E_f, \theta_f, T_s)$ gave rather uninformative broad lobular scattering with significant energy loss [158,159]. Comparison with equivalent experiments for rare gas scattering on Ag(111) suggests that there is a strong coupling of translation and rotation in the scattering, with loss of E_n into rotation (J_f). Definitive confirmation of this picture occurs via state-resolved direct scattering experiments [160,161] measuring $P(E_i, \theta_i, J_i \approx 0, \theta_f, J_f, T_s)$. The results show strong rotational excitation which scales approximately as $\langle E_J \rangle = a\,(E_n + \varepsilon_W) + bk_BT_s$, with $\varepsilon_W \approx W$ [162]. The conversion of E_n into E_J is most simply thought of as due to an anisotropic interaction potential $V(Z, \vartheta)$, where ϑ is the orientation of the NO axis relative to the surface normal. Figure 3.15 shows the rotational excitation at various E_n. When plotted in this manner, thermal rotational distributions are straight lines. Therefore, while the low J_f region is quasi-thermal, the high J_f region is decidedly non-Boltzmann and has been interpreted as arising from a rotational rainbow [160,163]. Rainbow scattering is simply a consequence of the phase space available for scattering into a particular range of states. For any $V(Z, \vartheta)$, the extent of rotational excitation J_f must depend on ϑ, with $J_f = 0$ at $\vartheta = 0, \pi/2$. Therefore, there must be some ϑ in between such that $J_f(\vartheta)$ is a maximum and the scattering intensity which is $\propto \left[dJ_f(\vartheta)/d\vartheta\right]^{-1}$ has a classical divergence. Of course, the divergence is smoothed when considering the other dynamical coordinates so that only the broad rainbow peak remains.

Initial theoretical efforts assumed that the anisotropy in the PES occurred principally in the hard repulsive wall of the PES and fit adjustable $V(Z, \vartheta)$ to the

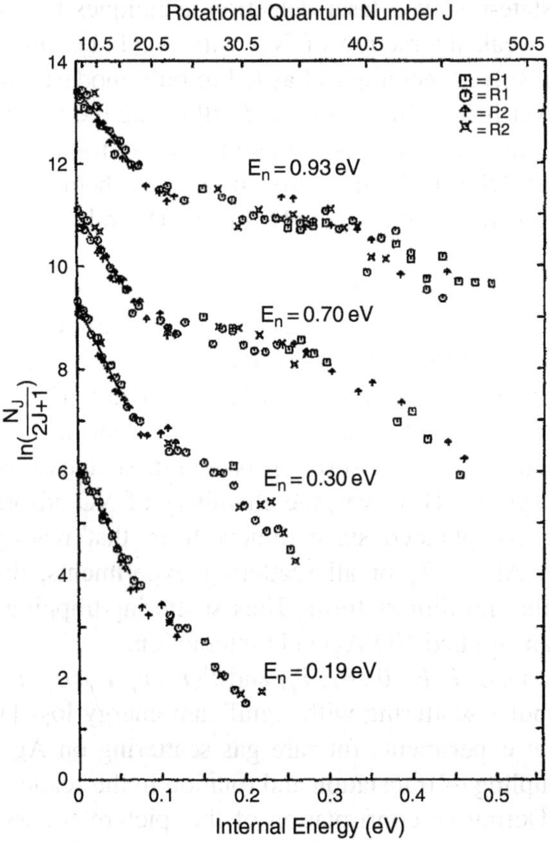

Figure 3.15. Rotational state distributions of NO produced in direct scattering from Ag(111) at $T_s \approx 600$ K as a function of incident normal energy E_n. Rotational populations N_J are plotted in such a way that a Boltzmann distribution characterized by a temperature T_J is a straight line. The different symbols correspond to rotational populations derived from the different rotational transitions as listed. From Ref. [160].

experiments [163,164]. Voges and Schinke [164] even suggested a strongly asymmetric PES for the two ends of the molecule so that a double rainbow could fit both the low and high J regions. On the other hand, Tully and collaborators assumed that the dominant anisotropy is in the attractive part of the PES. They were also able to fit the experiments with an empirical PES using multi-dimensional stochastic trajectories that included coupling to the lattice [165]. Thus, in the absence of a DFT PES, all one can say definitively from the rotational excitation is that there is anisotropy in the PES, which is hardly surprising.

Kleyn and collaborators have investigated the oriented scattering of NO from Ag(111) by passing a rotationally cold supersonic NO molecular beam through a hexapole field prior to scattering from the surface [166–169]. They observe that scattering with the N end preferentially towards the surface produces only the low J_f quasi-thermal part of the final rotational distribution, while scattering with the O end preferentially towards the surface produces both the high J_f and low J_f regions (see Figure 3.15). This partially justifies a two rainbow interpretation of the scattering based on anisotropy of the two ends of the NO molecule in the PES, although the nature of the PES that gives rise to this anisotropy is still open. One difficulty with a simple interpretation, however, is that the O end still produces a significant low J_f peak in addition to the high J_f peak.

In addition to the rotational excitation, strong rotational alignment is measured for the higher rotational states as shown in Figure 3.16 [170,171]. In this plot, the alignment is given in an equivalent manner to $A_0^{(2)}(J)$ as the ratio b_2/b_0, where b_i is the ith term in a Legendre expansion of the classical distribution of J vectors relative to the surface normal, $n(\vartheta_J) = \sum_i b_i P_i(\cos\vartheta_J)$. Perfect 'helicoptering' is $b_2/b_0 = +5$ and perfect 'cartwheeling' is $b_2/b_0 = -2.5$. As seen in Figure 3.16, there is a strong preference for 'cartwheeling', as anticipated since the forces are predominatly normal to the surface during the conversion of $E_n \rightarrow E_J$. The falloff in alignment at low J_f is most likely due to the finite initial J_i distribution in the beam [171]. The disappearance of alignment at the highest J_f is not well understood. All PES only of the form $V(Z, \vartheta)$ predict perfect

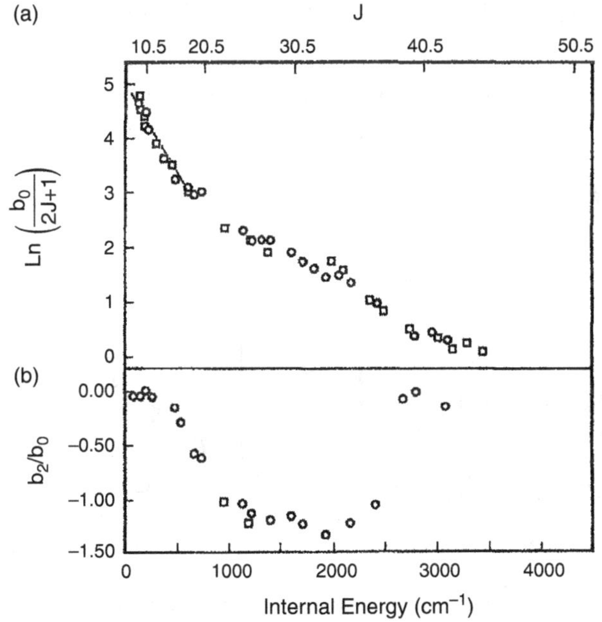

Figure 3.16. Population distribution b_0 (a) and rotational alignment b_2/b_0 (b) plotted vs. the internal energy. As in Figure 3.15, Boltzmann population distributions are linear in such plots. The different symbols correspond to different spin-orbit states of NO. From Ref. [170].

'cartwheeling', which is not observed. The multi-dimensional stochastic trajectories simulations [165] also overestimate the observed alignment.

NO is a free radical in a $^2\Pi$ electronic ground state and the laser spectroscopic detection also measures non-statistical partitioning into the electronic fine structure states for high J ($^2\Pi_{1/2}$ and $^2\Pi_{3/2}$) as well as into the two Λ-doublets [172]. It has been stressed by Alexander and co-workers [173,174] that because NO is a $^2\Pi$ state, its interaction with the Ag(111) surface gives two low-lying PES, V_+ and V_-, depending upon whether the singly occupied π orbital is pointing parallel to the surface or perpendicular to it, respectively. The two PES correlate differently to the asymptotic NO fine structure and Λ-doublet states and non-adiabatic transitions between the two PES occur during the scattering. Thus, the final partitioning is sensitive to details of $V_+ - V_-$ as well as of the average NO–Ag(111) PES.

Vibrational excitation of NO in direct scattering $P(E_i, \theta_i, v=0, v_f=1, T_s)$ is measured by summing measurements over all rotational states for $v = 0, 1$ and averaging over a wide range of angles θ_f [175,176]. Significant (~5%) vibrational excitation is observed at high E_n and high T_s as shown in Figure 3.17. The dependence on T_s is approximately Arrhenius, with an effective activation energy of roughly the NO vibrational frequency. One anticipates little vibrational excitation in direct scattering of a diatomic on a non-reactive PES [39]. Direct mechanical translation to intramolecular vibration energy transfer requires nearly head-on collisions with the surface and these represent a small fraction of all collisions with the surface. Phonon

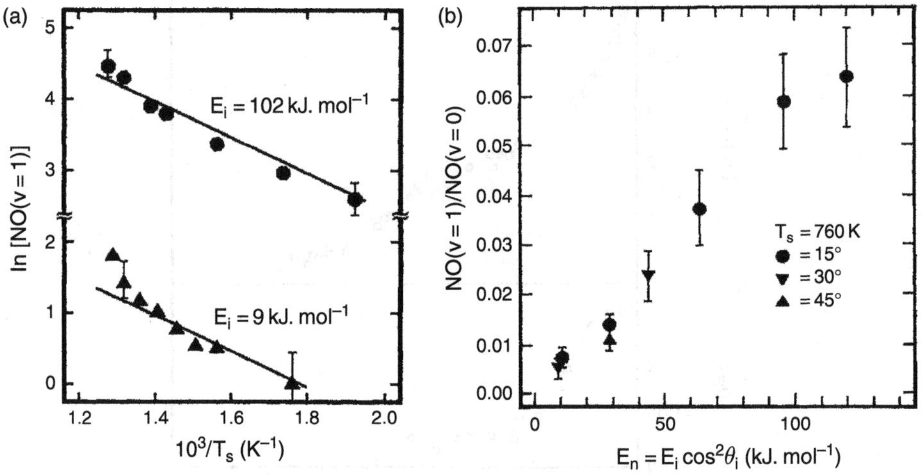

Figure 3.17. Vibrational excitation of NO ($v = 1$)/$v = 0$) in direct inelastic scattering from Ag(111). (a) as a function of the surface temperature T_s at two incident energies. The straight lines are Arrhenius fits with an effective activation energy of the NO vibrational frequency. The two sets of data at the different E_i are arbitrarily scaled relative to each other for clarity. (b) as a function of incident normal energy E_n at $T_s = 760$ K. From Ref. [175].

excitation of intramolecular vibration is weak because of the large frequency mismatch to the phonons. Therefore the ~5% vibrational excitation is quite surprising and the authors propose that vibrational excitation arises from resonant coupling to thermally excited e-h pairs [175,176]. This is an extension to scattering/excitation of the well accepted mechanism of intramolecular vibrational damping via e-h pairs [74,87]. A simple 1D non-adiabatic model based on electron transfer from the Ag to the NO $2\pi^*$ affinity level does fit the experiments [177]. However, purely adiabatic models with PES that include N−O bond stretching as a result of the electron transfer (and coupling to the lattice) can also fit the experiments as well because of enhanced translation to vibrational energy transfer on the reactive PES [178,179]. Thus, without a DFT PES, it is impossible to say whether the vibrational excitation observed is evidence for a breakdown in the BOA or not. One prediction of non-adiabatic coupling, however, is that vibrational de-excitation should increase significantly with E_n [78,178], and this has been observed for the related NO($v = 2$) scattering from Au(111) [180]. In addition, very strong multi-quantum vibrational de-excitation of NO($v = 15$) is observed in direct scattering from Au(111) [65]. It is hard to see how this massive vibrational de-excitation could have any other origin than a breakdown of the BOA. Non-adiabatic effects should be particularly large in this case because of the high nuclear velocities in $v = 15$ and because of the large variation in the electronic structure of the NO/Au PES sampled by $v = 15$. For example, the electron affinity of NO varies dramatically over the bond lengths sampled by $v = 15$ and this has prompted a qualitative description of the NO($v = 15$)/Au scattering experiment in terms of vibrationally promoted electron transfer [65].

Combining ion TOF techniques with REMPI gives measurements of $P(E_i, \theta_i, J_i \approx 0, E_f, \theta_f, J_f, T_s)$. This makes it possible to separate rotational excitation from energy loss to the lattice and how it varies with initial conditions [181]. Perhaps the most dramatic result is the anti-correlation between translational energy loss to the lattice and to rotational excitation as shown in Figure 3.18 [49]. This essentially reflects the competition for E_n in the scattering, and is rationalized as a change in the effective mass of the NO in classical energy transfer with angle ϑ, i.e., as $m(\vartheta)$ in the definition of μ in eq. (2.9). While the translational energy transfer to rotation scales well with $E_n = \cos^2\theta_i$, energy loss to the lattice scales with $E_i\cos\theta_i$, as shown in Figure 3.19. The magnitude of the energy transfer to the lattice is $> 50\,\%$ at high E_i and is therefore >> than energy transfer to rotation. The scaling in Figure 3.19 shows the breakdown of the conservation of parallel momentum in the lattice excitation. Another indication is that the energy transfer to the lattice is nearly independent of θ_f, similar to Figure 3.12 at high E_i. Stochastic trajectory simulations on an empirical PES are in semi-quantitative agreement with many of these new results, but only when the original PES [165] was varied to fit the new data [49].

With the exception of vibrational excitation, most of the behavior observed in NO/Ag(111) direct inelastic scattering has now been observed in the scattering of

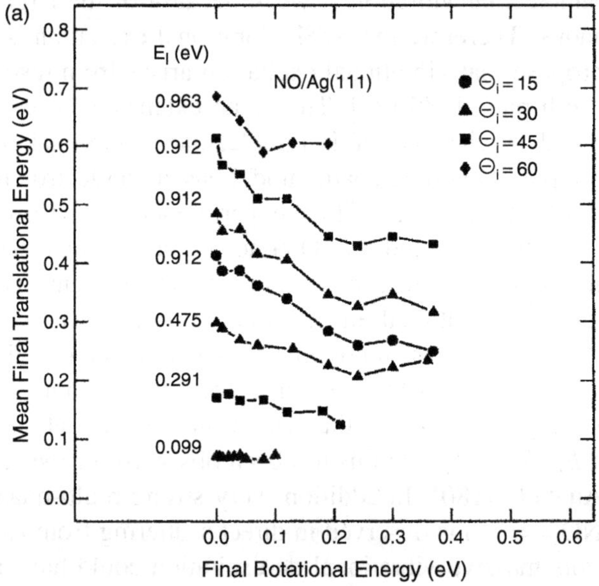

Figure 3.18. Variation of the mean translational energy E_f of NO in scattering from Ag(111) at the initial energies E_i and incident angles θ_i as noted on the figure. From Ref. [49].

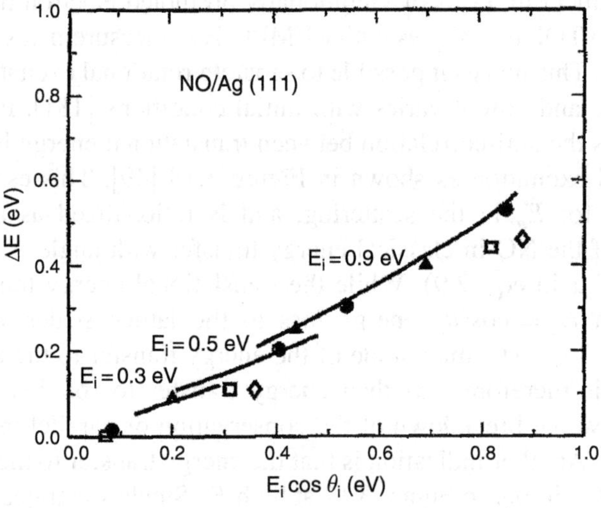

Figure 3.19. Variation of the energy transfer into the surface ΔE in scattering of NO from Ag(111) as a function of $E_e = E_i \cos\theta_i$. Solid lines and solid points are for rotationally elastic scattering $J_i = J_f = 0.5$ and the open points are for non-state-resolved scattering experiments (and therefore ΔE also contains a contribution from rotationally inelastic scattering). From Ref. [181].

other weak interaction systems, e.g., N_2/Ag(111) [182–186], N_2/Cu(110) [187], etc. Therefore, most of the findings here are probably quite general for weakly interacting systems.

Only limited studies of trapping and desorption have been performed for NO/Ag(111) [188,189]. A few measurements of $\alpha(E_i, \theta_i, T_s)$ observed in trapping-desorption suggests that trapping falls off with E_i roughly on an energy scale given by eq. (2.12) and scales either with E_i or $E_i\cos\theta_i$. The latter is the scaling observed for energy loss into the lattice. A steric effect is observed in trapping when using partially oriented beams, with more trapping for preferential O end surface collisions than N end surface collisions [189]. The authors suggest that this is due to the higher rotational excitation for O end collisions, i.e., from a contribution of $E_n \rightarrow E_J$ in the trapping. $\langle E_f \rangle < 2k_B T_s$ and an angular distribution broader than cos θ_f is observed for the desorbed NO, in agreement with detailed balance and the similar effects on Ar/Pt(111) discussed in Section 4.1.1. In addition, the rotational temperature characterizing the desorbing NO, $T_J = T_s$ for low T_s, but $T_J < T_s$ at higher T_s [190]. This is simply a consequence of the strong $E_n \leftrightarrow E_J$ coupling and detailed balance, and is discussed more fully in the next section on NO/Pt(111).

4.2.2. NO/Pt(111)

The interaction of NO with Pt single crystal surfaces has been much studied, in part to provide a fundamental background to understanding the reduction of NO in automotive exhaust catalysts (supported Pt/Rh). The adsorption of NO on Pt(111) is non-dissociative at terrace sites [191,192], at least for $E_i < 3$ eV and $T_s < 1000$ K [193], and therefore inactive for NO reduction. Dissociation readily occurs on Pt(100) [191] and on vicinal Pt(111) surfaces containing (100) steps [194]. The adsorption of NO on Pt(111) reflects modestly strong chemisorption, with the N end down at a variety of different adsorption sites (hollows, bridging, on-top), depending upon Θ_{NO} and T_s [195]. There is still controversy as to what binding sites exist under the different conditions. At low Θ_{NO} and for annealed surfaces, binding is at a single site and $W \approx 1.1$ eV [191,196]. The PES is anticipated to be extremely anisotropic and laterally corrugated because of the chemical nature of the bonding.

DFT calculations of the structure of the molecularly adsorbed NO are in reasonable agreement with experiments, but overestimate the binding energy [197,198]. A barrier of $\sim$2.1 eV to dissociation is predicted by DFT, with the NO at the transition state nearly parallel to the surface and N and O atoms in $\sim$ bridge sites [199]. This transition state geometry is similar to that of NO dissociation on other close-packed metal surfaces [200]. There is no global DFT PES so that all theoretical dynamics is based only on empirical model PES.

Modulated molecular beam studies show that while trapping-desorption is the dominant process at modest E_i and high T_s, there is also some direct or rather indirect inelastic scattering as well [192,201]. The dominance of trapping is anticipated for a system with a modestly deep chemical well. Because the binding energy of NO at steps is significantly greater than at terrace sites, these sites can dominate desorption at low Θ_{NO}, even though their density is low [196].

Measurement of $\alpha(E_i, \theta_i, T_s = 295\,K)$ are given in Figure 3.20 [193]. These results are not at all similar to predictions of the simple impulsive model, eq. (2.12), but are in qualitative agreement with classical dynamics simulations based on an empirical PES [193]. They are also qualitatively similar to α for other molecularly chemisorbed systems, e.g., CO on transition metals [50,202,203]. The gentle decrease in sticking with E_i is principally due to translation-rotation conversion from the strong anisotropy of the molecular well. This causes an increase in the number of round trips in the well due to rotational trapping, and this can enhance energy loss to the lattice for high E_i. On the other hand at low E_i, this conversion reduces energy transfer to the lattice during the initial impact with the surface (anti-correlation) and hence decreases α in this regime relative to an isotropic well. The net result is a generally monotonic fall-off in α over an appreciable E_i range. In a similar manner, the large corrugation in the PES causes facile interchange between perpendicular and parallel momentum. Because of the substantial energy scrambling of $E_{||}$, E_n and E_J in the well, α scales roughly with E_i. Rather surprisingly, oriented beam experiments show that trapping at low E_i is slightly higher with the O end towards the surface than the N end [204]. Naively, higher trapping is anticipated for the N

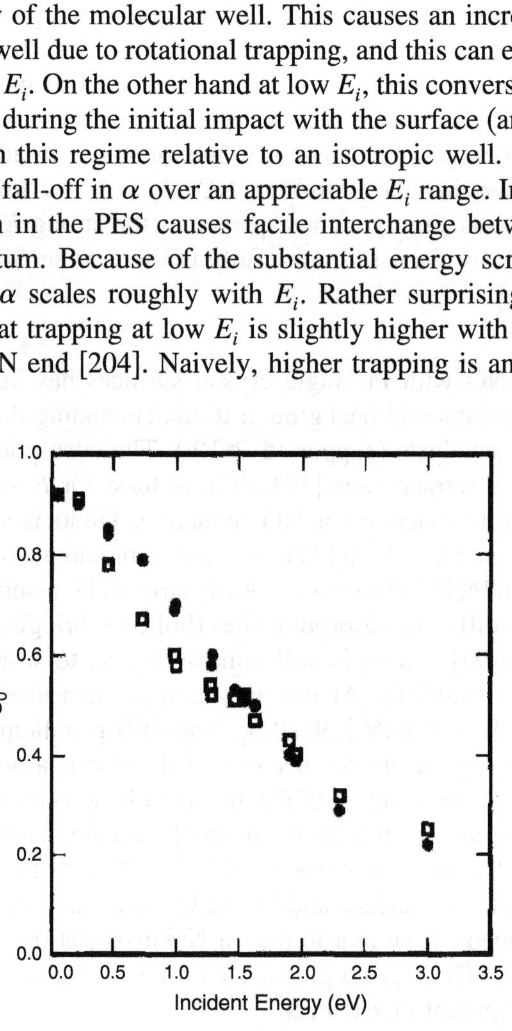

Figure 3.20. Molecular adsorption probability S_0 ($\equiv \alpha$ in the notation of this chapter) for NO on Pt(111) vs. incident translational energy E_i. Solid circles are for $\theta_i = 0°$ and open squares are for $\theta_i = 60°$. From Ref. [193].

end because of the deeper adsorption well for that end. It is suggested that the O end gives more rotational energy transfer, which aids trapping. The steric asymmetry in trapping is quite small however, only ~2.5 %, probably because of orientational steering in the adsorption.

Desorption of NO from Pt(111) is studied both in TPD and trapping-desorption. The former gives D_f at the lowest possible T_s, while trapping-desorption gives $D_f(T_s)$ at higher T_s and varying residence times τ_R. $D_f(J, T_s < 340\,\text{K})$ in TPD [205] gives NO rotationally equilibrated with the surface, i.e., $T_J \approx T_s$. However, $D_f(J, T_s)$ in trapping-desorption gives $T_J < T_s$ at higher T_s [206,207]. This dynamical effect, which seems quite general [208], is shown in Figure 3.21 and is readily understood in terms of detailed balance [165]. Large $E_n \leftrightarrow E_J$ coupling implies that $\alpha(E_n, J, T_s)$ will decrease with E_J as well as E_n. Therefore $D_f\,(E_n, J, T_s)$ must also have a rotational distribution colder than the thermal one at T_s to account for this. Because only low rotational states are populated at low T_s, the effect is minimal at low T_s. Significant quadrupolar alignment $A_0^{(2)}$ favoring helicopter motion is also observed in trapping-desorption [209]. This can also be rationalized via detailed balance since

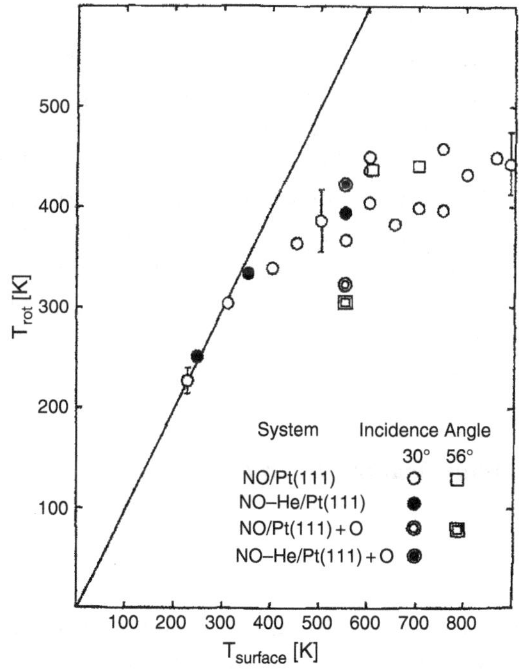

Figure 3.21. Rotational temperature T_{rot} (defined as T_J in this chapter) in trapping-desorption scattering of NO from Pt(111) and O covered Pt(111) as a function of the surface temperature T_s. Open points are for $E_i = 80$ meV and solid points are for $E_i = 220$ meV. The straight line is for $T_{\text{rot}} = T_s$. From Ref. [206].

cartwheel motion is better able to convert $E_J \rightarrow E_n$, therefore suppressing α relative to helicopter motion [210].

Molecular dynamics simulations suggest that there is little true direct scattering of the NO, but rather only indirect scattering where the NO makes several bounces off the surface before returning to the gas phase, with strong interchange between $E_{||}$, E_n and E_J in the chemisorption well [165,193,210,211]. The indirect nature of scattering is reflected in the final distributions. When compared with the scattering of other molecules and atoms from Pt(111), $P(E_i, \theta_i, E_f, \theta_f, T_s)$ for NO scattering shows very broad angular distributions and much more translational energy loss [212]. For most of the scattered NO, especially at hyperthermal E_i, rotational state distributions $P(E_i, \theta_i, J_f, \theta_f, T_s)$ are approximately Boltzmann, but with T_J increasing with E_i and greater than T_s [213]. Rotational state distributions are approximately statistical, but with the dynamical constraint of partial conservation of parallel momentum [214]. The implication is that $E_{||}$ mixes more slowly than E_n with E_J. Molecular dynamics simulations on empirical model PES give reasonable agreement with these scattering distributions [165] and even suggest the importance of 'chattering' collisions at high E_i [211]. Chattering occurs when an O end collision generates so much NO rotation that the N end also gets close to the surface before the NO has a chance to scatter away from the surface. Despite the complicated dynamics of scattering, preferential cartwheel alignment is still observed in the scattering distribution (at high J_f) similar to that observed in NO/Ag(111) rainbow scattering [215].

NO/Pt(111) is certainly the best experimentally studied example of trapping/scattering/desorption in the molecular chemisorption regime. Many other examples that have been studied in less detail show many similarities; e.g., CO trapping/scattering on Pt(111) [50], Ru(0001) [202] and Cu(100) [203], CO scattering from Ni(111) [216], etc. While details of the dynamics will undoubtedly vary from system to system depending upon the chemical details of the PES, the dynamic principles illustrated here are anticipated to be quite general to scattering/adsorption dynamics in the molecular chemisorption regime.

4.3. Direct dissociation/associative desorption

4.3.1. Activated dissociation

4.3.1.1. H_2–Cu(111)

The activated dissociation of H_2 (D_2) on Cu(111) and other single crystal Cu surfaces has played a special role in the development of reactive gas-surface dynamics. Early experiments and theory by Cardillo and collaborators [217–219] first demonstrated the power of molecular beam techniques to probe activated adsorption and the theoretical methodology developed by them (6D quasi-classical dynamics on a model PES) only differs from modern treatments in the use of DFT based PES.

They also first proposed the use of detailed balance to relate dissociative adsorption to associative desorption. In hindsight, both the experiments and the PES derived by them to fit the experiments were in significant quantitative error, but this in no way minimizes the major contribution of this early work to the development of reactive gas-surface dynamics.

The dissociation of H_2 on Cu(111) is nearly thermoneutral, with $\Delta H \approx 0.1$ eV. It is apparent from Figure 3.4(a) that activation is necessary for dissociation and that both translational and vibrational excitation can help surmount the barrier. Initially stimulated by theoretical predictions (based on a very crude DFT PES [220–222]), the H_2/Cu systems were re-investigated some 10–15 years after the initial experiments using seeded supersonic nozzle beams to separate translational excitation from vibrational excitation [223–225]. Figure 3.22 shows an example of this type of experiment for D_2 dissociation on Cu(111) [225]. Neglecting rotation and analyzing the results with eqs. (3.2) and (3.3) to give $S(E_i, \theta_i, v, T_s)$ gives a vibrational efficacy $\eta_v \approx 0.5$ from eq. (2.5). The dependence of S on θ_i is close to normal energy scaling, $E_e = E_i \cos^{1.8}\theta_i$. Similar results were obtained for H_2/Cu(111) [34]. There is only a modest isotope effect in the measured H_2/D_2 dissociation probabilities $S(E_i, \theta_i, T_v, T_s)$ on Cu surfaces, in apparent disagreement with the moderate

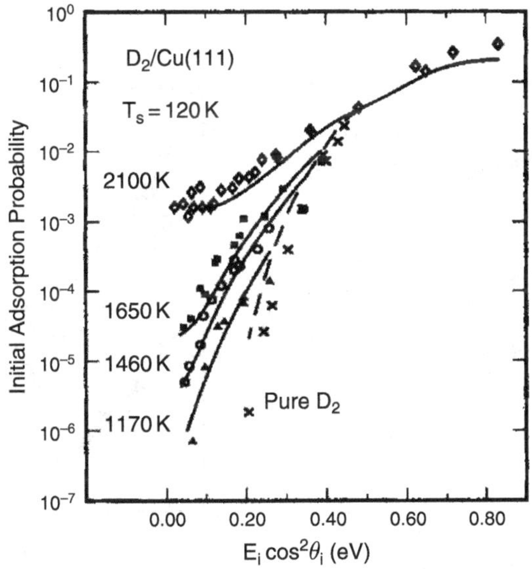

Figure 3.22. Dissociation probability S of D_2 on Cu(111) plotted logarithmically vs. the normal component of incident energy E_n for various vibrational temperatures ($T_v = T_n$) as listed in the figure. The × labeled pure D_2 is S for simply increasing T_v and E_n simultaneously of a pure D_2 beam by increasing T_n. From Refs. [33,225]. Lines are $S(E_n, T_v)$ calculated from associative desorption experiments measuring $S(E_n, v, J)$ and detailed balance.

influence of vibration on S. However, this results from an almost accidental compensation of the relative vibrational populations, zero points and relative thresholds for H_2/D_2 dissociation, all of which affect $S(E_i, \theta_i, T_v, T_s)$ [226]. Simple 2D dynamics on the DFT PES of Figure 3.4 (a) gives $\eta_v \approx 0.6$, in reasonable agreement with the experiments.

$D_f(E_f, \theta_f \approx 0, v, J, T_s)$ has been measured in permeation induced associative desorption of H_2 and D_2 from Cu(111) using (2+1) REMPI detection and ion time of flight techniques to measure E_f [33,34,227]. Figure 3.23 (a) shows $\langle E_f \rangle$ and Figure 3.23 (b) shows the flux-weighted population distribution spectrum for for the various v,J states produced in D_2 associative desorption from Cu(111). Note that most of the barrier energy ends up in translation rather than vibration, in semi-quantitative agreement with 2D dynamics on the PES of Figure 3.4 (a). Using detailed balance to convert $D_f(E_f, \theta_f \approx 0, v, J, T_s)$ to $S(E_i, \theta_i \approx 0, v, J = 2, T_s)$ via eq. (2.8), the derived results are also in semi-quantitative agreement with $S(E, \theta_i \approx 0, v, T_s)$ obtained by the analysis of Figure 3.22 via eq. (3.2). However, in the results derived from detailed balance, not all the $A(v)$ in eq. (3.3) are equal as assumed in the analysis of Figure 3.22.

A key new element in the associative desorption experiments is the ability to probe different rotational states and rotational alignment. Assuming that the dominant effect of rotation is to make the thresholds dependent upon rotational state,

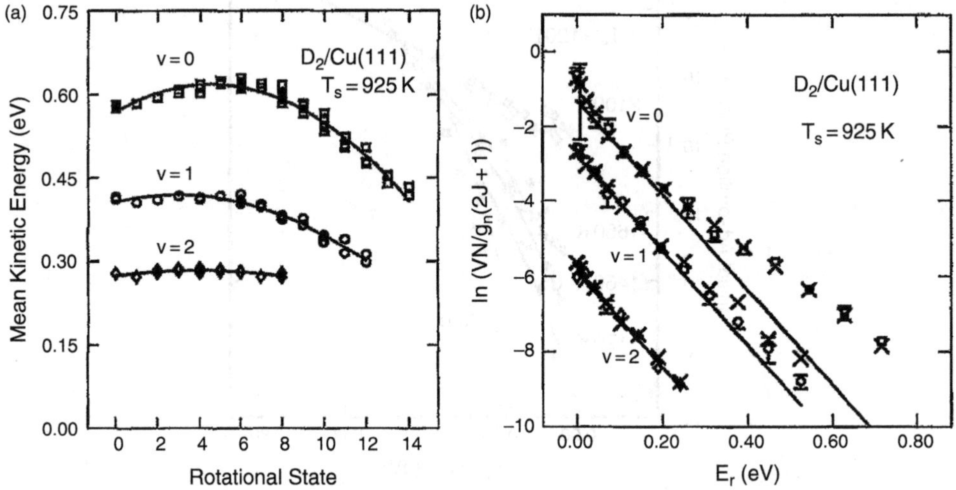

Figure 3.23. State-resolved associative desorption of D_2 from Cu(111). (a) average desorbing kinetic energy $\langle E_f \rangle$ as a function of v,J quantum state. (b) state-resolved desorbing flux $D_f(v, J, T_s = 925\ \mathrm{K})$ normalized by the rotational degeneracy and plotted in a manner such that a Boltzmann distribution is linear. The straight lines correspond to a rotational temperature $T_J = T_s$ for each v state. From Ref. [33].

i.e., as $E_0(v, J)$ in eq. (3.3), gives a strong dependence on J as well as v to E_0 in plots similar to that of Figure 3.5. This is also implicit in Figure 3.23(a). This dependence is complicated since $E_0(v, J)$ first decreases with J but then increases as J becomes larger. The explanation is that rotational hindering inhibits dissociation (by not allowing projections predominantly parallel to the surface) while the adiabatic conversion of rotational energy into E_n during the reactive collision assists dissociation [228]. The strong dependence of E_0 on J as well as v implies that the assumption inherent in eq. (3.2) is not rigorous, at least for H_2/Cu dissociation.

Using (1+1) REMPI, the quadrupole alignment $A_0^{(2)}$ $(E, v, J, T_s = 925\,K)$ has also been measured for D_2 associative desorption from Cu(111) [229] and some of the results are given in Figure 3.24. $A_0^{(2)}$ favors 'helicoptering' alignment, in agreement with a PES that favors dissociation of molecules preferentially parallel to the surface (and detailed balance). It increases substantially with J but decreases with E_f. The

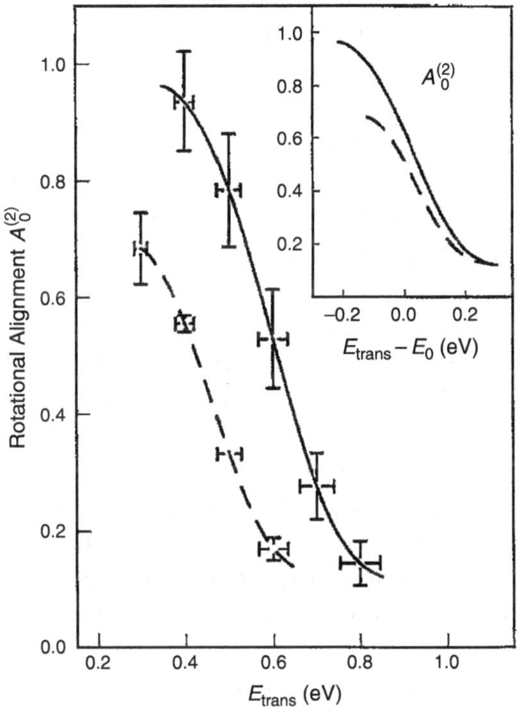

Figure 3.24. The rotational quadrupole alignment $A_0^{(2)}$ as a function of translational energy $E_{trans} = E_f$ of D_2 desorbing from Cu(111) in the $D_2(v = 0, J = 11)$ quantum state (points connected by solid line) and in the $D_2(v = 1, J = 6)$ quantum state (points connected by dashed lines). The inset shows the same results plotted against $E_f - E_0(v, J)$, where $E_0(v, J)$ is the center of the 'S'-shaped state-resolved translational excitation function obtained from unpolarized $D_f(E_f, v, J, T_s)$ measurements. From Ref. [421].

latter is interpreted via detailed balance as resulting from a barrier distribution $V^*(\vartheta)$ that allows wider D_2 angles to contribute to dissociation at the higher E_i [230].

In detailed comparisons of direct measurements of S with those based on D_f and detailed balance, it must be remembered that T_s is different in the two types of experiments; ~120 K in direct sticking and ~925 K in associative desorption. Using atom beam pulse techniques (and conventional TPD), $D_f(\theta_f, T_s)$ [231,232] and $D_f(E_f, v, J, T_s)$ [113] have been measured over a wide range of T_s. Using detailed balance, these results are evidence for a modest broadening in the 'S'-shaped $S(E_i, v, T_s)$ curves with T_s, with $W(v=0, J)$ broadening from ~0.14 eV at $T_s = 370$ K to ~0.22 eV at $T_s = 900$ K for both H_2 and D_2 and for $J = 0$–6. This broadening is evidence for a modest interaction with the lattice and is well described by eq. (3.5) with an $a_{v=0} = 0.00014$ eV/K. Since H_2 and D_2 are so light and the broadening is the same for H_2 and D_2, this T_s dependence is inconsistent with dynamic recoil, and it is suggested that the dominant effect is due to the thermal modulation of the barrier height, i.e., a thermal roughening of the surface [113].

Vibrational and rotational inelastic scattering has also been probed by combining molecular beam and (2+1) REMPI detection. At high incident E_i, significant vibrational excitation $P(v=0 \rightarrow v=1)$ is observed for both H_2 and D_2 scattering from Cu(111) [233,234]. At modest incident energy, there is significant vibrational de-excitation $P(v=1 \rightarrow v=0)$, especially for H_2 [234,235] and at low incident energy significant rotational excitation occurs for $v=1$, $P(v=1, J=0 \rightarrow v=1, J=2)$. The rotational excitation is quenched at higher energies by the onset of vibrational de-excitation.

Unfortunately, there is not at present a 'state-of-the-art' 6D DFT PES for this system to allow detailed comparison of 'first principles' adiabatic theory with experiment. There are only two different 6D PES based on assuming a London-Eyring-Polanyi-Sato (LEPS) form and fit to a limited set of DFT 2D slices [236]. The 6D quantum dynamics on these PES give only modest agreement with experiments [236,237]. Theoretical $S(E_i, v, J=0)$ predict higher vibrational efficacies than observed in the experimental fits. It should be pointed out, however, that the shape of the state-resolved excitation functions, eq. (3.3), assumed in the analysis of the experiments is not well corroborated by the theoretical calculations so this may also reflect a limitation in the analysis of the experiments rather than a problem with the theory. It seems more meaningful to directly compare $S(E_i, T_v, T_s=0)$ calculated from the theoretical $S(E_i, v, J)$ with the experiments. In addition, theoretical calculation of $P(v=1, J=0 \leftrightarrow v=0, J=0)$ vastly underestimates the vibrational excitation and de-excitation observed in experiments. The authors of Ref. [237] suggest that this is principally due to a failure of the LEPS fit form of the PES, as 6D quantum dynamics on a more accurate DFT PES for H_2/Cu(100) do predict significant vibrational excitation for collisions over the atop site [238]. It has also been suggested that vibrational de-excitation could result from non-adiabatic coupling to e-h pairs [78].

For the similar system H_2/Cu(100), measurements of $S(E_i, \theta_i, T_v, T_s)$ are not nearly as extensive as those for Cu(111) [223,226] and there are at present no measurements of $D_f(E_f, v, J, T_s)$. However, there are extensive measurements of 'state-to-state' inelastic scattering $P(E_i, \theta_i, v = 1, J, T_s \rightarrow E_f, \theta_f, v_f, J_f, T_s)$ using a combination of molecular beam techniques, Raman pumping of the initial state and REMPI detection of the final state [239–241]. These results show little rotational inelasticity but imply considerable pure vibrational de-excitation, especially for H_2 scattering. Two related 6D DFT PES have been described for H_2/Cu(100), their primary difference being that one of these was a more accurate fit to more extensive DFT data, and 6D quantum dynamic calculations have been performed to compare with the experiments [237]. 'First principles' 6D dynamics on the more accurate PES is only in fair quantitative agreement with these experiments, with the calculated $S(E,v)$ thresholds lower than those extracted from experiments [237]. In addition, the calculations overestimate rotational inelasticity and underestimate pure vibrational de-excitation inferred in the experiments. The authors suggest that the DFT barriers are too small and that this rationalizes the disagreements of the 'first principles' dynamics with experiment by allowing the H_2 to experience too much potential anisotropy. For H_2/Cu(110), only measurements of $S(E_i, \theta_i, T_n, T_s)$ are available [223,224,226,242] and this is in very good agreement with 6D first principles dynamics [26].

It is unclear at this time whether the lack of quantitative agreement in the 'first principles' theory for H_2/Cu(111) and H_2/Cu(100) with experiments is due to uncertainties in the DFT PES (i.e., due to standard DFT approximations such as the exchange-correlation functional) or in limitations of the 6D adiabatic dynamical model itself. The latter neglects lattice interactions and the role of non-adiabatic interactions. While there is evidence of modest lattice interaction from $S(T_s)$, the small H (D) mass argues against any stronger interaction via dynamic recoil. 2D MDEF (see Section 2.6.1), with both the 2D PES and the electronic friction tensor obtained from DFT calculations [68], indicate very little energy transfer to e-h pairs in associative desorption, $\Delta E_{eh} << V^*$. This is shown in Figure 3.25(a) and indicates that the BOA is a very good approximation to describe the dissociative adsorption/associative desorption. On the other hand, 'first principles' 2D MDEF for the scattering of vibrationally excited H_2 do suggest that significant pure vibrational de-excitation should occur in scattering due to non-adiabatic de-excitation [78]. Both the experimental magnitude of the vibrational de-excitation observed in experiments on H_2/Cu(100) and its dependence on E_i and isotope are in agreement with the MDEF, and this is taken as indirect evidence for the importance of non-adiabatic coupling to e-h pairs in the vibrational de-excitation observed in the scattering experiments [78].

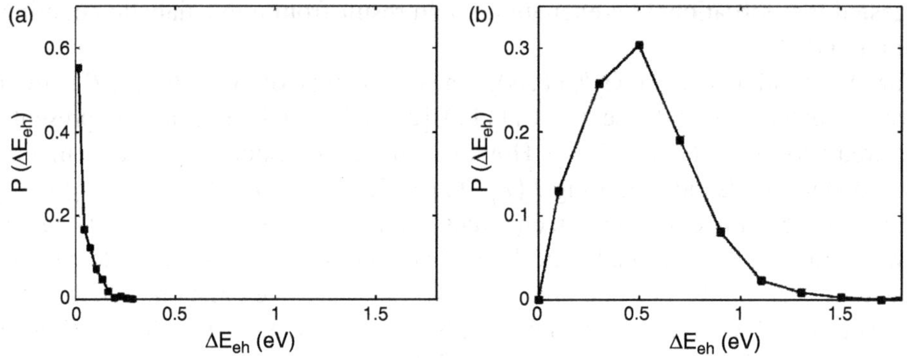

Figure 3.25. Probability of a given energy loss into e-h pairs of magnitude $\Delta E_{\rm eh}$ vs. $\Delta E_{\rm eh}$ occurring in associative desorption of a diatomic from a metal surface from 3D non-adiabatic dynamics. (a) is for H_2 associative desorption from Cu(111), with $\langle \Delta E_{\rm eh} \rangle \approx 0.02$ eV and (b) is N_2 associative desorption from Ru(0001), with $\langle \Delta E_{\rm eh} \rangle \approx 0.5$ eV. From Ref. [68].

4.3.1.2. N_2/Ru(0001)

Another direct activated adsorption system that has received much experimental and theoretical attention is N_2 dissociation on Ru(0001). In part, this interest arises because Ru is a better catalyst for NH_3 synthesis than the traditional Haber-Bosch catalyst, albeit more expensive. The dissociation of N_2 at free sites on Ru is the rate-limiting step in the catalysis, although it is now well documented that step sites are the active catalytic sites for NH_3 synthesis because the barrier to N_2 dissociation is much lower at these sites than the terrace sites [243] (see the chapter by Bligaard and Nørskov). In addition, there is fundamental interest in the dynamics of dissociative adsorption/associative desorption since the PES topology (Figure 3.4(b)) is quite different from that of H_2/Cu(111) (Figure 3.4(a)); the barrier is higher and is preferentially along the vibrational coordinate instead of the translational coordinate.

Molecular beam sticking experiments measuring $S(E_i, \theta_i = 0, T_v, T_s)$ for N_2 dissociation on Ru(0001) are given in Figure 3.26(a) [244]. This figure is a summary of measurements from three different groups [244–246]. It is possible that the dissociation at low E_i is dominated by steps/defects, but molecular beam experiments with the steps blocked by Au adsorption show that at $E_i > 1$ eV, S principally measures sticking on the terrace sites [246]. The sticking curve of Figure 3.26(a) looks quite different to that for D_2/Cu(111) in Figure 3.22; S increases quite gradually with E_i and for $E_i >> V^*$ saturates at $\sim 10^{-2}$ rather than nearly unity. Several suggestions have been proposed to account for this behavior; diabatic curve crossing and tunneling [245,247], energy loss to the lattice [244], damping into e-h pairs [126] and finally as a steric consequence of the late barrier PES [27,248]. Another interesting aspect of Figure 3.26(a) is the dependence on T_v.

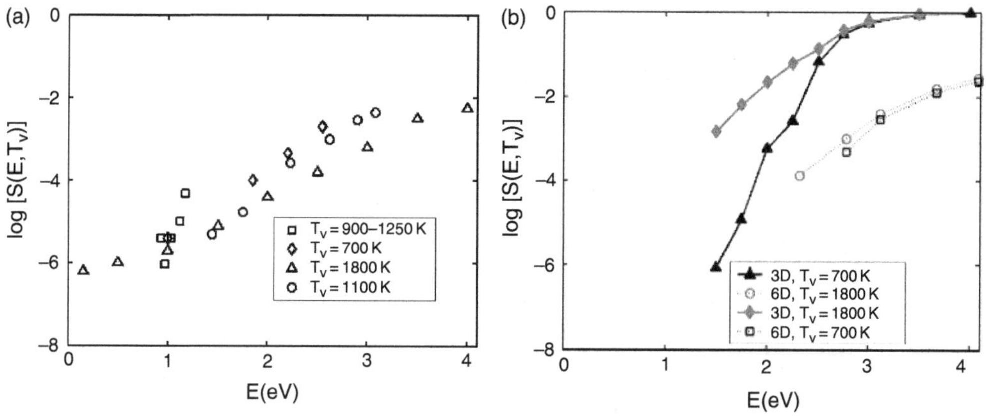

Figure 3.26. (a) The experimental dissociation probability S for N_2 on Ru(0001) plotted logarithmically vs. the incident normal energy $E = E_n$ for three different N_2 vibrational temperatures as noted in the legend. The squares varied both E_n and T_v simultaneously. From Ref. [244]. (b) 'First principles' predictions of the logarithim of the dissociation probability at two vibrational temperatures as noted in the legend. The solid points are from 3D (Z, R, q) quasi-classical dynamics and the open points are from 6D quasi-classical dynamics. The latter are from Ref. [27].

Since there is not an overabundance of experimental data, this was originally interpreted [126] as a rather weak dependence compared to that in Figure 3.22. However, a reanalysis reveals that $\eta_v > 1$ is in better agreement with the limited experimental data [248]. A high η_v is consistent with the exit channel (vibrational) nature of the barrier.

$D_f(\theta_f, T_s)$ in TPD is moderately peaked about the normal $\sim\cos^7\theta_f$ as expected for a system with a high barrier [249]. $D_f(E_f, \theta_f = 0, T_s)$ is also measured for different initial N surface coverages Θ_N using a laser-induced T-jump to induce associative desorption (LAAD technique) [38,250]. Raw experimental TOF are given in Figure 3.27(a) and show broad desorption peaks tailing out to E_f>2 eV. Since the highest E_f observed is a lower limit to V^*, the barrier must increase substantially with Θ_N [250,251]. DFT calculations of V^* and its dependence on Θ_N are in very good quantitative agreement [250,252]. A summary of the strong Θ_N dependence of V^* for this system is given in Figure 3.27(b). Because $\langle E_f \rangle << V^*$, it was originally suggested that the rest of the energy of associative desorption resided in N_2 vibration because of the vibrational nature of the barrier in the PES [250], and this suggestion is consistent with adiabatic dynamics on the PES of Figure 3.4(b) [126,253]. The first state-resolved experiments for associative desorption $D_f(E_f, <\theta_f>, v, J, T_s)$ were also interpreted in terms of a vibrational inversion between $v = 1$ and $v = 0$ [22], although later experiments did not support this conclusion [126]. In fact, this later work surprisingly found little vibrational excitation in the associatively desorbing N_2 as shown in Figure 3.28(a). Attempts to duplicate the initial experiments have

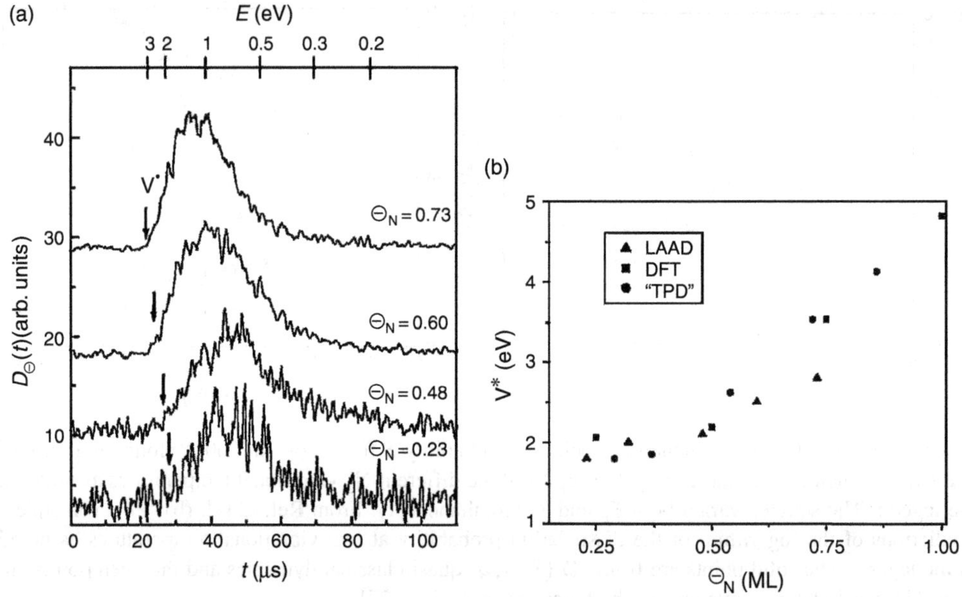

Figure 3.27. (a) N_2/Ru(0001) associative desorption density normal to the surface $D_\Theta(t)$ versus experimental time of flight t (and corresponding translational energy E) measured as a function of initial N coverage on the surface Θ_N. The arrows mark the adiabatic barrier V^* for dissociative adsorption extracted from the $D_\Theta(t)$. (b) Adiabatic barriers for N_2 dissociation on Ru(0001) as a function of initial N coverage Θ_N estimated from the different techniques listed in the legend; LAAD from analysis of associative desorption, i.e., (a) in this figure, DFT from DFT calculations and 'TPD' from shift of TPD peak for associative desorption combined with DFT calculations of N binding energy. From Ref. [38].

been unsuccessful [254], and in fact the results of Figure 3.28(a) are now observed by different N atom preparation techniques [38,126,254]. It is possible that the original experiments measured principally associative desorption at steps, while later experiments measured associative desorption on the terraces [38,126].

Since both adsorption and desorption experiments have been performed, it is possible to determine if the experiments satisfy detailed balance and probe the same phase space. They do not, probably because sticking at low E_i is dominated by dissociation at the steps while associative desorption at higher Θ_N principally measures desorption from the terraces [244]. There is also ambiguity as to whether energy loss to the lattice and e-h pairs is the same in the two different types of experiments.

State-resolved inelastic scattering for a wide range of incident conditions (E_i, θ_i) are measured for this system by combining molecular beam techniques with (2 + 1) ion TOF REMPI detection of the scattered molecules [58]. Energy transfer parallel to the surface is measured from the Doppler broadening of the REMPI spectra. Trapping

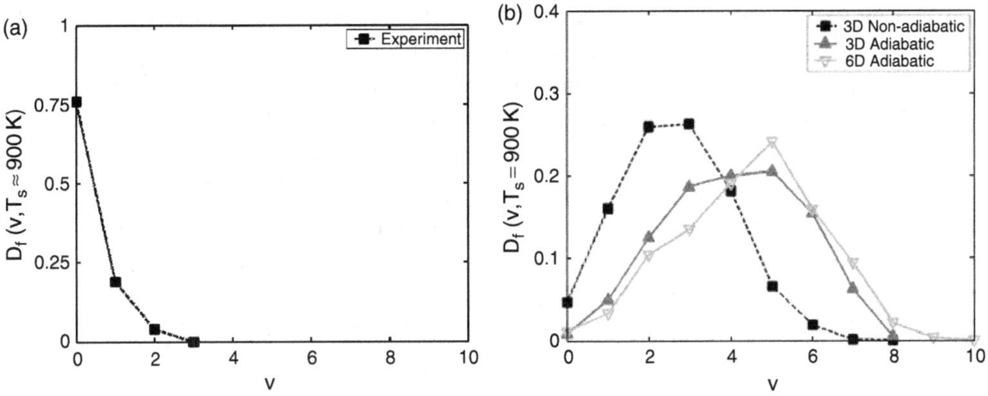

Figure 3.28. N_2 vibrational state distribution in associative desorption from Ru(0001). (a) Observed in experiment. From Ref. [126]. (b) From 3D (Z, R, q) 'first principles' quasi-classical dynamics, with the solid triangles pointing upward being adiabatic dynamics and the squares from molecular dynamics with electronic frictions also from DFT. Based on the PES and frictions of Ref. [68]. The open triangles pointing downward are the results of 6D 'first principles' adiabatic quasi-classical dynamics from Ref. [253].

is minimal under the conditions of the experiments, so that the scattering represents direct inelastic scattering. Since all N_2 degrees of freedom are measured, the energy loss to the surface can be inferred. The results show no (or minimal) vibrational excitation, a weak rotational excitation which is well described as a Boltzmann distribution that increases nearly linearly in T_J with E_i, a modest excitation of translational energy parallel to the surface, a strong energy loss to the surface that depends mostly on the normal component of incident energy E_n, and an anti-correlation between surface and rotational excitation. The net energy loss to the surface in scattering is $\sim$50 % at high E_i, in reasonable agreement with 3D classical dynamics based on the DFT PES of Figure 3.4(b) and assuming dynamic recoil as the mechanism of energy transfer to the lattice (see Section 2.5.2). Most experimental results for N_2/Ru(0001) scattering are qualitatively similar to other state-resolved studies of N_2 scattering on other *d*-band metals [255–258]. The only unusual aspect to the scattering of N_2/Ru(0001) seems to be the absence of any vibrational excitation at incident energies as high as 2.8 eV, especially when compared with the large vibrational excitation observed in H_2/Cu scattering. The 3D adiabatic dynamics (Z, R, q) on the PES of Figure 3.4(b) predict substantial vibrational excitation in scattering at $E_n > 2.2$ eV. 6D dynamics on a 6D DFT PES (but neglecting lattice coupling) also predict significant vibrational excitation at $E_n = 2.8$ eV [27]. The failure of adiabatic dynamics, i.e., that it predicts too much vibrational excitation in scattering, has led to the suggestion that non-adiabatic de-excitation of vibration is the cause [58,126].

The low excitation of N_2 vibration in associative desorption, absence of vibrational excitation in scattering at high E_i and the unusual $S(E_n, T_v)$ are all qualitatively inconsistent with 3D (Z, R, q) adiabatic dynamics on the DFT PES [126]. For example, $S(E_n, T_v)$ predicted by the 3D dynamics saturates at $S \sim 1$ for $E_n >> V^*$ as shown in Figure 3.26(b). The very large η_v predicted in Figure 26(b) for $E_n < 2$ eV is in part due to the role of lattice coupling. In addition, the 3D adiabatic dynamics predicts strong vibrational inversion in associative desorption as shown in Figure 3.28(b), with populations peaking at ca $v = 4 - 6$, depending upon the height of the barrier in the DFT PES (i.e., due to Θ_N). Because of the low excitation of vibration in the experiments, the total average energy of desorbing $N_2 \langle E_f + E_v + E_J \rangle \approx (1/3)V^*$, i.e. nearly 2/3 of the desorption energy is lost to the surface. It is certainly not anticipated that the high frequency N_2 vibration readily damps by exciting low frequency lattice excitations because of the large frequency mismatch. This conclusion is also consistent with the absence of large lattice relaxations at V^* in the DFT PES [259]. Adiabatic dynamics using many different models to couple the lattice to vibration [126] also indicates little vibrational de-excitation. It was therefore suggested that N_2 damps by exciting e-h pairs in the Ru. 3D nearly adiabatic dynamics calculations, MDEF (see Section 2.6.1), with electronic frictions treated as adjustable parameters could fit the observed vibrational de-excitation in associative desorption, but only if the electronic frictions were much larger than those typically associated with vibrational damping of chemisorbed molecules [126]. The MDEF calculations were also qualitatively consistent with the absence of $P(v = 0 \rightarrow v = 1)$ in scattering and the $S(E_n, T_v)$ observed. Since it was not clear whether this was reasonable, electronic frictions were calculated using DFT [68] (see Section 2.6.1). The frictions for N_2/Ru(0001) are nearly an order of magnitude larger than those for H_2/Cu(111) and peak near the transition state. The 3D MDEF using the DFT electronic frictions do predict substantial non-adiabatic energy loss in associative desorption as shown in Figure 3.25(b), and this implies a significant breakdown in the BOA in associative desorption. Most of the energy loss occurs from the vibrational coordinate as shown in Figure 3.28(b). However, this 'first principles' non-adiabatic energy loss is still only ~1/3 of that necessary to agree with the experiments. It is presently unclear if this is due to a breakdown in the friction approximation, the limited dimensionality of the non-adiabatic dynamics or a limitation in the classical approximation to describe vibrational damping. We do note, however, that application of eq. (2.16) shows that $\frac{1}{2}\pi\nu X_v \cdot \nabla_X \ln(\eta_{XX}) \sim 1$ so that the underestimate is most likely due to limitations of the friction approximation because of the rapid change in vibrational friction as the molecule desorbs, i.e., travels from the transition state to the desorption asymptote.

A significant limitation in both the adiabatic and non-adiabatic dynamics discussed above is that they are only 3D. Recently quasi-classical adiabatic dynamics calculations of S on a 6D DFT PES have shown that the predominant reason that $S << 1$

at $E >> V^*$ is simply a steric constraint of the late barrier PES topology [27,248] as shown in Figure 3.26(b). Therefore, the exit channel nature of the topology clearly leads to very different dissociation behavior than that for H_2/Cu(111). However, this 6D dynamics still predicts essentially the same highly inverted $D_f(v)$ distribution in associative desorption [253] as the 3D adiabatic dynamics as shown in Figure 3.28(b) and significant $P(v=0 \rightarrow v=1)$ in scattering at high E_n [27], neither of which is observed in experiments. These 6D calculations do not include energy loss to the lattice nor non-adiabatic effects so that quantitative comparison to experiments is still somewhat uncertain. 3D dynamics calculations suggest that energy loss to the lattice via dynamic recoil affects S appreciably [244] and that non-adiabatic coupling strongly affects associative desorption/scattering. Therefore, at present the dynamics story is still incomplete for this system. It is interesting that 6D adiabatic dynamics apparently gives a reasonable description of dissociative adsorption at $E_i \sim 2$ eV, while it fails to describe the vibrational quenching observed in associative desorption (and vibrationally inelastic scattering). This is probably related to the higher vibrational velocities produced in associative desorption relative to both the translational and vibrational velocities in dissociative adsorption.

4.3.1.3. CH_4/Ni(100)

The dynamics of the activated dissociation of CH_4 on Ni(100) and other transition metals has been actively studied for many years because of its central role in the steam reforming of natural gas [260]. Breaking the first C−H bond is generally the rate limiting step in the catalysis and this is also the step probed in molecular beam experiments [261,262]. The overall energy diagram for CH_4 dissociation on other transition metals, e.g., Ru(0001) [263], is similar to that for Ni (100) [264] and the measured dynamics of CH_4 activated dissociation appear quite similar for all transition metals [265]. Because of this similarity, the dynamic description of activated adsorption will not focus solely on Ni(100) when more detailed experiments exist for other systems. There are a few metal surfaces (Ir(111), Ir(110), and Pt(110)) where low energy paths to dissociation also exist. These are either parallel precursor-mediated paths or arise from dynamic steering to low barrier sites [266]. However, the nature of the low energy paths on these surfaces is still not well understood.

Figure 3.29 (a) shows measurements of $S(E, \theta_i = 0, T_v, T_s)$, again emphasizing the importance of both translation and vibration in activating dissociation [267,268]. Of course, there are now many different modes of vibration for CH_4 that are thermally excited and can contribute to the vibrational activation. Quantum chemistry [269] and DFT calculations [270] of the transition state and reduced dimensional dynamic models of the dissociation [59] suggest that the dominant vibrational effect is due to a single 'local' C−H stretch mode that points towards the surface. Fitting $S(E_i, \theta_i = 0, T_v, T_s)$ to activation by a single C−H stretch mode using eq. (3.2) requires

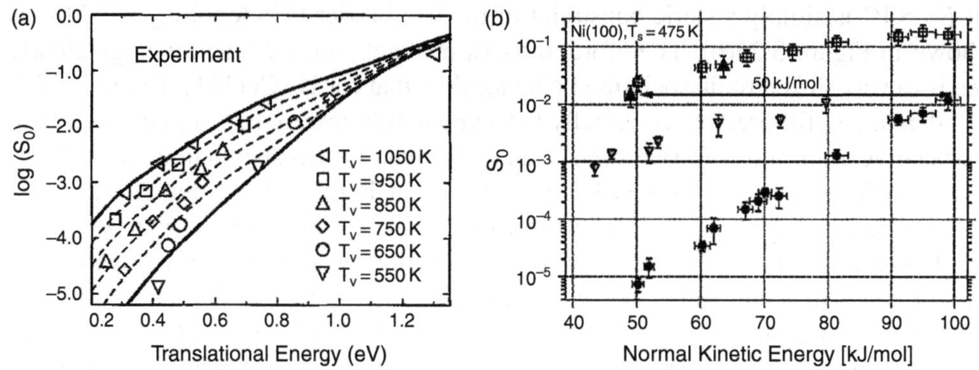

Figure 3.29. CH_4 dissociation probability S_0 (or S in this chapter) on Ni(100) plotted logarithmically vs. the normal kinetic energy E_n. (a) as a function of vibrational temperature T_v as noted in the figure. The lines are the fits of eq. (3.2) to the experiments. From Refs. [267,268]. (b) is for vibrationally state-resolved dissociation measurements, with CH_4 in the ν_1 (solid triangles facing upwards), $2\nu_3$ (open squares), ν_3 (open downward facing traingles) and $\nu = 0$ (solid circles) vibrational states. From Ref. [117].

$\eta_v > 1$. Although this was initially surprising, it could readily occur if external or other internal vibrational modes also contribute to the activation [268], when lattice coupling is important [112] or if the dissociation is strongly vibrationally non-adiabatic [116]. Molecular beam studies of CH_4 dissociation on many other metals, e.g., W(110) [271,272], Ni(111) [262], Pt(111) [273], and Ru(0001) [114,274], are all similar to those on Ni(100) and imply both translational and vibrational activation of similar but varying magnitudes.

Of course, vibrational activation in CH_4 adsorption is significantly more complicated than for a diatomic. There are four normal modes; a symmetric C−H stretch ν_1, an anti-symmetric C−H stretch ν_3, and two bending modes ν_2 and ν_4, and activation by each may be different. One anticipates that those modes, which have the largest projection onto the reaction coordinate, will have the largest activation. Recently, a series of beautiful experiments involving direct IR excitation or Raman pumping of the CH_4 molecular beams prior to dissociative adsorption on Ni(100) [115,117,118,275,276] and Ni(111) [115,116] has allowed the beginning of a detailed examination of mode-specific vibrational activation. Vibrationally resolved $S(E, \theta_i = 0, \nu_i, T_s)$ for sticking on Ni(100) [117] are directly measured and shown in Figure 3.29(b). These demonstrate that activation is significantly more pronounced for ν_1 than ν_3 and that activation for $\nu_3 > 3\nu_4$ [115]. In similar experiments on Ni(111), $\eta_v(3\nu_4) = 0.72$ and $\eta_v(\nu_3) = 1.25$. The observation of the state-resolved $\eta_v(\nu_3)>1$ is intriguing and could be due to a variety of phenomena [116]. Detailed theoretical analysis of these experiments is still not possible since no multi-dimensional PES exists. However, two key points are apparent. First, the state-resolved sticking experiments are all in strong conflict with a microcanonical unimolecular rate model [277–279] which has been proposed to account for the dissociation of molecules at surfaces (based on fits to experimental data with several adjustable parameters). In this case, the only important parameter in activation is the energy of the excitation, independent of the vibrational mode and even whether it is translation or vibration. While the possibility of such a model to explain surface reactivity is appealing because of its simplicity, there is really no reasonable theoretical basis for the assumptions inherent in the model [280]. Second, the state-resolved activation follows the pattern $v_1 > v_3 > v_4 > v = 0$. This is the same pattern as the normal mode projections onto a 'local' C−H stretch mode that is the closest approximation for describing the transition state [281], so that the reaction is most likely dominated by this local mode distortion as assumed in simple dynamical models [59,282]. However, since all vibrational modes contribute somewhat to activation, the dissociation is certainly more complicated than as described by the simple dynamical models.

A very large kinetic isotope effect, $S(CH_4) \approx 10 \times S(CD_4)$ is also observed in the molecular beam experiments [267,268]. Again, this large isotope effect seems rather general for CH_4/CD_4 dissociation on other transition metals as well. The observation of this large isotope effect originally stimulated the proposal of simple 1D tunneling models to describe the dissociation [271,283]. However, a later 3D dynamical model [59] suggests that this large isotope effect could also be simply a dynamical consequence of the strong vibrational activation of the dissociation combined with the initial conditions of the molecular beam experiments.

$S(E_i, \theta_i = 0, T_v, T_s)$ for CH_4/ Ni(100) also shows a modest dependence on T_s. Similar but more dramatic dependences on T_s have also been observed for dissociation on Pt(111) [273] and Ru(0001) [114], and it is probably only the different experimental conditions which prevent the Ni(100) system from exhibiting a stronger T_s dependence in S. The T_s dependence of S for CH_4 dissociation on Pt(111) [284] is given in Figure 3.30. This T_s dependence is clearly not Arrhenius (i.e., exponential in T_s and independent of E_i) as anticipated for a precursor-mediated process and is attributed to lattice coupling in the direct activated dissociation [59] (see Section 2.5.2). The T_s dependence is well approximated as a broadening of the 'S'-shaped excitation functions with T_s as in eq. (3.5) [114].

There is unfortunately no sensitive laser spectroscopic detection technique for CH_4 that can yield state-resolved associative desorption fluxes $D_f(E_f, v_f, J, T_s)$ of CH_4 as for H_2/Cu(111) and N_2/Ru(0001). However, $D_f(E_f, \theta_f = 0, T_s)$ has been measured for associative desorption from Ru(0001) using the technique of LAAD [114]. The results are in good agreement via detailed balance with molecular beam measurements of $S(E_i, \theta_i = 0, T_v, T_s)$ [114,274]. The adiabatic barrier $V^* = 0.8$ eV is obtained straightforwardly in these experiments as the high-energy threshold in the desorbed flux (extrapolated to $T_s = 0$). This barrier is in good agreement with the indirect value estimated by the molecular beam experiments from the threshold

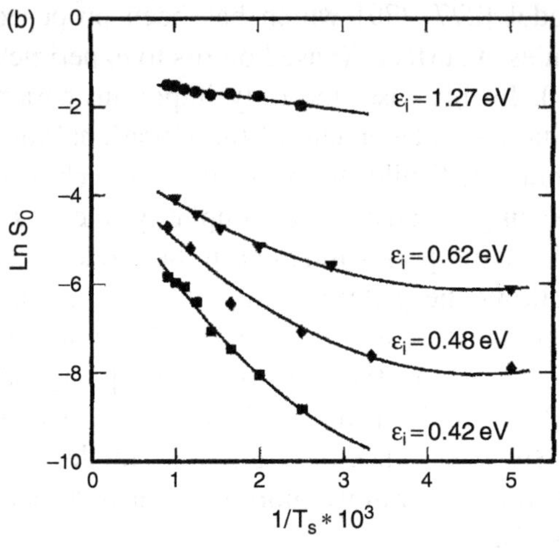

Figure 3.30. Natural logarithm of the CH_4 dissociation probability S_0 (≡ S) on Pt(111) vs. $1/T_s$ for different incident energies at normal incidence ($\varepsilon_i \equiv E_n$). If the T_s dependence was due to precursor-mediated dissociation, it would be Arrhenius (a straight line with slope independent of ε_i) and hence is due to lattice coupling.

in S (and corrected by models of dynamic recoil) via eq. (3.4). This value is also in good agreement with V^* estimated from DFT calculations with a frozen lattice. However, $E_a \approx 0.6$ eV in the isothermal rate [285] and this is less than the frozen lattice V^*. This emphasizes that some caution must be used in equating E_a with V^* when lattice coupling is significant. On average, slightly more than $\frac{1}{2}V^*$ is partitioned into E_f in associative desorption. While detailed vibrational state-resolved dynamics can not be inferred from the unstructured $D_f(E_f, \theta_f = 0, T_s)$, comparison with various models allows some inference, i.e., that the vibrations are not populated statistically (as predicted via the unimolecular rate model) and that both bends and stretches must be populated in associative desorption. Via detailed balance, this simply says that both stretch modes and bending modes must contribute somewhat to vibrational activation (as observed in the state-resolved dissociation experiments), although the relative importance of each can not be quantified from D_f.

It is easy to conclude that activated CH_4 dissociation on Ni(100) (and other transition metals) is dominated by direct rather than precursor-mediated processes under molecular beam conditions because of the strong dependence of S on E_i and T_v. However, it is not as easy to decide in thermal bulb experiments and there has been considerable controversy over which dominates the thermal CH_4 dissociation [59,285,286]. This is especially true since both lattice coupling in direct dissociation and a precursor-mediated dissociation with $E_c > E_d$ (see Section 2.3.2) can cause $k(T_g = 300\text{K}, T_s)$ to increase with T_s. One way to distinguish between these two possibilities is to compare isothermal rates $k(T_g = T_s)$ with non-isothermal rates $k(T_g \sim 300\,\text{K} < T_s)$ at the same T_s. $k(T_g = T_s) >> k(T_g \sim 300\,\text{K} < T_s)$ implies a direct activated mechanism since k increases with increasing initial CH_4 energy. Because $k(T_g = T_s) >> k(T_g \sim 300\,\text{K} < T_s)$ for CH_4 on Ni(100) [285], the dissociation is a direct activated process under thermal conditions as well as under molecular beam conditions. The temperature dependence under isothermal conditions gives an $E_a = 0.61$ eV. Averaging $S(E, \theta_i = 0, T_v, T_s)$ from the molecular beam experiments over thermal distributions appropriate to the rate experiments gives good agreement with the measured isothermal rate as well as to the value of E_a [267,285]. A similar comparison of molecular beam studies with thermal rates for CH_4 dissociation on Ru(0001) also showed excellent agreement [114].

At present, no DFT PES of any dimensionality exists for CH_4 dissociation on any transition metal. Only DFT calculations of the transition state and in some cases the minimum energy path for dissociation exist for various transition metal surfaces (Ni(100) [287], Ni(111) [62,270] and Ru(0001) [288]). All DFT calculations imply that the dominant CH_4 distortion at the transition state is extensive stretching of a single C−H bond that points towards the surface, although some bending distortion is also present. Without a PES, only models of the dynamics are possible. Because the experiments show strong activation by E, v, and T_s, a 3D dynamical model involving coordinates Z the distance the molecule is to the surface, R a 'local' vibrational coordinate for a single C−H bond and q a single phonon coordinate were initially included in the dynamics [59]. The coupling to the lattice assumed dynamic recoil (see Section 2.5.2), although later DFT calculations also show that lattice relaxation at the transition state can reduce the barrier moderately and also induce lattice coupling [61–63]. A reasonable, but totally arbitrary model PES, gives qualitative agreement of the 3D dynamics model with observed CH_4 dissociation experiments on a wide variety of transition metals [59]. The modeling of molecular beam experiments for Ni(100) tried to include other dynamical variables via the 'hole' model [268]. This 3D dynamical model has also been generalized to include a rotational degree of freedom [282] since DFT calculations imply tight orientational constraints at the transition state [270]. Although the general assumptions of these models seem reasonable in view of the DFT calculations and the many experimental results, they remain nevertheless very simplified low-dimensional dynamic models with an arbitrary PES. Therefore, the dynamic models should not be compared quantitatively with experiments to judge the 'correctness' of the models. They are at best only a qualitative rationale for parametric experimental dependences. Quantitative comparison of experiment with first principles theory must await a DFT PES, even one of limited dimensionality. However, we do point out that the fundamental assumptions of any dynamical model are antithetical to the assumptions of a statistical model, and the qualitative success of the dynamical models is further evidence that a statistical model is not the correct description of the activated dissociation of CH_4.

4.3.2. Weakly activated dissociation

4.3.2.1. H_2/Pt(111)

The dissociation of H_2 on Pt(111) was studied quite early in the evolution of surface dynamics because of the importance of Pt as a catalyst for hydrogenation reactions. It was first suggested that the H_2 and D_2 exchange reaction implied relatively weak H_2 (D_2) dissociation that occurs only at steps on Pt(111) [289], but later reanalysis of these experiments [290] and more extensive measurements [291] implied that H_2 dissociation occurs on the terraces, with only a modest enhancement of dissociation at the steps.

Detailed molecular beam measurements of $S(E_i, \theta_i, T_v, T_s)$ for D_2 (H_2) on Pt(111) probe the direct dissociation on the terraces [292,293]. The dependence of S on E_i and θ_i is shown in Figure 3.31 (a) for D_2 dissociation on Pt(111). Note that $S \approx 0$ at $E_i \approx 0$ but that it increases dramatically with E_i, especially at $\theta_i = 0$. This is interpreted as showing that $V^* = 0$ for the optimum steric configuration, but that V^* increases appreciably for other impact sites (X,Y) and angular orientations (ϑ, φ). Therefore the increase in S with E_i represents the opening up of a phase space hole for dissociation, i.e., as in the 'hole' model. The θ_i dependence of Figure 3.31 (a) implies that $E_{||}$ inhibits dissociation, i.e., that energetic corrugation is more important than geometric corrugation [24,294]. There is no dependence of S on either T_v or isotope, suggesting that when barriers exist they are principally in the entrance channel. In addition, there is a weak increase in S with T_s, consistent with direct dissociation, but implying small lattice coupling in the dissociation. Similar molecular beam experiments for D_2 dissociation on Pt(100) also observed a weak increase in S with T_s [60]. Co-adsorbing small amounts of K on the Pt(111)

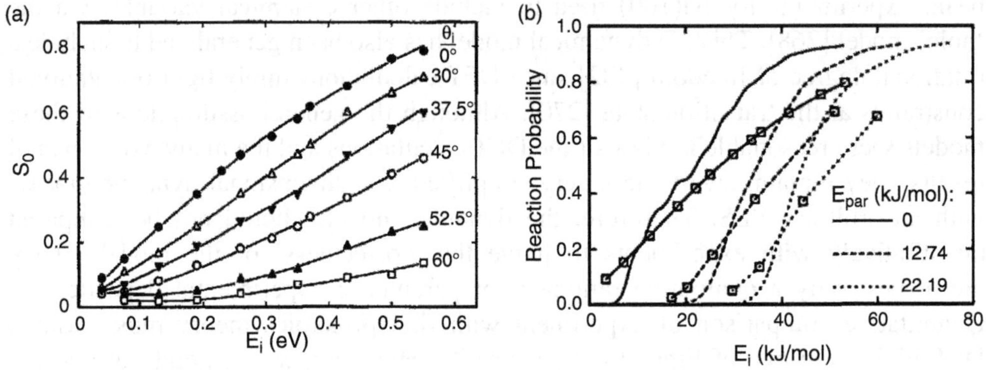

Figure 3.31. (a) Experimental dissociation probability $S_0(\equiv S)$ for D_2 on Pt(111) as a function of E_i and θ_i. From Ref. [292]. (b) Points connected by lines are some of the experimental results of (a) re-plotted at fixed parallel incident energy $E_{par}(\equiv E_{||})$. The pure solid, dashed and dotted lines are the equivalent results from 6D 'first principles' dynamics. From Ref. [300].

surface leads to very long-range (70–430 Å^2) poisoning of D_2 dissociation due to electrostatic perturbations to the barrier [295]. DFT calculations suggest that the effect is due to dipole-dipole repulsion between the adsorbed alkali and the H_2-metal transition state (where charge transfer from the metal to H_2 occurs at the transition state and induces a dipole) [296].

One puzzle in H_2/Pt(111) interactions is that the sticking experiments imply a strongly corrugated reactive PES, while early elastic scattering experiments (diffraction) for HD/Pt(111) seem to imply an uncorrugated physisorption PES [297]. This dichotomy has been much discussed in the literature [294,298,299], and the conflict recently resolved by a combination of theory and new experiments. Strong diffraction from the reactive PES only occurs out of the scattering plane [23] and this was not measured in the initial experiments. Subsequent diffraction experiments did measure the predicted strong out of plane diffraction [300].

Measurements of $S(E_i, \theta_i, T_v, T_s)$ for H_2 (D_2) on Pt(533) have clarified greatly the previously confusing role of steps in this dissociation [301]. Pt(533) is a Pt vicinal surface with (100) oriented steps separated by four rows of (111) terraces. In this case, S is composed of two components; one decreasing with E_i at low E_i due to the steps and a component increasing with E_i which is identical to that observed on Pt(111) and assigned to direct dissociation on the (111) terraces. The step component is attributed to a conventional precursor-mediated dissociation (by trapping in the physisorbed molecular state on the terraces) for the lowest E_i and to steering or dynamic trapping at somewhat higher E_i.

The H_2/Pt(111) dynamics has been extensively investigated by 'state-of-the-art' 6D dynamic theory. These calculations are based on a detailed 6D DFT PES obtained by accurate fits to many separate ab initio points [8,302] combined with 6D quantum dynamical calculations [23,300]. The latter avoids any ambiguity in the comparison with experiment due to the validity of the classical approximation in describing H_2 dissociation/scattering. Figure 3.31(b) compares theory with the experiments of Figure 3.31(a), but re-plotted in a way that the theoretical calculations were performed. The agreement is quite good and demonstrates the inhibition of dissociation by parallel momentum. However, the S curves from theory are narrower than those in experiment, and all attempts to resolve this discrepancy as due to experimental averaging, lattice coupling, etc. failed [303]. The authors of Ref. [303] speculate that this is just the limits of DFT theory and our current knowledge of the exchange-correlation functional. Diffraction intensities from the theory are also in very good agreement with experiment [300]. Thus, 'state-of-the-art' 6D theory seems to describe both H_2/Pt(111) dissociation and scattering dynamics well (although there is still some quantitative disagreement). In fact, because of the agreement of 6D adiabatic dynamics theory simultaneously to both dissociation *and* elastic scattering, the authors argue the general validity of the Born-Oppenheimer approximation to describe H_2/Pt(111) dynamics and suggest reasons why this should be general for H_2/metals [300]. While the BOA is undoubtedly a good approximation for H_2/Pt(111) dynamics, it is unclear to this author how this allows an estimate of the strength of non-adiabatic effects since no attempt has been made to include them in the theoretical dynamics and to determine their effect on the experiments discussed.

4.3.3. Non-activated dissociation

4.3.3.1. H_2/Pd(100)

There are a great many examples of non-activated dissociation at surfaces. H_2 dissociation on Pd(100) and the other single crystal faces of Pd are perhaps the best studied examples, both experimentally and theoretically. Part of the interest in these systems is the importance of Pd as a hydrogenation catalyst. The other major technological impetus is that bulk Pd readily absorbs large amounts of H, so is a prototype for metal-hydride hydrogen storage systems.

Molecular beam studies of dissociative adsorption $S(E_i, \theta_i = 0, T_s = 170\,\text{K})$ for H_2 on Pd(100) are given as the points [304] in Figure 3.32. There is high S throughout the energy range, suggesting non-activated dissociation. In fact, somewhat higher values of S were obtained by Rettner and Auerbach [12], and these authors speculate that surface contamination may have made the results in Ref. [304] artificially low. To account for the initial decrease in S with increasing E_n, Rendulic et al. [304]

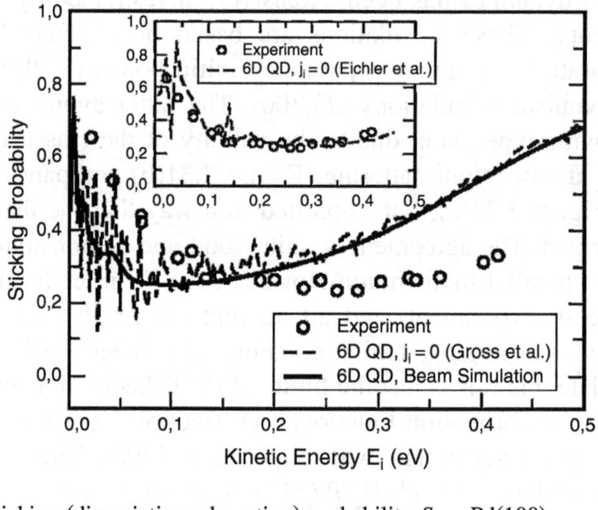

Figure 3.32. H_2 Sticking (dissociative adsorption) probability S on Pd(100) as a function of incident normal kinetic energy $E_i \equiv E_n$. Circles are experiment [304], dashed and solid line are 6D 'first principles' quantum dynamics with H_2 in the ground state and a thermal distribution appropriate to the experiments, respectively [109]. The inset is also 6D 'first principles quantum dynamics but based on a better PES [309]. From Ref. [2].

suggested that precursor-mediated dissociation dominated at low E_n and weakly activated dissociation at higher E_n. Similar experimental results and explanations were suggested for H_2 dissociation on W(100) as well [305]. However, the physisorbed H_2 state on these metals is too weakly adsorbed to act as the molecular precursor anticipated from eq. (2.12), and S is not a function of T_s as anticipated for a precursor [305].

Theoretical dynamics first suggested that steering in non-activated dissociation gives the behavior observed in $S(E_n)$ [109,306,307]. The 6D DFT PES for H_2 dissociation on Pd(100) shows that V varies appreciably with impact site on the surface (and orientation) [308]. For some X,Y the PES is purely attractive while other X,Y have small barriers. Therefore, when an H_2 is incident on the surface at low E_n, it can be 'steered' to the attractive site (high S). At higher E_n, H_2 does not have time to steer to the attractive zero barrier sites and hence S initially decreases with E_n. At even higher E_n, H_2 has enough energy to dissociate over the barrier at these sites and S therefore increases again. Quantum dynamics on 6D DFT PES [109] give very good agreement with the experiments as shown in Figure 3.32. The inset shows even better agreement from 6D quantum dynamics on a better PES [309]. However, some caution is in order in assessing the differences in the two calculations since the different experimental groups measuring $S(E_n)$ do not agree as well as the differences between the two calculations.

Not only is there lateral steering on the surface, but there is also orientational steering from the anisotropy in the PES. Therefore, S should also decrease with J_i as the higher rotational energies are not as effectively steered into the low barrier parallel orientation. This was originally suggested as a way to distinguish steering from a precursor-mediated channel [306]. State-resolved sticking experiments show that $S(E_n, J_i, T_s = 80\,\mathrm{K})$ does decrease ~20 % with $J_i = 0-3$ [12].

State-resolved associative desorption experiments $D_f(E_f, v, J, T_s)$ for H_2/Pd(100) find Boltzmann distributions in all coordinates, with $T = T_s$ for the E_f distribution [310] and slight rotational cooling and vibrational heating as given in Figure 3.33 [124]. 6D quantum dynamics on the DFT PES of state-resolved S combined with detailed balance give the theoretical estimates [124] of $D_f(E_f, v, J_f, T_s)$ in Figure 3.33 and these are in excellent agreement with experiments. The rotational cooling in adsorption is qualitatively consistent via detailed balance with the decrease in S with J_i discussed above. The vibrational heating is simply a consequence of the adiabatic lowering of the vibrational frequencies at the transition state, which is larger for $v = 1$ than $v = 0$. In addition, modest 'helicopter' rotational alignment occurs in associative desorption of H_2, with $A_0^{(2)} \approx 0.2$. Dynamical calculations are in only modest agreement with this aspect of the experiments [311].

Although not as well studied experimentally, H_2 dissociation on Pd(111) and Pd(110) is similar to that on Pd(100), i.e., $S(E_i, \theta_i = 0, T_s)$ shows an initial decrease

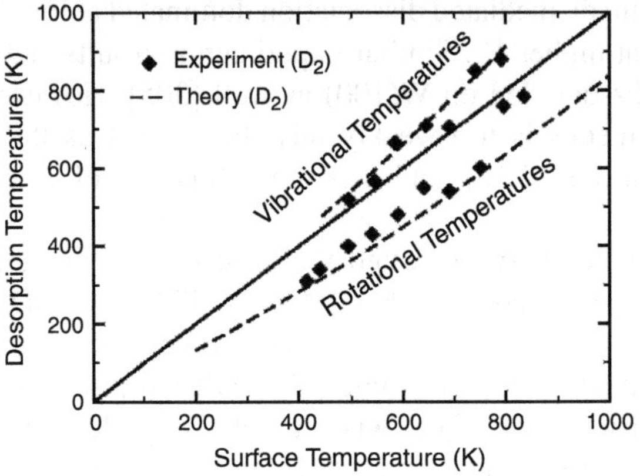

Figure 3.33. Vibrational and rotational temperatures (T_v and T_J) produced in associative desorption of D_2 from Pd(100) as a function of the surface temperature T_s. Points are experiments and dashed lines are from 6D 'first principles' quantum dynamics. The solid line corresponds to $T_v = T_J = T_s$. From Ref. [124].

with E_i [312] and a decrease in S with initial rotation at low E_i [313,314]. Extensive studies of 6D classical (and quantum) dynamics on DFT PES for these systems suggest an entirely different picture of H_2/Pd dissociation to that of steering and direct dissociation, i.e., a combination of dynamic trapping and direct dissociation [110,111,315]. In this scenario, the long range attractive nature of the PES accelerates the H_2 towards the surface until it reaches a region where the anisotropy and corrugation in the PES produce energy transfer from translation normal to the surface to other molecular degrees of freedom, $E_n \rightarrow E_J, E_{||}$. Molecules that do not directly dissociate rebound from the surface and may be prevented from returning to the vacuum by the energy transfer from E_n, remaining trapped near the surface, which ultimately favors dissociation. Dynamic trapping is similar to the process discussed earlier of the quasi-trapping in the molecular chemisorption regime, except that it ultimately results in indirect dissociation of the H_2 after multiple rebounds from the surface. When E_i increases, the dynamic trapping probability decreases but direct dissociation generally increases. The combination of the two processes results in the typical $S(E_i)$ observed for H_2/Pd surfaces. A moderately strong J dependence observed at low E_i [313,314] in $S(E_i, \theta_i = 0, J, T_s)$ is well described also in the dynamics [315] and suggests that $E_n \rightarrow E_J$ is the dominant energy transfer mechanism in dynamic trapping. Recently, the concept of dynamic trapping has received experimental support for H_2/Pd(110) because the scattered H_2 exhibits only a broad $\cos\theta_f$ distribution without diffraction peaks [316]. For this system, the 6D first principles dynamics predicts that some of the dynamically trapped fraction returns to the gas phase after loosing all memory of initial momentum.

In reality, the various mechanisms dominant for dissociation of H_2/Pd at low E_i form a continuum. Steering implies only modest molecular energy transfer ($E_n \rightarrow E_J, E_{||}$) prior to dissociation, dynamic trapping implies considerable molecular energy transfer ($E_n \rightarrow E_J, E_{||}$) so that dissociation is indirect and a precursor-mediated process implies not only an indirect interaction from ($E_n \rightarrow E_J, E_{||}$), but also thermalization with the lattice prior to dissociation [317].

4.4. Precursor-mediated dissociation/associative desorption

4.4.1. O_2/Pt(111)

The O_2 dissociation on Pt(111) is another of the well-studied paradigms in the dynamics of surface chemistry, both because of the richness of its dynamics and because Pt is an important oxidation catalyst, e.g., in automotive exhausts.

The O_2 exhibits complex chemical interactions with Pt(111) dominated by electron transfer from the surface to O_2 because of its large electron affinity, rather schematically described by Figure 3.34 [318–320]. Several molecularly adsorbed states are known, differing principally in the degree of electron transfer from the surface to the molecule; a physisorbed state (nominally O_2) at $T_s \leq 35$ K [30,321] and two molecularly chemisorbed states, a superoxo or O_2^- (at $T_s \leq 120$ K) and the peroxo or O_2^{2-} (at $T_s \leq 95$ K) [153,322,323]. The O_2^- state is centered at bridge sites while the O_2^{2-} state is centered over fcc threefold hollow sites [153]. The O_2^{2-} state predominates at low O_2 coverage, while the O_2^- state predominates at higher coverage [153]. Only the dissociated state is stable on the surface for $T_s \geq 130$ K, saturating at a $p(2 \times 2)$ overlayer with O in fcc threefold hollow sites. Binding energies for the various states are W $\approx$ 0.12 eV for the physisorbed state [30], $W \approx 0.4 - 0.5$ eV for O_2^- and for O_2^{2-} [320,322] and $W \approx 3.7$ eV for O [322].

Molecular beam studies of $S(E_i, \theta_i, T_s \geq 130$ K) measure dissociative sticking on the surface. The complex dependence of S on E_n and T_s in Figure 3.35 (a). suggests that there are two distinct mechanisms involved in the sticking [319]. The low E_n and T_s region is described as precursor-mediated dissociation because of the strong quenching of S with E_n and T_s (see Figure 3.11). In addition, the E_i, θ_i dependence of non-dissociative trapping into molecularly chemisorbed states at $T_s = 90$ K is nearly identical to that observed in low E_n and T_s dissociative $S(E_i, \theta_i, T_s \geq 130$ K). Thus, one or both molecularly chemisorbed states are a precursor to dissociation. One difficulty with this interpretation, however, is that the decrease in S with E_n suggests W ~0.1 eV instead of ~0.5 eV by eq. (2.12) [319,324].

The higher E_n regime shows a complex dependence on E_i, θ_i (Figure 3.35 (b)) somewhat characteristic of direct dissociation where energetic corrugation dominates over geometric corrugation. This regime was originally described as weakly

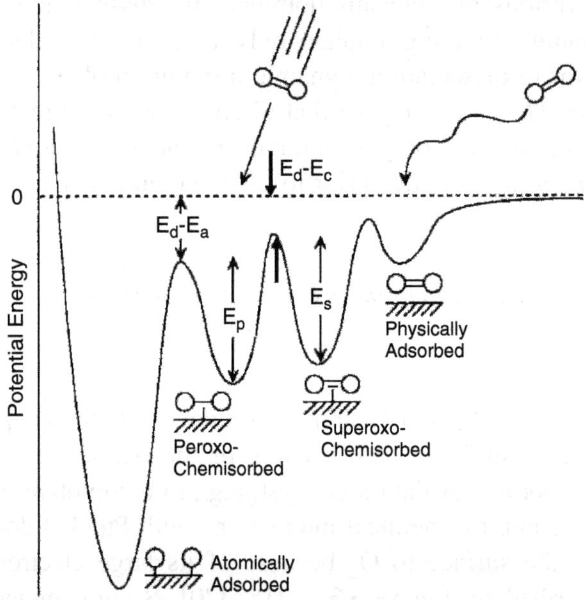

Figure 3.34. Schematic 1D PES and dynamics for O_2 dissociation on Pt(111). High incident energies allow adsorption directly into the molecularly chemisorbed states, which then act as precursors to dissociation. At lower incident energies, O_2 first adsorbs in the physisorption well and then proceeds through sequential precursors to dissociation. From Ref. [320].

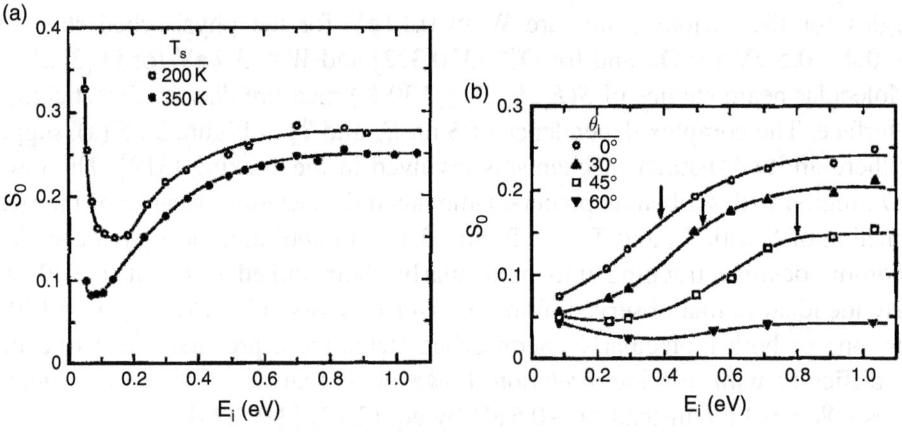

Figure 3.35. Dissociative adsorption S_0 ($\equiv S$) of O_2 on Pt(111). (a) is for normal incidence $\theta_i = 0 (E_i = E_n)$ and for two different T_s as labeled. (b) is for different θ_i and E_i at $T_s = 350$ K. The arrows mark where the center of the 'S'-shaped curve should occur for E_n scaling. From Ref. [319].

activated quasi-direct dissociation because even at high E_n there is still a significant T_s dependence. Chemical titration by reaction with CO [324] and vibrational spectroscopy [320] demonstrate that there is in fact no dissociation at high E_n and low T_s. Under these conditions only the O_2^{2-} molecular state is produced and at low E_n (and $T_s = 90\,\mathrm{K}$) both the O_2^{2-} and O_2^- molecular states are produced [320]. Therefore, the high E_n dissociation channel is suggested to be due to an activated precursor, i.e., activated adsorption into one or both molecularly adsorbed states which then acts as precursor to thermal dissociation [324].

Spectroscopic measurements of O_2 adsorption at very low T_s (and low E_n) demonstrate that only the physisorbed state is populated under this condition, and this converts thermally to chemisorbed O_2 at $T_s = 35\,\mathrm{K}$ [30,325]. It was therefore proposed that the dissociation of O_2 at low E_n is due to sequential precursors; the physisorbed state acts as a precursor to molecular chemisorption and the molecularly chemisorbed state is a precursor to dissociation [30]. This can be written as the following kinetic equation:

$$\mathrm{O_2}\,(g) \underset{k_d}{\overset{\alpha}{\rightleftarrows}} \mathrm{P} \underset{k_{-c}}{\overset{k_c}{\rightleftarrows}} \mathrm{C} \overset{k_a}{\rightarrow} 2\mathrm{O}\,(a)\,,$$

where $O_2(g)$ is the gas-phase O_2, P is the physisorbed species, C is the molecularly chemisorbed species and O(a) is the adsorbed atomic species. This mechanism lumps both molecularly chemisorbed states together, so is something of an oversimplification. But it does readily describe the rapid fall off in S with E_n in Figure 3.35(a) as due to the physisorbed state rather than the chemisorbed state. At higher E_n, direct trapping into the molecularly chemisorbed state occurs and the kinetics is best described via

$$\mathrm{O_2}\,(g) \underset{k_d}{\overset{\alpha_c}{\rightleftarrows}} \mathrm{C} \overset{k_a}{\rightarrow} 2\mathrm{O}\,(a)\,.$$

Kinetic analyses of these two schemes qualitatively rationalize all aspects of the experiments. Schematically, the dissociation channels of O_2 are given in Figure 3.34. There are obviously many thermal barriers and wells. Kinetic analysis of the sticking [319,320,324], analysis of thermal desorption [322,323] and T_s dependent spectroscopy [319,320] can 'fit' all available experiments to give the energy differences labeled in Figure 3.34, as well as the W for the various adsorbed states. These are as follows; $E_d - E_a \approx 0.13\,\mathrm{eV}$, $E_p \approx 0.29\,\mathrm{eV}$ ($E_p \approx E_s$) and $E_d - E_c \approx 0.01\,\mathrm{eV}$. In addition, all ratios of pre-exponentials in the Arrhenius rates for processes going towards the gas phase, i.e., v_d/v_a and $v_d/v_c > 1$. One should not consider the parameters obtained via indirect kinetic methods as especially accurate or unique. They likely depend significantly on O_2 coverage, as well as defect densities of the surfaces. Nevertheless, qualitatively at low O_2 coverage, $E_d - E_c > 0$ and $E_d - E_a > 0$

because the physisorbed state converts entirely to the chemisorbed state rather than desorbing and the molecular chemisorbed state dissociates rather than desorbing. At higher O_2 coverage, desorption is also observed from both states, so kinetic parameters change with coverage as anticipated.

STM experiments are in general agreement with the picture above. STM directly observes the two different molecular precursor states at low T_s and observes their dissociation by thermal annealing, photochemistry and via tunneling electrons [153,326]. The molecularly adsorbed states and dissociated O show a strong T_s dependent clustering at low T_s, and this is evidence for a very mobile physisorbed precursor since the molecularly chemisorbed states are not significantly mobile at these T_s [326,327].

Since there is no good internal state detector for O_2, there have been no detailed dynamic studies of associative desorption. Measurements of $D_f(\theta_f, T_s)$ for associative desorption do show modest peaking about the normal [328], and this peaking increases with T_s [329]. It therefore appears that at higher T_s, the associative desorption samples part of the barrier causing the activated precursor that is apparent in S at higher E_i. On the other hand, TPD from the molecularly chemisorbed states at $T_s = 140$ K gives $D_f \propto \cos\theta_f$ [329].

Scattering experiments $P(E_i, \theta_i, E_f, \theta_f, T_s)$ give a rather different and simple picture of the O_2/Pt(111) interaction [330]. Rather narrow lobular scattering slightly shifted towards the subspecular direction is observed, much as predicted from cube models of scattering. The energy transfer to the lattice is ~40% and is roughly independent of θ_f. Thus, the scattering looks similar to that of non-reactive scattering of Ar from Pt(111) (Section 4.1.1). The only hint that there is significant chemical interaction with the surface is an increase in the angular widths of the scattering and increased energy transfer to the lattice at $E_i > 1$ eV.

DFT calculations [17,331,332] of O_2/Pt(111) find the two different molecular precursors defined in the discussions above and observed by STM [153,326]. However, DFT calculations of O_2 dissociation on surfaces are problematical and two different standard approximations for the exchange correlation functional (PW91 and RPBE) get somewhat different results. The DFT calculations are in only fair agreement with experiments. Most troubling is that all DFT barriers to dissociation from both molecular precursors are significant, 0.6–0.9 eV relative to the O_2+ Pt(111) asymptote. This is qualitatively inconsistent with experiment, which shows that thermal dissociation is energetically favored over desorption.

The qualitative picture that emerges of O_2/Pt(111) dynamics is slightly puzzling in some respects. First, barriers for O_2 dissociation from the kinetics measurements are relatively small and O_2 readily dissociates thermally. However, there is *no* dissociation using only translational energy, even at $E_i > 1$ eV. Second, the reactive PES shows evidence of strong anisotropy and lateral corrugation. However, the scattering is similar to Ar scattering from Pt(111). Tight-binding molecular dynamics calculations have attempted to resolve these issues and give a 'unified picture' of O_2/Pt(111) adsorption [333,334]. Tight-binding molecular dynamics is an approximate method of ab initio molecular dynamics, in which electronic energies are approximated by a tight-binding Hamiltonian, with parameters fit to DFT calculations. The electronic energies are, however, far less accurate than obtained in accurate analytical fitting procedures to the PES. The great advantage of the technique over more conventional 6D dynamics on an accurate fit PES is that the coupling to the lattice can be approximately included without resorting to a preconceived model, and this is an essential ingredient to describe trapping into the molecular chemisorption well and subsequent thermal dissociation. These calculations show that direct dissociation at high E_i is simply blocked by steric constraints, i.e., that the O_2 cannot reach the highly elongated barrier between the molecularly chemisorbed state and the dissociated state before energy loss to the lattice causes trapping in a molecularly adsorbed state. They also find that the O_2 scattering looks similar to that of Ar/Pt(111) scattering because only the uncorrugated repulsive tails of the PES lead to scattering trajectories. Closer encounters, where the PES is strongly corrugated, ultimately trap. The calculated trapping into the molecularly chemisorbed state $S(E_n, T_s)$ falls with E_n at low E_n due to steering, and they suggest that this is the reason for the initial decrease in S observed in the experiments rather than due to an initial precursor in the physisorbed state. This suggestion is, however, in conflict with many experimental observations noted above, e.g., the strong Arrhenius T_s dependence in S [319] and STM experiments [153,326,327]. The DFT calculations were incapable of describing the physisorbed state.

4.5. Direct and precursor-mediated dissociation

4.5.1. N_2/W(100)

The interaction of N_2 with transition metals is quite complex. The dissociation is generally very exothermic, with many molecular adsorption wells, both oriented normal and parallel to the surface and at different sites on the surface existing prior to dissociation. Most of these, however, are only metastable. Both vertically adsorbed (γ^+) and parallel adsorption states (γ^-) have been observed in vibrational spectroscopy for N_2 adsorbed on W(100), and the parallel states are the ones known to ultimately dissociate [335]. The dissociation of N_2 on W(100) has been well studied by molecular beam techniques [336–339] and these studies exemplify the complexity of the interaction. $S(E_i, \theta_i, T_s)$ for this system [339] in Figure 3.36 (a) is interpreted as evidence for two distinct dissociation mechanisms; a precursor-mediated one at low E_i and T_s and a direct activated process at higher E_i. These results are similar to those of Figure 3.35 for O_2/ Pt(111), except that there is no T_s

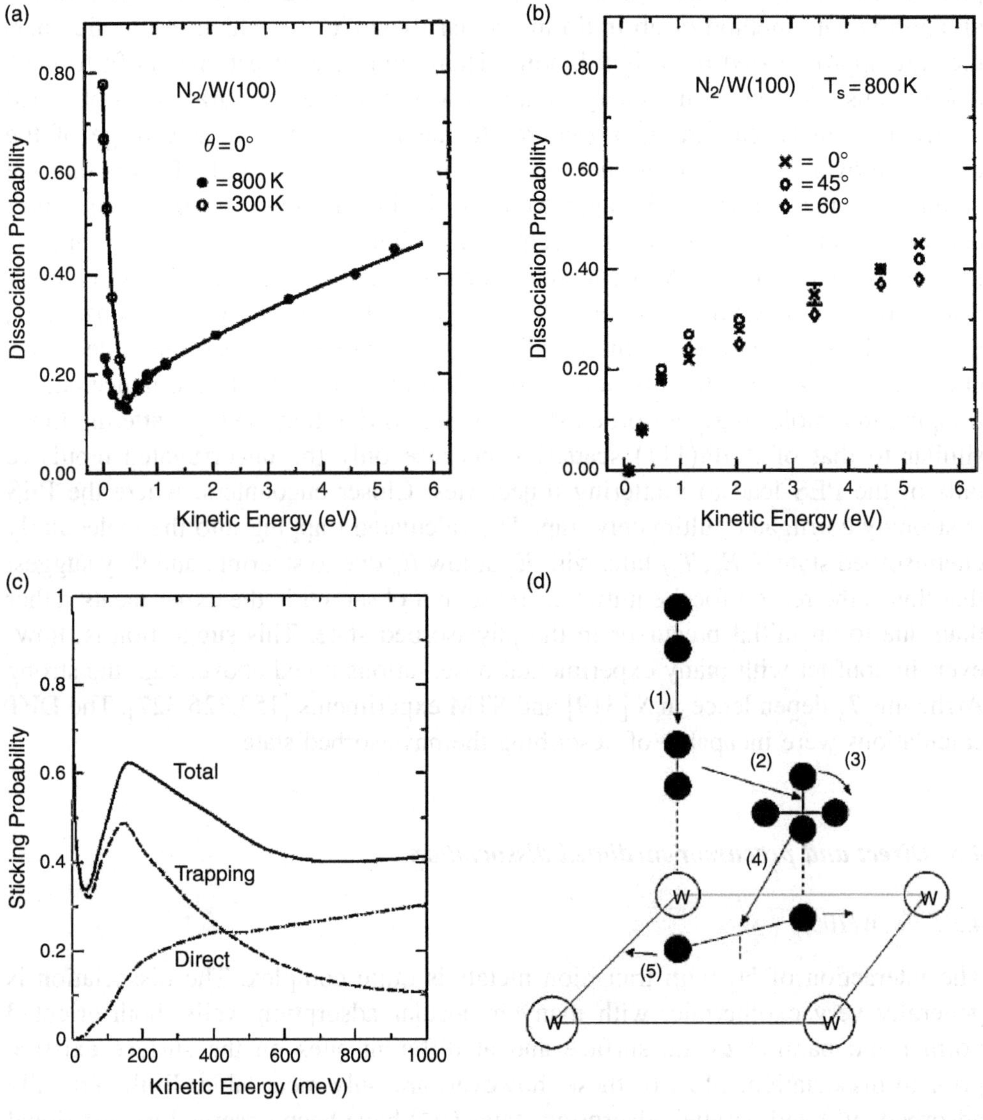

Figure 3.36. Nitrogen dissociation on W(100). (a) Experimental measurements of the dissociation probability S as a function of E_n and T_s. (b) Experimental measurements of only the direct component of dissociation probability S as a function of E_i and θ_i. (a) and (b) from Ref. [339]. (c) Dissociation probability S from 'first principles' classical dynamics, separated into a dynamic trapping fraction and a 'direct' dissociation fraction. (d) Approximate reaction path for dynamic trapping mediated dissociation from the 'first principles' dynamics. The numbers indicate the temporal sequence. (c) and (d) from Ref. [343].

dependence at high E_i for N_2/W(100). It is this fact that causes the assignment to a direct rather than activated precursor channel.

By measuring $P(E_i = 0.088\,\text{eV}, \theta_i = 60°, E_f, \theta_f, T_s)$ to separate trapping-desorption and direct inelastic scattering simultaneously with $S(E_i = 0.088\,\text{eV}, \theta_i = 60°, T_s)$ gives a measure for the trapping probability $\alpha(E_i, \theta_i, T_s)$ in the energy regime where precursor-mediated dissociation is anticipated to occur. The $\alpha(E_i, \theta_i, T_s)$ obtained in this manner as a function of T_s shows that <10% of the T_s dependence in Figure 3.36 (a) is due to the T_s dependence of α [338]. The rest is simply due to the competition between desorption of the precursor back into the gas phase with dissociation. Fitting the T_s dependence of S in this regime to eq. (2.7) gives $E_d - E_c = 0.16\,\text{eV}$ and $\nu_d/\nu_c = 18$. S in the precursor regime, and presumably also α, are almost independent of θ_i and therefore scale approximately with E_i rather than E_n. The E_i range for trapping is too great for a physisorbed state to be the molecular precursor, and in fact XPS experiments monitoring adsorption at 20 K do not find any evidence for the existence of a separate physisorbed state in this system [18]. Therefore, a molecularly chemisorbed precursor is interpreted as responsible for the E_i dependence.

As shown in Figure 3.36 (b), the direct activated dissociation (obtained by subtracting the small precursor-mediated component at low E_i) also scales roughly with E_i rather than E_n, in a manner similar to that for N_2 activated dissociation on W(110) [340]. For N_2 dissociation on W(110), however, there is no accompanying non-activated path to dissociation. The extracted V^* for direct activated adsorption is ~0.1 eV on W(100) and ~0.5 eV on W(110) [339]. Early molecular dynamics calculations on empirical PES (for W(110)) suggested that the E_i scaling resulted from temporary trapping in a weak adsorption well and a barrier that contains tight steric constraints in the exit channel that causes significant scrambling of E_n and $E_{||}$ prior to dissociation [341].

Direct inelastic scattering measurements $P(E_i, \theta_i, E_f, \theta_f, T_s)$ exhibit ~45–50% energy loss to the surface that is roughly independent of θ_f, as for many other surface scattering systems [339]. However, the angular halfwidths in θ_f are quite narrow at modest E_i (0.5–2 eV), broadening at very low E_i (0.088 eV) due to thermal broadening and at high E_i (5 eV) due to the emergence of structure scattering. The observed narrow angular halfwidths in much of the scattering regime implys a relatively non-corrugated PES and this is in conflict with the E_i dependence of trapping which naively implies a corrugated PES. This is the same dilemma noted earlier for O_2/Pt(111) (Section 4.4.1) and its resolution may be similar.

DFT calculations [342,343] demonstrate that the only stable molecularly adsorbed N_2 state is for an atop vertically oriented molecule with $W = 1.1$ eV, although a shallow metastable well parallel to the surface also is predicted. Classical calculations on a 6D DFT PES observe dynamic trapping and non-activated dissociation throughout the energy range studied ($E_i < 1\,\text{eV}$) (Figure 3.36(c)), although the dissociation is arbitrarily broken up into two fractions depending upon the number of rebounds with the surface prior to dissociation, and this varies with E_i [343]. The dominant path leading to dissociation is complex and shown in Figure 3.36(d), although all of the steps most likely do not occur in a single unit cell. This path is similar to that proposed for N_2 dissociation on Fe(111) [344]. These dynamics calculations suggest that there are not two entirely separate mechanisms, only two shades of the same one and this interpretation is therefore in considerable conflict with the experimental interpretations. The key limitations in the first principles dynamics is that the PES and dynamics do not include lattice coupling and these are undoubtedly very strong for this system. For example, lattice relaxation in the PES could make parallel metastable wells much more strongly bound than in a frozen lattice calculation. It is also possible that 6D dynamics with lattice coupling could cause the dynamic trapping to evolve into a true precursor, at least at low E_i. However, one difficulty in this interpretation is that in the DFT calculations to date only the atop vertical molecularly adsorbed state is an obvious candidate for a precursor, and this is predicted to have $E_d - E_c = 0.9\,\text{eV}$, in poor agreement with experiment. This suggests that inclusion of lattice coupling in the DFT PES and dynamics is critical to reasonable first principles dynamics for this system.

4.5.2. NH_3/Ru(0001)

The dissociation of NH_3 on Ru(0001) is the competing back reaction to NH_3 synthesis on Ru(0001). In addition, NH_3 has been proposed as a viable hydrogen storage/transport medium in a hydrogen economy [345] so that its dissociation liberating N_2 and H_2 are also important.

NH_3 adsorbs molecularly on Ru(0001), with a well depth $W = 0.93\,\text{eV}$ and N end down and normal to the surface at low Θ_{NH_3} [346]. At higher Θ_{NH_3}, several new adsorption states exist. Molecular beam experiments $S(E_i, \theta_i, T_s)$ show two mechanisms for dissociative adsorption; a direct mechanism increasing with E_i and independent of T_s at $E_i > 0.5\,\text{eV}$ and a precursor-mediated mechanism at low E_i with a very unusual T_s dependence [347]. Rather than the usual exponential decrease (or increase) in S with T_s, the low E_i regime shows a peaking in S at $T_s \approx 450$ K with a width (FWHM) of ~200 K. The contribution of the precursor component increases by an order of magnitude when the defect density (created by light sputtering) is increased an order of magnitude ($\Theta_{\text{defect}} = 0.25\,\% \rightarrow 2\,\%$). This demonstrates that the precursor channel is associated with the defects. On the other hand, the large S at high E_i implies that the direct activated dissociation occurs on the terrace sites. This picture is consistent with the known dissociation rate of NH_3 on stepped and highly corrugated Ru surfaces [348,349]. Kinetic modeling suggests that the decrease in the precursor-mediated $S(E_i, \theta_i, T_s)$ for $T_s > 450\,\text{K}$ is simply that $E_c - E_d < 0$ at the steps, while the decrease for $T_s < 400\,K$ is due to site blocking of the steps by the dissociation products (the experiments average a small but finite dose of NH_3 >> density of steps).

Two DFT studies have calculated the stepwise addition of adsorbed H to adsorbed N on Ru(0001) terraces to form NH_3. The last step is the reverse of NH_3 dissociation on the terraces. One study finds $W = 1.3\,eV$ and $V^* = 1.4\,eV$ (for NH_3 dissociation) [350,351] while the other finds $W = 0.89$ eV and $V^* = 0.8$ eV [352]. The second study is in better agreement with experiments in the direct dissociation regime. However, only Refs. [350,351] study dissociation of NH_3 at step sites and find $W = 1.76$ eV and $E_c - E_d = -0.5$ eV. This agrees reasonably with experiment for the precursor channel. Therefore, DFT studies do qualitatively support the picture suggested by the experiments.

The scenario described here, i.e., activated dissociation at terrace sites and precursor-mediated dissociation at step or defect sites, is likely to be a very general one since barriers to dissociation are generally much lower at step sites [353]. There are already many known examples; C_2H_6 dissociation on Ir(111) [354], CH_3OH dissociation on Pt(111) [355] and neopentane dissociation on Pt(111) [356], etc.

4.6. Langmuir-Hinschelwood chemistry

Certainly most surface chemistry occurs as adsorbates come together as a result of thermal diffusion on the surface. When both reagents are in thermal equilibrium with the surface before reacting, the surface chemistry is described as a Langmuir-Hinschelwood (LH) mechanism. Even most gas-surface reactions occur via this mechanism. However, when the product of the reaction also remains on the surface, no dynamic information is available. Therefore, the only LH reactions discussed in this chapter are when the product of the reaction is a gas phase species. One example already discussed extensively is associative desorption. Here, another well-studied example is considered.

4.6.1. (O + CO)/Pt(111)

The LH recombination of O + CO to form CO_2 on Pt(111) has been extensively studied over the years, in part because this reaction is the basic oxidation process occurring in the automotive exhaust catalyst. The reaction on other crystal faces of Pt, e.g., Pt(110), exhibit a rich variety of nonlinear dynamics [357]. Even for the simpler reaction on Pt(111), the oxidation is complex because it shows a strong dependence on the relative morphology of the two adsorbates on the surface, and hence on a wide variety of experimental parameters. Although the simple associative desorption of a single atomic species adsorbed on a surface is also LH recombination, this type of complexity is not present for these systems.

The overall conversion of CO and O_2 in the gas phase to CO_2 in the gas phase via the LH reaction on Pt(111) is exothermic by ~2.9 eV. Modulated molecular beam studies of the kinetics of $(O_2 + CO)$ on Pt(111) demonstrate the LH mechanism and show that the LH barrier of the adsorbed species O and CO, V^*_{LH}, depends on Θ_O, from ~1 eV at low Θ_O to ~0.5 eV at high Θ_O [358]. This large variation in V^*_{LH} with Θ_O is most likely due to coverage dependent variations in the $CO_a + O_a$ asymptote rather than any fundamental change in the transition state. CO_2 desorption characteristics, although labeled by $D_f(\Theta_O, \Theta_{CO}, T_s, E_f, \theta_f)$, also depend on details of the surface morphology and preparation of the mixed layers. In many experiments, two separable desorption components are observed; one with a highly peaked $\cos^n \theta_f$ distribution with $n = 7-10$ and $\langle E_f \rangle = 0.3-0.4$ eV and the other being fully accommodated to the surface, i.e., with a $\cos \theta_f$ distribution and $\langle E_f \rangle = 2k_B T_s$ [359–361]. In addition, IR emission experiments of the nascent CO_2 show that it is both vibrationally and rotationally excited [362,363]. Since $v = 0$ is not observed in IR emission, it is generally assumed that the full distributions can be described by effective temperatures. These results show that $T_v >> T_s$ for the antisymmetric stretch vibration and for the bending mode of CO_2. Thus, the highly peaked and energy-rich desorption component is qualitatively compatible with a bent transition state. The origin of the accommodated component is not entirely clear. Its relative contribution to the overall CO_2 production depends on Θ_O, Θ_{CO}, T_s, step density, etc. and it is not observed in all experiments [364]. Some authors believe it is related to a similar reaction at steps [359], while others believe it involves an entirely different transition state on the terraces than that which produces the hot CO_2 product [360].

The details of the mechanism of the reaction depend on the adsorbate morphology. STM experiments show that at high Θ_O and high Θ_{CO}, reaction occurs principally at the perimeters of segregated O and CO islands, and this shows the same V^*_{LH} as observed in the macroscopic kinetics experiments [365]. On the other hand, time-resolved NEXAFS experiments suggest that another reaction path also exists between diffusing isolated O and CO [366]. This path dominates at low Θ_O and low Θ_{CO} in experiments co-adsorbing CO and O_2 at high T_s. The similarity of $D_f(E_f, \theta_f, T_s)$ for the two reaction conditions suggests that the transition state is nearly identical in the two cases.

In addition to the fully LH reaction of O + CO, CO_2 apparently produced by 'hot O atoms' is also observed by thermal dissociation [367], photochemical dissociation [368,369] or collisional dissociation [370] of O_2 co-adsorbed with CO. Dosing a CO adsorbed layer with gas phase O atoms also appears to give the same hot atom reaction, i.e., reaction prior to the O atom fully equilibrating with the surface [371]. In all cases, the oxidation occurs at T_s far below the thermal LH reaction of adsorbed O + CO at $T_s = 330$ K, and the $D_f(E_f, \theta_f, T_s)$ is even slightly more energetic (more peaked in angle and higher $\langle E_f \rangle$) than that of the LH reaction. In addition, there is no fully accommodated component in $D_f(E_f, \theta_f, T_s)$. The CO_2 produced by $O_2 + CO$ layers could also in principle be an activated bimolecular reaction [372]. However, the similarity of $D_f(E_f, \theta_f, T_s)$ to that produced by gas phase O atom dosing, the occurrence of O_2 dissociation under identical conditions to CO_2 formation and DFT studies of the transition state (see below) all argue for a mechanism based on hot O atom reaction with CO.

Microcalorimetry studies [373] of the heat remaining in the substrate for CO(g) reaction with Pt(111)/O ($\Theta_O = 0.25$) indicate some energy release into the nascent CO_2. Unfortunately, these studies were done at $T_s = 300$ K and most of the oxidation to CO_2 at these experimental conditions occurs at $T_s = 330$ K. It is likely that the authors only produced a small amount of CO_2 and can therefore quote only a modest lower bound to the energy release into CO_2. The total energy release into CO_2 is an important number to judge whether the experiments described below are correct.

Somorjai and collaborators [374,375] have recently observed large chemicurrents from the reaction of O_2 and CO at high pressures on Pt thin films on GaN or TiO_2. Because the Schottky barriers are ~1.4 V for these systems, they have interpreted the chemicurrent as due to the formation of very high energy e-h pairs via the chemical formation of CO_2, and speculated that this is an efficient method to convert chemical energy into electrical energy. However, this is unlikely for several reasons. If the reaction is LH, then there is not enough exothermicity to generate such high energy e-h pairs. In addition, if the system is like most other moderate exothermicity reactions on transition metals, the reaction should be in the nearly adiabatic limit, i.e., only many low-energy e-h pairs are created. It is possible that the chemicurrent measured by them is due to pinholes in the thin films and therefore a result of only very low-energy electron generation.

In DFT calculations of the reaction path [376,377], the CO migrates to an atop site adjacent to an O atom, forcing it from the hollow to a bridge site at the transition state. The transition state is therefore a tilted and asymmetrically stretched nascent CO_2, in agreement with suggestions from experiments. The barrier to reaction arises because it costs energy to break some of the O bonding to the surface when it is forced to the bridge site. Therefore O atom mobility is the key limit in the reaction. Barriers of 0.8–1 eV are calculated for the reaction. In addition, the reaction path from co-adsorbed $O_2 + CO$ to form CO_2 is also calculated [377]. In this case, the lowest barrier to reaction occurs if first, the molecular O_2 bond is broken and one of the atoms forms a transition state similar to that calculated for O + CO, while the other O atom is adsorbed in a hollow site. In addition, the barrier to desorption of O_2 is lower than that for reaction, so O_2 desorption, formation of adsorbed O and reaction to form CO_2 should be more or less simultaneous when raising T_s as observed in experiments.

4.7. Eley-Rideal/Hot atom chemistry

The Eley-Rideal (ER) mechanism of surface chemistry is the antithesis of the LH mechanism. In this case, an incoming gas phase species directly reacts with an adsorbate equilibrated on the surface. Of course, if a species impinges on the surface near an adsorbate, it can make a few bounces on the surface before reacting with the adsorbate. When it retains most of its initial energy of adsorption prior to reacting in this fashion, it is called a hot atom (HA) mechanism. There is no rigid division between ER and HA, so they will be considered together. In the discussion of H + H/Cu(111), the difference between the two will be clarified.

Although the division of surface reaction mechanisms into LH or ER dates to the early days of catalysis, ER/HA surface reactions have only been demonstrated recently and only for strongly reactive atomic gas phase species, e.g., H, O. There are many differences between the ER/HA mechanism and the LH mechanism that can be used to separate them experimentally. For example, ER/HA reactions of reactive incident atoms are very exothermic relative to the equivalent LH reaction, typically by several eV. Much of this released energy should end up in the gas-phase product molecule. ER/HA are direct non-activated reactions whose final state properties depend on the initial conditions of the incoming atom and not T_s. This is of course the exact opposite of LH properties.

4.7.1. H+H/Cu(111)

The abstraction of H by incident H on a H covered Cu(111) surface is the best studied example of an ER/HA reaction. The direct reaction is exothermic by ~2.3 eV. In contrast, if the incident H atom first equilibrates with the surface and reacts via associative desorption (LH mechanism), the reaction is nearly thermoneutral, albeit over a barrier of ~0.5 eV (Section 4.3.1.1). The first evidence for this ER/HA reaction was in molecular beam experiments detecting HD produced by H reacting with D/Cu(111) ($\Theta_D = 0.5$) and D reacting with H/Cu(111) ($\Theta_H = 0.5$), both at $T_s = 100$ K [378]. A very broad translational energy distribution $F(E_f)$ was observed for the HD product with $\langle E_f \rangle \sim 1$ eV. Angular distributions of the nascent HD $F(\theta_f)$ were sharply peaked and displaced only slightly from the surface normal in the specular direction. This displacement shows sensitivity to the incident E_i of the atom beam, implying memory of the initial parallel momentum in the reaction. There were small but measurable differences in both $\langle E_f \rangle$ and $F(\theta_f)$ depending upon the isotopes of the experiment, i.e., H atom impinging on D/Cu or D atom on H/Cu. Many aspects of the experiment show that the HD must result from the direct ER/HA reaction; i.e., reaction occurs at a T_s much lower than associative desorption, there is large energy release into the product and there is a sensitivity to initial atom energy and momentum of the gas phase atom.

Detecting the nascent HD with laser state-resolved techniques gives a very detailed picture of energy release in this ER/HA reaction [141,379]. The state-resolved reaction fluxes are defined as $F_f(E_i, \theta_i, \langle E_f\rangle, \langle\theta_f\rangle, v, J, T_s = 100\text{ K})$, where i refers to incident atom properties and f to HD properties. Figure 3.37 is the energy partitioning observed into HD (v, J) states [141]. A very wide distribution of internal states is observed, with some states at internal energies up to the reaction exothermicity of ~2.3 eV. There are slight differences in the two isotopic cases, but the trends are very similar. Overall, $\langle E_v\rangle = 0.6$ eV (H on D/Cu) or 0.68 eV (D on H/Cu) and $\langle E_J\rangle \approx 0.36$ eV, so that $\langle E_f + E_v + E_J\rangle \approx 2$ eV and most of the available reaction energy ends up in the HD product.

The qualitative picture of the abstraction reaction that emerges from the energy partitioning is that as the H(D) approaches the adsorbed D(H), it feels the strong

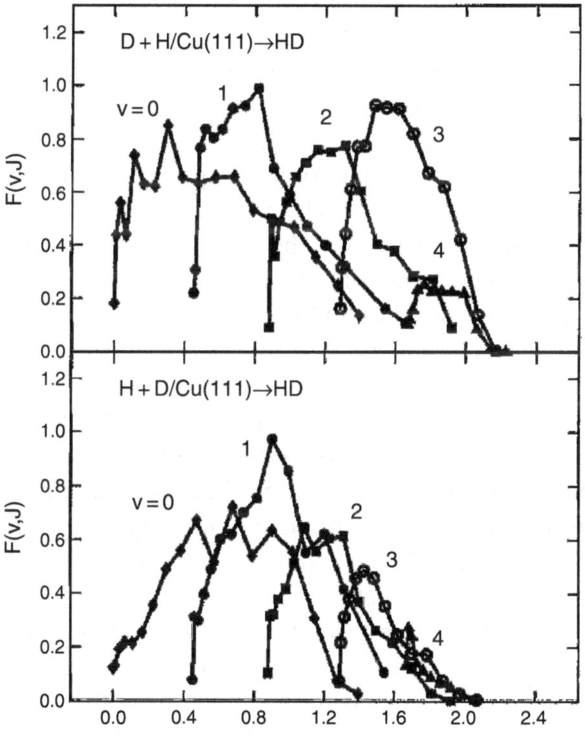

Figure 3.37. HD (v,J) flux $F(v,J)$ produced by the Eley-Rideal reaction of D+H/Cu(111) and H+D/Cu(111) as a function of the vibrational state v and rotational energy $E_r(E_J)$. From Ref. [422].

attractive potential for formation of HD. Most of the energy released ends up in the nascent HD. Because significant vibrational excitation occurs, the reactive PES must be attractive, i.e., release a large fraction of the energy in the entrance channel [1]. The large rotational excitation also implies that the abstraction is not predominantly collinear.

The chemical rate of formation of HD from H + D/Cu(111) (and D + H/Cu(111)) is well fit by a simple quasi first-order ER type rate law

$$\frac{d[HD]}{dt} = \sigma\Phi_H [D]_0 \exp(-\sigma\Phi_H t), \tag{3.6}$$

where σ is the cross-section for the abstraction reaction, Φ_H is the flux of H atoms onto the D/Cu(111) surface and $[D]_0$ is the initial D coverage on the surface [142]. $\sigma = 2$ Å^2 from a fit to this kinetic data [142] and is estimated as ~5 Å^2 by Rettner et al. [141]. No abstraction occurs from H in the subsurface sites, which are also populated by atom dosing (see Section 4.1.2). In addition to HD, a small fraction (1–2%) of D_2 is also produced from H + D/Cu(111), but with a relative abundance quadratic in $[D]_0$. In addition, for the H that ultimately trap on the surface, only ~50% end up abstracting D, i.e., the reaction probability for trapped atoms is $p_r \approx 0.5$ [141,142]. The trapped but non-reactive fraction $p_s \approx 0.5$ is observed by subsequent LH production of HD [141].

Although the kinetic rate and energy partitioning are qualitatively consistent with a pure ER process, other aspects of the experiments and most of the theory (see discussion below) imply that the abstraction is more properly described as a combination of ER and HA reactions. The large σ for abstraction is inconsistent with theoretical studies of a pure ER process as this requires a direct hit of the incoming H(D) with the adsorbed D(H) [380,381]. There is also no way to reconcile formation of homonuclear products with a pure ER process. In addition, similar kinetic experiments on other metals, e.g., Ni(100) [146], Pt(111) [147,382], etc., are not even in qualitative agreement with the simple ER rate law above. In those cases, it is necessary to develop more sophisticated HA kinetic mechanisms to describe the kinetics experiments [383–385]. The key parameter of these kinetic models is the ratio of reaction to non-reactive trapping, p_r/p_s. For $p_r/p_s = 1$, the HA kinetics looks very much like the simple ER case, and this is the reason H(D) + D(H)/Cu(111) has such simple kinetics.

There has been a long history in theoretical efforts to understand H + H/Cu(111) and its isotopic analogs because it represents the best studied prototype of an ER/HA reaction. These have evolved from simple 2D collinear quantum dynamics on model PES [386] to 6D quasi-classical dynamics on PES fit to DFT calculations [380,387,388], and even attempts to include lattice motion on ER/HA reactions [389]. These studies show that there is little reflection of incident H because of the deep well and energy scrambling upon impact, i.e., $\alpha \approx 1$. Although some of the incident atoms react directly with the adsorbate via an ER process, most initially trap on the surface as hot atoms with energies up to 2.3 eV, either by the corrugation of the surface as discussed in Section 4.1.2 or by scattering from an existing adsorbate which also causes energy loss/energy scrambling [388]. The rate of energy loss to the lattice is slow because of small mass ratio (see Section 2.5.1). If the HA reacts with adsorbates before it dissipates the excess energy, energy-rich HA products are produced in the gas phase. Some of the HA do not react and therefore ultimately thermalize with the surface. Also, some adsorbates may be excited by the HA to become HA themselves. This is facilitated by the fact that $\mu \sim 1$ for adsorbate and incoming atom so that energy transfer is efficient. This can produce homonuclear HA products. Thus, the theory is in qualitative agreement with all experimental results.

One way to measure the relative importance of ER to HA in the simulations is to look at the distribution of total HD energy produced in the nascent molecule. This is shown in Figure 3.38 (a) for two different PES, both of which are equally well fit to the DFT calculations [388]. The sharp peak at the highest energy ~2.4 eV is due to ER, while the straggling to lower energies is due to HA reactions. The latter dominate the total reaction. In these simulations, no energy transfer to the lattice is included so that all energy straggling is due to energy transfer to other

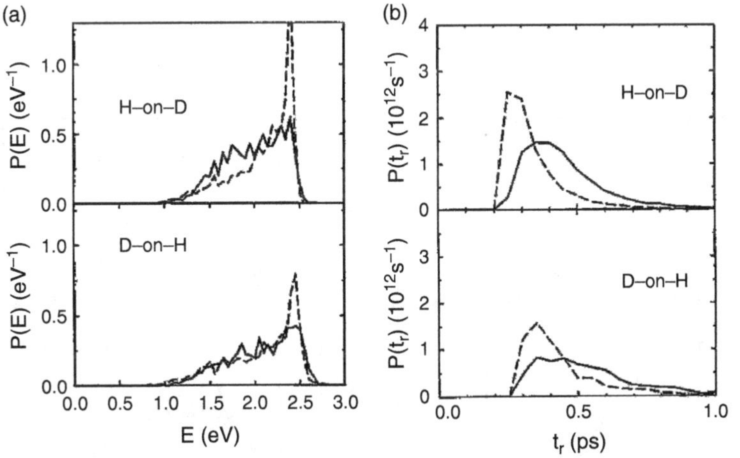

Figure 3.38. (a) Probability distribution for the asymptotic total product energy for two different PES fit to DFT calculations (solid and dashed lines) for the Eley-Rideal/ hot atom reactions of H+D/Cu(111) and D+H/Cu(111). The sharp peak at ~2.4 eV corresponds to ER reaction, while the straggling to lower energies result from HA reaction. (b) the probability distribution of reaction times t_r in molecular dynamics simulations of the Eley-Rideal/hot atom reactions on the same two PES as in (a). The initial portion at t_r~0.35 ps correspond to ER reaction and the straggling to longer times to HA reaction. From Ref. [423].

adsorbates on the surface. Another way to estimate the ER vs. HA fraction is to look at the distribution of times for reaction as shown in Figure 3.38 (b) [388]. Only reaction at the earliest interaction times, ~0.35 psec, are ER. Note that the two PES give somewhat different estimates for ER/HA by both criteria. These simulations also demonstrate why some of the trapped HA do not always react readily with adsorbates, often requiring many collisions. The chief reason is that the H–H interaction is repulsive when the atoms are close to the surface and only becomes attractive or reactive when the atoms move away from the surface. In fact, the non-reactive scattering is so efficient that ~50% (experimentally) ultimately thermalize with the surface rather than reacting with the adsorbates.

Only some aspects of the simulations are in semi-quantitative agreement with the experiments. Vibrational energy partitioning is slightly underestimated in the simulations, while the rotational state distributions are in good agreement with experiments. These comparisons are shown in Figure 3.39. In addition angular distributions and the anti-correlation of rotation with vibrational state are also well described. However, total theoretical abstraction cross-sections are only ~0.5 Å^2, ~5–10 times smaller than experiment. Presumably, the difficulty in quantitative agreement is in the accuracy of the PES or to effects of lattice coupling. Even though the H + H/Cu(111) reaction is very exothermic and attractive, the surface causes competing trapping and relaxation, so that the net rate of reaction may be very sensitive to details of the PES and lattice coupling.

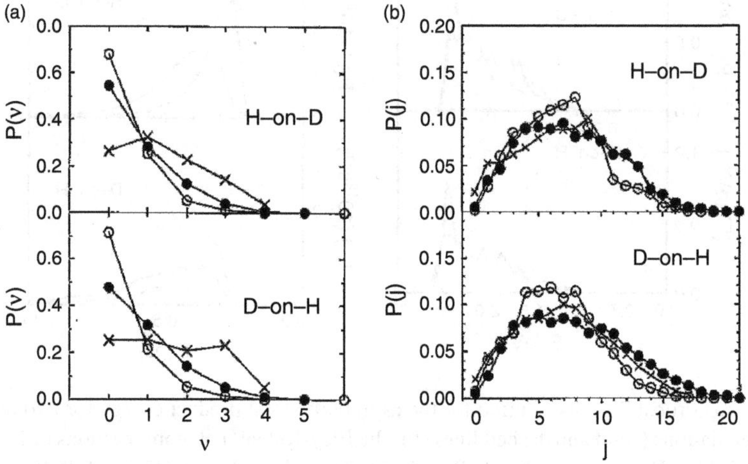

Figure 3.39. Vibrational (a) and rotational (b) state distributions from the ER/HA reactions of H + D/Cu(111) and D + H/Cu(111). The crosses are from experiment and the open circle and closed circle symbols are results from dynamic simulations based on two different PES that fit equally well DFT calculations. From Ref. [423].

4.8. Hot electron chemistry

4.8.1. Photochemistry/femtochemistry

4.8.1.1. NO/Pt(111)

The photon induced desorption of NO from Pt(111) has been actively studied by measurements of $Y(\hbar\omega, F_\omega, \theta_i, P_\omega, E_f, \theta_f, v, J)$ by several groups using ns lasers [390–393]. The pulsed laser-induced photodesorption of saturated NO adsorption at $T_s \approx$ 100 K consists of two separate components, a thermal one arising from the T-jump on the surface and a 'fast' non-thermal component arising from photodesorption [391]. Annealing the surface to $T_s \geq 220$ K (with NO assumed to be only in atop sites with $\Theta_{NO} \leq 0.25$) produces only a fast non-thermal desorption component. Under these conditions, measurements of $Y(\hbar\omega, F_\omega, \theta_i, P_\omega)$ indicate that the photophysics is substrate-mediated and linear in the Pt(111) absorbed flux. This implies that the mechanism for photo-desorption is much as described in Figure 3.8 and Figure 3.9. The relevant $2\pi^*$ for NO/Pt(111) is a resonance centered ~1.5 eV above ε_F [394]. Y is quite small, increasing from $\sim 5 \times 10^{-9}$ at 1.17 eV to 5×10^{-8} at 3.49 eV. In fact Y increases almost exponentially with photon energy for NO adsorbed on a variety of different metals but with a threshold at ca 1.1 eV, the NO binding energy to the surface [395]. If photochemistry occurs only from nascent hot electrons, Y should peak (or at least level off) at some $\hbar\omega > 1.5$ eV. Therefore, contributions to photodesorption from both nascent hot electrons and a hot electron cascade are most likely important.

Detailed (E_f, v, J) distributions have been measured for $\hbar\omega = 1.17-6.4$ eV [391,393]. Boltzmann E_f distributions with temperatures $T_f >> 2T_s$ are obtained in all cases, with T_f increasing significantly with the rotational energy E_J. In addition, T_f also increases with $\hbar\omega$ and with T_s. The desorption is highly peaked normal to the surface $Y \sim \cos^{11}\theta_f$. The NO desorbs vibrationally hot (relative to T_s) for all $\hbar\omega > 1.17$ eV, increasing somewhat with $\hbar\omega$, i.e., $T_v = 840$ K at 2.33 eV to $T_v = 950$ K at 6.4 eV. Rotational state distributions are almost constant at low J but ~ Boltzmann for high J. There is also a significant inversion in the spin-orbit states.

Although first principles theory is currently impossible for DIET processes, most of the above experiments have been qualitatively rationalized or fit (by varying excited state PES and lifetime) to the type of diabatic models outlined in Section 2.6.2 [96]. The hot E_f and T_v distributions arise from wavepacket propagation in the 2D excited state PES with lifetime t_R [95]. Of course quantitative comparison with experiment requires averaging over the distribution of excited state lifetimes $f(t_R, \tau^*)$ and the initial geometries of the adsorbed NO [96]. Averaging over a distribution of t_R predicts a Boltzmann distribution in E_f as observed [97,396]. Rotational excitation in photodesorption is minimal since rotational periods are very long compared to t_R. In a model of a sudden unhindered rotor, averaging over the initial librational states of the adsorbate at T_s gives a Boltzmann distribution of J

states in desorption [397]. It is possible that a tilted NO adsorption geometry causes the downward curvature in the Boltzmann rotational plot at low J and the inversion in the spin-orbit states (since the higher spin orbit state of NO adiabatically correlates to the lower occupied 2π state of NO) [397]. The correlation of T_f with E_J principally arises because the excitation of each degree of freedom is higher when t_R is longer [97,398].

Using fs laser excitation at 620 nm, a 2PC in Y of $\sim$0.5 ps [399] implicates hot electrons, probably thermalized at T_e, as the mechanism for desorption induced by the fs laser (Section 2.6.2). Rotational state distributions are nearly Boltzmann characterized by T_J. The 2PC of internal state distributions was also obtained. Rather surprisingly, significant differences in these 2PC were obtained for T_J and the state-resolved yield for the two spin-orbit states and this was qualitatively rationalized by a DIMET picture [399]. Where overlap in experiments exist, the qualitative results are similar to those for fs laser induced desorption of NO/Pd(111) [400,401]. For this latter system, the absolute yield $Y \sim F_\omega^3$ is large at typical fluencies used in the experiments and a very hot vibrational distribution was observed ($T_v = 2900$ K).

4.8.1.2. O_2/Pt(111)

There have been many studies of photochemistry in the single excitation (DIET) regime for O_2 adsorbed on Pt(111) [402] and for CO/O_2/Pt(111) [369] and the related adsorbates on Pd(111) [403]. For O_2/Pt(111) and CO/O_2/Pt(111), both photodesorption and photodissociation are observed with roughly the same yields [404]. In Section 4.6.1, it was argued that the photochemical production of CO_2 is due to hot O atom reaction with CO (or at least via a similar transition state) so that the CO_2 production is a consequence of the photodissociation of O_2 in CO/O_2/Pt(111). Thresholds for photodesorption and photodissociation/ CO_2 formation are $\sim$2.8 eV, although the thermal activation energy E_a for each process is only $\sim$0.4 eV. The mechanistic interpretation for the processes (and some of the experimental results) is decidedly different between different groups. Based on measurements of $Y(\hbar\omega, \theta_i, P_\omega)$ and analogy to electronic excitations observed in EELS, Mieher and Ho [369] suggest that both photodesorption and photodissociation are caused by direct excitation of the adsorbate-substrate complex, with excitation to the $3\sigma_u{}^*$ state being the most likely candidate [405]. However, the experimental results are equally well in agreement with a mechanism of substrate mediated hot electron excitation of the $3\sigma_u{}^*$ resonance [406]. On the other hand, other authors suggest that photodesorption and photodissociation involve different mechanisms, i.e., that photodesorption involves direct excitation but that photodissociation is principally substrate mediated [407]. The uncertainty in the mechanisms for the photophysics simply reflects the general lack of knowledge of the excited states of the adsorbates.

Both photodesorption of O_2 from O_2/Pt(111) and photodesorption and photo-oxidation of CO in CO/O_2/Pt(111) have also been well studied using fs lasers [408,412]. All experiments exhibit typical attributes of femtochemistry; much higher yields than in the single excitation regime, non-linear dependence of Y on F_ω, $\langle E_f \rangle$ that depends on F_ω, ~1 ps 2PC, etc. Because the femtochemistry occurs for $\hbar\omega$ less than the threshold for DIET, it is believed to arise from multiple excitations to the $2\pi^*$ resonance of the adsorbate (in the DIMET language). Figure 3.40(d) shows the photodesorption of O_2 from O_2/Pt(111) as a function of F_ω induced by a ~100 fs laser pulse. These results show the transition between the DIET (linear in F_ω) and DIMET regimes ((non-linear in F_ω). Y in the fs DIET regime is in good agreement with that from ns laser desorption. Because $Y \propto F_\omega^{6.3}$ is the same at 310 and 620 nm, it is suggested that the reaction is due to fully thermalized hot electrons defined by an electron temperature T_e. Similar results for O_2 desorption and CO_2 produced in CO/O_2/Pt(111) femtochemistry are given in Figure 3.40(a) and (b). Figure 3.40(c) shows that the branching ratio $Y(O_2)/Y(CO)$ increases dramatically with F_ω over that of the DIET regime, again emphasizing that the femtochemistry is a different mechanism than that of DIET photochemistry. However, it appears

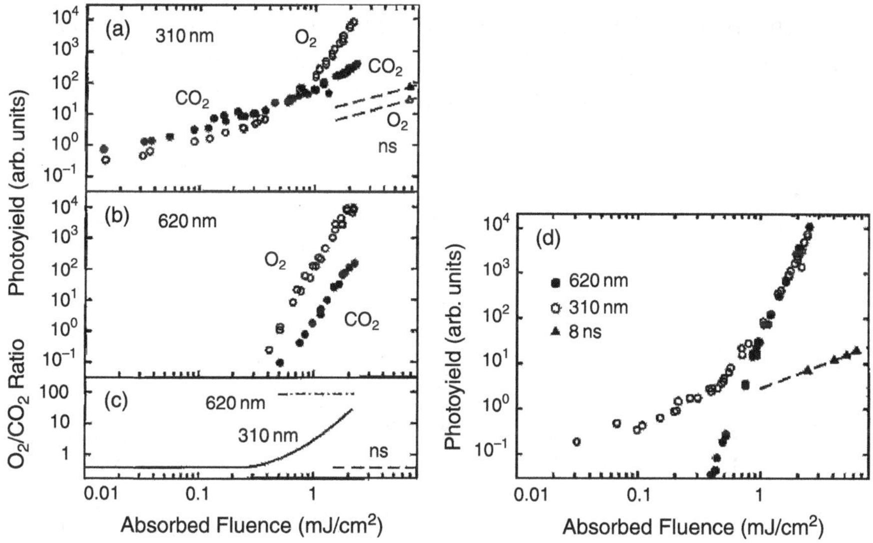

Figure 3.40. (a) Photoyield vs. absorbed laser fluence for O_2 and CO_2 desorption from a saturated CO/O_2/Pt(111) surface using 110 fs 310 nm laser pulses (circles) and 8 ns 355 nm laser pulses (triangles). (b) Same as in (a) except using 90 fs 620 nm laser pulses. (c) the line is the O_2/CO_2 branching ratio at 310 nm as a function of laser fluence, the dashed curve is the branching ratio for 355 nm ns laser pulses and the dot dashed line is that at 620 nm laser pulses. (d) Photoyield for O_2 desorption from a saturated O_2/Pt(111) surface from 620 nm 90 fs laser pulses, 310 nm 110 fs laser pulses and 355 nm 8 ns laser pulses as noted in the figure. From Ref. [411].

that in this case the Y dependence on F_ω for CO_2 desorption is not identical at 310 and 620 nm. Other studies show that the absolute Y and dependence on F_ω changes with $\hbar\omega$, especially for CO_2 formation and suggests that femtochemistry in $CO/O_2/Pt(111)$ is due to non-thermalized hot electrons [412]. However, there is a note of caution in this conclusion since several corrections are necessary in data analysis to extract $Y(\hbar\omega, F_\omega)$ properly, e.g., to treat non-homogeneous spatial distributions in the beam (as yield-weighted fluences) [413] and to include varying penetration depths of the lasers with $\hbar\omega$ (that affect T_e) [414]. It is unclear how these aspects have been included in the analysis of existing experiments.

4.8.1.3. H_2/Ru(0001)

Although there is no single-excitation DIET photochemistry for H adsorbed on Ru(0001), fs laser excitation at 830 nm of Ru(0001)(1×1)H induces associative desorption of H_2 and this has been extremely well studied experimentally [413,415,416]. With sufficient laser fluence, the yield of associative desorption can be quite high, nearly unity. Figure 3.41(a) shows Y measured for H_2 and D_2 as a function of F_ω. Y is nonlinear, i.e., $\propto F_\omega^{\ 2.8}$ for H_2 and $F_\omega^{\ 3.2}$ for D_2 and there is an enormous isotope effect with $Y(H_2)/Y(D_2) \approx 5-20$ depending upon F_ω. The 2PC for H_2 desorption is given in Figure 3.42(a). Both the large isotope effect and the 2PC implicate a hot electron mechanism for the associative desorption. TOF measurements indicate that $\langle E_f \rangle$ increases with F_ω and that $\langle E_f \rangle$ for H_2 is greater than that for D_2 at the same F_ω. State-resolved experiments show that $\langle E_f \rangle/2 > \langle E_v \rangle > \langle E_J \rangle$.

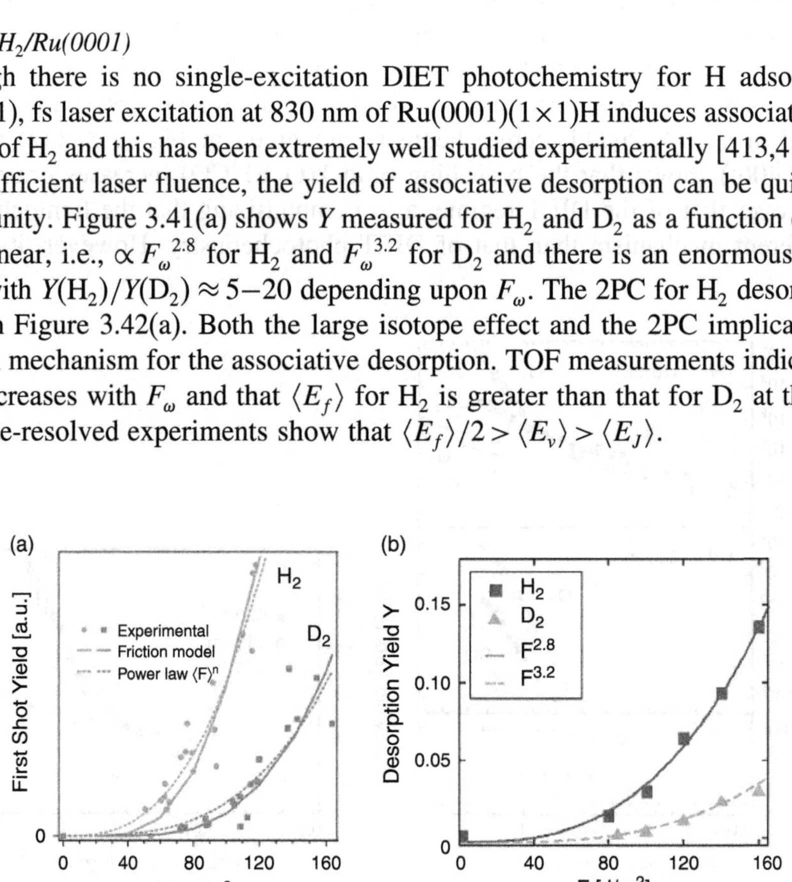

Figure 3.41. Photoyield Y for H_2 and D_2 associative desorption from Ru(0001)(1×1)H and Ru(0001)(1×1)D as a function of absorbed 800 nm 130fs laser pulse fluence F. (a) experimental results with circles for H_2 desorption and squares for D_2 desorption. Solid lines are fits to a 1D friction model and dashed lines are fits to power law expressions, $Y \propto F^{2.8}$ for H_2 and $Y \propto F^{3.2}$ for D_2. From Ref. [413]. (b) Equivalent photoyields for associative desorption from 3D 'first principles' molecular dynamics with electronic frictions. From Ref. [101].

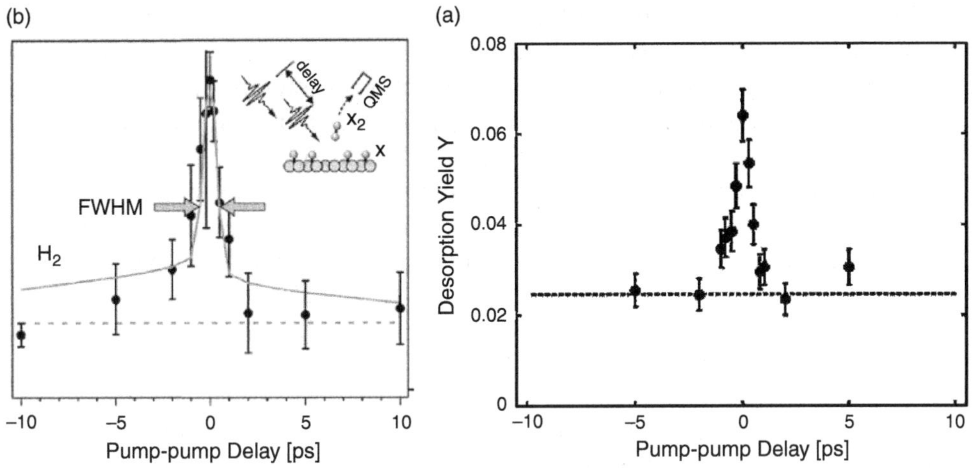

Figure 3.42. Two pulse correlation (2PC) for H_2 associative desorption induced by two nearly equal 800 nm 130 fs laser pulses as outlined in the inset to figure (b). The desorption yield is plotted as a function of delay between the two pulses. (b) Experimental results (points). The line is a fit of a 1D friction model to the experiments. From Ref. [413]. (a) 2PC from 3D 'first principles' molecular dynamics with electronic frictions. From Ref. [101].

The solid lines in Figure 3.41(a) and Figure 3.42(b) are obtained by fitting a 1D electronic friction model mentioned in Section 2.6.2 to the experiments. Best fits were obtained for a well depth $V_0 = 1.35$ eV and $(\eta/m) = (1/180)$ fs^{-1} for H_2 and $(\eta/m) = (1/360)$ fs^{-1} for D_2. However, it is not at all obvious why a 1D model is appropriate to these experiments or how to interpret parameters since associative desorption is at least a 2D dynamical problem in terms of Z and R. For example, the E_a for thermal associative desorption is 0.9 eV and there is no obvious reason in a 1D model why $V_0 \neq E_a$.

Results from the 3D dynamical model outlined in Figure 3.7 are given in Figure 3.41(b) and Figure 3.42(a) to compare with the corresponding experiments [101]. They are based on MDEF, with the PES from DFT and the electronic friction tensor also from DFT via eq. (2.15). $T_e(t)$ and $T_q(t)$ are obtained from the two-temperature model for conditions appropriate to the experiment [413]. Therefore, these theoretical results are truly 'first principles' with no adjustable parameters. The agreement with experiment is excellent. The MDEF calculations also semi-quantitatively predict the increase of $\langle E_f \rangle$ with F_ω, that $\langle E_f \rangle$ for H_2 is greater than that for D_2 and $\langle E_f \rangle/2 > \langle E_v \rangle$. The latter is simply due to a small translational barrier in the exit channel for desorption. The good agreement with the experiments justifies the nearly adiabatic frictional description of the femtochemistry. By analyzing trajectories, these calculations also give insight into why the 1D model can be used to fit the experiments of inherently multi-dimensional dynamics. Excitation of nuclei occurs from the fluctuating forces $F_i(t)$ when $T_e(t)$ is large. However, there is very rapid interchange between the Z and R coordinates as the H−H (or D−D) climbs out of the adsorption well towards desorption. Therefore, it is only the total energy or adsorbate temperature $T_{ads} = \langle E_{H2} \rangle / 2k_B$ that is important in desorption. However, not all molecules with $E_{H_2} > E_a$ desorb because of phase space constraints. There is effectively an 'entropic barrier' in multi-dimensional dynamics that forces a 1D fitting model to have $V_0 > E_a$. Figure 3.43 shows the time evolution of T_{ads}, T_e, T_q and the rate of desorption, and this is quite similar to equivalent pictures using the 1D model [413].

4.8.2. *Single molecule chemistry*

Single molecule chemistry induced by an STM is one of the new frontiers in surface chemistry, and it is beyond the scope of this chapter to treat this in any meaningful way. Only a brief discussion of one system is given to illustrate the relationship to many of the dynamic principles discussed here. The reader is referred to an excellent review article for recent status in this field [417].

4.8.2.1. *O_2/Pt(111)*

As discussed in Section 4.4.1, O_2 molecularly adsorbs onto Pt(111) into two molecularly adsorbed states; an O_2^- state centered at the bridge site and an O_2^{2-} state centered at the fcc hollow site. When an STM tip is positioned over an isolated O_2^{2-} adsorbate, tunneling electrons from the tip induce dissociation of only that molecule without affecting the surrounding [153,326]. The instant of dissociation is readily

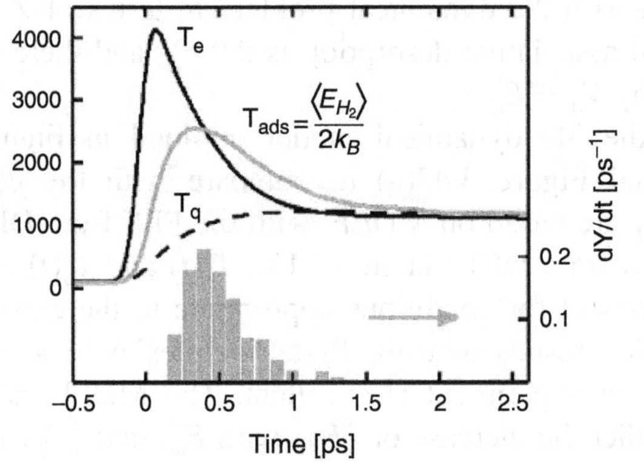

Figure 3.43. The time dependent electronic temperature T_e, lattice temperature T_q, and adsorbate temperature defined as $T_{ads} = \langle E_{H_2} \rangle / 2k_B$ following a 130 fs laser pulse with absorbed laser fluence of 120 J/m^2 centered at time $t = 0$. The bar graph is the rate of associative desorption dY/dt as a function of t. T_e and T_q are from the conventional two temperature model and T_{ads} and dY/dt are from 3D 'first principles' molecular dynamics with electronic frictions. From Ref. [101].

monitored by the STM as a sharp step in the tunneling current. Repeated measurements for different isolated molecules allows a measurement of the dissociation rate R_d as a function of tip bias and tunneling current. Experimental results are shown as the solid lines in Figure 3.44(c). $R_d \propto I^n$ with $n = 0.8 \pm 0.2$, 1.8 ± 0.2 and 2.9 ± 0.3 for tip biases of 0.4, 0.3 and 0.2 V respectively. No simultaneous desorption was induced by tunneling electrons. For the O_2^- state, tunneling electrons also induce dissociation *and* desorption, but only at higher tip biases, ~0.5 eV.

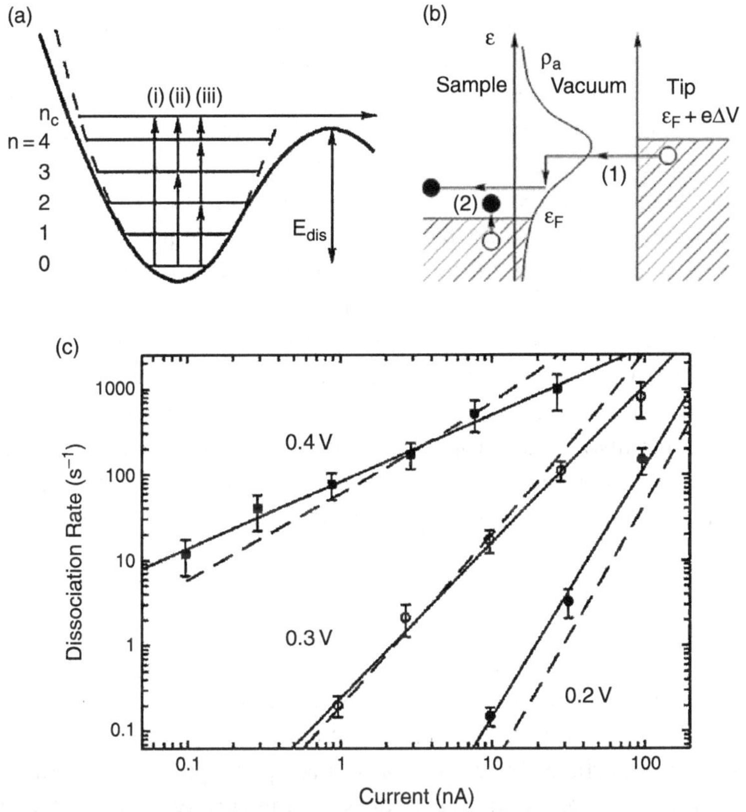

Figure 3.44. Dissociation of O_2 adsorbed on Pt(111) by inelastic tunneling of electrons from a STM tip. (a) Schematic 1D PES for chemisorbed O_2^{2-} dissociation and illustrating different types of excitations that can lead to dissociation. (b) Schematic picture of inelastic electron tunneling to an adsorbate-induced resonance with density of states ρ_a inducing vibrational excitation (1) competing with non-adiabatic vibrational de-excitation that creates e-h pairs in the substrate (2). (c) Dissociation rate R_d for O_2^{2-} as a function of tunneling current I at the three tip bias voltages labeled in the figure. Solid lines are fits of $R_d \propto I^N$ to the experiments with N = 0.8, 1.8, and 3.2 for tip biases of 0.4, 0.3, and 0.2 V, respectively and correspond to the three excitation conditions in (a). Dashed lines are results of a theoretical model incorporating the physics in (a) and (b) and a single fit parameter. From Ref. [153].

The theoretical model developed to explain these experiments is based on inelastic tunneling of electrons from the tip into the $2\pi^*$ adsorbate resonance that induces vibrational excitation in a manner similar to that of the DIMET model (Figure 3.44(b)). Of course, in this case, the chemistry is induced by specific and variable energy hot electrons rather than a thermal distribution at T_e. Another significant difference is that STM induced currents are low so that vibrational excitation rates are smaller than vibrational de-excitation rates via e-h pair damping. Therefore, coherent vibrational ladder climbing dominates over incoherent ladder climbing, i.e., dissociation take place by the minimum number of vibrational transitions. On the other hand, hot electron densities in fs laser induced femtochemistry are so large that incoherent ladder climbing is believed to be the dominant ladder climbing mechanism in this case. The theoretical model is outlined in Figure 3.44(a) and treats the vibrational ladder climbing in a 1D PES (truncated harmonic oscillator) via a Master equation. The key inputs are the (current independent) vibrational relaxation rate taken from experiment and a tunneling electron-vibration coupling constant. The latter is simply a fitting variable and is the equivalent to the unknown excited state PES and lifetime in the DIMET model. The model gives good agreement with experiments (Figure 3.44(c)) and suggests that the different n observed for the different bias voltages simply reflect the minimum number of excitation events required to dissociate at a given bias voltage because coherent multi-quantum excitations dominate the ladder climbing. One caveat to the above model is that the tunneling electrons also induce rotation of the O_2^{2-} and with a rate several orders of magnitude larger than that for dissociation [418,419]. Therefore, it is not entirely obvious that a 1D model is entirely appropriate to describe dissociation.

5. Summary and outlook

This chapter has tried to summarize what is the current state of knowledge (or at least the author's) in the fall of 2006 of the dynamics of bond making and breaking at the gas-surface interface. Of course, any attempt to provide such a summary is like hitting a moving target since our understanding of any field evolves with time. Hopefully, future readers will not hold the author too accountable if concepts change as our knowledge base grows in this field. The modus operandi of this chapter were to first introduce general concepts necessary to understand the dynamics of chemical change at surfaces, a few of the experimental techniques currently in use and then to treat in some detail a few selected examples of the different types of chemical change at surfaces that are summarized in Figure 3.1. Most examples were chosen because of extensive experimental *and* theoretical work on these particular systems. Special emphasis has been placed on trying to access how well 'first principles' dynamics can now quantitatively interpret detailed experiments.

Not surprisingly, the field of gas-surface chemical dynamics has evolved and matured significantly during the past three decades. In the early pre-DFT years, detailed experiments were the only way to gain any insight into the dynamics and ultimately the PES topology, although theory played the important role in assimilating this knowledge. Now, it seems that roles have reversed, at least at the level of 6D 'first principles' dynamics. This first principles dynamics now provides insight significantly more easily (and more cheaply) than detailed dynamics experiments. However, there are still major challenges in generalizing the theory beyond that of H_2 + metals, e.g., how to include lattice coupling reasonably and even understanding when the breakdown in the BOA is important.

In this author's opinion, the experimental study of gas-surface dynamics has unfortunately dropped off precipitously in the last several years. There are probably many reasons for this; funding vagaries, the dominance of the STM in surface science today, the success of 'first principles' dynamics, etc. However, it is absolutely essential to have the most detailed experiments to benchmark first principles theory. Comparisons for even the simple systems considered here demonstrate that theory is far from fully understood. Hopefully, a new generation of experimentalists/theorists will accept this challenge so that the next status report will make this one obsolete.

The author wishes to thank his many collaborators and colleagues in this field, almost too numerous to mention. However, special thanks to the collaborators and colleagues at both IBM and Denmark and to Mats Persson, Geert-Jan Kroes, and Bret Jackson who have read parts of this review and offered comments. Nevertheless, the opinions presented here are the sole responsibility of the author. Discussions and arguments with many of the participants in this field, both in print and over many a bottle of wine/beer, and in many parts of the world, over the past 25 years have led to the opinions presented in this review. It has been a fun journey of discovery. Of course since this is still an active research field, some of the concepts/opinions are not yet etched in stone, but hopefully time will prove those presented here to be more fact than fiction. The author also wishes to thank the Alexander von Humboldt foundation for support of visits to Berlin where this chapter was initiated and Bonnie for her patience while this chapter was written.

References

[1] R.D. Levine and R.B. Bernstein, Molecular Reaction Dynamics and Chemical Reactivity, Oxford University Press, New York, 1987.

[2] A. Gross, Theoretical Surface Science: A Microscopic Perspective, Springer, Berlin, 2003.

[3] J. C. Tully, in Dynamics of Molecular Collisions, Vol. B (W. H. Miller, ed.), Plenum, NY, 1976, p. 217.

[4] G. R. Darling and S. Holloway, Rep. Prog. Phys. 58 (1995) 1595.

[5] M. C. Payne, M. P. Teter, D. C. Allan, T. A. Arias, and J. D. Joannopoulos, Rev. Modern Phys. 64 (1992) 1045.
[6] A. Devita, I. Stich, M. J. Gillan, M. C. Payne, and L. J. Clarke, Phys. Rev. Lett. 71 (1993) 1276.
[7] H. F. Busnengo, A. Salin, and W. Dong, J. Chem. Phys. 112 (2000) 7641.
[8] C. Crespos, M. A. Collins, E. Pijper, and G. J. Kroes, J. of Chem. Phys. 120 (2004) 2392.
[9] S. Lorenz, M. Scheffler, and A. Gross, Phy. Rev. B 73 (2006).
[10] G. J. Kroes, A. Gross, E. J. Baerends, M. Scheffler, and D. A. McCormack, Accts. Chem. Res. 35 (2002) 193.
[11] G. J. Kroes and M. F. Somers, J. Theor. Comput. Chem. 4 (2005) 493.
[12] C. T. Rettner and D. J. Auerbach, Chem. Phys. Lett. 253 (1996) 236.
[13] E. Zaremba and W. Kohn, Phys. Rev. B 13 (1976) 2270.
[14] N. Esbjerg and J. K. Nørskov, Phys. Rev. Lett. 45 (1980) 807.
[15] J. E. Lennard-Jones, Trans. Faraday Soc. 28 (1932) 333.
[16] S. Holloway and J. W. Gadzuk, J. Chem. Phys. 82 (1985) 5203.
[17] A. Eichler, F. Mittendorfer, and J. Hafner, Phys. Rev. B 62 (2000) 4744.
[18] M. J. Grunze, J. Fuhler, M. Neumann, C. R. Brundle, D. J. Auerbach, and J. Behm, Surf. Sci. 139 (1984) 109.
[19] W. Brenig, in Kinetics of Interface Reactions (M. Grunze and H. J. Kreuzer, eds.), Springer, Berlin, 1987, p. 19.
[20] A. C. Luntz and P. Kratzer, J. Chem. Phys. 104 (1996) 3075.
[21] B. Hammer, M. Scheffler, K. W. Jacobsen, and J. K. Nørskov, Phys. Rev. Lett. 73 (1994) 1400.
[22] M. J. Murphy, J. F. Skelly, A. Hodgson, and B. Hammer, J. Chem. Phys. 110 (1999) 6954.
[23] E. Pijper, G. J. Kroes, R. A. Olsen, and E. J. Baerends, J. Chem. Phys. 117 (2002) 5885.
[24] G. R. Darling and S. Holloway, Surf. Sci. 304 (1994) L461.
[25] K. Gundersen, K. W. Jacobsen, J. K. Nørskov, and B. Hammer, Surf. Sci. 304 (1994) 131.
[26] A. Salin, J. Chem. Phys. 124 (2006) 104704.
[27] C. Diaz, J. K. Vincent, G. P. Krishnamohan, R. A. Olsen, K. Honkala, J. K. Nørskov, and G. J. Kroes, J. Chem. Phys. (2006) 114706.
[28] C. T. Campbell, Y. K. Sun, and W. H. Weinberg, Chem. Phys. Lett. 179 (1991) 53.
[29] M. Grunze, M. Golze, W. Hirschwald, H. J. Freund, H. Pulm, U. Seip, M. C. Tsai, G. Ertl, and J. Kueppers, Phys. Rev. Lett. 53 (1984) 850.
[30] A. C. Luntz, J. Grimblot, and D. E. Fowler, Phys. Rev. B: Condens. Matter 39 (1989) 12903.
[31] C. T. Rettner and H. Stein, Phys. Rev. Lett. 59 (1987) 2768.
[32] J. C. Tully, Surf. Sci. 111 (1981) 461.
[33] H. A. Michelsen, C. T. Rettner, D. J. Auerbach, and R. N. Zare, J. Chem. Phys. 98 (1993) 8294.
[34] C. T. Rettner, H. A. Michelsen, and D. J. Auerbach, J. Chem. Phys. 102 (1995) 4625.
[35] C. T. Rettner, E. K. Schweizer, and C. B. Mullins, J. Chem. Phys. 90 (1989) 3800.
[36] T. B. Grimley and S. Holloway, Chem. Phys. Lett. 161 (1989) 163.
[37] J. C. Tully, Surf. Sci. 299–300 (1994) 667.
[38] L. Diekhöner, H. Mortensen, A. Baurichter, and A. C. Luntz, J. Chem. Phys. 115 (2001) 3356.
[39] R. R. Lucchese and J. C. Tully, J. Chem. Phys. 80 (1984) 3451.
[40] E. Watts and G. O. Sitz, J. Chem. Phys. 111 (1999) 9791.
[41] S. A. Adelman and J. D. Doll, J. Chem. Phys. 64 (1976) 2375.
[42] J. C. Tully, J. Chem. Phys. 73 (1980) 1975.
[43] N. Pineau, H. F. Busnengo, J. C. Rayez, and A. Salin, J. Chem. Phys. 122 (2005) 214705.
[44] N. Perron, N. Pineau, E. Arquis, J. C. Rayez, and A. Salin, Surf. Sci. 599 (2005) 160.

[45] J. Harris, in Dynamics of Gas–Surface Interactions (C. T. Rettner and M. N. R. Ashfold, eds.), Royal Soc. Chem., Cambridge, UK, 1991.
[46] E. K. Grimmelmann, J. C. Tully, and M. J. Cardillo, J. Chem. Phys. 72 (1980) 1039.
[47] J. A. Barker and D. J. Auerbach, Chem. Phys. Lett. 67 (1979) 393.
[48] J. C. Tully, J. Chem. Phys. 92 (1990) 680.
[49] J. Kimman, C. T. Rettner, D. J. Auerbach, J. A. Barker, and J. C. Tully, Phys. Rev. Lett. 57 (1986) 2053.
[50] J. Harris and A. C. Luntz, J. Chem. Phys. 91 (1989) 6421.
[51] M. Persson and J. Harris, Surf. Sci. (1987) 0039.
[52] D. Farias and K. H. Rieder, Reports on Prog. Phys. 61 (1998) 1575.
[53] H. Hoinkes, Rev. Modern Phys. 52 (1980) 933.
[54] S. Andersson, L. Wilzen, M. Persson, and J. Harris, Phys. Rev. B: Condens. Matter 40 (1989) 8146.
[55] H. Schlichting, D. Menzel, T. Brunner, W. Brenig, and J. C. Tully, Phys. Rev. Lett. 60 (1988) 2515.
[56] H. Schlichting, D. Menzel, T. Brunner, and W. Brenig, J. Chem. Phys. 97 (1992) 4453.
[57] M. Hand and J. Harris, J. Chem. Phys. 92 (1990) 7610.
[58] H. Mortensen, E. Jensen, L. Diekhöner, A. Baurichter, A. C. Luntz, and V. Petrunin, J. Chem. Phys. 118 (2003) 11200.
[59] A. C. Luntz and J. Harris, Surf. Sci. 258 (1991) 397.
[60] A. T. Pasteur, S. J. Dixon-Warren, Q. Ge, and D. A. King, J. Chem. Phys. 106 (1997) 8896.
[61] G. Henkelman and H. Jonsson, Phys. Rev. Lett. 86 (2000) 664.
[62] H. S. Bengaard, in Departmen t of Physics, Technical University of Denmark, Lyngby, Denmark, 2001.
[63] S. Nave and B. Jackson, Phys. Rev. Lett. 98 (2007) 173003.
[64] A. Amirav and M. J. Cardillo, Phys. Rev. Lett. 57 (1986) 2299.
[65] Y. H. Huang, C. T. Rettner, D. J. Auerbach, and A. M. Wodtke, Science 290 (2000) 111.
[66] B. Kasemo, E. Tornqvist, J. K. Nørskov, and B. I. Lundqvist, Surf. Sci. 89 (1979) 554.
[67] A. Hellman, B. Razaznejad, and B. I. Lundqvist, Phys. Rev. B 71 (2005) 205424.
[68] A. C. Luntz and M. Persson, J. Chem. Phys. 123 (2005) 074704.
[69] E. G. Dagliano, P. Kumar, W. Schaich, and H. Suhl, Phys. Rev. B 11 (1975) 2122.
[70] M. Head-Gordon and J. C. Tully, J. Chem. Phys. 103 (1995) 10137.
[71] B. Hellsing and M. Persson, Phys. Scr. 29 (1984) 360.
[72] J. R. Trail, D. M. Bird, M. Persson, and S. Holloway, J. Chem Phys 119 (2003) 4539.
[73] N. Lorente and M. Persson, Faraday Discuss. (2000) 277.
[74] D. C. Langreth and M. Persson, in Laser Spectroscopy and Photochemistry on Metal Surfaces, Vol. 5 (H.-L. Dai and W. Ho, eds.), World Scientific, River Edge, New Jersey, 1995, p. 498.
[75] M. Head-Gordon and J. C. Tully, Phys. Rev. B: Condens. Matter 46 (1992) 1853.
[76] M. Persson, Philosophical Transactions of the Royal Society of London Series a-Mathematical Physical and Engineering Sciences 362 (2004) 1173.
[77] V. Krishna and J. C. Tully, J. Chem. Phys. 125 (2006) 054706.
[78] A. C. Luntz, M. Persson, and G. O. Sitz, J. Chem. Phys. 124 (2006) 091101.
[79] J. T. Kindt, J. C. Tully, M. Head-Gordon, and M. A. Gomez, J. Chem. Phys. 109 (1998) 3629.
[80] J. R. Trail, M. C. Graham, D. M. Bird, M. Persson, and S. Holloway, Phys. Rev. Lett. 88 (2002) 166802.
[81] V. Krishna, J. Chem. Phys. 125 (2006) 034711.
[82] J. K. Nørskov, J. Vac. Sci. Technol. 18 (1981) 420.
[83] A. Hellman, B. Razaznejad, and B. I. Lundqvist, J. Chem. Phys. 120 (2004) 4593.
[84] J. Behler, B. Delley, K. Reuter, and M. Scheffler, Phys. Rev. B 75 (2007) 115409.

[85] J. D. White, J. Chen, D. Matsiev, D. J. Auerbach, and A. M. Wodtke, Nature 433 (2005) 503.
[86] H. Nienhaus, Surf. Sci. Reports 45 (2002) 3.
[87] J. C. Tully, M. Gomez, and M. Head-Gordon, J. Vac. Sci. Technol. A 11 (1993) 1914.
[88] S. M. Li and H. Guo, J. Chem. Phys. 117 (2002) 4499.
[89] Y. Yourdshahyan, B. Razaznejad, and B. I. Lundqvist, Phys. Rev. B 65 (2002) 075416.
[90] J. Behler, B. Delley, S. Lorenz, K. Reuter, and M. Scheffler, Phys. Rev. Lett. 94 (2005).
[91] P. Feulner and D. Menzel, in Laser Spectroscopy and Photo-chemistry on metal surfaces (H.-L. Dai and W. Ho, eds.), World Scientific, Singapore, 1995, p. 627.
[92] D. Menzel and R. Gomer, J. Chem. Phys. 41 (1964) 3311.
[93] P. A. Redhead, Can. J. Phys. 42 (1964) 886.
[94] P. R. Antoniewicz, Phys. Rev. B 21 (1980) 3811.
[95] J. W. Gadzuk, L. J. Richter, S. A. Buntin, D. S. King, and R. R. Cavanagh, Surf. Sci. 235 (1990) 317.
[96] F. M. Zimmermann and W. Ho, Surf. Sci. Reports 22 (1995) 129.
[97] E. Hasselbrink, Chem. Phys. Lett. 170 (1990) 329.
[98] N. Del Fatti, R. Bouffanais, F. Vallee, and C. Flytzanis, Phys. Rev.Lett. 81 (1998) 922.
[99] S. I. Anisimov, B. L. Kapeliovich, and T. L. Perel'man, Sov. Phys. JETP 39 (1974) 375.
[100] J. A. Misewich, T. F. Heinz, and D. M. Newns, Phys. Rev. Lett. 68 (1992) 3737.
[101] A. C. Luntz, M. Persson, S. Wagner, C. Frischkorn, and M. Wolf, J. Chem. Phys. 124 (2006) 244702.
[102] D. M. Newns, T. F. Heinz, and J. A. Misewich, Progress of Theoretical Physics Supplement (1991) 411.
[103] M. Brandbyge, P. Hedegard, T. F. Heinz, J. A. Misewich, and D. M. Newns, Phys. Rev. B 52 (1995) 6042.
[104] G. Scoles (ed.), Atomic and Molecular Beam Methods, Oxford University Press, Cary, North Carolina, 1988.
[105] G. O. Sitz, Reports on Progress in Physics 65 (2002) 1165.
[106] M. Silva, R. Jongma, R. W. Field, and A. M. Wodtke, Ann. Rev. Phys. Chem. 52 (2001) 811.
[107] C. T. Rettner, H. A. Michelsen, and D. J. Auerbach, Faraday Discuss. 96 (1993) 17.
[108] M. Kay, G. R. Darling, S. Holloway, J. A. White, and D. M. Bird, Chem. Phys. Lett. 245 (1995) 311.
[109] A. Gross, S. Wilke, and M. Scheffler, Phys. Rev. Lett. 75 (1995) 2718.
[110] C. Crespos, H. F. Busnengo, W. Dong, and A. Salin, J. Chem. Phys. 114 (2001) 10954.
[111] M. A. Di Cesare, H. F. Busnengo, W. Dong, and A. Salin, J. Chem. Phys. 118 (2003) 11226.
[112] A. C. Luntz, J. Chem. Phys. 113 (2000) 6901.
[113] M. J. Murphy and A. Hodgson, J. Chem. Phys. 108 (1998) 4199.
[114] H. Mortensen, L. Diekhöner, A. Baurichter, and A. C. Luntz, J. Chem. Phys. 116 (2002) 5781.
[115] L. B. F. Juurlink, R. R. Smith, D. R. Killelea, and A. L. Utz, Phys. Rev. Lett. 94 (2005) 208303.
[116] R. R. Smith, D. R. Killelea, D. F. DelSesto, and A. L. Utz, Science 304 (2004) 992.
[117] P. Maroni, D. C. Papageorgopoulos, M. Sacchi, T. T. Dang, R. D. Beck, and T. R. Rizzo, Phys. Rev. Lett. 94 (2005) 246104.
[118] R. D. Beck, P. Maroni, D. C. Papageorgopoulos, T. T. Dang, M. P. Schmid, and T. R. Rizzo, Science 302 (2003) 98.
[119] J. B. Miller, H. R. Siddiqui, S. M. Gates, J. N. Russell, Jr., J. T. Yates, Jr., J. C. Tully, and M. J. Cardillo, J. Chem. Phys. 87 (1987) 6725.
[120] K. D. Rendulic and A. Winkler, Surf. Sci. 299-300 (1994) 261.
[121] K. H. Allers, H. Pfnuer, P. Feulner, and D. Menzel, Surf. Sci. 291 (1993) 167.
[122] G. Comsa and R. David, Surf. Sci. Rep. 5 (1985) 145.
[123] K. H. Allers, H. Pfnuer, P. Feulner, and D. Menzel, Surf. Sci. 286 (1993) 297.

[124] D. Wetzig, M. Rutkowski, H. Zacharias, and A. Gross, Physical Review B 63 (2001) 205412.
[125] M. J. Murphy, J. F. Skelly, and A. Hodgson, Chem. Phys. Lett. 279 (1997) 112.
[126] L. Diekhöner, L. Hornekaer, H. Mortensen, E. Jensen, A. Baurichter, V. Petrunin, and A. C. Luntz, J. Chem. Phys. 117 (2002) 5018.
[127] J. E. Hurst, C. A. Becker, J. P. Cowin, K. C. Janda, L. Wharton, and D. J. Auerbach, Phys. Rev. Lett. 43 (1979) 1175.
[128] M. G. Tenner, E. W. Kuipers, W. Y. Langhout, A. W. Kleyn, G. Nicolasen, and S. Stolte, Surf. Sci. 236 (1990) 151.
[129] X. Y. Zhu, J. M. White, M. Wolf, E. Hasselbrink, and G. Ertl, Chemical Physics Letters 176 (1991) 459.
[130] L. J. Richter, S. A. Buntin, D. S. King, and R. R. Cavanagh, Chem. Phys. Lett. 186 (1991) 423.
[131] J. A. Prybyla, H. W. K. Tom, and G. D. Aumiller, Phys. Rev. Lett. 68 (1992) 503.
[132] E. H. G. Backus, A. Eichler, A. W. Kleyn, and M. Bonn, Science 310 (2005) 1790.
[133] Y. Andersson, D. C. Langreth, and B. I. Lundqvist, Phys. Rev. Lett. 76 (1996) 102.
[134] M. Head-Gordon, J. C. Tully, C. T. Rettner, C. B. Mullins, and D. J. Auerbach, J. Chem. Phys. 94 (1991) 1516.
[135] D. Kulginov, M. Persson, C. T. Rettner, and D. S. Bethune, J. Phys. Chem. 100 (1996) 7919.
[136] R. J. W. E. Lahaye, S. Stolte, A. W. Kleyn, R. J. Smith, and S. Holloway, Surf. Sci. 307–309 (1994) 187.
[137] J. E. Hurst, L. Wharton, K. C. Janda, and D. J. Auerbach, J. Chem. Phys. 78 (1983) 1559.
[138] J. E. Hurst, Jr., L. Wharton, K. C. Janda, and D. J. Auerbach, J. Chem. Phys. 83 (1985) 1376.
[139] C. B. Mullins, C. T. Rettner, D. J. Auerbach, and W. H. Weinberg, Chem. Phys. Lett. 163 (1989) 111.
[140] J. Stromquist, L. Bengtsson, M. Persson, and B. Hammer, Surf. Sci. 397 (1998) 382.
[141] C. T. Rettner and D. J. Auerbach, J. Chem. Phys. 104 (1996) 2732.
[142] T. Kammler and J. Kuppers, J. Chem. Phys. 111 (1999) 8115.
[143] G. Lee and E. W. Plummer, Surf. Sci. 498 (2002) 229.
[144] F. Greuter and E. W. Plummer, Solid State Comm. 48 (1983) 37.
[145] A. D. Johnson, K. J. Maynard, S. P. Daley, Q. Y. Yang, and S. T. Ceyer, Phys. Rev. Lett. 67 (1991) 927.
[146] T. H. Kammler, J. Lee, and J. Kuppers, J. Chem. Phys. 106 (1997) 7362.
[147] S. Wehner and J. Kuppers, Surf. Sci. 411 (1998) 46.
[148] D. V. Shalashilin and B. Jackson, J. Chem. Phys. 109 (1998) 2856.
[149] J. Harris and B. Kasemo, Surf. Sci. 105 (1981) L281.
[150] C. L. A. Lamont, B. N. J. Persson, and G. P. Williams, Chemical Physics Letters 243 (1995) 429.
[151] H. Nienhaus, H. S. Bergh, B. Gergen, A. Majumdar, W. H. Weinberg, and E. W. McFarland, Phys. Rev. Lett. 82 (1999) 446.
[152] B. Gergen, H. Nienhaus, W. H. Weinberg, and E. W. McFarland, Science 294 (2001) 2521.
[153] B. C. Stipe, M. A. Rezaei, W. Ho, S. Gao, M. Persson, and B. I. Lundqvist, Phys. Rev. Lett. 78 (1997) 4410.
[154] G. Wahnstrom, A. B. Lee, and J. Stromquist, J. Chem. Phys. 105 (1996) 326.
[155] R. J. Behm and C. R. Brundle, Journal of Vacuum Science & Technology A – Vacuum Surfaces And Films 2 (1984) 1040.
[156] S. K. So, R. Franchy, and W. Ho, J. Chem. Phys. 91 (1989) 5701.
[157] M. P. Jigato, D. A. King, and A. Yoshimori, Chem. Phys. Lett. 300 (1999) 639.
[158] H. Asada and T. Matsui, Japanese Journal Of Applied Physics Part 1-Regular Papers Short Notes & Review Papers 21 (1982) 259.

[159] H. Asada, Surf. Sci. 110 (1981) 270.
[160] A. W. Kleyn, A. C. Luntz, and D. J. Auerbach, Phys. Rev. Lett. 47 (1981) 1169.
[161] G. M. McClelland, G. D. Kubiak, H. G. Rennagel, and R. N. Zare, Phys. Rev. Lett. 46 (1981) 831.
[162] G. D. Kubiak, J. E. Hurst, H. G. Rennagel, G. M. McClelland, and R. N. Zare, J. Chem. Phys. 79 (1983) 5163.
[163] J. A. Barker, A. W. Kleyn, and D. J. Auerbach, Chem. Phys. Lett. 97 (1983) 9.
[164] H. Voges and R. Schinke, Chem. Phys. Lett. 100 (1983) 245.
[165] C. W. Muhlhausen, L. R. Williams, and J. C. Tully, J. Chem. Phys. 83 (1985) 2594.
[166] M. G. Tenner, F. H. Guezebroek, E. W. Kuipers, A. E. Wiskerke, A. W. Kleyn, S. Stolte, and A. Namiki, Chem. Phys. Lett. 168 (1990) 45.
[167] M. G. Tenner, E. W. Kuipers, A. W. Kleyn, and S. Stolte, J. Chem. Phys. 94 (1991) 5197.
[168] A. W. Kleyn, Surf. Rev. Lett. 1 (1994) 157.
[169] F. H. Geuzebroek, A. E. Wiskerke, M. G. Tenner, A. W. Kleyn, S. Stolte, and A. Namiki, J. Phys. Chem. 95 (1991) 8409.
[170] A. C. Luntz, A. W. Kleyn, and D. J. Auerbach, Phys. Rev. B: Condens. Matter 25 (1982) 4273.
[171] A. W. Kleyn, A. C. Luntz, and D. J. Auerbach, Surf. Sci. 152-153 (1985) 99.
[172] A. C. Luntz, A. W. Kleyn, and D. J. Auerbach, J. Chem. Phys. 76 (1982) 737.
[173] S. Gregurick, M. H. Alexander, and A. E. DePristo, J. Chem. Phys. 100 (1994) 610.
[174] A. E. DePristo and M. H. Alexander, J. Chem. Phys. 94 (1991) 8454.
[175] C. T. Rettner, F. Fabre, J. Kimman, and D. J. Auerbach, Phys. Rev. Lett. 55 (1985) 1904.
[176] C. T. Rettner, J. Kimman, F. Fabre, D. J. Auerbach, and H. Morawitz, Surf. Sci. 192 (1987) 107.
[177] D. M. Newns, Surface Science 171 (1986) 600.
[178] A. Gross and W. Brenig, Chemical Physics 177 (1993) 497.
[179] G. A. Gates, G. R. Darling, and S. Holloway, J. Chem. Phys. 101 (1994) 6281.
[180] Y. Huang, A. M. Wodtke, H. Hou, C. T. Rettner, and D. J. Auerbach, Phys. Rev. Lett. 84 (2000) 2985.
[181] C. T. Rettner, J. Kimman, and D. J. Auerbach, J. Chem. Phys. 94 (1991) 734.
[182] G. O. Sitz, A. C. Kummel, and R. N. Zare, J. Chem. Phys. 87 (1987) 3247.
[183] A. C. Kummel, G. O. Sitz, R. N. Zare, and J. C. Tully, J. Chem. Phys. 89 (1988) 6947.
[184] G. O. Sitz, A. C. Kummel, R. N. Zare, and J. C. Tully, J. Chem. Phys. 89 (1988) 2572.
[185] G. O. Sitz, A. C. Kummel, and R. N. Zare, J. Chem. Phys. 89 (1988) 2558.
[186] A. C. Kummel, G. O. Sitz, R. N. Zare, and J. C. Tully, J. Chem. Phys. 91 (1989) 5793.
[187] J. L. W. Siders and G. O. Sitz, J. Chem. Phys. 101 (1994) 6264.
[188] E. W. Kuipers, M. G. Tenner, M. E. M. Spruit, and A. W. Kleyn, Surf. Sci. 205 (1988) 241.
[189] E. W. Kuipers, M. G. Tenner, A. W. Kleyn, and S. Stolte, Chem. Phys. 138 (1989) 451.
[190] J. C. Tully, in Kinetics of Interface Reactions;Springer Ser. Surf. Sci., Vol. 8 (M. Grunze and H. J. Kreuzer, eds.), Springer, Berlin, 1987, p. 37.
[191] R. J. Gorte, L. D. Schmidt, and J. L. Gland, Surface Science 109 (1981) 367.
[192] C. T. Campbell, G. Ertl, and J. Segner, Surf. Sci. 115 (1982) 309.
[193] J. K. Brown and A. C. Luntz, Chem. Phys. Lett. 204 (1993) 451.
[194] E. H. G. Backus, A. Eichler, M. L. Grecea, A. W. Kleyn, and M. Bonn, Journal Of Chemical Physics 121 (2004) 7946.
[195] P. Zhu, T. Shimada, H. Kondoh, I. Nakai, M. Nagasaka, and T. Ohta, Surface Science 565 (2004) 232.
[196] J. A. Serri, J. C. Tully, and M. J. Cardillo, J. Chem. Phys. 79 (1983) 1530.
[197] Q. Ge and D. A. King, Chem. Phys. Lett. 285 (1998) 15.
[198] D. C. Ford, Y. Xu, and M. Mavrikakis, Surf. Sci. 587 (2005) 159.

[199] A. Bogicevic and K. C. Hass, Surf. Sci. 506 (2002) L237.
[200] M. Mavrikakis, L. B. Hansen, J. J. Mortensen, B. Hammer, and J. K. Nørskov, in ACS Symp. Ser., Vol. 721, 1999, p. 245.
[201] J. A. Serri, M. J. Cardillo, and G. E. Becker, J. Chem. Phys. 77 (1982) 2175.
[202] S. Kneitz, J. Gemeinhardt, and H. P. Steinruck, Surf. Sci. 440 (1999) 307.
[203] J. T. Kindt and J. C. Tully, Surf. Sci. 477 (2001) 149.
[204] E. W. Kuipers, M. G. Tenner, A. W. Kleyn, and S. Stolte, Phys. Rev. Lett. 62 (1989) 2152.
[205] D. A. Mantell, R. R. Cavanagh, and D. S. King, J. Chem. Phys. 84 (1986) 5131.
[206] J. Segner, H. Robota, W. Vielhaber, G. Ertl, F. Frenkel, J. Haeger, W. Krieger, and H. Walther, Surf. Sci. 131 (1983) 273.
[207] M. Asscher, W. L. Guthrie, T. H. Lin, and G. A. Somorjai, J. Chem. Phys. 78 (1983) 6992.
[208] J. A. Barker and D. J. Auerbach, Surf. Sci. Rep. 4 (1985) 1.
[209] D. C. Jacobs, K. W. Kolasinski, R. J. Madix, and R. N. Zare, J. Chem. Phys. 87 (1987) 5038.
[210] D. C. Jacobs and R. N. Zare, J. Chem. Phys. 91 (1989) 3196.
[211] R. J. W. E. Lahaye, S. Stolte, S. Holloway, and A. W. Kleyn, J. Chem. Phys. 104 (1996) 8301.
[212] A. E. Wiskerke and A. W. Kleyn, J. Phys.: Condens. Matter 7 (1995) 5195.
[213] A. E. Wiskerke, C. A. Taatjes, A. W. Kleyn, R. J. W. E. Lahaye, S. Stolte, D. K. Bronnikov, and B. E. Hayden, J. Chem. Phys. 102 (1995) 3835.
[214] C. A. Taatjes, A. E. Wiskerke, and A. W. Kleyn, J. Chem. Phys. 102 (1995) 3848.
[215] D. C. Jacobs, K. W. Kolasinski, S. F. Shane, and R. N. Zare, J. Chem. Phys. 91 (1989) 3182.
[216] M. A. Hines and R. N. Zare, J. Chem. Phys. 98 (1993) 9134.
[217] M. Balooch, M. J. Cardillo, D. R. Miller, and R. E. Stickney, Surf. Sci. 46 (1974) 358.
[218] M. J. Cardillo, M. Balooch, and R. E. Stickney, Surf. Sci. 50 (1975) 263.
[219] A. Gelb and M. Cardillo, Surf. Sci. 59 (1976) 128.
[220] J. Harris and S. Andersson, Phys. Rev. Lett. 55 (1985) 1583.
[221] J. Harris, S. Holloway, T. S. Rahman, and K. Yang, J. Chem. Phys. 89 (1988) 4427.
[222] M. R. Hand and S. Holloway, J. Chem. Phys. 91 (1989) 7209.
[223] G. Anger, A. Winkler, and K. D. Rendulic, Surf. Sci. 220 (1989) 1.
[224] B. E. Hayden and C. L. A. Lamont, Phys. Rev. Lett. 63 (1989) 1823.
[225] C. T. Rettner, D. J. Auerbach, and H. A. Michelsen, Phys. Rev. Lett. 68 (1992) 1164.
[226] H. A. Michelsen and D. J. Auerbach, J. Chem. Phys. 94 (1991) 7502.
[227] H. A. Michelsen, C. T. Rettner, and D. J. Auerbach, Phys. Rev. Lett. 69 (1992) 2678.
[228] G. R. Darling and S. Holloway, J. Chem. Phys. 101 (1994) 3268.
[229] H. Hou, S. J. Gulding, C. T. Rettner, A. M. Wodtke, and D. J. Auerbach, Science 277 (1997) 80.
[230] J. Q. Dai and J. C. Light, J. Chem. Phys. 108 (1998) 7816.
[231] C. T. Rettner, H. A. Michelsen, D. J. Auerbach, and C. B. Mullins, J. Chem. Phys. 94 (1991) 7499.
[232] H. A. Michelsen, C. T. Rettner, and D. J. Auerbach, Surf. Sci. 272 (1992) 65.
[233] C. T. Rettner, D. J. Auerbach, and H. A. Michelsen, Phys. Rev. Lett. 68 (1992) 2547.
[234] A. Hodgson, J. Moryl, P. Traversaro, and H. Zhao, Nature 356 (1992) 501.
[235] A. Hodgson, P. Samson, A. Wight, and C. Cottrell, Phys. Rev. Lett. 78 (1997) 963.
[236] S. Nave, D. Lemoine, M. F. Somers, S. M. Kingma, and G. J. Kroes, J. Chem. Phys. 122 (2005) 214709.
[237] M. F. Somers, R. A. Olsen, H. F. Busnengo, E. J. Baerends, and G. J. Kroes, J. Chem. Phys. 121 (2004) 11379.
[238] G. J. Kroes, E. J. Baerends, and R. C. Mowrey, J. Chem. Phys. 107 (1997) 3309.
[239] E. Watts and G. O. Sitz, J. Chem. Phys. 114 (2001) 4171.

[240] E. Watts, G. O. Sitz, D. A. McCormack, G. J. Kroes, R. A. Olsen, J. A. Groeneveld, J. N. P. Van Stralen, E. J. Baerends, and R. C. Mowrey, J. Chem. Phys. 114 (2001) 495.
[241] L. C. Shackman and G. O. Sitz, J. Chem. Phys. 123 (2005) 064712.
[242] B. E. Hayden and C. L. A. Lamont, Chem. Phys. Lett. 160 (1989) 331.
[243] S. Dahl, A. Logadottir, R. C. Egeberg, J. H. Larsen, I. Chorkendorff, E. Tornqvist, and J. K. Nørskov, Phys. Rev. Lett. 83 (1999) 1814.
[244] L. Diekhöner, H. Mortensen, A. Baurichter, E. Jensen, V. Petrunin, and A. C. Luntz, J. Chem. Phys. 115 (2001) 9028.
[245] L. Romm, G. Katz, R. Kosloff, and M. Asscher, J. Phys. Chem. B 101 (1997) 2213.
[246] R. C. Egeberg, J. H. Larsen, and I. Chorkendorff, Phys. Chem. Chem. Phys. 3 (2001) 2007.
[247] M. Asscher, O. M. Becker, G. Haase, and R. Kosloff, Surf. Sci. 206 (1988) L880.
[248] C. Diaz, J. K. Vincent, G. P. Krishnamohan, R. A. Olsen, G. J. Kroes, K. Honkala, and J. K. Nørskov, Phys. Rev. Lett. 96 (2006) 096102.
[249] T. Matsushima, Surf. Sci. 197 (1988) L287.
[250] L. Diekhöner, H. Mortensen, A. Baurichter, A. C. Luntz, and B. Hammer, Phys. Rev. Lett. 84 (2000) 4906.
[251] L. Diekhöner, H. Mortensen, A. Baurichter, and A. C. Luntz, Journal of Vacuum Science & Technology a-Vacuum Surfaces and Films 18 (2000) 1509.
[252] B. Hammer, Phys. Rev. B 63 (2001) 205423.
[253] C. Diaz, A. Perrier, and G. J. Kroes, Chem. Phys. Lett 434 (2007) 231.
[254] A. Hodgson, (private communication)
[255] T. F. Hanisco and A. C. Kummel, J. Chem. Phys. 99 (1993) 7076.
[256] T. F. Hanisco and A. C. Kummel, J. Vac. Sci. Technol. A 11 (1993) 1907.
[257] K. R. Lykke and B. D. Kay, J. Chem. Phys. 90 (1989) 7602.
[258] C. M. Matthews, F. Balzer, A. J. Hallock, M. D. Ellison, and R. N. Zare, Surf. Sci. 460 (2000) 12.
[259] J. J. Mortensen, Y. Morikawa, B. Hammer, and J. K. Nørskov, J. Catal. 169 (1997) 85.
[260] J. R. Rostrup-Nielsen, J. Sehested, and J. K. Nørskov, in Advances In Catalysis, Vol 47, Vol. 47, 2002, p. 65.
[261] M. B. Lee, Q. Y. Yang, S. L. Tang, and S. T. Ceyer, J. Chem. Phys. 85 (1986) 1693.
[262] M. B. Lee, Q. Y. Yang, and S. T. Ceyer, J. Chem. Phys. 87 (1987) 2724.
[263] I. M. Ciobica, F. Frechard, R. A. van Santen, A. W. Kleyn, and J. Hafner, Chem. Phys. Lett. 311 (1999) 185.
[264] R. M. Watwe, H. S. Bengaard, J. R. Rostrup-Nielsen, J. A. Dumesic, and J. K. Nørskov, J. Catal. 189 (2000) 16.
[265] J. H. Larsen and I. Chorkendorff, Surf. Sci. Rep. 35 (1999) 163.
[266] C. T. Reeves, D. C. Seets, and C. B. Mullins, J. Mol. Catal. A – Chemical 167 (2001) 207.
[267] P. M. Holmblad, J. Wambach, and I. Chorkendorff, J. Chem. Phys. 102 (1995) 8255.
[268] A. C. Luntz, J. Chem. Phys. 102 (1995) 8264.
[269] H. Yang and J. L. Whitten, J. Chem. Phys. 96 (1992) 5529.
[270] P. Kratzer, B. Hammer, and J. K. Nørskov, J. Chem. Phys. 105 (1996) 5595.
[271] C. T. Rettner, H. E. Pfnuer, and D. J. Auerbach, Phys. Rev. Lett. 54 (1985) 2716.
[272] C. T. Rettner, H. E. Pfnur, and D. J. Auerbach, J. Chem. Phys. 84 (1986) 4163.
[273] A. C. Luntz and D. S. Bethune, J. Chem. Phys. 90 (1989) 1274.
[274] J. H. Larsen, P. M. Holmblad, and I. Chorkendorff, J. Chem. Phys. 110 (1999) 2637.
[275] M. P. Schmid, P. Maroni, R. D. Beck, and T. R. Rizzo, J. Chem. Phys. 117 (2002) 8603.
[276] L. B. F. Juurlink, P. R. McCabe, R. R. Smith, C. L. DiCologero, and A. L. Utz, Phys. Rev. Lett. 83 (1999) 868.
[277] V. A. Ukraintsev and I. Harrison, J. Chem. Phys. 101 (1994) 1564.

[278] H. L. Abbott, A. Bukoski, and I. Harrison, J. Chem. Phys. 121 (2004) 3792.
[279] A. Bukoski, D. Blumling, and I. Harrison, J. Chem. Phys. 118 (2003) 843.
[280] A. C. Luntz, Science 302 (2003) 70.
[281] L. Halonen, S. L. Bernasek, and D. J. Nesbitt, J. Chem. Phys. 115 (2001) 5611.
[282] M.-N. Carre and B. Jackson, J. Chem. Phys. 108 (1998) 3722.
[283] H. F. Winters, J. Chem. Phys. 64 (1976) 3495.
[284] J. Harris, J. Simon, A. C. Luntz, C. B. Mullins, and C. T. Rettner, Phys. Rev. Lett. 67 (1991) 652.
[285] B. O. Nielsen, A. C. Luntz, P. M. Holmblad, and I. Chorkendorff, Catal. Lett. 32 (1995) 15.
[286] R. A. Campbell, J. Szanyi, P. Lenz, and D. W. Goodman, Catal. Lett. 17 (1993) 39.
[287] W. Z. Lai, D. Q. Xie, and D. H. Zhang, Surf. Sci. 594 (2005) 83.
[288] I. M. Ciobica, F. Frechard, R. A. van Santen, A. W. Kleyn, and J. Hafner, J. Phys. Chem. B 104 (2000) 3364.
[289] S. L. Bernasek and G. A. Somorjai, J. Chem. Phys. 62 (1975) 3149.
[290] I. E. Wachs and R. J. Madix, Surf. Sci. 58 (1976) 590.
[291] K. Christmann, G. Ertl, and T. Pignet, Surf. Sci. 54 (1976) 365.
[292] A. C. Luntz, J. K. Brown, and M. D. Williams, J. Chem. Phys. 93 (1990) 5240.
[293] P. Samson, A. Nesbitt, B. E. Koel, and A. Hodgson, J. Chem. Phys. 109 (1998) 3255.
[294] E. Pijper, G. J. Kroes, R. A. Olsen, and E. J. Baerends, J .Chem. Phys. 113 (2000) 8300.
[295] J. K. Brown, A. C. Luntz, and P. A. Schultz, J. Chem. Phys. 95 (1991) 3767.
[296] B. Hammer, K. W. Jacobsen, and J. K. Nørskov, Surf. Sci. 297 (1993) L68.
[297] J. P. Cowin, C.-F. Yu, S. J. Sibener, and J. E. Hurst, J. Chem. Phys. 75 (1981) 1033.
[298] J. Harris, Faraday Discuss. 96 (1993) 1.
[299] A. C. Luntz, Phys. Scr. 35 (1987) 193.
[300] P. Nieto, E. Pijper, D. Barredo, G. Laurent, R. A. Olsen, E. J. Baerends, G. J. Kroes, and D. Farias, Science 312 (2006) 86.
[301] A. T. Gee, B. E. Hayden, C. Mormiche, and T. S. Nunney, J. Chem. Phys. 112 (2000) 7660.
[302] R. A. Olsen, H. F. Busnengo, A. Salin, M. F. Somers, G. J. Kroes, and E. J. Baerends, J. Chem. Phys. 116 (2002) 3841.
[303] J. K. Vincent, R. A. Olsen, G. J. Kroes, and E. J. Baerends, Surf. Sci. 573 (2004) 433.
[304] K. D. Rendulic, G. Anger, and A. Winkler, Surf. Sci. 208 (1989) 404.
[305] D. A. Butler, B. E. Hayden, and J. D. Jones, Chem. Phys. Lett. 217 (1994) 423.
[306] A. Gross, S. Wilke, and M. Scheffler, Surf. Sci (1996) 0039.
[307] G. R. Darling, M. Kay, and S. Holloway, Surf. Sci. 400 (1998) 314.
[308] S. Wilke and M. Scheffler, Phys. Rev. B: Condens. Matter 53 (1996) 4926.
[309] A. Eichler, J. Hafner, A. Gross, and M. Scheffler, Phys. Rev. B 59 (1999) 13297.
[310] L. Schroeter, C. Trame, R. David, and H. Zacharias, Surf. Sci. 272 (1992) 229.
[311] A. Gross and M. Scheffler, Prog. Surf. Sci. 53 (1997) 187.
[312] C. Resch, H. F. Berger, K. D. Rendulic, and E. Bertel, Surf. Sci. 316 (1994) L1105.
[313] M. Gostein and G. O. Sitz, J. Chem. Phys. 106 (1997) 7378.
[314] M. Beutl, M. Riedler, and K. D. Rendulic, Chem. Phys. Lett. 247 (1995) 249.
[315] H. F. Busnengo, E. Pijper, G. J. Kroes, and A. Salin, J. Chem. Phys. 119 (2003) 12553.
[316] D. Barredo, G. Laurent, C. Diaz, P. Nieto, H. F. Busnengo, A. Salin, D. Farias, and F. Martin, J. Chem. Phys. 125 (2006) 051101.
[317] H. F. Busnengo, W. Dong, and A. Salin, Phys. Rev. Lett. 93 (2004) 236103.
[318] S. Holloway and J. W. Gadzuk, J. Chem. Phys. 82 (1985) 5203.
[319] A. C. Luntz, M. D. Williams, and D. S. Bethune, J. Chem. Phys. 89 (1988) 4381.
[320] P. D. Nolan, B. R. Lutz, P. L. Tanaka, J. E. Davis, and C. B. Mullins, J. Chem. Phys. 111 (1999) 3696.

[321] W. Wurth, J. Stohr, P. Feulner, X. Pan, K. R. Bauchspiess, Y. Baba, E. Hudel, G. Rocker, and D. Menzel, Phys. Rev. Lett. 65 (1990) 2426.
[322] J. L. Gland, B. A. Sexton, and G. B. Fisher, Surf. Sci. 95 (1980) 587.
[323] H. Steininger, S. Lehwald, and H. Ibach, Surf. Sci. 123 (1982) 1.
[324] C. T. Rettner and C. B. Mullins, J. Chem. Phys. 94 (1991) 1626.
[325] J. Grimblot, A. C. Luntz, and D. E. Fowler, J. Electron Spectrosc. Relat. Phenom 52 (1990) 161.
[326] B. C. Stipe, M. A. Rezaei, and W. Ho, J. Chem. Phys. 107 (1997) 6443.
[327] T. Zambelli, J. V. Barth, J. Wintterlin, and G. Ertl, Nature 390 (1997) 495.
[328] C. T. Campbell, G. Ertl, H. Kuipers, and J. Segner, Surf. Sci. 107 (1981) 220.
[329] A. N. Artsyukhovich, V. A. Ukraintsev, and I. Harrison, Surf. Sci. 347 (1996) 303.
[330] A. E. Wiskerke, F. H. Geuzebroek, A. W. Kleyn, and B. E. Hayden, Surf. Sci. 272 (1992) 256.
[331] A. Eichler and J. Hafner, Phys. Rev. Lett. 79 (1997) 4481.
[332] Z. Sljivancanin and B. Hammer, Surf. Sci. 515 (2002) 235.
[333] A. Gross, A. Eichler, J. Hafner, M. J. Mehl, and D. A. Papaconstantopoulos, Surface Science 539 (2003) L542.
[334] A. Gross, A. Eichler, J. Hafner, M. J. Mehl, and D. A. Papaconstantopoulos, J. Chem. Phys. 124 (2006) 174713.
[335] A. Sellidj and J. L. Erskine, Surf. Sci. 220 (1989) 253.
[336] C. T. Rettner and E. K. Schweizer, Surf. Sci. 203 (1988) L677.
[337] C. T. Rettner, H. Stein, and E. K. Schweizer, J. Chem. Phys. 89 (1988) 3337.
[338] C. T. Rettner, E. K. Schweizer, H. Stein, and D. J. Auerbach, Phys. Rev. Lett. 61 (1988) 986.
[339] C. T. Rettner, E. K. Schweizer, and H. Stein, J. Chem. Phys. 93 (1990) 1442.
[340] D. J. Auerbach, H. E. Pfnur, C. T. Rettner, J. E. Schlaegel, J. Lee, and R. J. Madix, J. Chem. Phys. 81 (1984) 2515.
[341] A. Kara and A. E. Depristo, Surf. Sci. 193 (1988) 437.
[342] M. Serrano and G. R. Darling, Surf. Sci. 532 (2003) 206.
[343] G. Volpilhac and A. Salin, Surf. Sci. 556 (2004) 129.
[344] J. J. Mortensen, L. B. Hansen, B. Hammer, and J. K. Nørskov, J. Catal. 182 (1999) 479.
[345] P. J. Feibelman, Phys. Today 58 (2005) 13.
[346] C. Benndorf and T. E. Madey, Surf. Sci. 135 (1983) 164.
[347] H. Mortensen, L. Diekhöner, A. Baurichter, E. Jensen, and A. C. Luntz, J. Chem. Phys. 113 (2000) 6882.
[348] Y. Wang, A. Lafosse, and K. Jacobi, Surf. Sci. 507 (2002) 773.
[349] K. Jacobi, Y. Wang, C. Y. Fan, and H. Dietrich, J. Chem. Phys. 115 (2001) 4306.
[350] A. Logadottir and J. K. Nørskov, J. Catal. 220 (2003) 273.
[351] K. Honkala, A. Hellman, I. N. Remediakis, A. Logadottir, A. Carlsson, S. Dahl, C. H. Christensen, and J. K. Nørskov, Science 307 (2005) 555.
[352] C. J. Zhang, M. Lynch, and P. Hu, Surf. Sci. 496 (2002) 221.
[353] J. K. Nørskov, T. Bligaard, A. Logadottir, S. Bahn, L. B. Hansen, M. Bollinger, H. Bengaard, B. Hammer, Z. Zljivancanin, M. Mavrikakis, Y. Xu, S. Dal, and J. C. J. H., J. of Catal. 209 (2002) 275.
[354] D. F. Johnson and W. H. Weinberg, J. Chem. Phys. 101 (1994) 6289.
[355] L. Diekhöner, D. A. Butler, A. Baurichter, and A. C. Luntz, Surf. Sci. 409 (1998) 384.
[356] J. F. Weaver, M. A. Krzyzowski, and R. J. Madix, Surf. Sci. 393 (1997) 150.
[357] S. Jakubith, H. H. Rotermund, W. Engel, A. Vonoertzen, and G. Ertl, Phys. Rev. Lett. 65 (1990) 3013.
[358] C. T. Campbell, G. Ertl, H. Kuipers, and J. Segner, J. Chem. Phys. 73 (1980) 5862.
[359] J. Segner, C. T. Campbell, G. Doyen, and G. Ertl, Surf. Sci. 138 (1984) 505.

[360] K. H. Allers, H. Pfnuer, P. Feulner, and D. Menzel, J. Chem. Phys. 100 (1994) 3985.
[361] E. Poehlmann, M. Schmitt, H. Hoinkes, and H. Wilsch, Surf. Sci. 287 (1993) 269.
[362] G. W. Coulston and G. L. Haller, J. Chem. Phys. 95 (1991) 6932.
[363] K. Nakao, S. Ito, K. Tomishige, and K. Kunimori, J. Phys. Chem. B 109 (2005) 24002.
[364] C. A. Becker, J. P. Cowin, L. Wharton, and D. J. Auerbach, J. Chem. Phys. 67 (1977) 3394.
[365] J. Wintterlin, S. Volkening, T. V. W. Janssens, T. Zambelli, and G. Ertl, Science 278 (1997) 1931.
[366] I. Nakai, H. Kondoh, K. Amemiya, M. Nagasaka, A. Nambu, T. Shimada, and T. Ohta, J. Chem. Phys. 121 (2004) 5035.
[367] T. Matsushima, Surf. Sci. 127 (1983) 403.
[368] W. D. Mieher and W. Ho, J. Chem. Phys. 91 (1989) 2755.
[369] W. D. Mieher and W. Ho, J. Chem. Phys. 99 (1993) 9279.
[370] C. Aakerlund, I. Zoric, and B. Kasemo, J. Chem. Phys. 104 (1996) 7359.
[371] C. B. Mullins, C. T. Rettner, and D. J. Auerbach, J. Chem. Phys. 95 (1991) 8649.
[372] V. A. Ukraintsev and I. Harrison, J. Chem. Phys. 96 (1992) 6307.
[373] Y. Y. Yeo, L. Vattuone, and D. A. King, J. Chem. Phys. 106 (1997) 392.
[374] X. Z. Ji and G. A. Somorjai, J. Phys. Chem. B 109 (2005) 22530.
[375] X. Z. Ji, A. Zuppero, J. M. Gidwani, and G. A. Somorjai, J. Am. Chem. Soc. 127 (2005) 5792.
[376] A. Alavi, P. J. Hu, T. Deutsch, P. L. Silvestrelli, and J. Hutter, Phys. Rev. Lett. 80 (1998) 3650.
[377] A. Eichler and J. Hafner, Surf. Sci. 435 (1999) 58.
[378] C. T. Rettner, Phys. Rev. Lett. 69 (1992) 383.
[379] C. T. Rettner and D. J. Auerbach, Phys. Rev. Lett. 74 (1995) 4551.
[380] S. Caratzoulas, B. Jackson, and M. Persson, J. Chem. Phys. 107 (1997) 6420.
[381] B. Jackson and D. Lemoine, J. Chem. Phys. 114 (2001) 474.
[382] S. Wehner and J. Kuppers, J. Chem. Phys. 108 (1998) 3353.
[383] T. Kammler, S. Wehner, and J. Kuppers, J. Chem. Phys. 109 (1998) 4071.
[384] T. Kammler, D. Kolovos-Vellianitis, and J. Kuppers, Surf. Sci. 460 (2000) 91.
[385] B. Jackson, X. W. Sha, and Z. B. Guvenc, J. Chem. Phys. 116 (2002) 2599.
[386] B. Jackson and M. Persson, Surf. Sci. 269–270 (1992) 195.
[387] M. Persson, J. Stromquist, L. Bengtsson, B. Jackson, D. V. Shalashilin, and B. Hammer, J. Chem. Phys. 110 (1999) 2240.
[388] D. V. Shalashilin, B. Jackson, and M. Persson, J. Chem. Phys. 110 (1999) 11038.
[389] Z. B. Guvenc, X. W. Sha, and B. Jackson, J. Phys. Chem. B 106 (2002) 8342.
[390] S. A. Buntin, L. J. Richter, R. R. Cavanagh, and D. S. King, Phys. Rev. Lett. 61 (1988) 1321.
[391] S. A. Buntin, L. J. Richter, D. S. King, and R. R. Cavanagh, J. Chem. Phys. 91 (1989) 6429.
[392] R. Schwarzwald, A. Modl, and T. J. Chuang, Surf. Sci. 242 (1991) 437.
[393] K. Fukutani, Y. Murata, R. Schwarzwald, and T. J. Chuang, Surf. Sci. 311 (1994) 247.
[394] P. D. Johnson and S. L. Hulbert, Phys. Rev. B 35 (1987) 9427.
[395] W. Ho, in Laser spectroscopy and photo-chemistry on metal surfaces, Vol. 2 (H.-L. Dai and W. Ho, eds.), World Scientific, Singapore, 1995, p. 1047.
[396] F. M. Zimmermann and W. Ho, J. Chem. Phys. 100 (1994) 7700.
[397] F. M. Zimmermann and W. Ho, Phys. Rev. Lett. 72 (1994) 1295.
[398] F. M. Zimmermann and W. Ho, J. Chem. Phys. 101 (1994) 5313.
[399] T. Yamanaka, A. Hellman, S. W. Gao, and W. Ho, Surf. Sci. 514 (2002) 404.
[400] J. A. Prybyla, T. F. Heinz, J. A. Misewich, M. M. T. Loy, and J. H. Glownia, Phys. Rev. Lett. 64 (1990) 1537.
[401] J. A. Prybyla, T. F. Heinz, J. A. Misewich, and M. M. T. Loy, Surf. Sci. 230 (1990) L173.
[402] X. Y. Zhu, S. R. Hatch, A. Campion, and J. M. White, J. Chem. Phys. 91 (1989) 5011.

[403] E. Hasselbrink, H. Hirayama, A. De Meijere, F. Weik, M. Wolf, and G. Ertl, Surf. Sci. 269–270 (1992) 235.
[404] C. E. Tripa, C. R. Arumaninayagam, and J. T. Yates, Jr., J. Chem. Phys. 105 (1996) 1691.
[405] A. W. E. Chan, R. Hoffmann, and W. Ho, Langmuir 8 (1992) 1111.
[406] F. Weik, A. Demeijere, and E. Hasselbrink, J Chem Phys 99 (1993) 682.
[407] S. R. Hatch, X. Y. Zhu, J. M. White, and A. Campion, J. Phys. Chem. 95 (1991) 1759.
[408] F. J. Kao, D. G. Busch, D. Cohen, D. G. Dacosta, and W. Ho, Phys. Rev. Lett. 71 (1993) 2094.
[409] F. J. Kao, D. G. Busch, D. G. Dacosta, and W. Ho, Phys. Rev. Lett. 70 (1993) 4098.
[410] D. G. Busch, S. W. Gao, R. A. Pelak, M. F. Booth, and W. Ho, Phys. Rev. Lett. 75 (1995) 673.
[411] D. G. Busch and W. Ho, Phys. Rev. Lett. 77 (1996) 1338.
[412] T. H. Her, R. J. Finlay, C. Wu, and E. Mazur, J. Chem. Phys. 108 (1998) 8595.
[413] D. N. Denzler, C. Frischkorn, M. Wolf, and G. Ertl, J. Phys. Chem. B 108 (2004) 14503.
[414] S. Wagner, H. Ostrom, A. Kaebe, M. Krenz, M. Wolf, A. C. Luntz, and C. Frischkorn, (to be published)
[415] D. N. Denzler, C. Frischkorn, C. Hess, M. Wolf, and G. Ertl, Phys. Rev. Lett. 91 (2003) 226102.
[416] S. Wagner, C. Frischkorn, M. Wolf, M. Rutkowski, H. Zacharias, and A. C. Luntz, Phys. Rev. B 72 (2005) 205404.
[417] W. Ho, J Chem. Phys. 117 (2002) 11033.
[418] B. C. Stipe, M. A. Rezaei, and W. Ho, Phys. Rev. Lett. 81 (1998) 1263.
[419] D. Teillet-Billy, J. P. Gauyacq, and M. Persson, Phys. Rev. B 62 (2000) R13306.
[420] H. E. Pfnur, C. T. Rettner, J. Lee, R. J. Madix, and D. J. Auerbach, J. Chem. Phys. 85 (1986) 7452.
[421] H. Hou, S. J. Gulding, C. T. Rettner, A. M. Woodtke, and D. J. Auerbach, Science 277 (1997) 80.
[422] C. T. Rettner and D. J. Auerbach, J. Chem. Phys. 104 (1996) 2732.
[423] D. V. Shalashilin, B. Jackson, and M. Persson, Faraday Discuss. 110 (1998) 287.

Chapter 4

Heterogeneous Catalysis

T. Bligaard and J. K. Nørskov

Center for Atomic-scale Materials Design, Department of Physics, NanoDTU, Technical University of Denmark, DK-2800 Lyngby, Denmark

1. Introduction

The ability of solid surfaces to make and break bonds to molecules, from the surroundings, is the basis for the phenomenon of heterogeneous catalysis. Many chemical reactions are catalyzed by solid surfaces. This includes most large scale chemical processes in industry and a number of environmental protection processes such as the exhaust gas cleaning in automobiles. The energy sector is particularly dependent on heterogeneous catalysis; the gasoline at a fueling station has, for instance, seen on the order of 10 catalysts during its journey through the refinery. Future energy technologies based on hydrogen or other energy carriers, will also require heterogeneous catalysis – in fact, many of the future challenges connected with hydrogen production and fuel cells are closely related to finding new catalysts and catalytic processes.

A solid surface has three closely coupled functions when it works as a catalyst for a chemical reaction. First, it adsorbs the reactants and cleaves the required bonds. Next it holds the reactants in close proximity so that they can react, and finally the surface lets the products desorb back into the surrounding phase. Understanding the adsorption bond is therefore crucial to the understanding of the way surfaces act as catalysts. If we can understand, which factors determine if a surface is a good catalyst for a given chemical reaction, we will have the concepts needed to guide us to better and more efficient catalysts.

In the following, we will develop a set of simple concepts that allow us to understand variations in the reactivity from one surface to the next. These variations are essential to the understanding of heterogenous catalysis, because they hold the key to distinguishing between a good and a bad catalyst. We will concentrate in the present Chapter on transition metal surfaces, but many of the concepts should be applicable to oxides, sulfides and other catalytically interesting materials.

2. Factors determining the reactivity of a transition metal surface

The description of bonding at transition metal surfaces presented here has been based on a combination of detailed experiments and quantitative theoretical treatments. Adsorption of simple molecules on transition metal surfaces has been extremely well characterized experimentally both in terms of geometrical structure, vibrational properties, electronic structure, kinetics, and thermo-chemistry [1–3]. The wealth of high-quality experimental data forms a unique basis for the testing of theoretical methods, and it has become clear that density functional theory calculations, using a semi-local description of exchange and correlation effects, can provide a semi-quantitative description of surface adsorption phenomena [4–6]. Given that the DFT calculations describe reality semi-quantitatively, we can use them as a basis for the analysis of catalytic processes at surfaces.

In Chapter 2, it was demonstrated how the experiment-theory combination has provided a very detailed insight into the nature of the surface chemical bond [7]. In the present chapter, we will exploit this insight to build an understanding of variations in bond energies form one metal to the next. We will show that it is through an understanding of these variations, that we extract concepts that allow us to determine why a particular surface will be a good catalyst for a given reaction, while other surfaces are not. One of the aims of this Chapter is to introduce *descriptors* that can be used to characterize the catalytic properties of a surface. Such descriptors may be used to rationalize experimental data, consequently bringing order to the known literature concerning the catalytic rates of different materials. Furthermore, they also have the potential to become an important ingredient in designing new catalysts.

The aim of the following is to develop an understanding of the differences in reactivity between one elemental metal, and another. The concepts will be extended to describe what happens if two metals are alloyed, and the subsequent role, in the reactivity of adsorbed species. This understanding leads to the realization of two general classes of phenomena, which determine the reactivity. The first is sometimes referred to as the *electronic effect* in heterogeneous catalysis [2], and we will go on to demonstrate which electronic effects are dominant. The second class of phenomena is the influence of surface structure, the so-called *geometrical effect* in heterogeneous catalysis [2]. The discussion will start by considering simple adsorption systems and proceed to variations in activation energies for surface reactions.

3. Trends in adsorption energies on transition metal surfaces

As an introduction to the problem of understanding trends in adsorption energies on metal surfaces, consider the adsorption of atomic oxygen on a range of late transition metal surfaces. Figure 4.1 shows calculated energies as a function of the

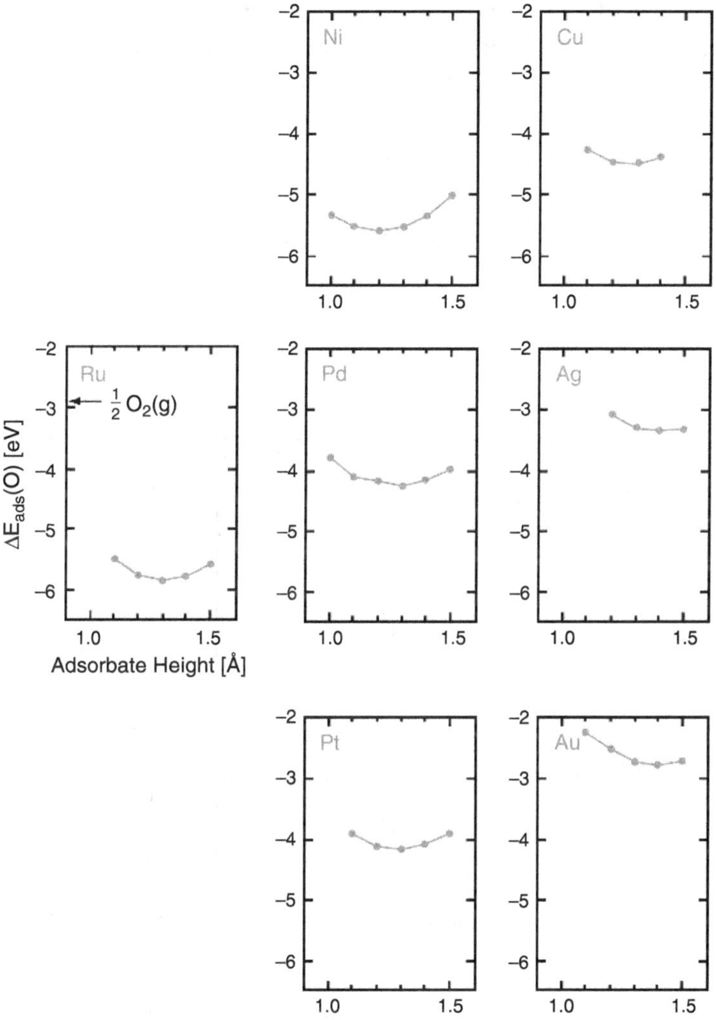

Figure 4.1. Calculated adsorption energy for atomic oxygen as a function of distance of the O atom above the surface for a range of close-packed transition metal surfaces (ordered according to their position in the periodic table). In the box showing results for Ru, the energy per O atom in O_2 is shown for comparison. Only metals where the minimum in the adsorption energy function is below this value will be able to dissociate O_2 exothermally. Adapted from Ref. [4].

distance of the O atom above the surface for a small section of the periodic table. It can be seen that O binds most strongly to the metals to the left in the transition metal series and stronger to the 3d than to the 4d and 5d metals. This is in excellent agreement with experimental findings [1]; Ru bonds much stronger than Pd and Ag, and Au is very noble with a bond energy per O atom less than that in O_2, Ag is just able dissociate O_2 exothermically, and Cu forms quite strong bonds.

To understand these variations it is instructive to look at the variations in the electronic structure of O adsorbed on the different metals, see Figure 4.2. As shown in Chapter 2, bonding of an atom like O to a transition metal results in the formation of bonding and anti-bonding states below and above the metal d bands. It can be seen how the anti-bonding states for O on Ru are less filled than on Pd and Ag,

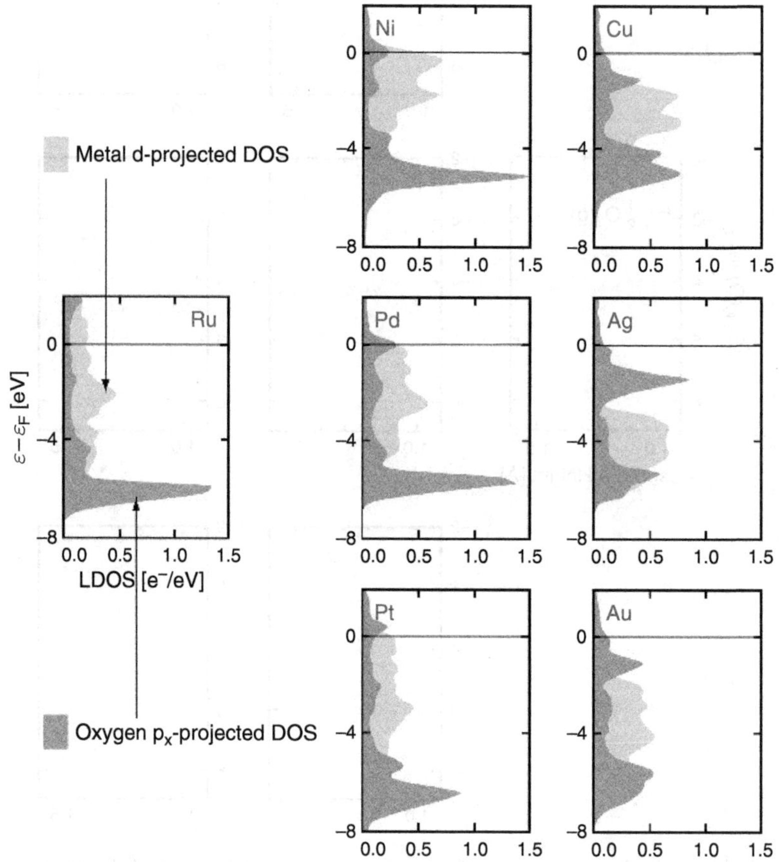

Figure 4.2. The density of states projected onto the d states of the surface atoms for the surfaces considered in Figure 4.1 (grey). Also shown (black) is the oxygen $2p_x$ projected density of states for O adsorbed on the same surfaces. The formation of bonding and anti-bonding states below and above the metal d states is clearly seen. Adapted from Ref. [4].

explaining why the bonding becomes weaker in that order. This picture cannot explain why 3d metals bond stronger than 4d and 5d metals, and it is not clear how to extend it to look at more complex systems. We will therefore develop a more extensive model to explain these effects.

4. The d-band model

In density functional theory the total energy is usually written as:

$$E_{tot}[n] = T_{HK}[n] + F[n] = T_{HK}[n] + \tfrac{1}{2}\int\int \tfrac{n(r)n(r')}{|r-r'|} + \int v(r)n(r) + E_{nn} + E_{xc}[n]. \quad (1)$$

Here

$$\begin{aligned} T_{HK}[n] &= \sum_{i\,occ} <\psi_i| -\tfrac{1}{2}\nabla^2|\psi_i> \\ &= \sum_{i\,occ} \varepsilon_i - \int v_{eff}(r)n(r) \end{aligned} \quad (2)$$

is the kinetic energy of a non-interacting electron gas moving in an effective potential $v_{eff}(r)$ chosen so that the non-interacting system has the same electron density as the real system. $F[n]$ is the sum of the average electrostatic energy, the interaction of the electrons with the external potential set up by the nuclei, the nucleus–nucleus interaction, E_{nn}, and the exchange-correlation energy, $E_{xc}[n]$.

Developing a model to describe the trends in bond energies from one transition metal to the next, based on the density of one-electron states poses a fundamental question. How can we get bond energies directly from the Kohn-Sham one-electron energies given that eqs. (1) and (2) show that the total energy is *not* equal to the sum of the one-electron energies in density functional theory? This problem is present in all simple one-electron descriptions of bonding in molecules, and we address it by showing that, if calculated correctly, *changes* in the Kohn-Sham one-electron energies can give *changes* in bond energies. This is enough for our analysis, since we are mainly concerned with understanding the changes in bonding from one system to the next.

The following analysis follows that of Hammer and Nørskov [4,8].

4.1. One-electron energies and bond energy trends

Consider an adsorbate a outside a metal surface m. We want to estimate the change in adsorption energy of the adsorbate when the metal is modified slightly to $\tilde{m}$. The modification could for instance be that another atom or molecule was adsorbed on m close to a, or that m was exchanged for another metal close to m in the Periodic Table.

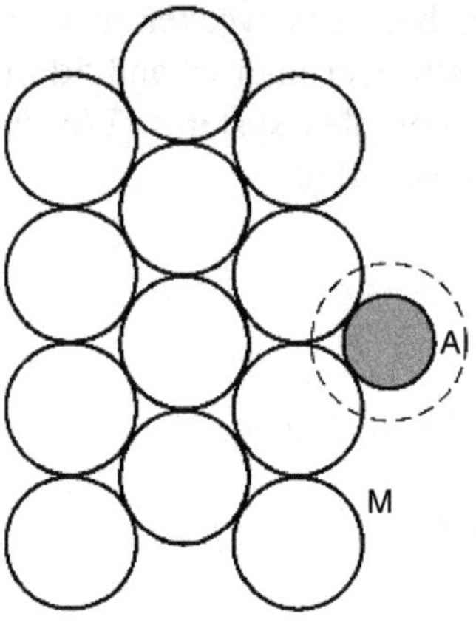

Figure 4.3. Illustration of the near-adsorbate region, A, and the near-metal region, M.

We are interested in the difference in adsorption energy in the two cases:

$$\begin{aligned}\delta E_{ads} &= \Delta E[\tilde{m}] - \Delta E[m] \\ &= (E_{tot}[\tilde{m}+a] - E_{tot}[\tilde{m}] - E_{tot}[a]) - (E_{tot}[m+a] - E_{tot}[m] - E_{tot}[a]) \\ &= E_{tot}[\tilde{m}+a] - E_{tot}[\tilde{m}] - (E_{tot}[m+a] - E_{tot}[m]). \end{aligned} \quad (3)$$

We now divide space into two regions, A and M, as illustrated in Figure 4.3. In the near-adsorbate region, A, the electron density and one-electron potential will only be affected slightly by the change of m to $\tilde{m}$. Likewise, in the near metal region, M, the effect of the adsorbate will be weak. We can exploit this in connection with the generalized variational principle of density functional theory, which says that a change in the electron density *and* in the one-electron potential will only give rise to changes in the total energy to second-order [9]. Because region A is dominated by the adsorbate we choose the same density and potential in this region irrespective of the metal. Likewise, we let the density and potential in region M be independent of the presence of the adsorbate. Freezing the density and potential in this way will only lead to second-order errors in δE_{ads}.

Consider first the contribution of the F[n] term, eq. (1), to δE_{ads}. In the generalized gradient approximation used throughout this work, $E_{xc}[n]$ is a local function of position in space. This is not generally true of the electrostatic energy contributions. If, for example, the adsorbate has a dipole moment, this will give rise to an electrostatic potential in region M. For the present we shall neglect such non-local electrostatic interactions between regions A and M. In that case, we can then divide F into contributions from the two regions $F = F_A + F_M$ and write:

$$\begin{aligned}\delta F_{local} &= F_A[\tilde{m}+a] - F_A[\tilde{m}] - (F_A[m+a] - F_A[m]) + F_M[\tilde{m}+a] - F_M[\tilde{m}] \\ &\quad - (F_M[m+a] - F_M[m]) \\ &= (F_A[\tilde{m}+a] - F_A[m+a]) - (F_A[\tilde{m}] - F_A[m]) + (F_M[\tilde{m}+a] - F_M[\tilde{m}]) \\ &\quad - (F_M[m+a] - F_M[m])\end{aligned}$$

With the frozen density ansatz all terms in parenthesis in the last equation will be zero. The only contribution from F to the adsorption energy difference is therefore the non-local electrostatic energy,

$$\delta E_{es,A-M} = \int_A \int_M \frac{\rho(\mathbf{r})\rho(\mathbf{r}')}{|\mathbf{r}-\mathbf{r}'|} d\mathbf{r} d\mathbf{r}'.$$

In the same way, the frozen potential and density ansatz renders the net contribution from the nv integrals in the kinetic energy difference zero, and only the difference in the one-electron energies calculated with the frozen potentials outlined above will contribute to the kinetic energy contribution:

$$\begin{aligned}\delta T_{HK} = \delta E_{1el} &= \sum \varepsilon \{v_M[\tilde{m}], v_A[a]\} - \sum \varepsilon \{v_{M+A}[\tilde{m}]\} \\ &\quad - (\sum \varepsilon \{v_M[m], v_A[a]\} - \sum \varepsilon \{v_{M+A}[m]\}).\end{aligned}$$

The difference in adsorption energy is therefore given by the one-electron energy difference plus the difference in electrostatic interaction between the surface and the adsorbate in the two situations:

$$\delta E_{ads} = \delta E_{1el} + \delta E_{es,A-M}. \tag{4}$$

This is a very interesting result. It shows that energy *differences* are given by an inter-atomic electrostatic energy and the sum of the one-electron energies, provided they are calculated in the right way, using frozen potentials. Equation (4) gives a theoretical background for the use of arguments based on the one-electron spectra. It shows that such arguments are particularly useful for making comparisons between similar systems, not for calculating the total binding energy.

Using the knowledge that the non-self-consistent, one-electron energy differences do contain information about bonding trends, we can exploit the understanding of the electronic structure of adsorbates, outside metal surfaces, developed in Chapter 2. Generally, an adsorbate's energy level interacting with a broad band, such as the free-electron-like sp bands of metals, experiences shifts and a broadening, see Figure 4.4. When the coupling to the narrow d bands of transition metals is subsequently switched on, bonding and anti-bonding states appear. This is exactly what can be observed in Figure 4.2. The question we will address in the following is what happens to the interaction when we move from one metal to the next, or if we change the surroundings of the metal atoms to which the adsorbate bonds.

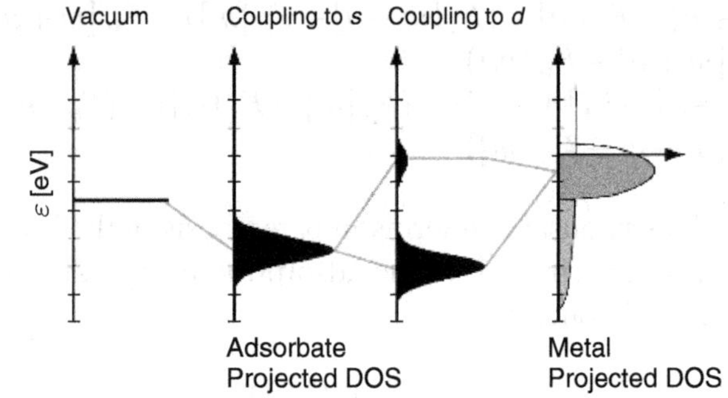

Figure 4.4. Schematic illustration of the formation of a chemical bond between an adsorbate valence level and the s and d states of a transition metal surface. Adapted from Ref. [4].

Since all transition metals have a half filled, broad s band, we will assume, in the spirit of the preceding section, that this part is independent of the metal in question. We therefore write the adsorption energy as [4,8,10]:

$$\Delta E = \Delta E_0 + \Delta E_d, \tag{5}$$

where ΔE_0 is the bond energy contribution from the free-electron-like sp electrons and ΔE_d is the contribution from the extra interaction with the transition metal d electrons.

One of the basic assumptions of the d band model is that ΔE_0 is independent of the metal. This is not a rigorous approximation. It will for instance fail when metal particles get small enough that the sp levels do not form a continuous (on the scale of the metal-adsorbate coupling strength) spectrum. It will also fail for metals where the d-states do not contribute to the bonding at all. The other basic assumption is that we can estimate the d contribution as the non-self-consistent one-electron energy change as derived above:

$$\Delta E_d \cong \int \varepsilon(\Delta n'(\varepsilon) - \Delta n(\varepsilon))d\varepsilon.$$

Here $\Delta n'(\varepsilon)$ and $\Delta n(\varepsilon)$ are the adsorbate-induced densities of states with and without the d coupling included, respectively. We will show that in spite of the approximate nature of the model it can describe a large number of trends. In the following we introduce the model which we will use to estimate ΔE_d.

4.2. The Newns-Anderson model

The simplest model that includes all the essential ingredients to describe the coupling between an adsorbate state and a band of metal states is the Newns-Anderson model [11–15]. In the following, we introduce the basics of the model and simple methods utilizing it to estimate bond energies.

Consider a metal surface with one-electron states $|k\rangle$ with energies ε_k, and an adsorbate with a single valence state $|a\rangle$ of energy ε_a. When the adsorbate approaches the surface from far away, to a position just outside, the two sets of states are coupled by matrix elements $V_{ak} = \langle a|\mathbf{H}|k\rangle$, where $\mathbf{H}$ is the Hamiltonian of the combined system. If we expand the solutions $|i\rangle$ of $\mathbf{H}$ in terms of the free adsorbate and surface solutions:

$$|i\rangle = c_{ai}|a\rangle + \sum_k c_{ki}|k\rangle,$$

and neglect the overlap $\langle a|k\rangle$, then the Schrödinger equation can be written

$$\mathbf{H}\mathbf{c}_i = \varepsilon_i \mathbf{c}_i,$$

where $H_{aa} = \varepsilon_a$, $H_{kk} = \varepsilon_k$, and $H_{ak} = V_{ak}$.

The projection of the density of states on the adsorbate state can be written:

$$n_a(\varepsilon) = \sum_i |\langle i|a\rangle|^2 \delta(\varepsilon - \varepsilon_i),$$

where the sum is over the eigenstates of the full Hamiltonian. We can rewrite this as

$$n_a(\varepsilon) = -\frac{1}{\pi} Im \sum_i \frac{\langle a|i\rangle\langle i|a\rangle}{\varepsilon - \varepsilon_i + i\delta} = -\frac{1}{\pi}\mathrm{Im}\ G_{aa}(\varepsilon)$$

with $\delta = 0^+$. Here $G_{aa}(\varepsilon)$ is the projection on the adsorbate state of the single particle Green function,

$$G(\varepsilon) = \sum_i \frac{|i\rangle\langle i|}{\varepsilon - \varepsilon_i + i\delta},$$

defined by the formal equation:

$$(\varepsilon - \mathbf{H} + i\delta)\mathbf{G}(\varepsilon) = 1.$$

To get $n_a(\varepsilon)$ we therefore just need the imaginary part of the $|a\rangle$ projection of $\mathbf{G}(\varepsilon)$. These equations can be solved for G_{aa} to yield:

$$G_{aa}(\varepsilon) = \frac{1}{\varepsilon - \varepsilon_a - q(\varepsilon)},$$

where the self-energy,

$$q(\varepsilon) = \Lambda(\varepsilon) - i\Delta(\varepsilon),$$

is given by real and imaginary parts:

$$\Delta(\varepsilon) = \sum_k |V_{ak}|^2 \delta(\varepsilon - \varepsilon_k) \tag{6}$$

and

$$\Lambda(\varepsilon) = \frac{P}{\pi} \int \frac{\Delta(\varepsilon')}{\varepsilon - \varepsilon'} d\varepsilon'.$$

This immediately gives the projected density of states:

$$n_a(\varepsilon) = \frac{1}{\pi} \frac{\Delta(\varepsilon)}{(\varepsilon - \varepsilon_a - \Lambda(\varepsilon))^2 + \Delta(\varepsilon)^2}. \tag{7}$$

It can be seen that the chemisorption function, $\Delta(\varepsilon)$, can be regarded as a local projection of the metal density of states around the adsorbate.

Knowing $\Delta(\varepsilon)$ (and therefore also $\Lambda(\varepsilon)$), we can calculate the change in the sum of one-electron energies associated with chemisorption as:

$$\Delta E = 2\left(\frac{1}{\pi} \int_{-\infty}^{\varepsilon_F} Arc\tan \frac{\Delta(\varepsilon)}{\varepsilon - \varepsilon_a - \Lambda(\varepsilon)} d\varepsilon - \varepsilon_a\right). \tag{8}$$

In the following, we will use ΔE calculated in this way to estimate the d-contribution, ΔE_d, to the bonding. This amounts to using a (fictitious) transition metal with no d electrons as the metal surface m in the preceding section, and using the real transition metal including the d electrons as the metal $\tilde{m}$. By using the unperturbed values of the energy of the adsorbate state and the metal d states as done in the simplest form of the Newns-Anderson model, we implicitly invoke the frozen density and potential approximation needed for the one electron energy difference, ΔE_d, to give the d contribution to the bonding.

The analysis of the Newns-Anderson model becomes particularly elegant by introducing the group orbital:

$$|d\rangle = \frac{1}{V} \sum_k V_{ak} |k\rangle, \quad V = \sqrt{\sum_k |V_{ak}|^2}. \tag{9}$$

Introducing the matrix element $V_{ad} = \langle a|H|d\rangle$, we can rewrite the chemisorption function as:

$$\Delta(\varepsilon) = \pi V_{ad}^2 n_d(\varepsilon)$$
$$n_d(\varepsilon) = \sum |\langle k|d\rangle|^2 \delta(\varepsilon - \varepsilon_k).$$

The group orbital is the combination of metal states coupling directly to the adsorbate state, and it is therefore the projection of the substrate density of states onto this state that defines the chemisorption function. The Newns-Anderson model will be used in the following to describe the coupling of an adsorbate orbital to the d states of a surface. In this case the group orbital is a single localized d orbital, or a linear combination of a few d orbitals on the transition metal atoms bonding directly to the adsorbate. In the following, we will make the approximation that the first moment of the relevant group orbital is given by the average of all the d states of the relevant atoms. It has been shown that in some cases an improved description can be obtained by explicitly using the d states of the proper symmetry [16].

The group-orbital projected metal density of states, $n_d(\varepsilon)$, can be characterized by its moments:

$$M_n = \int_{-\infty}^{\infty} \varepsilon^n n_d(\varepsilon) d\varepsilon \Big/ \int_{-\infty}^{\infty} n_d(\varepsilon) d\varepsilon. \qquad (10)$$

The first moment is the center, $M_1 = \varepsilon_d$, and the second moment is the width, $M_2 = W$, of the local projected band, and we note that the moments of $n_d(\varepsilon)$ and $\Delta(\varepsilon)$ are the same.

Figure 4.5 shows solutions to the Newns-Anderson model using a semi-elliptical model for the chemisorption function. The solution is shown for different surface projected density of states, $n_d(\varepsilon)$, with increasing d band centers ε_d. For a given metal the band width and center are coupled because the number of d electrons must be conserved.

It is seen in Figure 4.5 that as the d band shifts up in energy, ΔE_d becomes increasingly negative (meaning stronger bonds). This is a significant result, which cannot be obtained using simple rules from gas phase chemistry. To see this, compare the result of Figure 4.5 with the solution to the similar problem where a molecular state with energy ε_a interacts with another molecular state with energy ε_d with a coupling matrix element V_{ad}. If the state $|a>$ is completely filled as in Figure 4.5 and the state $|d>$ is fractionally filled, then the bond energy will be given by:

$$\delta E = -2(1-f)\omega,$$

where f is the filling fraction and ω is the absolute value of the down-shift of the bonding and up-shift of the anti-bonding state:

$$\omega = \sqrt{V_{ad}^2 + \left(\frac{\varepsilon_d - \varepsilon_a}{2}\right)^2} - \frac{\varepsilon_d - \varepsilon_a}{2}.$$

In the case where the coupling is weak ($|V_{ad}| << \varepsilon_d - \varepsilon_a$) we get the usual perturbation result,

$$\omega \cong \frac{V_{ad}^2}{\varepsilon_d - \varepsilon_a}.$$

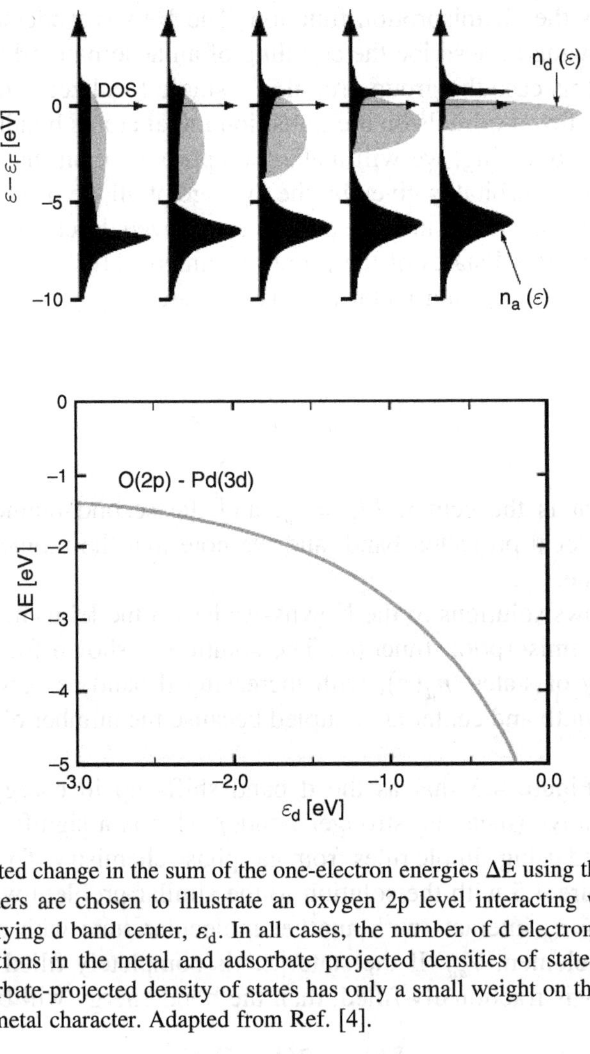

Figure 4.5. Calculated change in the sum of the one-electron energies ΔE using the Newns-Anderson model. The parameters are chosen to illustrate an oxygen 2p level interacting with the d states of palladium with a varying d band center, ε_d. In all cases, the number of d electrons is kept fixed. The corresponding variations in the metal and adsorbate projected densities of states are shown above. Notice that the adsorbate-projected density of states has only a small weight on the antibonding states since it has mostly metal character. Adapted from Ref. [4].

Both expressions have ω decreasing as ε_d shifts up in energy contrary to the result seen in Figure 4.5.

The difference between the gas phase bonding and the bonding at a metal surface is due to the presence of a Fermi energy ε_F. This means that the occupancy of the anti-bonding state is not determined solely by the filling of the d states, f, but rather by the position of the anti-bonding states relative to ε_F. It is clearly shown in Figure 4.5 that as the d band shifts up in energy, the number of anti-bonding states above ε_F increases, so leading to a stronger bond.

We note that if the adsorbate level is above the d band (and hence above the Fermi level) then we always get stronger bonding when the d states move up in energy. We therefore conclude that we would expect, that an up-shift in the d states should lead to a strengthening of the bonding of the adsorbate to the transition metal surface.

5. Trends in chemisorption energies

In the following, we first consider trends in chemisorption energies. Similar trends in activation barriers will be dealt with in the next section.

5.1. Variations in adsorption energies from one metal to the next

As a first example of the use of the d band model, consider the trends in dissociative chemisorption energies for atomic oxygen on a series of 4d transition metals (Figure 4.6). Both experiment and DFT calculations show that the bonding becomes

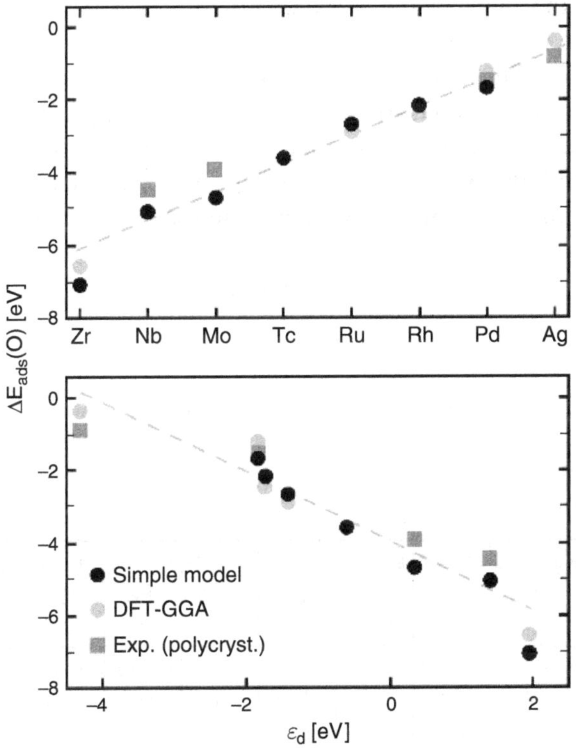

Figure 4.6. Variations in the O adsorption energy along the 4d transition metal series. The results of full DFT calculations are compared to those from the simple d band model and to experiments. Below the same data are plotted as a function of the d band center. Adapted from Ref. [4].

stronger (ΔE more negative) as we move left in the periodic table. The same is observed for the 3d and 5d series, as seen in Figure 4.1, and for a number of other adsorbates [17]. We note that the d band model with ΔE_d calculated using the Newns-Anderson model describes the trends quite well. It is also clear that the trends in the adsorption energy correlate well with $\varepsilon_d - \varepsilon_F$. The variations from one metal to the next along such a series reflect the fact that as we go left in the transition metal series, the number of d electrons decreases and hence the d bands must shift up in energy (c.f. Table 4.1).

In Figure 4.6, we have focused on metals for the 4d series of transition metals. We have done so because in such a series the matrix elements only vary modestly, see Table 4.1, making $\varepsilon_d - \varepsilon_F$ the most important variable. For the 3d and 5d series similar plots can be made, but $\varepsilon_d - \varepsilon_F$ will not *a priori* characterize variations down the columns in the periodic table, since the matrix elements vary considerably here. We will briefly suggest a way to understand these variations.

Let us focus on oxygen adsorption on Cu, Ag and Au. These metals are simple because the d bands are so low-lying that the anti-bonding states formed between the O atoms and the metal d states are all filled and contribute little to the bond energy, see Figure 4.2. There is, however, one contribution to ΔE_d, that we have neglected until now. This is the Pauli repulsion between the adsorbate states and the metal d states. The Pauli repulsion arises from the energy cost of orthogonalizing the adsorbate state to the metal states, and it is proportional to the overlap between the two sets of orbitals [8]. Since the overlap matrix elements will roughly scale with the coupling matrix elements, we can write the Pauli repulsion as:

$$\Delta E_d^{orthorg} = \alpha |V_{ad}|^2. \tag{11}$$

Table 4.1
Parameters describing the electronic structure of the transition metals. Adapted from Ref. [44].

Idealized *d*-band filling V_{ad}^2 (relative to Cu) 0.3 4.65 Ti 1.50 ε_d [eV]

0.1 20.8 Ca	0.2 7.90 Sc	0.3 4.65 Ti 1.50	0.4 3.15 V 1.06	0.5 2.35 Cr 0.16	0.6 1.94 Mn 0.07	0.7 1.59 Fe –0.92	0.8 1.34 Co –1.17	0.9 1.16 Ni –1.29	1.0 1.00 Cu –2.67	1.0 0.46 Zn
0.1 36.5 Sr	0.2 17.3 Y	0.3 10.9 Zr 1.95	0.4 7.73 Nb 1.41	0.5 6.62 Mo 0.35	0.6 4.71 Tc –0.60	0.7 3.87 Ru –1.41	0.8 3.32 Rh –1.73	0.9 2.78 Pd –1.83	1.0 2.26 Ag –4.30	1.0 1.58 Cd
0.1 41.5 Ba	0.2 17.1 Lu	0.3 11.9 Hf 2.47	0.4 9.05 Ta 2.00	0.5 7.72 W 0.77	0.6 6.04 Re –0.51	0.7 5.13 Os	0.8 4.45 Ir –2.11	0.9 3.90 Pt –2.25	1.0 3.35 Au –3.56	1.0 2.64 Hg

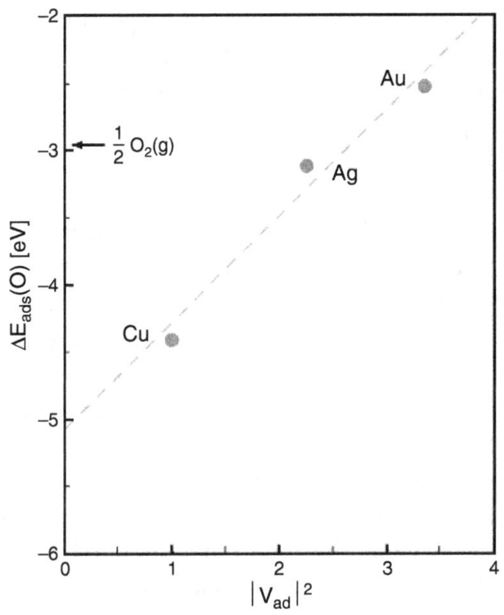

Figure 4.7. Variations in the O adsorption energy (from Figure 4.1) with the size of the coupling matrix element (from Table 4.1) for Cu, Ag and Au. Adapted from Ref. [4].

Table 4.1 shows $|V_{ad}|^2$ to increase rapidly down the columns in the periodic table, and in Figure 4.7 it is shown how the DFT calculations from Figure 4.1 scale nicely with $|V_{ad}|^2$ for Cu, Ag and Au, as expected from Eq. (11).

This picture shows why Au is so inert with respect to oxygen. Oxygen bonding is weakest for the metals with low-lying, filled d bands, because here the anti-bonding adsorbate-metal d states are filled. Of the metals with filled d bands ($f = 1$ in Table 4.1), Au is the metal with the largest matrix element $|V_{ad}|^2$, and hence the metal with the largest Pauli repulsion and the weakest oxygen chemisorption bond. Au is followed by Hg and Ag. Of the metals having a filled d shell, included in Table 4.1, Zn should be the one with the strongest oxygen bond, and this is, indeed, found to be the case.

The d band model, including Pauli repulsion, can therefore be used to understand variations in oxygen binding energies in the periodic table. It turns out that a similar description can be used for a number of other adsorbates [4,18].

5.2. *Ligand effects in adsorption – changing the d band center*

We will now consider a number of systems where the energy of the adsorbate state(s), ε_a, and the coupling matrix element, V_{ad}, are essentially constant. We will look at situations where an adsorbate bonds to a specific kind of transition metal atom, and the surroundings (or ligands) of the relevant metal atoms contributing to the coupling, c.f. Eq. (9), are changed. In these cases, we would expect from the analysis above that the position of the average energy of the d electrons relative to the Fermi level, $\varepsilon_d - \varepsilon_F$ should to a first approximation determine the variations in the interaction energy.

It should be noted that apart from the d band center, the interaction energy will depend on the shape and width of the projected d density of states, $n_d(\varepsilon)$. These variations are, however, often coupled to the d band center variations, and can therefore be lumped into that dependence. To illustrate the latter point, consider a situation where the width (the second moment, W) of $n_d(\varepsilon)$ is decreased for some reason – it could be because the surface layer is strained so that the coupling, V_{dd}, of the metal d states to the neighbors is smaller ($W \sim |V_{dd}|$) or because the number of metal neighbors (the coordination number, N_M) is decreased by creating a step or a kink on the surface ($W \sim N_M^{0.5}$). Changing W for a fixed $\varepsilon_d - \varepsilon_F$, would change the number of d electrons. It is generally found that the number of d electrons does not change for a given kind of metal, and the system compensates for this by shifting the d states up in energy as illustrated in Figure 4.8.

We will consider three classes of trends determined largely by variations in $\varepsilon_d - \varepsilon_F$.

5.2.1. *Variations due to changes in surface structure*

For one kind of transition metal atoms, the d band center can be varied by changing the structure. As mentioned above, the band width depends on the coordination number of the metal and this leads to substantial variations in the d band centers [19]. Atoms in the most close-packed (111) surface of Pt have a coordination number of 9. For the more open (100) surface it is 8 and for a step or for the (110)

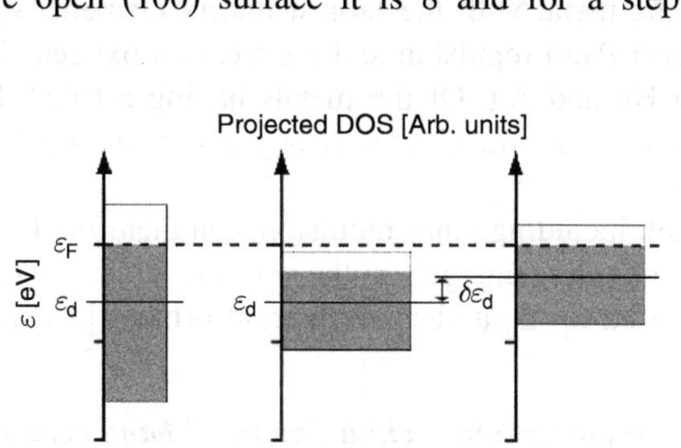

Figure 4.8. Schematic illustration of the coupling between bandwidth and d band center for a band with a fixed number of d electrons. When the bandwidth is decreasing the only way of maintaining the number of d electrons is to shift up the center of the band.

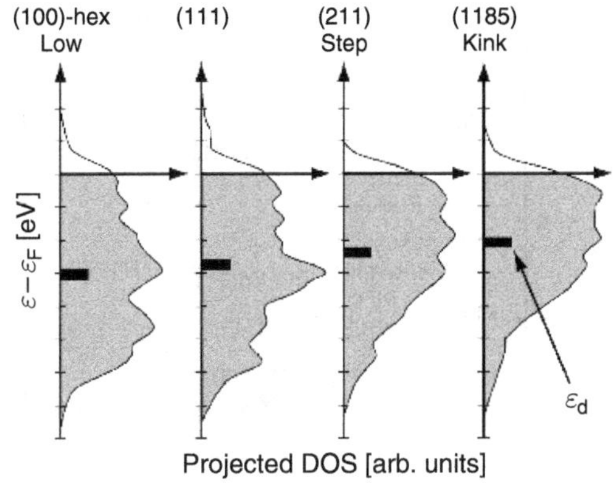

Figure 4.9. Projected densities of states onto the d states of the surface atoms for different Pt surfaces with decreasing atom density: The hexagonally reconstructed (100) surface, the close-packed (111) surface, the step atoms on a (211) surface and the kink atoms on a (11 8 5) surface. Adapted from Ref. [19].

surface it is 7. At a kink the coordination number is as low as 6. Figure 4.9 shows that this can lead to variations in the d band center of almost 1 eV, and the chemisorption energy of CO varies by a similar amount, see Figure 4.10. The fact that steps bind CO stronger than the flat surfaces is in excellent agreement with experiments [1,20].

Pt surfaces tend to restructure into overlayers with an even higher density of Pt atoms than the close-packed (111) surface [21]. The Pt atoms are closer to each other on the reconstructed surfaces than in the (111) surface. The overlap matrix elements and hence the bandwidth are therefore larger, the d bands are lower and consequently these reconstructed surfaces bind CO even weaker than the (111) surface. The reconstructed Pt surfaces are examples of strained overlayers. The effect of strain can be studied theoretically by simply straining a slab. Examples of continuous changes in the d band center and in the stability of adsorbed CO due to strain are included in Figure 4.10. The effect due to variations in the number of layers of a thin film of one metal on another can also be described in the d band model [22,23].

For atomic chemisorption, similar structural effects are found (see the middle panel of Figure 4.10). As for molecular chemisorption, low-coordinated atoms at steps bind adsorbates stronger and have lower barriers for dissociation than surfaces with high coordination numbers and lower d band centers. The d band model thus explains the many observations that steps form stronger chemisorption bonds than flat surfaces [1,20,24–28]. The finding that the correlation with the d band center is independent of the adsorbate illustrates the generality of the d band model.

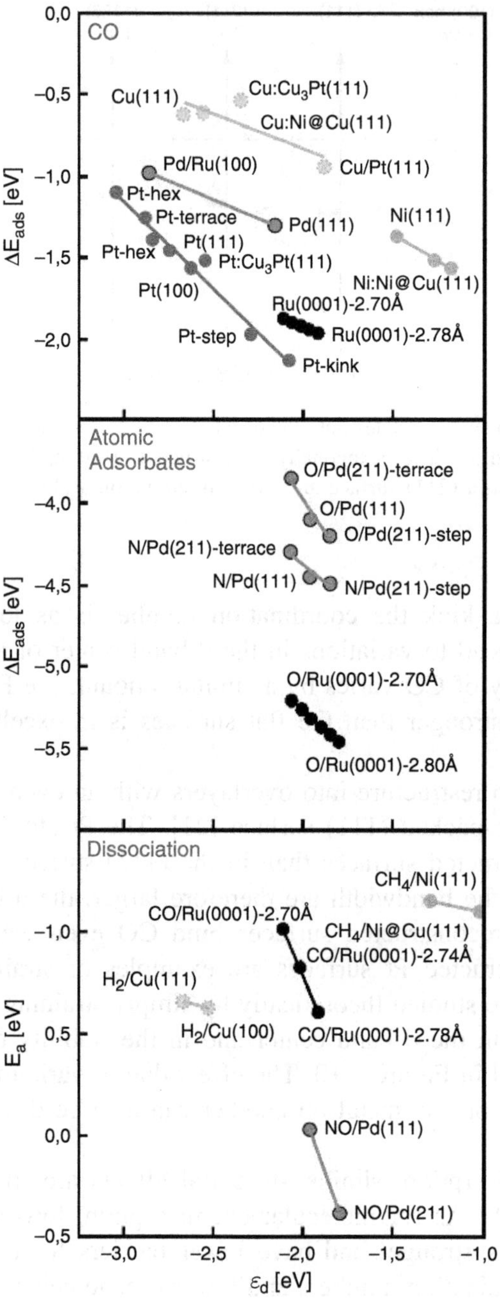

Figure 4.10. Calculated chemisorption energies for CO (top panel) and for different atomic adsorbates (middle panel), together with transition state energies for dissociation reactions (bottom panel) which are shown as a function of the average energy of the d states projected onto the surfaces atoms to which the adsorbates form bonds. Adapted from Ref. [29].

5.2.2. Variations due to alloying

The effect of alloying can also be understood in terms of *d* band shifts, which can be deduced from Figure 4.10. In Figure 4.11, this is shown in even more detail. By considering a Pt(111) surface where a series of different 3d metals have been sandwiched between the first and second layer, the effect of the intercalated layer of atoms on the reactivity of a Pt(111) overlayer can be studied. Such near-surface alloys [30], or 'skins' have been extensively studied as oxygen reduction catalysts in PEM fuel cells [31–32]. The d states of the surface Pt atoms are shifted down in energy as the second layer metal is chosen further to the left in the periodic table. The O and H adsorption energies show the same trends: as the d band center is shifted up in energy towards the Fermi level the bond becomes stronger and stronger. For the near surface alloys the bandwidth changes by the hybridization of the d states of the surface Pt atoms with the second layer atoms. Such an indirect interaction can also be termed a ligand effect – the metal ligands of the surface atoms are changed.

Similar effects can be found for metal overlayers, where a monolayer of one metal is deposited on top of another metal. Here there is an additional effect relating to the fact that the overlayer usually takes the lattice constant of the substrate. For metal overlayers we therefore find a combination of ligand and strain effects.

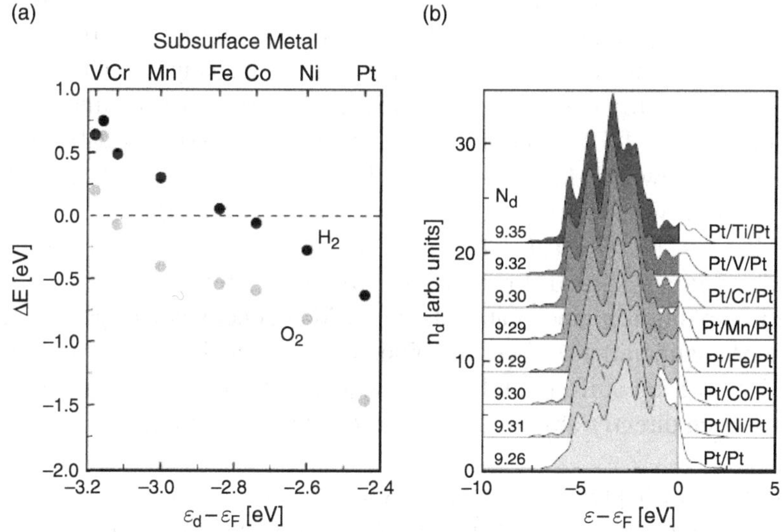

Figure 4.11. Calculated changes in the adsorption energy of atomic H and O on a series of Pt(111) surfaces, where the second layer has been replaced by a layer of 3d transition metals. To the right the variations in the d-projected density of states for the Pt surface atoms are shown. Adapted from Ref. [33].

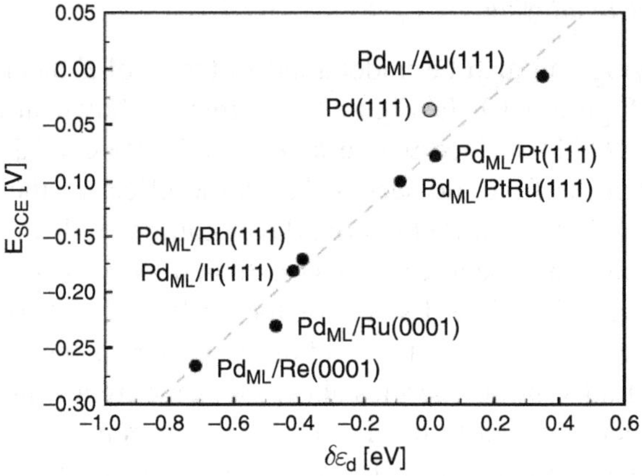

Figure 4.12. Electrochemically measured changes in the hydrogen adsorption energy for Pd overlayers on a number of metals shown as a function of the calculated shift of the d-band center. Adapted from Ref. [38].

Again the d band centers are found to describe changes in adsorption energies quite well [34–37]. This is illustrated in Figure 4.12, through the electrochemically determined variations in the hydrogen adsorption energy, for different Pd overlayers as a function of the calculated d band shifts [38].

The reactivity of a given metal can be varied substantially by depositing it on another metal. This provides an elegant way of controlling the reactivity of a given metal. Knowing how the d band center for a given metal changes when it is deposited on top of another provides a good starting point for choosing interesting metal combinations. Again DFT calculations can be used to systematically explore the possibilities as shown in Figure 4.13. If, for instance, one wants a surface that bonds CO a little weaker than Pt(111), then Figure 4.13 suggests that putting Pt on top of Fe, Co, Ni, Cu, Ru, Rh, and Ir should give the desired result because the d bands of Pt are shifted down relative to Pt(111) for these substrates. This is important in looking for anode catalysts for PEM fuel cells where poisoning by CO is a severe problem, and a surface that binds CO weaker (but still dissociates H_2) is desirable. The d-band shifting and the subsequent control of the adsorbate binding energy have been directly observed in single-crystal experiments [39,40] and in fuel cells [41–43].

As described in Chapter 2, a number of spectroscopic surface methods give information relating to d band shifts [45]. Ross, Markovic and coworkers have developed synchrotron-based high resolution photoemission spectroscopy to directly measure d band centers giving results in good agreement with the DFT calculations [46]. Another possibility is to exploit the fact that in some cases a shift in the d states can be measured as a core-level shift, as the d states and the core levels shift

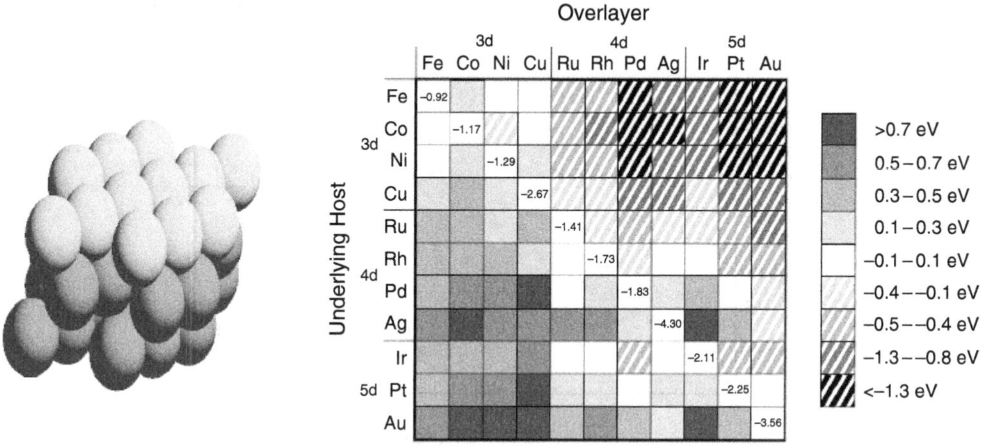

Figure 4.13. Calculated shifts in the d band centers for a number of overlayer structures. The shifts are calculated relative to the d band center for the pure overlayer metal surface. The shifts therefore reflect the change in reactivity of the overlayer relative to the pure metal. Adapted from ref. [44].

together [47,48]. This can explain the correlations between surface core level shifts found by Rodriques and Goodman [49] for a number of metal overlayers and similar results [50] for fuel cell catalysis by Watanabe and coworkers [51].

5.3. Ensemble effects in adsorption – the interpolation principle

Most of the effects considered until now have been indirect in the sense that the adsorbate has interacted with one kind of metal atom with varying surroundings or ligands. We now turn to the important case where a single adsorbate interacts with an ensemble of different metal atoms [52]. We will show that the chemisorption energy at a site with several different metal neighbors (i.e., the adsorption energy in the mixed site) to a first approximation is an appropriate average of the contributions from the individual components. This 'interpolation' principle turns out to be useful in thinking about new catalysts, as we will discuss later [53].

Figure 4.14 illustrates the interpolation principle [54,55]. It shows DFT calculations of the oxygen binding energy for a large number of alloys. The DFT values are compared to two different levels of model. Both models use the interpolation principle; in one case the bonding to a site with two kinds of neighbors, A and B, on the same substrate B is calculated as

$$\Delta E(A_xB_{1-x}/B) = x\Delta E(B/B) + (1-x)\Delta E(A/B).$$

Such an estimate gives typical errors of the order 0.1 eV in the prediction of the chemisorption energy. A cruder model would be to use chemisorption energies

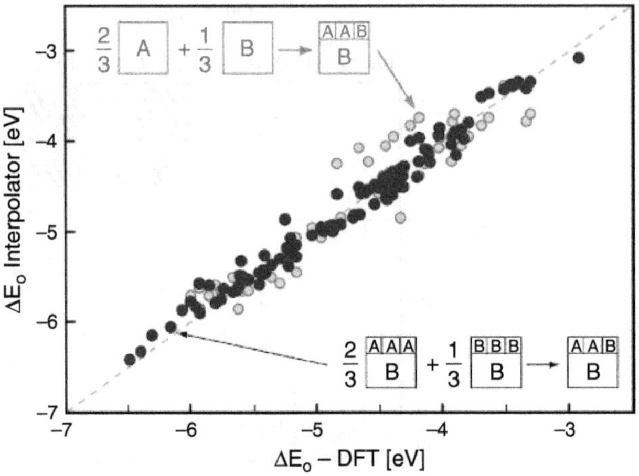

Figure 4.14. Illustration of the interpolation principle. Full DFT calculations for oxygen chemisorption energies are compared to two simple interpolation models for a series of surface alloys. Adapted from Ref. [54].

calculated for the pure metal surfaces as the starting point, thus neglecting the strain and ligand effects, for an overlayer of one metal on another:

$$\Delta E(A_x B_{1-x}/B) = x\Delta E(B/B) + (1-x)\Delta E(A/A).$$

This gives a slightly larger error; the average error for the data set in Figure 4.14 is 0.17 eV, however, its extreme simplicity means that the adsorption energy, on any alloy, can be estimated from a data base of adsorption energies calculated for the pure metals. In Table 4.2 we show such a data base [56].

While the interpolation model is far from perfect it gives a fast way of estimating the adsorption energies for alloys. Given the simplicity of the model it is surprising how well it works. The d band model can be used to indicate why this is the case.

Under the assumption that ΔE_0 in Eq. (5) is independent of the metal considered, all effects due to having several metal components, are to be found in the ΔE_d term. We have seen above that ΔE_d is a function of the d band center, and for small variations in ε_d, the relationship must be linear:

$$\Delta E_d(\varepsilon_d) = \Delta E_d(\varepsilon_{d0}) + \Delta E_d'(\varepsilon_d - \varepsilon_{d0}). \tag{12}$$

The d band center is given by the first moment of the chemisorption function, Eq. (10). We therefore need to understand qualitatively how $\Delta(\varepsilon)$ behaves for a multi-component system. To see this it is useful to expand the metal wave functions

Table 4.2
Dissociative chemisorption energies calculated by density functional theory for various molecules on a number of stepped transition metal surfaces. All values are given in eV per molecule. Positive and negative values signify endothermal and exothermal chemisorption reactions, respectively.

	H_2	OH	N_2	CO	NO	O_2	H_20	$H_2O \rightarrow$ $OH^*+0.5H_2$	CO_2	NH_3	CH_4	$CH_4 \rightarrow$ C^*+2H_2
Fe	−1.15	−4.14	**−1.27**	−2.53	−4.66	−6.30	− 1.98	−0.86	−2.51	−1.45	−1.07	1.24
Co	−0.78	−3.43	−0.38	**−1.51**	−3.63	−5.07	− 0.99	−0.65	−0.83	−0.43	0.09	1.65
Ni	−0.82	−2.77	−0.10	**−1.05**	−2.87	−3.90	− 0.45	−0.49	0.17	−0.37	−0.13	1.52
Cu	−0.29	−1.81	2.88	1.77	−0.68	−2.51	0.78	−0.07	3.69	1.92	3.06	3.64
Mo	−0.92	−4.61	−2.76	−3.61	−5.99	−7.48	−2.33	− 1.20	−4.18	−1.84	−1.09	0.74
Ru	−1.09	−3.27	**−0.84**	**−1.62**	−3.60	−4.62	−1.08	−0.64	−0.77	−1.14	−0.88	1.30
Rh	−0.79	−2.82	−0.70	**−1.12**	**−3.23**	−4.03	−0.48	−0.27	0.03	−0.61	−0.06	1.51
Pd	−0.78	−1.40	1.78	0.38	−0.58	**−1.20**	0.95	0.36	2.96	0.64	0.04	1.60
Ag	0.53	−0.48	5.86	4.32	1.73	−0.65	2.52	0.52	7.16	4.63	6.31	5.26
W	−1.29	−5.37	−4.33	−4.73	−7.34	−8.62	−3.27	− 1.45	−5.87	−3.18	−2.37	0.20
Ir	−1.26	−3.37	−0.59	−1.07	−3.49	−4.65	−1.26	− 0.35	−0.23	−1.27	−0.65	1.87
Pt	−1.12	−2.06	1.37	0.37	**−1.27**	**−2.17**	0.12	0.25	2.45	−0.08	−0.18	2.07
Au	0.18	−0.05	5.89	4.58	2.34	0.54	2.77	0.92	8.02	4.12	5.28	4.92

in localized basis sets $|j\rangle$ and $|j'\rangle$ consisting of d states on the individual atoms of the surface [57]:

$$\Delta(\varepsilon) = \sum_{j,j'} \sum_{k} \langle a|H|j\rangle \langle j \mid k\rangle \langle k \mid j'\rangle \langle j'|H|a\rangle\, \delta(\varepsilon - \varepsilon_k).$$

We have assumed here that the d states form complete sets for the metal states. This assumption should be sufficiently accurate for the determination of qualitative trends.

The first moment of $\Delta(\varepsilon)$ is then:

$$\varepsilon_d = \frac{\sum_{j,j'} \sum_{k} \langle a|H|j\rangle \langle j \mid k\rangle \langle k \mid j'\rangle \langle j'|H|a\rangle\, \varepsilon_k}{\sum_{j,j'} \sum_{k} \langle a|H|j\rangle \langle j \mid k\rangle \langle k \mid j'\rangle \langle j'|H|a\rangle} = \frac{\sum_{j,j'} \langle a|H|j\rangle \langle j|H_{metal}|j'\rangle \langle j'|H|a\rangle}{\sum_{j,j'} \langle a|H|j\rangle \langle j \mid j'\rangle \langle j'|H|a\rangle}.$$

Noting that $\langle j|H_{metal}|j\rangle = \varepsilon_j = \int \varepsilon \sum_k |\langle j|k\rangle|^2 \delta(\varepsilon - \varepsilon_k) d\varepsilon$ we have:

$$\varepsilon_d = \frac{\sum_j |V_{aj}|^2 \varepsilon_j + \sum_j \sum_{j' \neq j} \langle a|H|j\rangle \langle j|H_{metal}|j'\rangle \langle j'|H|a\rangle}{\sum_j |V_{aj}|^2 + \sum_j \sum_{j' \neq j} \langle a|H|j\rangle \langle j \mid j'\rangle \langle j'|H|a\rangle},$$

where

$$|V_{aj}|^2 = \langle a|H|j\rangle \langle j|H|a\rangle.$$

If the metal d-states are sufficiently localized, the second terms in both the numerator and denominator will be small, and we have

$$\varepsilon_d \approx \frac{1}{V}\sum_j \left|V_{aj}\right|^2 \varepsilon_j; \quad V^2 = \sum_j \left|V_{aj}\right|^2.$$

Together with Eq. (12) this shows ΔE_d to be an appropriately weighted average of the contributions from the different metal atoms to which the adsorbate bonds (where the matrix element is non-negligible).

6. Trends in activation energies for surface reactions

To describe catalytic reactions on a metal surface, adsorption energies of the reactants, intermediates and products are essential and so are the activation energies separating different intermediate steps. Figure 4.15 illustrates a full potential energy diagram for a catalytic reaction; the synthesis of ammonia $N_2 + 3H_2 \rightarrow 2NH_3$.

The activation energies are defined as the energy of the transition state for a given reaction relative to the initial state:

$$E_a = \Delta E_{TS} - \Delta E_{IS}.$$

We therefore also need to understand the variation of the interactions of molecules in their transition states, with the metal surface in order to understand differences in reactivity. In the following discussion, we will show that the trends are qualitatively the same as those governing the chemisorption energies.

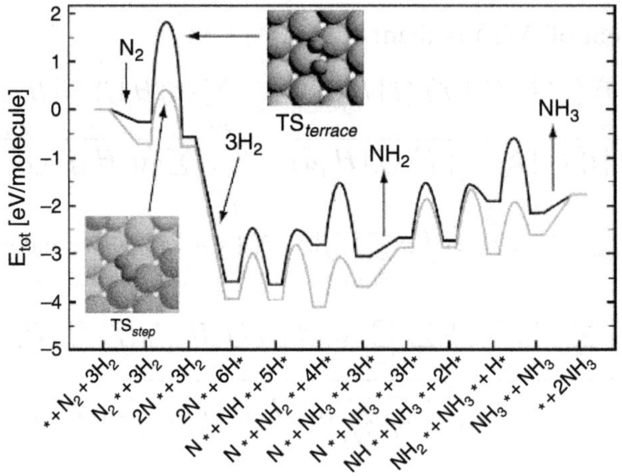

Figure 4.15. Calculated potential energy diagram for ammonia synthesis over a close-packed and a stepped Ru surface. Adapted from Ref. [58].

6.1. Electronic effects in surface reactivity

The arguments behind the d band model are quite general and should apply to the interactions in the transition state as well as in the initial and final (adsorbed) states of the process. We therefore expect correlations between the d band center and transition state energies to be the same as for chemisorption energies. This is illustrated in the bottom panel of Figure 4.10. Figure 4.16 shows in detail how the activation energy for methane on different Ni surfaces scales with the center of the d bands projected onto the appropriate metal states to which the transition state couples.

It is seen that both effects of alloying (NiAu) and structural effects (compare Ni(111), Ni(211) and the effect of strain) are described by the d band center variations. In fact, the alloying effects can be observed directly in molecular beam scattering experiments monitoring the methane sticking probability as a function of the Au coverage on a Ni(111) surface, see Figure 4.17, and such insight has formed the basis for the design of catalysts with new properties [59].

The structural effects can also be observed experimentally. Again we illustrate it for the methane activation over Ni surfaces. Figure 4.16 shows that the stepped Ni(211) surface has a considerably lower barrier for dissociation than the flat (111) surface. This has been confirmed in an elegant set of experiments [59]. It turns out that sulfur atoms preferentially adsorb at steps. In this way sulfur can be used to titrate the step sites, and since it also increases the barrier for methane activation considerably (see Figure 4.16), it can be used to block the step sites selectively. Figure 4.18 shows experimental data for the carbon uptake of a Ni(14 13 13) surface

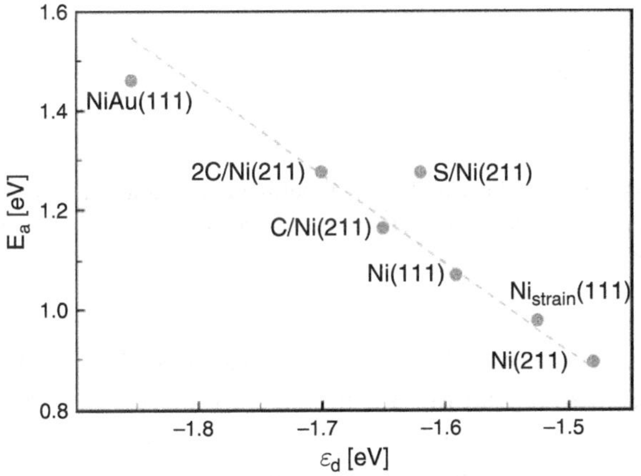

Figure 4.16. Calculated variations in the activation energy for methane dissociation over a number of different surfaces. The results are shown as a function of the energy of the d states coupling to the transition state methane molecule. Adapted from Ref. [57].

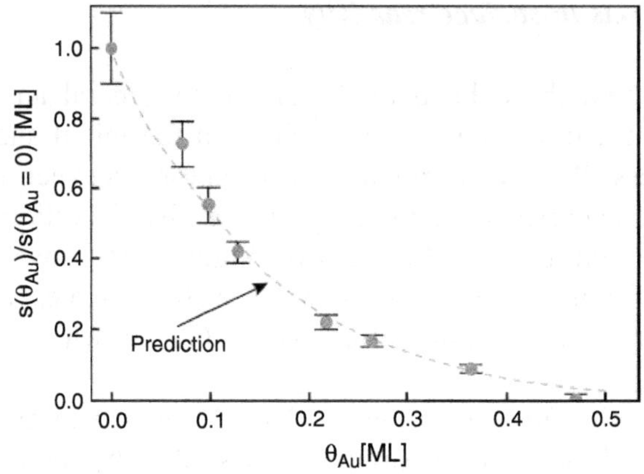

Figure 4.17. Measured sticking probability (relative to the clean surface) of a methane molecular beam on Ni(111) surfaces with varying amounts of Au alloying into the surface. The result of a model (prediction) based on DFT calculations of the change in the activation energy due to the addition of Au atoms is also shown. The beam data primarily measures methane sticking on the facets. Adapted from Ref. [59].

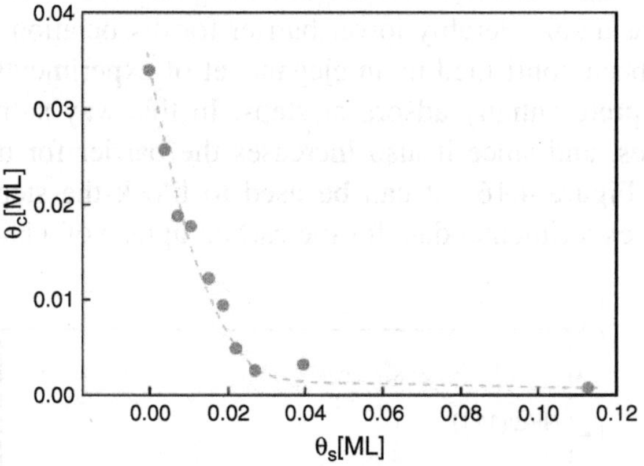

Figure 4.18. Experimental data showing how the thermal dissociation rate for methane (as measured by the C uptake) decreases rapidly as S atoms cover the 4% of steps on a Ni(14 13 13) surface. Adapted from Ref. [60].

which contains 4% step atoms. If 2% of S is adsorbed, which corresponds to a half covered step where all step atoms are blocked by having one S neighbor, the rate of methane dissociation is decreased substantially. This observation shows directly, how the step atoms can dissociate methane much faster than the terrace atoms.

Figure 4.16 also illustrates a case where the indirect interaction of one adsorbate with another can be described by the d band model (the effects of pre-adsorbed C atoms). If two adsorbates interact with the same metal atom, they are often found to repel each other. This can be viewed in the following way: Imagine a metal surface where one atom is adsorbed, consider for example a Ni(211) surface with C atoms adsorbed. The adsorbed C atoms will affect the d states of the neighboring Ni atoms making them less reactive for a second atom or molecule. Similar effects have been observed for a number of adsorbates [61].

Figure 4.16 also shows that there are additional effects due to direct interactions between adsorbates that are not described using the d band model. A large adsorbate like S, will have a sizable overlap to the valence orbitals of the incoming molecule, giving rise to a repulsion which is larger than what can be readily explained by the indirect interaction through d band shifts.

6.2. Geometrical effects in surface reactivity

Methane activation happens over a single Ni atom and the effect of changing the structure can be attributed primarily to changes in the electronic structure in the vicinity of the active site. There is, however, an additional effect of surface structure on the reactivity of transition metal surfaces that is of a purely geometrical nature. The effect is illustrated in Figure 4.15 [62]. The stepped Ru surface has a much lower barrier for dissociation than the close-packed (0001) surface. Part of the reason for the lower barrier at the step is due to the electronic effect: the Ru step atoms have higher-lying d states and thus the transition state energy is lower compared to the close-packed surface. The electronic effect can also be observed for atomic N adsorption in Figure 4.15. This is, however, not enough to explain the more than 1 eV difference in barrier. The purely geometrical effect is related to the fact that at the step more metal atoms can take part in the stabilization of the transition state than on the flat surface [62,63]. Figure 4.15 shows the transition state structures on the two surfaces. On the flat surface only four Ru atoms take part and one of them has to contribute to the stabilization of both N atoms. This gives rise to extra repulsion as discussed above. At the step the two N atoms are stabilized by five Ru atoms and none of them is 'shared'. The reactive step site involving five atoms was first speculated to be particularly important in catalysts by Van Hardevelt in 1966 [26].

There is a long history of geometrical effects in heterogeneous catalysis dating back at least to Taylor in 1925 [1,64–67]. Theoretical calculations have helped giving an idea about the order of magnitude of the effect and recent detailed experiments have shown this to be correct. The difference in barrier for N_2 dissociation on Ru steps and close-packed surfaces, for instance, is so large that the small number of steps on even the best (0001) single-crystals would completely dominate the rate of N_2 dissociation. By decorating the few steps of a Ru(0001) surface with gold atoms it was shown that the reactivity of the step atoms was at least 9 orders of magnitude larger than the terrace atoms [62], see Figure 4.19.

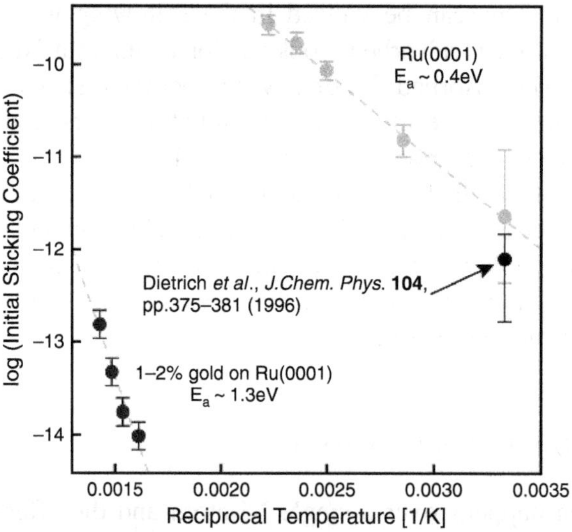

Figure 4.19. Arrhenius plot of measured thermal sticking coefficients of N_2 on a clean Ru(0001) surface and the same surface covered with 0.01–0.02 ML of gold. The point taken from Dietrich et al. [68] is the result from a similar measurement at room temperature. Adapted from Ref. [62].

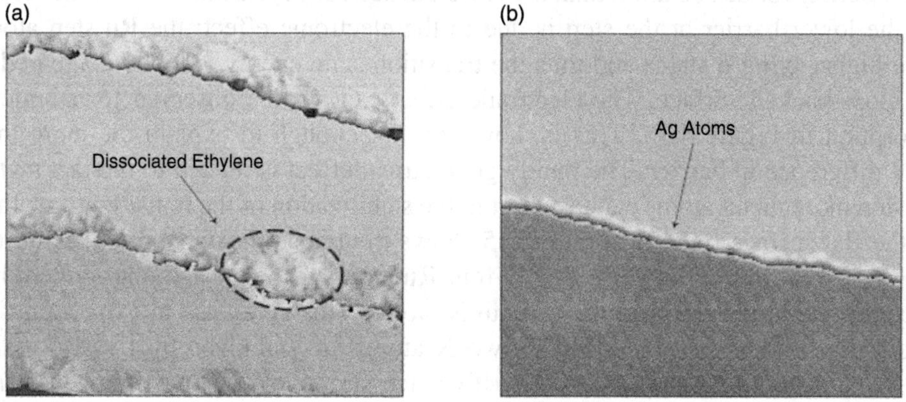

Figure 4.20. (a) STM image (200 × 200 Å^2) of a Ni(111) surface after exposure to ethylene (10^{-8} torr; 100 s) at room temperature. A brim of decomposed ethylene is formed along the step edges. (b) STM image (400 × 400 Å^2) of a Ni(111) surface with the step edges blocked by Ag atoms. No decomposition of ethylene is observed on this modified surface. Adapted from Ref. [70]. © 2005 Nature Publishing Group.

As we will discuss later the step effects are found to be quite general, and the role of steps in dissociation reactions has also been observed directly for ethylene activation on Ni surfaces, see Figure 4.20 [69,70].

7. Brønsted-Evans-Polanyi relationships in heterogeneous catalysis

Given that the variations in adsorption energies and transition state energies are governed by the same basic physics, it is not surprising that variations in adsorption energies of different molecules and transition state energies are found to be correlated. Such relationships are extremely important in building an understanding of heterogeneous catalysis. They can be viewed as a simple way of estimating activation energies on the basis of adsorption energies. Furthermore, at a more fundamental level they provide guidance in building kinetic models to understand trends in catalytic activity.

7.1. Correlations from DFT calculations

Linear correlations between activation (free) energies and reaction (free) energies are widespread in chemistry dating back to Brønsted in 1928 [71] and Evans and Polanyi ten years later [72]. Such relations have been assumed to hold in heterogeneous catalysis [73], however, it was not until DFT calculations became sufficiently accurate to make reliable predictions that it became possible to establish activation energies and reaction energies over a sufficient range to establish such correlations. Figures 4.21 and 4.22 show some of the first such Brønsted-Evans-Polanyi (BEP) relationships published [74].

The slopes of the BEP relations depend on the reaction studied. For dissociative adsorption processes involving simple diatomic molecules, the slope is often close

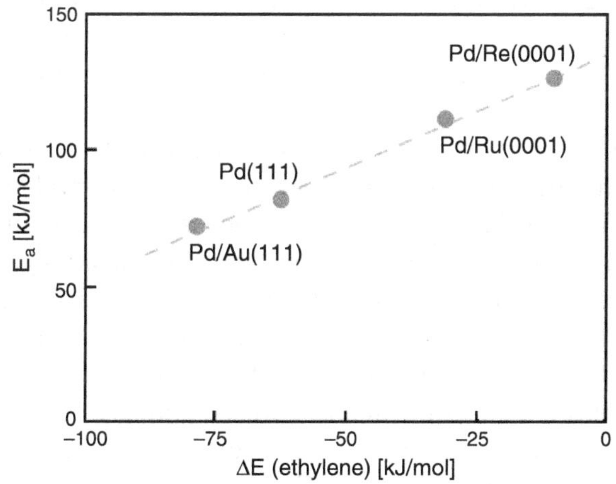

Figure 4.21. DFT-calculated activation barrier for ethyl C—H bond breaking, as a function of ethylene adsorption energy on pseudomorphic Pd overlayers. Adapted from Ref. [75].

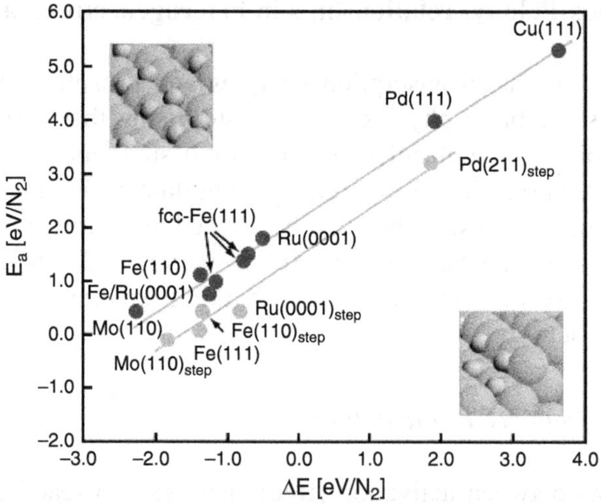

Figure 4.22. Calculated transition state energies for N_2 dissociation shown as a function of the dissociative N_2 chemisorption energy for both close-packed and stepped metal surfaces. Adapted from Ref. [74].

to 1 implying that the electronic structure of the transition state is similar to that of the final state. This is indicative of a late transition state, something that can be observed directly in the transition state structures in Figure 4.22. By the same token associative desorption shows very little dependence on the reaction energy so long as the dissociation process is activated. This has been observed directly in thermal desorption experiments. Recombinative desorption of N_2 from the Fe(111) surface, where dissociation is barely activated, happens at 750 K [76], while on Cu(110) where it is impossible to dissociate N_2 thermally, the recombinative desorption peaks at 750 K [77].

The DFT calculations show that the geometrical effect discussed above for N_2 dissociation on Ru surfaces is found for all the metals considered. In general, it is difficult to distinguish the electronic and geometrical effects in the reactivity of molecules. A step of course exhibits a new geometrical arrangement of the surface, but it also has an electronic effect due to the higher-lying d states at the low-coordinated atoms at the edge of the step. The electronic effect will move both ΔE_{TS} and ΔE, and the movement should be along the BEP line. The shift of the line is therefore a measure of the purely geometrical effect. We note that in principle there can be a new line for each considered surface geometry, so one should think of a family of BEP lines. For N_2 dissociation where a large number of geometries have been investigated [58], the close-packed surface and the step shown in Figure 4.22, seem to be close to the extrema, although that would

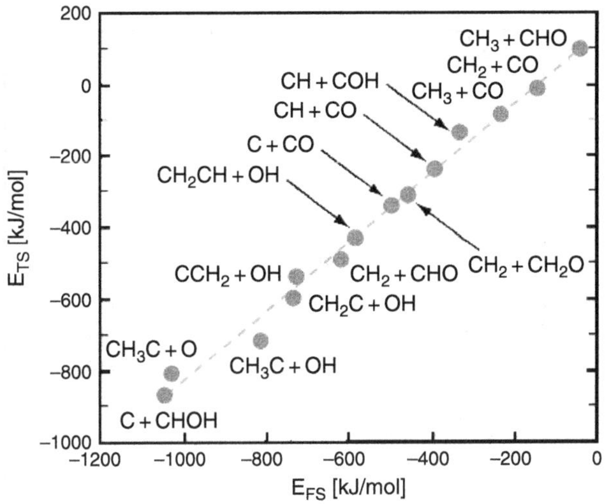

Figure 4.23. BEP plot for C−O and C−C bond cleavage reactions calculated from DFT. Final state (E_{FS}) and transition state (E_{TS}) energies are relative to initial state gas-phase energies. Adapted from Ref. [78].

in principle have to be investigated for each reaction. A guideline for identifying low-lying lines is that they correspond to a surface geometry where the two fragments of the reaction can both be stabilized without too many 'shared' metal atoms.

Another type of BEP relation in surface chemistry is shown in Figure 4.23. Here transition state energies are shown for the same surface (Pt(111)), but for a number of different C−O and C−C bond cleavage reactions that may occur in connection with the reforming of ethanol [78]. This is an interesting reaction which finds application in hydrogen production from renewable bio-resources [79].

7.2. Universal relationships

It turns out that if one compares dissociation of a number of similar molecules, their transition state energies scale with the reaction energy in much the same way as the relationships discussed previously; see Figure 4.24 [80]. This is a remarkable result indicating that the nature of the relationship between the final state and the transition state for dissociation of these molecules is quite similar. This is also borne out by direct comparison of the structures, see Figure 4.25. For the large number of systems considered in Figure 4.24 essentially all transition states look the same for a given surface structure.

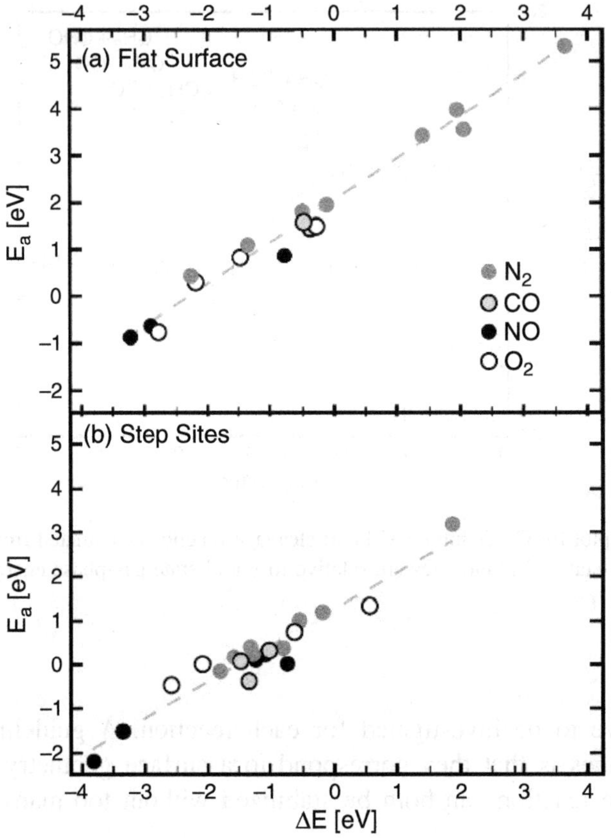

Figure 4.24. BEP plots for dissociation of a number of simple diatomic molecules. For a given surface geometry, the data cluster around the same 'universal' line. Adapted from Ref. [80].

We note in Figure 4.24 that the geometrical effect discussed for N_2 dissociation holds for all the adsorbates considered. This means that CO, NO, and O_2 dissociation should also be much faster at steps than at the most close-packed surface [81–83]. As noted above this is in agreement with a growing body of experimental evidence [83–85].

We note that other systems not resembling the simple diatomic molecules considered here may follow a different BEP relationship [86]. There may be other classes of reactions, dehydrogenation or C−C bond breaking that may follow other similar relationships and thus form another universality class. We also note that there are exceptions to the BEP relations, most notably for H_2 dissociation on near-surface alloys [87]. These deviations from the rules are still describable within the d band model, though [87].

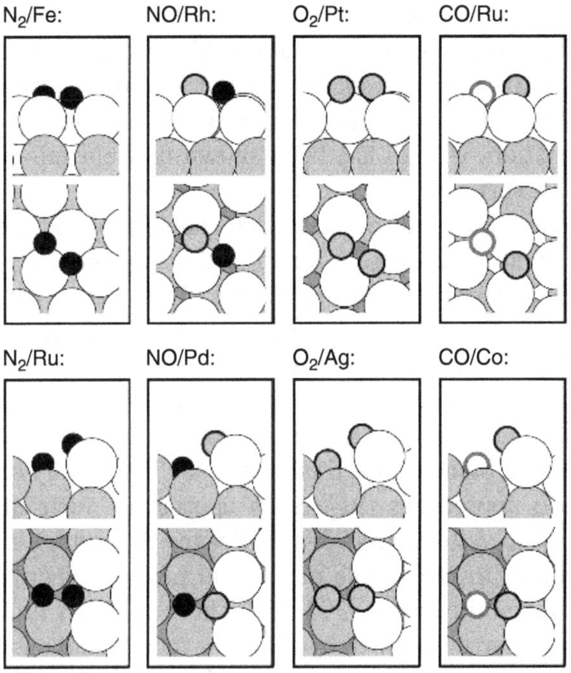

Figure 4.25. Side and top views of the calculated transition state structures for different diatomic molecules dissociating on different close-packed (top) and stepped (bottom) metal surfaces. Adapted from Ref. [80].

8. Activation barriers and rates

As shown in the previous sections, much can be learned about the energetics of adsorption systems by performing electronic structure calculations on relevant species. Knowing the adsorption energies and reaction barriers of the relevant molecular species is a good starting point for determining the actual rates of catalytic turnover for a given reaction. One however still has to go through a series of steps to reliably evaluate absolute rates. One could in principle use the electronic structure calculations to calculate reaction rates directly, by performing molecular dynamics calculations on the relevant system for enough time to obtain reliable statistics on the reaction rate. This, however, is rarely a practical approach. The main problem with performing full molecular dynamics to observe reactions is that many interesting reactions (if not all) happen on a much longer time-scale than the atomic vibration periods. In order to reliably integrate the equations of motion for the atomistic system in question, one has to sample time-intervals in the dynamics smaller than an atomic vibration period, as information about the dynamics is simply lost if longer time-steps are used.

Fortunately, the reaction rates of many important processes can be obtained without a full molecular dynamics simulation. Most reaction rate theories for elementary processes build upon the ideas introduced in the so-called transition state theory [88–90]. We shall focus on this theory here, particularly because it (and its harmonic approximation, HTST) has been shown to yield reliable results for elementary processes at surfaces.

Having obtained rate constants describing the key elementary processes for a catalytic reaction, these rate constants can be utilized to calculate the total catalytic rate. The coupling of elementary rates to yield the rate of an over-all catalytic reaction can also be performed in different ways, depending on the detail in which the elementary reaction rates are known, the importance of adsorbate–adsorbate interactions, and the degree to which the adsorbate coverage structure can be expected to be important for the total rate. The full solution of the reaction equations including neighbor interactions up to some predetermined neighbor shell is usually solved using lattice kinetic Monte Carlo approaches [91]. Such approaches have also been used for heterogeneous catalytic systems [58,92–94]. For the description of trends in catalysis, it is often adequate to use simple mean-field models, where surface coverages are averaged over the entire surface [95–98]. These methods often entirely neglect adsorbate-adsorbate interaction, thereby ignoring most details about the adsorbate-structure, islanding phenomena etc. They have, however, had a large impact on the understanding of heterogeneous catalytic processes [99]. A benefit of these models is that they can often be made to give analytic expressions, thereby yielding easily interpreted results and more fundamental insights into the mechanisms of the catalytic surfaces. They can also to varying degree be made to include lateral interactions [99]. We will return to some applications of these mean-field models after first introducing transition state theory.

8.1. Transition state theory

Transition state theory (TST) as derived in the 1930s [88,89,100] approaches the problem of calculating a reaction rate for a rarely occurring elementary reaction by separating space into two regions called the reactant region (RR) and the product region (PR). The reactant region defines the general region in which the system can be found before reacting, and the product region defines what is thought of as a product of the elementary reaction in question. The border between the two regions is referred to as the transition state (TS). The lowest energy configurations in the reactant and product regions are often referred to as the initial state (IS) and final state (FS), respectively.

If the transition state is assumed to have zero thickness, it is of a dimensionality of one lower than the configuration space of the system. If entropy effects were to be neglected, the natural choice would be to let the transition state follow the

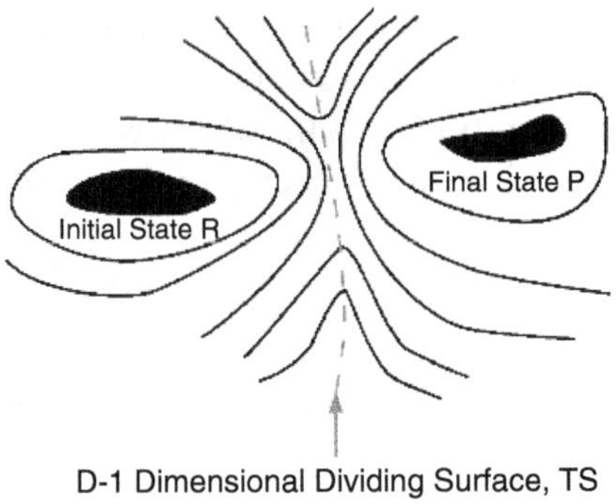

Figure 4.26. Separation of the configuration space according to transition state theory.

energy ridge between the reactant and product regions (as drawn in Figure 4.26). The configuration space is the space spanned by the displacements of the positions for all relevant atoms. The division of configuration space in transition state theory can also be formulated in terms of a more general configuration space, which is constituted by both the positions and the momenta of the involved atoms. This formulation is mostly used for more advanced applications of TST. The derivation of the TST rate constant below shall take advantage of this more general configuration space.

When applying transition state theory to potential energy surfaces, quantum-tunneling effects are assumed negligible and the Born-Oppenheimer approximation is invoked. Besides these, two central assumptions are made in transition state theory. The first is that the atoms in the reactant state have energies that are Boltzmann distributed. This should be satisfied if the system has had enough time to thermally equilibrate. An equivalent assumption, which is usually applied in the case that the initial state is unbounded, is therefore that an incoming flux of reactants should be thermally equilibrated. This assumption is certainly not always fulfilled. One well-studied case where the assumption breaks down is for molecular beam experiments, where the incoming flux can only be made to represent a thermal ensemble if extreme care is taken. The other assumption behind transition state theory is that once the system attains the transition state, with a velocity towards the product configuration, it will not reenter the initial state region again. This is a serious approximation, which one has to be aware of, since it leads to severe constraints on where a transition state can be chosen. One can, however, effectively correct for this assumption as we shall sketch later.

Under these assumptions transition state theory states that the rate constant of an elementary reaction (the rate assuming that one has the reactant in the initial state) is given by:

$$k_{TST} = P_{TS} \cdot r_{c,TS}, \tag{13}$$

where P_{TS} is the probability of finding the system in the transition state region, and $r_{c,TS}$ is the rate by which the transition state region is crossed. We assume here that the transition state region has a finite width, which we can let approach zero. Using the assumption of the system having its energy Boltzmann distributed:

$$E(\vec{x},\vec{v}) = E_{pot} + E_{kin} = V(\vec{x}) + \sum_i \frac{1}{2} m_i v_i^2 \tag{14}$$

the probability of finding the system in a given region of configuration space $([\vec{x}, \vec{x}+d\vec{x}], [\vec{v}, \vec{v}+d\vec{v}])$ can be evaluated:

$$\begin{aligned} P(\vec{x},\vec{v})d\vec{x}d\vec{v} &= Ae^{-E(\vec{x},\vec{v})/k_BT} d\vec{x}d\vec{v} \\ &= A_x e^{-V(\vec{x})/k_BT} \cdot A_v e^{-\sum_i \frac{1}{2} m_i v_i^2 / k_BT} d\vec{x}d\vec{v} \\ &= P_x(\vec{x})d\vec{x} \cdot P_v(\vec{v})d\vec{v}, \end{aligned} \tag{15}$$

Where A, A_x, and A_v are normalization constants. Since the potential and kinetic energies are additive, the probability of finding the system in the configuration interval $([\vec{x}, \vec{x}+d\vec{x}], [\vec{v}, \vec{v}+d\vec{v}])$ turns out to be separable into probabilities of position and velocity. We can now determine the probability of finding the system in the infinitesimal vicinity of thickness, δx, around the transition state to be:

$$P_{TS} = \delta x \cdot \frac{\int_{TS} e^{-V(\vec{x})/k_BT} d\vec{x}}{\int_{RR} e^{-V(\vec{x})/k_BT} d\vec{x}} = \delta x \cdot \frac{Z_{TS}}{Z_{RR}}, \tag{16}$$

where Z_{TS} and Z_{RR} are called the configuration integrals over the transition state (dividing surface) and the reactant region, respectively. The rate at which the system crosses the infinitesimal region of thickness, δx, around the transition state towards the final state is:

$$r_{c,TS} = \frac{\langle v_\perp \rangle}{\delta x}, \tag{17}$$

where $\langle v_\perp \rangle$ is the velocity perpendicular to the transition state in the direction towards the final state see Figure 4.27.

This velocity can be evaluated directly from the Boltzmann distribution:

$$\langle v_\perp \rangle = \frac{\int_0^\infty v_\perp \cdot e^{-\sum_i \frac{1}{2} m_i v_i^2 / k_BT} dv_\perp}{\int_{-\infty}^\infty e^{-\sum_i \frac{1}{2} m_i v_i^2 / k_BT} dv_\perp} = \sqrt{\frac{k_BT}{2\pi\mu}}. \tag{18}$$

The variable μ introduced at this point is an effective mass of the system in motion at the transition state. This can be evaluated for the specific application [101], but often it is apparent from the context of the application which mass is to be used.

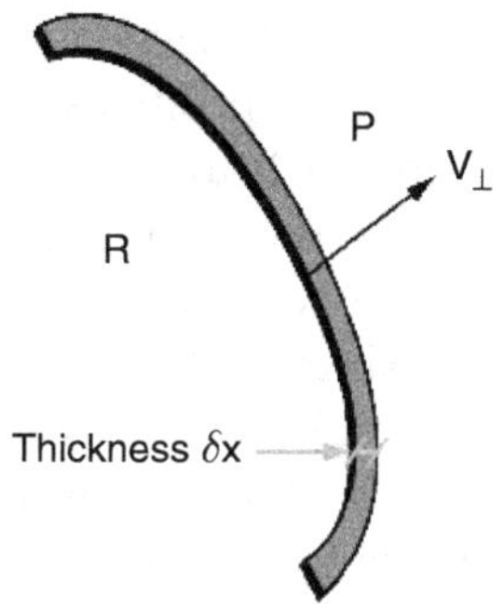

Figure 4.27. Definition of the vicinal region of the transition state and the perpendicular forward velocity. Here 'R' designates the reactant region and 'P' the product region.

The central result of transition state theory is then the definition of the transition state theory rate constant, which is obtained by combining the equations (13, 16, 17, and 18):

$$k_{TST} = P_{TS} \cdot r_{c,TS} = \sqrt{\frac{k_B T}{2\pi\mu}} \cdot \frac{Z_{TS}}{Z_{RR}}. \tag{19}$$

8.2. Variational transition state theory and recrossings

It is assumed in transition state theory that configurations, which are found in the transition state and have a velocity towards the product region will eventually end up in the product region. This means that cases where the supposed product crosses back into the reactant region are miscounted (see Figure 4.28).

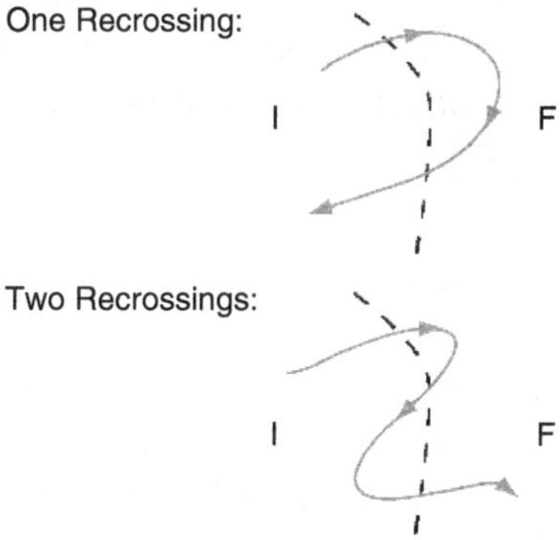

Figure 4.28. Various recrossings, all leading to an overestimation of the rate.

Since some trajectories turn back to the reactant region, these lead to an overestimation of the TST rate constant as 'false positives'. Other reaction trajectories will cross the transition state once, turn back to the reactant region and subsequently cross the transition state region again to end up in the final state. These trajectories also lead to an overestimation of the rate, since their single reactive event leads to a double counting of the forward crossings of the transition state. In fact, it is observed that no matter how many times a trajectory crosses the transition state before it turns into products or reactants, the TST rate constant will be overestimated, since it assumes that every crossing of the transition state towards the product region is one reactive event.

The overestimation of the TST rate constant leads to a variational principle for the optimization of the position of the dividing surface constituting the transition state. In general, one can write:

$$k_{TST} \geq k_{\text{exact}}, \tag{20}$$

with equality only holding in the probably unattainable case where there are no recrossings. It is therefore natural to choose the transition state in such a way that the rate constant is minimized. Such a transition state is by definition the 'best' transition state to be used for a TST evaluation of the rate [102,103]. In general, however, there are considerable difficulties in representing the dividing surface for systems of many atoms, and more effort is needed in this area to yield generally applicable and computationally tractable methods for larger systems [101,104,105].

Further improvements can be obtained by going beyond transition state theory and including dynamical corrections to the TST rate constant [106–109]. This is done by evaluating the dynamics of a Boltzmann ensemble of trajectories starting out on the transition state surface. These trajectories are only followed for a very short time compared to the rare reaction event, but long enough to give an estimate of how many times they cross the transition state before being either thermally equilibrated in either the reactant or product region. The inclusion of dynamical corrections to TST gives superior estimations of the rate constant, but can be rather computationally demanding [91].

8.3. Harmonic transition state theory (HTST)

Transition state theory is very often used in its harmonic approximation. The harmonic approximation is applicable under the normal assumptions of transition state theory, but further demands that the potential energy surface is smooth enough for a harmonic expansion of the potential energy to make sense. Since the harmonic expansion is performed in the initial state and in a first-order saddle point on the potential energy surface, it is also necessary that the potential is well-represented by its second-order Taylor expansion around both these configurations.

The procedure for determining the HTST rate constant thus follows a series of well-defined steps. First an initial state is determined as the lowest energy point in the reactant region, e.g., by direct structural optimization. Then a first-order saddle point on the potential energy surface needs to be determined. A first-order saddle point is a stationary point on the potential energy surface (like a local minimum or maximum), but in which one eigenmode is negative and all the other eigenmodes are positive. Many methods exist for performing such saddle point searches. Of special interest for future development are those methods which do not require information about the final state of the reaction, since these methods in principle allow for long time-scale simulations to be carried out without predefined event tables [110]. For extended systems relevant for heterogeneous catalysis it is often impractical to perform an analytical calculation of the Hessian matrix. One is therefore limited to saddle point methods, which only employ first derivatives. The majority of current studies in theoretical catalysis focus on predefined reactions, for which both the initial and final states are known, and the objective of the saddle point search is to find the saddle point in between. For problems of this type most often the Nudged Elastic Band (NEB) method [111–113] is applied or one of the many algorithms derived from the NEB [114,115]. The NEB algorithm establishes an ensemble of 'images' of the system along the minimal energy path (MEP) from the initial to the final state of the reaction. To employ this algorithm, one thus needs to *a priori* determine the final state as well. The saddle point is determined as the maximum energy configuration along the MEP. The transition state in HTST is chosen as the uniquely defined dividing surface, which is the hyperplane going through the saddle point and which is perpendicular to the reaction coordinate in the saddle point.

By performing a normal mode analysis in the initial state and in the saddle point, it is then possible to obtain the harmonic expansion of the potential in the reactant region:

$$V_{RR}(\vec{x}) = V(\vec{x}_{IS}) + \sum_{i=1}^{D} \frac{1}{2} k_{i,IS}\, q_{i,IS}^2 \tag{21}$$

and of the potential in the transition state:

$$V_{TS}(\vec{x}) = V(\vec{x}_{SP}) + \sum_{i=1}^{D-1} \frac{1}{2} k_{i,SP}\, q_{i,SP}^2. \tag{22}$$

The q_i's are the coordinates along the normalized eigenvectors, $\vec{n}_i$, in the initial state and in the saddle point, respectively, so that $\vec{x} = \sum_i q_i \vec{n}_i$. The k_i's are force constants obtained from the normal mode analysis by diagonalizing the mass-weighted Hessian matrix [116]. The expressions (21) and (22) can be inserted into the expression for the rate constant (19), and the configuration integrals (of equation (16)) can be evaluated explicitly. In terms of frequencies, $v_i = \frac{1}{2\pi} \cdot \sqrt{k_i/\mu_i}$, of the eigenmode i corresponding to an effective mass of μ_i, the HTST rate constant then becomes (at $k_B T >> h v_i$ for all i):

$$k_{\text{HTST}} = \frac{\prod_{i=1}^{D} v_{i,IS}}{\prod_{i=1}^{D-1} v_{i,SP}} e^{-\Delta E_{\text{barrier}}/k_B T}, \tag{23}$$

where $\Delta E_{\text{barrier}} = V(\vec{x}_{SP}) - V(\vec{x}_{IS})$. The expression for the harmonic TST rate constant thus has the form of an Arrhenius expression:

$$k_{\text{Arrh}} = v e^{-\Delta E/k_B T}. \tag{24}$$

Since in eq. (23) there is one frequency more in the numerator than in the denominator it is often interpreted as an attempt frequency of the reactant system multiplied by a Boltzmann factor corresponding to the energy barrier between the initial state and the saddle point in the transition state. The transition state theory result is often written in the form (see page 110 of [3]):

$$k = \frac{k_B T}{h} e^{-\Delta G/k_B T} = \frac{k_B T}{h} e^{\Delta S/k_B} \cdot e^{-\Delta H/k_B T}, \tag{25}$$

where ΔG, ΔS, and ΔH are differences in Gibbs free energy, entropy, and enthalpy between the initial and transition state, respectively. This form is often very convenient for applications, since it allows the inclusion of calculated partition functions or tabulated entropies in a straight-forward fashion. Often the change in entropy between reactant and transition state is relatively small. This is particularly true for systems where there are no changes in free translational degrees of freedom between initial and transition states, since free translations lead to the largest contributions to the entropy (e.g., as in gases). At catalytically relevant temperatures, $\frac{k_B T}{h}$ is on the order of 10^{13} s^{-1}. In the absence of larger entropic effects the prefactor will thus be on the order of magnitude of 10^{13} s^{-1}, which is also what is being observed experimentally [97]. For the study of trends in catalysis, one often looks at a given reaction, and varies the catalytic surface. Here the variations in entropy from system to system are usually very small, and give rise to variations in the rate by maximally a couple of orders of magnitude [117]. This should be compared to the changes in rates induced by varying the energy barriers of different reactions. A change in barrier of 1 eV will at room temperature, e.g., lead to a variation in the Boltzmann factor of approximately 10^{20}. Since the variation in activation barriers between neighboring metals in the Periodic Table is on the order of perhaps half an eV, it is often sufficient for the study of trends in heterogeneous catalysis to disregard effects of varying prefactor and concentrate only on the variations in the reaction energetics.

Many adsorbate and bulk systems fulfill the criteria for HTST to apply. For strongly bonded systems, the potential energy surface is often sufficiently harmonic in the initial state and in the saddle point. This is perhaps due to the rather large energy barriers involved in most surface processes. Often the most important processes are slow. This can be seen as natural, since strong bonds leads to attempt frequencies around 10^{13} s^{-1}. If the energy barrier is not substantial, the rate automatically turns out very high, and the initial and transition states would then turn out to be in thermal equilibrium. In Figure 4.29, results from a HTST calculation of the rate constant for N_2 dissociation over a ruthenium surface are compared to results obtained using the more advanced and much more demanding 'Multi Configuration Hartree Approach'. The latter approach takes quantum-tunneling effects into account. It is observed that the two lines are essentially identical down

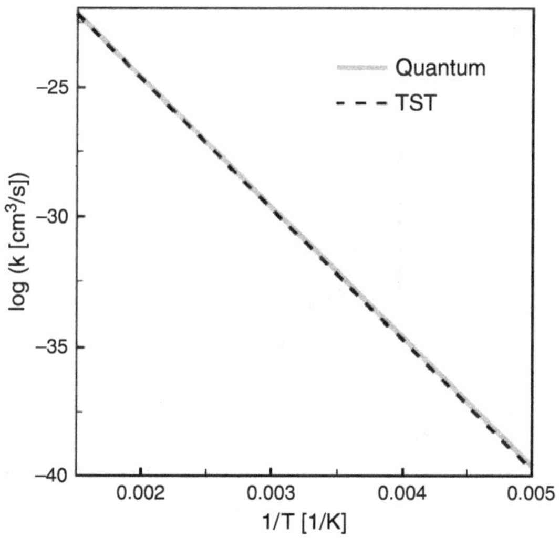

Figure 4.29. Difference between TST and Multi Configuration Hartree Approach. Adapted from Ref. [118].

to very low temperatures ($1/T = 0.005$ 1/K corresponds, of course, to $T = 200$ K. At low-enough temperatures the quantum tunneling rate will be dominant, but this occurs at even lower temperatures).

Some important systems, which certainly do not fulfill the assumptions of harmonic transition state theory are gas phase reactions. In the gas phase, there are zero-modes such as translation and rotation, and these lead to totally different configuration integrals than those obtained from a normal mode analysis. For these species one can in a simple manner modify the terms going into the HTST rate by incorporating the molecular partition functions [3,119].

Both the kinetic Monte Carlo and mean-field models for solving the reaction equations of a heterogeneously catalytic system are usually built upon elementary rates obtained from the harmonic transition state theory approach. In Figure 4.30 is shown an application of the kinetic Monte Carlo method to describe CO oxidation over a $RuO_2(110)$ surface purely on the basis of activation energies calculated using DFT [94]. The agreement with detailed surface science experiments is very good [94]. The same is true for a similar attempt to calculate the rate of ammonia synthesis over Ru, see Figure 4.31. Here there are high pressure data for comparison, and the agreement is again surprisingly good [58].

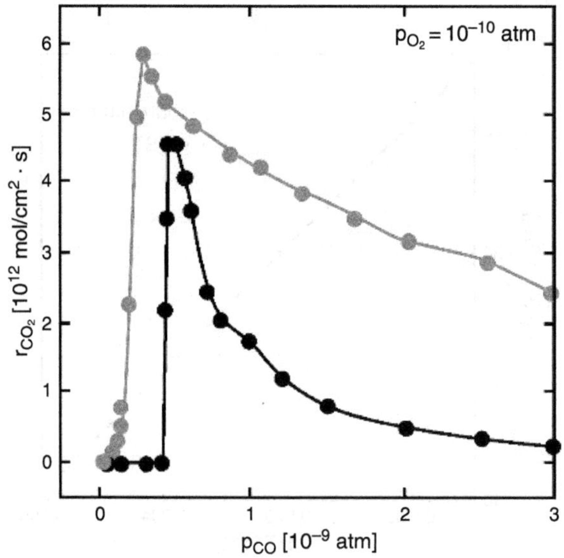

Figure 4.30. Comparison of calculated and experimental rates for CO oxidation over a $RuO_2(110)$ surface. The top curve is experimental and the bottom curve is calculated. Adapted from Ref. [94].

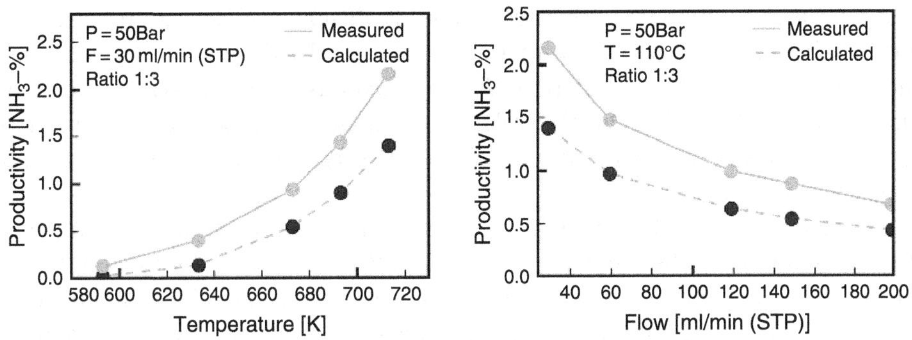

Figure 4.31. Comparison of calculated and experimental rates for ammonia synthesis of a nanoparticle Ru catalyst under industrial conditions. Adapted from Ref. [58].

9. Variations in catalytic rates – volcano relations

As described above, elaborate kinetic methods are available for performing the detailed description of the rate of a given heterogeneous reaction. Here we shall instead focus on the more general description of trends in catalysis. Mean-field microkinetic models are in many cases adequate for quantitatively describing the reaction rate [120,121]. An example is shown in Figure 4.32. The mean-field models have some distinct advantages when studying trends, since the introduction of a few additional assumptions (such as inclusion of a rate-determining reaction and the steady-state approximation) will often result in the model becoming entirely analytical.

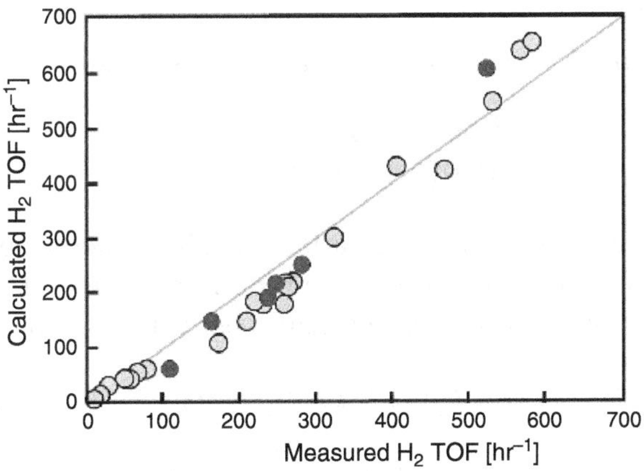

Figure 4.32. Methanol decomposition over Pt as determined from a microkinetic model. Adapted from Ref. [120].

The microkinetic models in this section are built upon BEP-relations of the type described above. It will be shown that an underlying BEP-relation in general leads to the existence of a volcano relation. We shall also use the microkinetic models in combination with the universal BEP-relation to explain why good catalysts for a long range of reactions lie in a surprisingly narrow interval of dissociative chemisorption energies.

A simple tool is described, which provides a conceptual framework for analyzing microkinetic models of heterogeneous reactions. We refer to this tool as the 'Sabatier Analysis'. The Sabatier Analysis of the microkinetic models developed in this section suggests that the clustering of good catalysts can be explained by the combination of the universal BEP-relation and activated re-adsorption of synthesis products onto the catalyst.

9.1. Dissociation rate-determined model

We commence by creating a simple model for a heterogeneously catalyzed reaction, which proceeds from a diatomic reactant and necessitates the cleavage of a strong molecular bond. A large number of important reactions belong to this category. One of the most simple such reactions that can be envisioned is the reaction:

$$A_2 + 2B \rightarrow 2AB, \qquad \{1\}$$

where the dissociation of the diatomic molecule, A_2, happens at the same time as the molecule is adsorbed upon the surface. The other reactant, B, reacts with the adsorbed species, A, without prior adsorption of B on the surface, and with direct desorption of the product, AB, to the gas phase. The reaction scheme of elementary steps is thus:

Model 1: Dissociative chemisorption as the rate-determining step.

$$A_2 + 2^* \rightarrow 2A^* \qquad \{2\}$$
$$A^* + B \rightarrow AB + ^*, \qquad \{3\}$$

where an asterisk represents an active surface site. The microkinetic model for this system [1,97] allows us to express the turnover frequency (TOF), which is the frequency of net creation of product molecules per active site on the catalyst:

$$r(T, P_x) = 2 \cdot k_1 \cdot P_{A_2} \cdot \theta_*^2(1 - \gamma) \qquad (26)$$

Here k_1 is the temperature dependent rate constant for the forward direction of reaction step 1, which is assumed to follow an Arrhenius expression with activation energy of E_T of Figure 4.33, P_{A_2} is the pressure of the reactant A_2, θ_* is the fraction

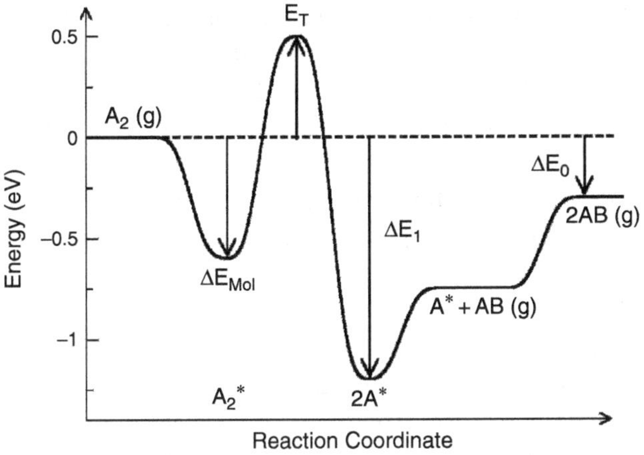

Figure 4.33. Energy diagram for the microkinetic Model 1.

of the active sites that are free, and γ is the approach to equilibrium (sometimes also referred to as the reversibility of the reaction) [96,122,123]. For the expression to be valid in general, γ should be the approach to equilibrium for reaction step 1, but if the dissociative chemisorption is the rate-determining step, which we shall assume for now (later this assumption will be removed), then the γ in equation (26) is the overall gas phase approach to equilibrium:

$$\gamma = \frac{P_{AB}^2}{K_{eq} \cdot P_{A_2} \cdot P_B^2}, \tag{27}$$

where K_{eq} is the equilibrium constant for the reaction, P_B is the pressure of the reactant, B, and P_{AB} is the pressure of the product, AB. Under the assumption that dissociative chemisorption is the rate-determining step, the coverage of free sites can be determined analytically:

$$\theta_* = \frac{1}{1+\frac{\theta_A}{\theta_*}} = \frac{1}{1+\frac{P_{AB}}{K_2 P_B}} = \frac{1}{1+\sqrt{K_1 P_{A_2} \gamma}} \tag{28}$$

where $K_1 = \exp(-\Delta G_1/k_B T)$ is the equilibrium constant for reaction step 1 with standard reaction Gibbs energy $\Delta G_1 = \Delta E_1 - T\Delta S_1$. Under the given assumptions it is thus possible to obtain an analytical expression for the turnover frequency, which is the primary reason for assuming that a given reaction step is rate-determining. The dissociative chemisorption energy ΔE_1 determines how very reactive surfaces (very negative ΔE_1) will poison the reaction in the sense that there will be very few free sites on which to dissociate onto. The turnover frequency will thus decrease as $\Delta E_1 \to -\infty$. The BEP-relation relates the transition state energy of dissociation to the dissociative chemisorption energy on a given surface. The activation barrier will be large on less reactive surfaces (ΔE_1 numerically small or even positive), and the turnover frequency will thus also decrease as $\Delta E_1 \rightarrow \infty$. In the intermediate range, the turnover frequency goes through a maximum that often resembles a volcano, and the turnover frequency is thus said to follow a volcano curve.

To analytically determine such volcano curves for the simple model reaction, we need to make some further assumptions (the assumptions are realistic at least for the case of NH_3 synthesis):

- The entropy of gas phase A_2 is 200 J/(mol K), which is reasonably close to the correct value for many diatomic molecules [124].
- The entropy of A atoms adsorbed on the surface is assumed to be negligible. This is not generally true, but often a sufficiently good approximation [97].
- The entropy of the reactant, B, in the gas phase is equal to the entropy of the product AB in the gas phase. This is sometimes a poor approximation, but the turnover frequency rarely depends strongly on this value.
- The transition state for the dissociative chemisorption and for desorption is strongly constrained. This assumption allows us to set the partition function in the transition state equal to 1, and the prefactors in the Arrhenius expressions for the rate constants of desorption of AB and re-desorption of A_2 thus become $\frac{k_B T}{h}$, where k_B is the Boltzmann constant, T is the temperature and h is the Planck constant [125].

Using these assumptions, we avoid the more involved treatment of the detailed adsorption processes. Specific variations in sticking coefficients or steering effects can be calculated theoretically [126,127], but often these lead to relatively small variations compared to the trends induced by varying the energetics. Under these assumptions, the universal BEP-relation allows the analytical calculation of the turnover frequency when dissociation is rate-determining.

Various reactions which proceed via the dissociation of diatomic molecules can have drastically different over-all reaction energies, ΔE_0. This demonstrates how useful it is to write the microkinetics in terms of the approach to equilibrium instead of using the product pressure and reaction energy to describe the given reaction. Reactions with drastically different reaction energies all follow the same microkinetic model for a given approach to equilibrium, but a given approach to equilibrium will correspond to very different product pressures for the various reactions. The volcano-curves obtained at different approaches to equilibrium for the generalized microkinetic model with rate-determining dissociation are shown in Figure 4.34.

It is seen in Figure 4.34 how reaction conditions close to equilibrium ($\gamma \rightarrow 1$) require a catalyst, which is nobler than further away from equilibrium. This is

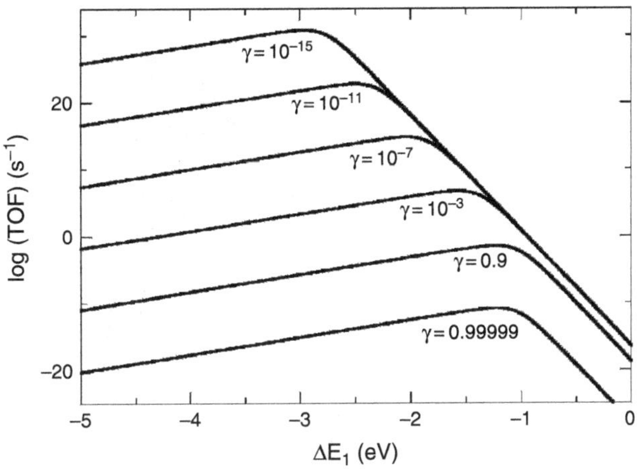

Figure 4.34. Turnover frequencies at various approaches to equilibrium.

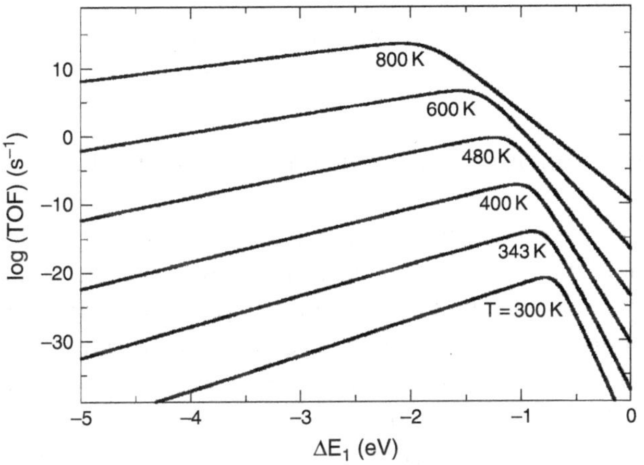

Figure 4.35. Temperature dependence of the optimal catalyst.

observed also under industrial conditions in, e.g., the ammonia synthesis process. In that process, iron is the optimal catalyst far from equilibrium, but to reach an approach to equilibrium close to one in the last part of the reactor bed, ruthenium is a more suitable (and more noble as well as much more expensive) catalyst [128].

The same microkinetic model can be used to investigate how the reactivity of the optimal catalyst changes with other reaction conditions such as temperature or pressure. In Figure 4.35, the dependence of the turnover frequency on temperature is shown. For high temperatures, the optimal catalyst moves out towards the more reactive surfaces. Figure 4.36 shows the dependence of the turnover frequency on the pressure of the more important reactant. The position of the optimal catalyst for

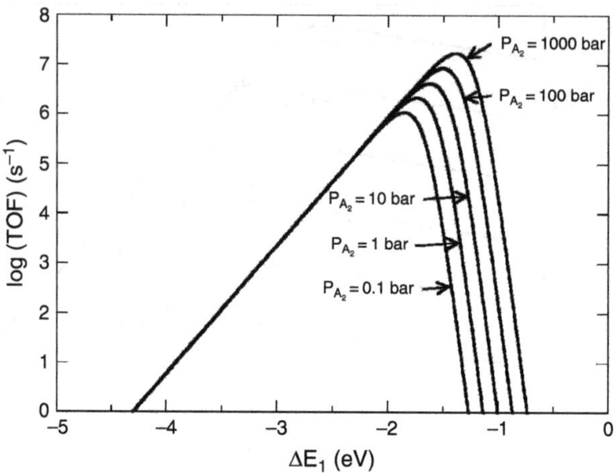

Figure 4.36. Pressure dependence of the optimal catalyst.

a given reaction varies less with pressure than the possible variations due to changes in the approach to equilibrium and the temperature.

9.2. A Le Chatelier-like principle for heterogeneous catalysis

It is perhaps not clear that the optimal catalyst should have a dependence on the reaction conditions as described here. For developing new catalysts it would be useful, if there was access to a conceptual framework from which the effect of changing reaction conditions could directly give insight into the position of the maximum of the volcano-curve describing the catalytic turnover frequency. It is possible to develop such a conceptual framework and thus a more intuitive understanding through an analysis of the surface coverage at the optimum of the volcano-curve, at least in the case of a dissociation rate-determined kinetics.

To determine the analytical expression for the coverage of free sites on the optimal catalyst, we simply differentiate the turnover frequency given in equation (26) with respect to ΔE_1 and set this equal to zero. This determines the optimal ΔE_1. With the universal BEP-relation substituted into the forward rate constant k_1, and using that the equilibrium constant K_1 also depends exponentially on ΔE_1 (through the Gibbs' relations for equilibrium constants), it can be shown after some straightforward algebra that the coverage of free sites at the optimal ΔE_1 always is [129]:

$$\theta_* = 1 - \alpha \tag{29}$$

for the given model. Here α is the slope of the universal BEP-relation. This result is surprisingly independent of the reaction conditions and suggests that the optimal catalyst under given reaction conditions as a reactivity leading to an intermediate coverage of the key reactants. Because of the conservation of the coverage of free sites, a Le Chatelier-like principle of heterogeneously catalyzed reactions can be developed for reactions that are dissociation rate-determined and proceed by the dissociation of a diatomic molecule.

When changing reaction conditions, the new optimal catalyst will be a surface of reactivity that counteracts the change of the coverage of free sites, which is induced by the new reaction conditions.

It is observed that this principle correctly describes the trends observed in Figures 4.34–4.36. When the approach to equilibrium is decreased, less product intermediates are present on the surface, and the optimum will move towards the more reactive surfaces to compensate. A temperature increase will remove adsorbates from the surface, due to the entropic push towards the gas phase, and the optimal catalyst must thus be a more reactive surface to compensate. An increase in the reactant pressure will increase the surface coverage. The optimal catalyst under the higher pressure must therefore be a nobler surface to conserve the optimal coverage.

9.3. Including molecular precursor adsorption

A strongly bound precursor could change the reactivity of the catalyst, and this effect cannot be analyzed within the model that has been used so far. The reaction system in *Model 1* will therefore now be expanded slightly, to facilitate this analysis:

Model 2: Including the effects of a strongly bound precursor.

$$
\begin{array}{ll}
A_2 + {}^* \rightarrow A_2^* & \{1a\} \\
A_2^* + {}^* \rightarrow 2A^* & \{1b\} \\
A^* + B \rightarrow AB + {}^* & \{2\}
\end{array}
$$

In this model, A_2 molecules are first adsorbed on the surface non-dissociatively. The A_2 molecular precursor might dissociate if there is a free active site adjacent to it, and if it is capable of climbing the dissociation energy barrier due to thermal excitation, or the precursor could be thermally activated to desorb as A_2 into the gas phase again. It is still assumed that the dissociation (now from the precursor state and not from the gas phase) is the rate-determining step. If the reaction proceeds to a steady-state, but the over-all gas phase reactants and products are kept out of equilibrium, the precursor state will be in equilibrium with the gas phase reactant, but not with the dissociated state. This model will have a turnover frequency given by:

$$r(T, P_x) = 2 \cdot k_{1b} \cdot \theta_{A_2} \cdot \theta_*(1-\gamma), \tag{30}$$

where k_{1b} is the forward rate constant of the (rate-determining) reaction step 1b and γ is the approach to equilibrium for the over-all gas phase reaction. The analytically determined coverage of free sites is now:

$$\theta_* = \frac{1}{1+\frac{P_{AB}}{K_2 \cdot P_B} + K_{1a} \cdot P_{A_2}}, \qquad K_{1a} = \exp(-(\Delta E_{1a} - T\Delta S_{1a})/k_B T) \tag{31}$$

and the coverage of the molecular precursor:

$$\theta_{A_2} = \frac{K_{1a} \cdot P_{A_2}}{1+\frac{P_{AB}}{K_2 \cdot P_B} + K_{1a} \cdot P_{A_2}}. \tag{32}$$

In these relations, K_i denotes the equilibrium constant of reaction step i. For the numerical evaluation of the model, it is assumed that the backward reaction of step 1b has the same transition state as the transition state for the re-desorption of A_2 in *Model 1*, and that the entropy of the molecular precursor on the surface is negligible. The results are shown in Figure 4.37. It is observed that the model predicts that catalysts of much larger reactivity (more negative ΔE_1) will be optimal for reactions where the diatomic molecule is strongly bound to the surface before the dissociation.

In the other cases discussed above, the optimal catalyst is relatively close to the narrow region of dissociative chemisorption energies from −2 to −1 eV. It does, however, appear that the models developed so far could also have a problem describing why some high temperature and very exothermic reactions (with corresponding small approaches to equilibrium) also lie within the narrow window of chemisorption energies. To remove these discrepancies we shall relax the assumption of one rate-determining step, but retain an analytic model, by use of a least upper bound approach.

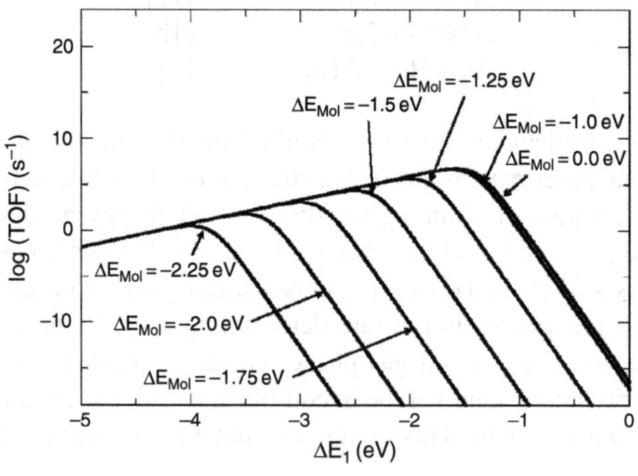

Figure 4.37. Variation of volcano plots with the molecular precursor adsorption energy.

9.4. Sabatier analysis

The analysis given until this point is primarily valid for heterogeneously catalyzed reactions, which are dissociation rate-determined when using an optimal catalyst. Some reactions are known to be rate-determined by the dissociative chemisorption reaction step, as is the case for ammonia synthesis [130,131]. For other less studied reactions, it is perhaps not clear whether there is a rate-determining step, or which step this is. It is often less practical having to find the exact numerical solution of the given microkinetic model, and then as a post-treatment calculate whether the dissociation step is rate-determining or not. Often it is preferable to have an analytical solution based on an assumption of a rate-determining step, since the lack of analytical expressions for the results hinders the subsequent development in the understanding of the reaction in question. It is therefore desirable to have an analytical tool available for analyzing when the assumption of rate-determination breaks down.

To develop such tools, the microkinetic *Model 1* is first evaluated exactly in a numerical manner, now without the assumption of a rate-determining step. The resulting volcano curves are shown in Figure 4.38. On the right-hand side, the volcano-curves are identical to the solutions presented in Figure 4.34, which were obtained using the assumption of a rate-determining dissociation step. On the left-hand side, the volcano-curves for reaction conditions close to equilibrium also closely follow the behavior in Figure 4.34. The volcano-curves far from equilibrium, however, show a drastically different behavior. In Figure 4.38 the solution for $\gamma = 10^{-15}$ is actually also shown, but it cannot be distinguished from the

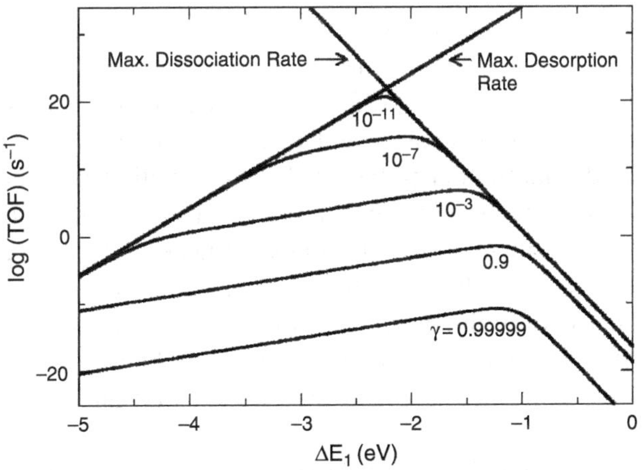

Figure 4.38. Sabatier volcano-curve: The limiting case of the exact numerical solution of the microkinetic Model 1.

volcano-curve with $\gamma = 10^{-11}$. The lines representing the maximal desorption and maximal dissociation rates clarify the situation. These maximal rates are simply calculated as the rates of desorption or dissociation assuming an optimal coverage for the given reaction step. When the microkinetic model is assumed to be dissociation rate-determined, the turnover frequency will automatically be bounded by the line corresponding to the maximal possible dissociation rate. For very reactive surfaces or very small approaches to equilibrium, the solution obtained with the assumption of a rate-determining step will eventually violate the other exact bound on the turnover frequency, that the turnover frequency cannot be larger than the maximal desorption rate. The exact solution looks like the solution obtained with the rate-determining step, until that solution approaches the maximal desorption rate, which it will join smoothly (through a suitable adjustment of the coverages).

Several important conclusions can be drawn from Figure 4.38. It appears that in general a simple catalytic reaction, which includes the dissociation of a diatomic molecule, will have this dissociation as the rate-determining step, when the reaction takes place under conditions close to equilibrium. This agrees well with the ammonia synthesis being dissociation rate-determined, as this process is the prototype of an equilibrium-limited reaction [128]. When the reaction is taking place far from equilibrium, the actual approach to equilibrium becomes unimportant, and the volcano plot very closely follows the volcano defined by the minimum value among the maximal possible rates for all reaction steps.

For the general case, the limiting Sabatier Volcano-Curve can be defined as:

$$TOF_{\text{Sabatier}} = \min(\max R_1, \max R_2, \ldots, \max R_N) \tag{33}$$

Here $\max R_i$ is the maximal rate of reaction step i, which is calculated by assuming optimal coverages for that reaction step. This (usually multi-dimensional) volcano-curve we shall refer to as the Sabatier volcano-curve, as it is intimately linked to the original Sabatier principle [132,133]. This principle states that desorption from a reactive metal catalyst is slow and will increase on less reactive metals. On very noble metals the large energy barrier for dissociation will, however decrease the dissociation rate. The best catalyst must be a compromise between the two extremes. As has been shown above, this does not necessarily mean that the optimal compromise is obtained exactly where the maximal desorption and dissociation rates are competing. That is only the case far from equilibrium. Close to equilibrium the maximum will often be attained while dissociation is the rate-determining step, and the maximum of the volcano-curve will then be reached due to a lack of free sites to dissociate into.

We will refer to the method of using the maximal possible reaction rates for all the reaction steps of a heterogeneously catalyzed reaction as 'Sabatier Analysis'. This analytical method might prove useful for various purposes in the future. Here its

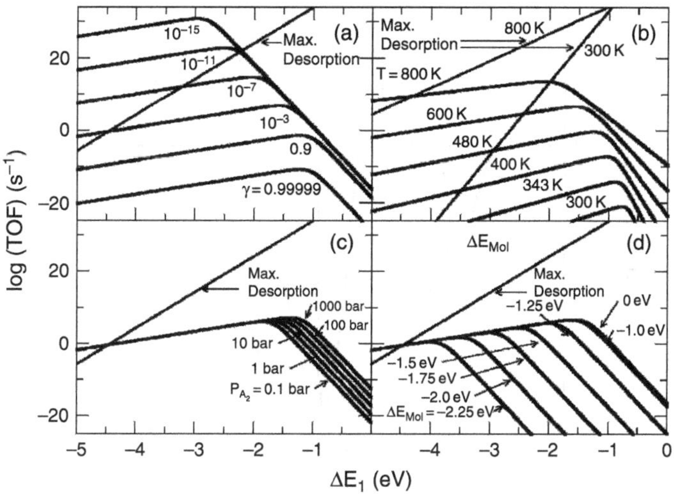

Figure 4.39. Sabatier desorption-limitation of the dissociation rate-determined kinetics.

use will be demonstrated to assess the validity of the results obtained earlier in this chapter. It was shown in Figures 4.34–4.37 how changes in reaction conditions and the chemisorption energy of the molecular precursor could significantly move the optimum of the turnover frequency volcano-curves. These results were based on the assumption of dissociation being the rate-determining step. In Figure 4.39, the same figures are shown again. This time a line, which represents the maximal desorption rate is drawn on each figure. The maximal desorption rate is independent of γ, P_{A_2} and ΔE_{Mol}, but it does have a dependence on T. Therefore two different maximal desorption rate-lines are drawn on Figure 4.39(b), which shows the volcano-curves at various temperatures. One line represents the maximal desorption rate at 300 K, the other at 800 K.

This Sabatier Analysis shows that the assumption of rate-determining dissociation was valid to a large extent: We would have to modify the results obtained from the simple analysis based on a rate-determining step only at relatively small approaches to equilibrium, extremely low pressures or quite strongly bound molecular precursor states.

9.5. A realistic desorption model

By analyzing energy barriers for product desorption under ammonia synthesis, CO hydrogenation, and NO reduction by CO, we can refine the models further. For these three processes, the reaction conditions are very different. The ammonia synthesis process is weakly exothermic, whereas the CO hydrogenation reaction has

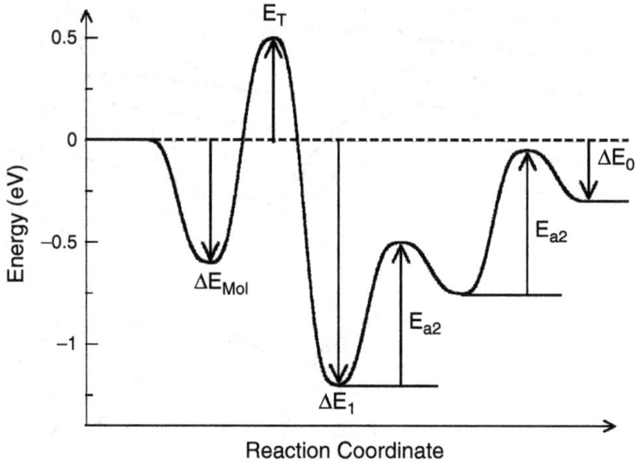

Figure 4.40. Energy diagram including a generalized desorption barrier.

an intermediate exothermicity and the NO reduction process is extremely exothermic. In the analysis of the energy diagrams in [56] it is shown that the desorption process in all three cases is further activated than just the direct desorption to the level of the over all exothermicity. In *Model 1* of the previous section, the reaction had no energy barrier for re-adsorption of the product AB onto the surface (following the energy diagram in Figure 4.33). If such a re-adsorption barrier is introduced, the desorption process will have a higher barrier, now following the energy diagram represented in Figure 4.40. With the barrier for the generalized desorption process being increased, the Sabatier Analysis of the dissociation rate-determined microkinetic models presented in the previous section will now give a stronger desorption constraint on the volcano-curves. This is shown in Figure 4.41.

In Figure 4.41, the maximal desorption rate has been determined under the assumption that the generalized desorption barrier E_{a2} follows a linear relation in ΔE_1, in such a way that the barrier grows for more reactive surfaces.

$$E_{a2} = \alpha_2 \cdot \Delta E_1 + \beta_2 \tag{34}$$

From the analysis of the ammonia synthesis, CO hydrogenation and NO reduction by CO in [56] we deduce that the barrier for generalized desorption is on the order of 1.5 eV for surfaces with dissociative chemisorption energies in the optimal range. If we choose the α_2-parameter to be 0.5 and set $\beta_2 = 0.75$ eV, a barrier corresponding to the one mentioned above is obtained. It is seen in Figure 4.41(a) that the maximal desorption rate now cuts off the volcano-curves much closer to equilibrium. This lends further credit to the Sabatier Analysis tool. If the re-adsorption of product molecules is generally an activated process, and the generalized desorption barrier is

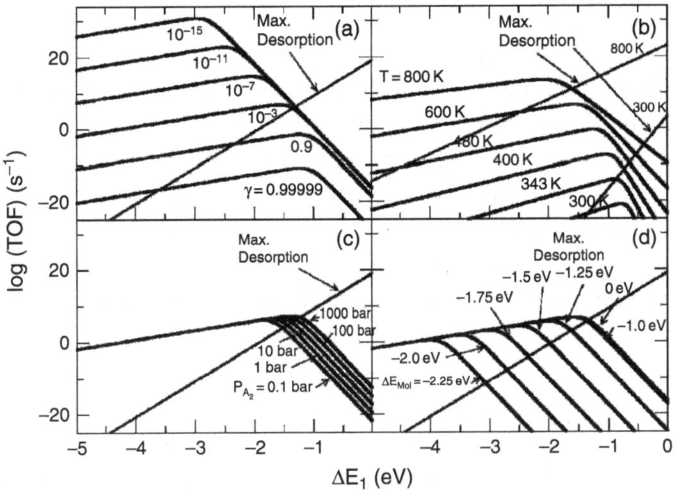

Figure 4.41. Sabatier desorption-limitation of the dissociation rate-determined kinetics with a generalized desorption barrier.

on the order of 1.5 eV, then the Sabatier-limit could be relevant for the large class of reactions which are carried out further from equilibrium than $\gamma \approx 10^{-3}$. It is observed in Figure 4.41(b) that the volcano curves at high temperature can have their optimum moved towards the less reactive surfaces due to the generalized desorption barrier. At 800 K, the optimum of the dissociation rate-determined volcano-curve will clearly be constrained, while the 300 K maximal desorption curve does not constrain the optimum of the volcano-curve at 300 K. The pressure dependence of the optima of the volcano-curves is likewise constrained. This is shown in Figure 4.41(c). In all these cases, the Sabatier Analysis shows that the optimal catalyst is constrained by the desorption process to lie within a narrow window of dissociative chemisorption energies. The only apparent exception from this rule is that very strongly bound precursor states dictate that the optimal catalytic surface will be a more reactive one.

The one remaining problem is now the strong dependence of the optimal catalyst on the molecular chemisorption energy. The increased generalized desorption barrier introduced through the Sabatier Analysis does not appear to constrain the maxima of the volcano-curves of the kinetics which include a strongly bound precursor to the narrow window in question. It is interesting that rhodium is used as a car catalyst to remove NO, and that the dissociative chemisorption energy of NO on a stepped rhodium surface is lower than −3.2 eV. This would appear to be well in line with the analysis above, as the NO molecule binds with approximately 2 eV to the stepped rhodium surface. Platinum is, however, also used as a catalyst for NO activation and the dissociative chemisorption energy of NO on platinum is only ≈−1.3 eV. The coverage on the rhodium catalyst is probably quite high under working conditions due to its high reactivity. In the analysis presented here the possibility of coverage-dependent adsorption energies has not been included. This could potentially explain part of the only remaining discrepancy between the model presented here and the observation of clustering of good catalysts. The discrepancy is that those catalysts, which have strongly adsorbed precursors are not confined as strongly as expected from the experimental evidence. It is noted that the value of α_2 chosen here could be too low. If the re-adsorption of the (usually larger) product molecules also follows a BEP-relation between their re-adsorption barrier and dissociative chemisorption energy, and if the slope of this relation is larger than 0.5 (as for the universal BEP-relation for stepped surfaces with a slope of 0.87), then the desorption confinement will be enhanced.

The Sabatier analysis has been shown here to be a very effective tool for interpreting microkinetic models based on a rate-determining step. The limiting Sabatier volcano-curve of eq. (33) will in some cases need to be augmented to retain its usefulness. Close to equilibrium, which is an interesting case in its own right, and a very important case for many industrial processes, the Sabatier volcano of eq. (33) becomes too weak a bound on the turnover frequency for Sabatier Analysis to retain its usefulness in that form. It was shown in [56] how each of the maximal rates going into the limiting Sabatier volcano can be adjusted by a factor of $1 - \gamma^{1/k_i}$, where k_i is determined from the dependence of the approach to equilibrium of the gas phase reaction on the approach to equilibrium for reaction step i:

$$\gamma = \gamma^{1/k_1} \cdot \gamma^{1/k_2} \cdot \ldots \cdot \gamma^{1/k_N} \tag{35}$$

This augmentation factor expands the use of Sabatier analysis to reactions close to equilibrium. In other cases, it might for other reasons not be relevant to apply Sabatier analysis directly in the form of the bounding volcano of eq. (33). Often microkinetic models include one or more pre-determined equilibria between various surface species. There could for example be an implicitly assumed equilibrium between adsorbed hydrogen species and free sites on a surface of an ammonia catalyst [74], such that the number of free sites never gets close in number to the optimal coverage used in eq. (33) for determination of the maximal dissociation rate. In such cases, the individual maximal rates of eq. (33) can be modified by a factor that takes this into account. In all cases, however, the cruder version of the Sabatier volcano in eq. (33) will give an exact upper bound to the real turnover frequency of the microkinetic model. This is even true for a non-stationary state of the catalytic process. In the analysis given above it has implicitly been assumed throughout that the catalyst was working in the stationary state. Because the maximal rates are determined at optimal coverages there does not exist a solution of the microkinetic model that has a higher turnover frequency. The local variations in temperature and pressure induced by an oscillatory or chaotic state of the heterogeneous reaction could however make it difficult to choose correct reaction conditions for the model.

9.6. Database of chemisorption energies

It has been shown above that there is plenty of evidence that good catalysts for heterogeneously catalyzed reactions that depend on the dissociation of diatomic molecules lie in a narrow window of dissociative chemisorption energies. In the previous sections, we have analyzed through the development of microkinetic models, why this would be the case. The clustering principle can be used directly to find new good catalysts. If a catalyst turns out to lie in the narrow region of dissociative chemisorption energies for a given process, the volcano relation will assure that it is a reasonably good catalyst. A problem that arises is that it is difficult to obtain reliable experimental chemisorption data for given catalytic reactions over surfaces with drastically varying activity. A database of such chemisorption data has thus been established from density functional theory. In Table 4.2, dissociative chemisorption energies for various small molecules on various surfaces are shown.

A database of molecularly adsorbed species on various surfaces is also included (see Table 4.3). In all cases, the chemisorption energies have been calculated on stepped surfaces using density functional theory (see [56] for details). The metals have been modeled by slabs with at least three close-packed layers. The bcc metals are modeled by the bcc(210) surface and the fcc and hcp metals have been modeled by the fcc(211) surface. A small discrepancy between the adsorption on the hcp metals in the fcc(211) structure is thus expected when the results are compared to the adsorption energies on the correct stepped hcp structure instead. When mixing

Table 4.3
Molecular chemisorption energies calculated by density functional theory for various molecules on a number of stepped transition metal surfaces. All values are given in eV per molecule.

	OH	N_2	CO	NO
Fe	−3.60	−0.35	−1.52	−2.34
Co	−3.48	−0.47	−1.50	−2.13
Ni	−3.23	−0.47	−1.66	−2.10
Cu	−2.81	0.07	−0.62	−0.71
Mo	−3.94	−0.24	−1.60	−2.59
Ru	−3.37	−0.61	−1.77	−2.35
Rh	−3.00	−0.56	−1.79	−2.16
Pd	−2.38	−0.25	−1.74	−1.79
Ag	−2.22	0.04	−0.06	−0.08
W	−4.19	−0.50	−2.02	−2.81
Ir	−3.08	−0.69	−1.96	−2.32
Pt	−2.49	−0.24	−1.89	−1.91
Au	−1.81	0.05	−0.35	−0.22

fcc and hcp metals they will often stabilize in the fcc structure, and the database might therefore be particularly suitable for being used together with the recently proposed principle of interpolation in the Periodic Table [53].

Using the data base we can test some of the conclusions in the preceding sections. It was shown that for reactions where a molecule is dissociated and the surface intermediates are then reacted away, the optimum rate for the reaction occurs for a dissociative chemisorption energy for the dissociating molecule in the range −1 to −2 eV. If we look at ammonia synthesis (N_2 dissociation) the best catalysts of the surfaces considered in Table 4.2 are Fe and Ru [134] with N_2 chemisorption energies according to Table 4.2 of −1.27 and −0.84 eV. For methanation and Fischer-Tropsch synthesis (CO dissociation), the best catalysts are Co, Ru, and Ni [135], with CO dissociative chemisorption energies of −1.51, −1.62, and −1.05 eV. For oxidation reactions (O_2 dissociation) the best catalysts are Pd, Pt, and for partial oxidation Ag [135], with associated O_2 chemisorption energies of −1.20, −2.17, and −0.65 eV. All of which is in good agreement with the general discussion above. For NO reduction Pt and Rh are used and while Pt lies in the −1 to −2 eV range Rh is well below (−3.23 eV) [137]. This is however not so strange since of the molecules considered here NO has the strongest molecularly adsorbed state (−2.35 eV for Rh), see Table 4.3. According to Figure 4.41 this should lead to optimum catalysts with stronger dissociative adsorption energies. Given the simplicity of the kinetics arguments, the agreement with observations is surprisingly good, and it indicates that we have the concepts necessary to understand trends in overall reactivity for this class of surface catalyzed reactions.

10. The optimization and design of catalyst through modeling

This chapter has so far attempted to explain the experimentally observed variations in adsorption energetics, and introduced how reaction rates of catalytic reactions can be modeled on the basis of the adsorption energetics. The real test of the quality of the theory is given in the attempt to predict what is not already known. Is the modeling today at such a level, that it can start competing with modern experimental design approaches [138] in usefulness as a tool for the design of new heterogeneous catalysts, or will we have to move the modeling precision much further towards 'chemical accuracy' before we can even contemplate design? One argument pointing in favor for design by 'Catalysis Informatics' is certainly the constantly increasing availability of computational power. A power that we can translate directly into better accuracy as well as larger databases of computed materials. Below two examples are given, which could lead to some optimism for the role of theory as a supplementary tool in future catalyst design.

10.1. The low-temperature water gas shift (WGS) reaction

One reaction for which it has been useful to have a consistent set of theoretically obtained adsorption energies, as those described in the previous section, is for the water gas shift (WGS) reaction [139]. The WGS reaction:

$$H_2O(g) + CO(g) \rightarrow H_2(g) + CO_2(g) \qquad \{WGS\}$$

is employed in several industrial processes. One of the more important is in combination with steam reforming for the production of hydrogen from hydrocarbons. The proposed reaction mechanism for the water gas shift reaction is rather complex:

$$
\begin{array}{ll}
H_2O(g) + {}^* \rightarrow H_2O^* & \{WGS1\} \\
H_2O^* + {}^* \rightarrow H^* + OH^* & \{WGS2\}^* \\
2OH^* \rightarrow H_2O^* + O^* & \{WGS3\} \\
OH^* + {}^* \rightarrow O^* + H^* & \{WGS4\}^* \\
2H^* \rightarrow H_2(g) + 2^* & \{WGS5\} \\
CO(g) + {}^* \rightarrow CO_* & \{WGS6\} \\
CO^* + O^* \rightarrow CO_2^* + {}^* & \{WGS7\}^* \\
CO_2^* \rightarrow CO_2(g) + {}^* & \{WGS8\} \\
H^* + CO_2^* \rightarrow HCOO^* + {}^* & \{WGS9\}
\end{array}
$$

In addition, one would therefore perhaps assume that such a reaction would need a large set of parameters to describe variations in reactivity through the transition metals. On the basis of BEP-relations and adsorption energy correlations it was however possible to reduce the number of descriptors to only two. The chosen descriptors were the adsorption energy of oxygen and of carbon monoxide on the transition metal surfaces. The resulting model is in good qualitative agreement with experiments, showing that copper is the most active of the late and noble transition metals (see Figure 4.42). More importantly, this modeling of the trends in reactivity behind the WGS reaction predicts a way to improve the industrial catalyst by making it slightly more reactive with respect to both oxygen and carbon monoxide adsorption [139].

10.2. Methanation

The methanation reaction is primarily used to remove any traces of CO and CO_2 in the hydrogen feed gas for ammonia synthesis. The reaction has been known for more than a century [141]. This reaction has found renewed interest in connection with the transformation of coal to natural gas. The hydrogen for ammonia synthesis is here normally produced by steam reforming with subsequent water gas

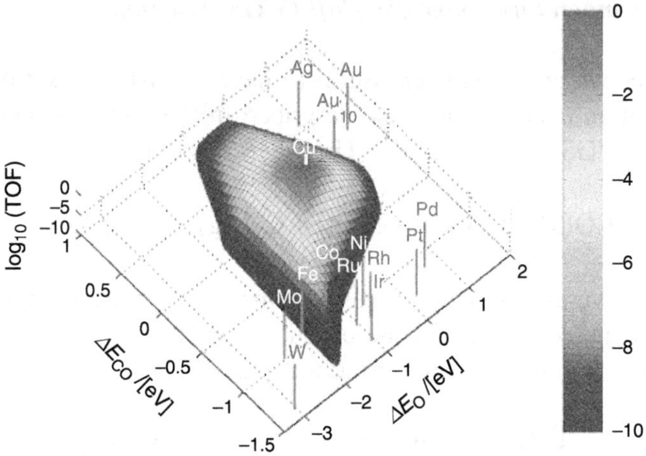

Figure 4.42. The turnover frequencies for the low-temperature WGS reaction as a function of adsorption energies of oxygen and carbon monoxide. The positions of the step sites on noble and late transition metals are shown. As observed experimentally only copper appears to be a suitable pure metal catalyst for the process. Adapted from [139].

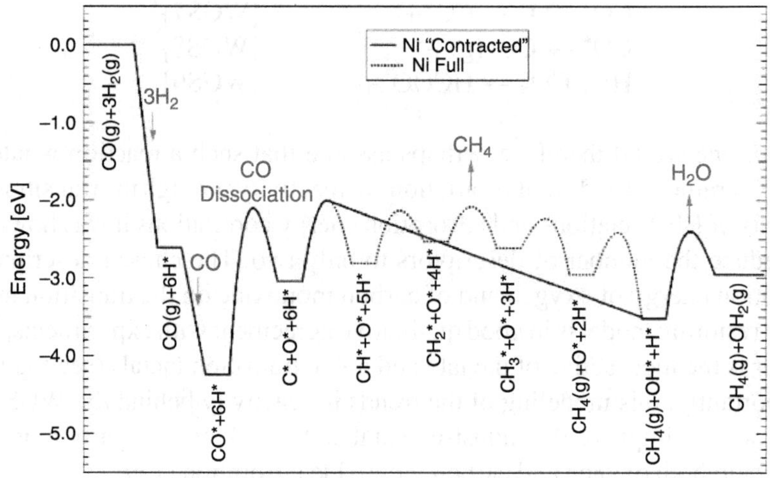

Figure 4.43. Energy diagram for CO methanation over Ni. Adapted from [140].

shift as described above. Unreacted CO is then removed by the carbon monoxide methanation process:

$$CO(g) + 3H_2(g) \rightarrow CH_4(g) + H_2O(g) \qquad \{\text{Methanation}\}$$

This process proceeds by the dissociative chemisorption of CO and H_2, and the subsequent recombination of adsorbed species to form CH_4 and H_2O, which finally

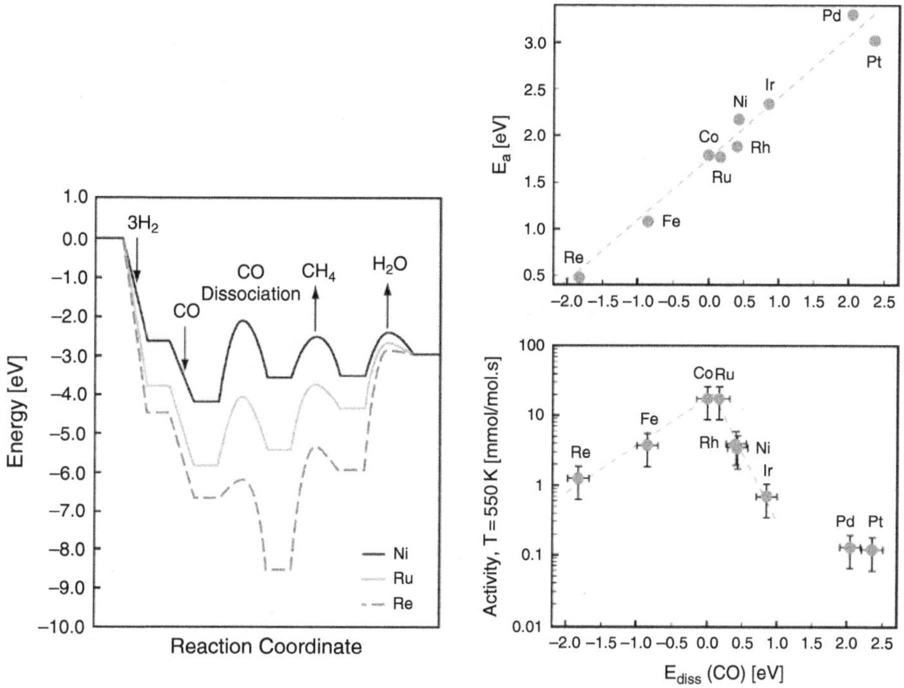

Figure 4.44. Contracted energy diagrams for CO methanation over Ni, Ru, and Re (Left). BEP-relation for CO dissociation over transition metal surfaces (right-top) and the corresponding volcano-relation for the turnover frequency (right-bottom). Adapted from [55,140].

desorb [142]. The energy diagram for the reaction can be calculated in detail (see Figure 4.43), and a simplified (contracted) energy diagram for the process over Ni, Ru, and Re is shown in Figure 4.44. It is observed that over Re the barrier for dissociating CO is small compared to the CO dissociation barrier over Ni and Ru. Over Re, however, the barriers for desorbing methane and water are high. The opposite is the case for Ni. Here the barrier for dissociating CO is high, but the barriers for desorbing water and methane are lower than the corresponding barriers over Ru and Re. Ru constitutes the best compromise, as on this metal none of the barriers are very high. So this is well in line with the Sabatier Principle. From Figure 4.24 it is known that CO dissociation obeys a BEP-relation when the dissociation barrier is correlated against the dissociative chemisorption energy. In Figure 4.44(right-top) such a BEP-relation is shown, taking the strongly adsorbed CO precursor as reference instead of CO in the gas phase. When the experimentally measured activity is plotted against this dissociation energy, a very well-behaved volcano appears (see Figure 4.44(right-bottom)).

The quality of this volcano turns out to be so high that using it together with the interpolation principle (see Section 5.3) it is possible to predict alloys which show a higher methanation activity at a lower price of the constituents than the

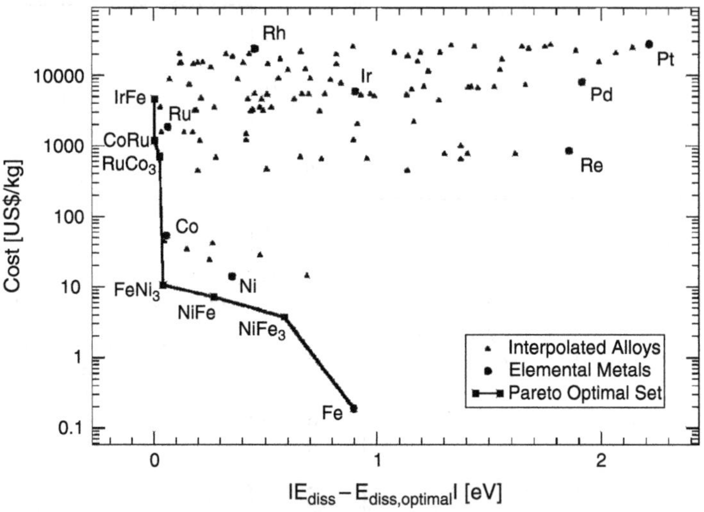

Figure 4.45. Pareto plot of interpolated catalysts predicted to be good compromises with respect to cost and activity for methanation. The positions of the interpolated catalysts are determined by the cost of their constituent elements vs. their distance from the optimal dissociative chemisorption energy for CO with respect to the experimentally observed optimum (see Figure 4.44 right-bottom). Adapted from Ref. [55].

industrially used (Ni) catalyst [55]. Figure 4.45 shows the 'Pareto-optimal' set of interpolated alloys with respect to low cost and short distance to the optimal dissociative chemisorption energy. It is observed that some Fe−Ni alloys are predicted to yield better activity than Ni and at a reduced cost, since Fe is a much cheaper material than Ni. Subsequent experiments have corroborated these theoretical predictions [55]. Curiously, a simultaneous state-of-the-art high-throughput experimentation study was not successful in predicting the utility of Fe–Ni alloys as improved low-cost catalysts for the methanation reaction [143].

11. Conclusions and outlook

In the present chapter, we have attempted to illustrate how surface bonding and catalytic activity are closely related. One of the main conclusions is that adsorption energies of the main intermediates in a surface catalyzed reaction is often a very good *descriptor* of the catalytic activity. The underlying reason is that we find correlations, Brønsted-Evans-Polanyi relations, between activation barriers and reaction energies for a number of surface reactions. When combined with simple kinetic models such correlations lead to volcano-shaped relationships between catalytic activity and adsorption energies.

We have shown that the variation in adsorption energies from one transition metal to the next and the variation from one surface geometrical structure to the next can be understood on the basis of a model describing the coupling between the adsorbate states and the transition metal *d* states. One important finding is that variations in the reactivity of a given metal when the surroundings are changed are governed to a large extent by the local value of the average energy of the *d* states. In this way, we have provided notions to understand some of the electronic and geometrical factors governing catalytic rates.

For the simplest catalytic reactions on transition metal surfaces it is now possible to understand the basic descriptors well enough that the insight can be used to help identifying new catalysts. This we take as evidence that our fundamental understanding of transition metal catalysis has reached a useful level. The prospects for catalyst development are enormous. If insight and computational methods can be used to narrow down the number of possible catalysts for a given reaction this will make the design of new catalysts substantially simpler than today. A large number of challenges are, however, still to be solved. Up to now most work has been concentrating on transition metal surfaces and on very simple reactions.

We need to develop methods to understand trends for complex reactions with many reaction steps. This should preferentially be done by developing models to understand trends, since it will be extremely difficult to perform experiments or DFT calculations for all systems of interest. Many catalysts are not metallic, and we need to develop the concepts that have allowed us to understand and develop models for trends in reactions on transition metal surfaces to other classes of surfaces: oxides, carbides, nitrides, and sulfides. It would also be extremely interesting to develop the concepts that would allow us to understand the relationships between heterogeneous catalysis and homogeneous catalysis or enzyme catalysis. Finally, the theoretical methods need further development. The level of accuracy is now so that we can describe some trends in reactivity for transition metals, but a higher accuracy is needed to describe the finer details including possibly catalyst selectivity. The reliable description of some oxides and other insulators may also not be possible unless the theoretical methods to treat exchange and correlation effects are further improved.

References

[1] G.A. Somorjai. Introduction to Surface Chemistry and Catalysis, Wiley, New York (1994).

[2] J.M. Thomas and W.-J. Thomas. Principle and Practice of Heterogeneous Catalysis, Wiley-VCH, Weinheim (1997).

[3] I. Chorkendorff and J.W. Niemantsverdriet. Concepts of Modern Catalysis and Kinetics, Wiley-VCH, Weinheim (2003).

[4] B. Hammer and J.K. Nørskov, Adv. Catal. **45** (2000) 71.

[5] J. Greeley, J.K. Nørskov, and M. Mavrikakis, Ann. Rev. Phys. Chem. **53** (2002) 319.
[6] R.A. van Santen and M. Neurock. Molecular Heterogeneous Catalysis, Wiley-VCH, Weinheim (2006).
[7] A. Nilsson and L.G.M. Pettersson, Surf. Sci. Rep. **55** (2004) 49.
[8] B. Hammer and J.K. Nørskov. Theory of adsorption and surface reactions in NATO ASI Series E **331**, Eds. R. Lambert and G. Pacchioni, Kluwer Academic Publishers, Dordrecht (1997).
[9] K.W. Jacobsen, J.K. Nørskov, and M.J. Puska, Phys. Rev. B **35** (1987) 7423.
[10] B.I. Lundqvist, O. Gunnarsson, H. Hjelmberg, and J.K. Nørskov, Surf. Sci. **89** (1979) 196.
[11] P.W. Anderson, Phys. Rev. **124** (1961) 41.
[12] D.M. Newns, Phys. Rev. **178** (1969) 1123.
[13] T.B. Grimley, J. Vac. Sci. Technol. **8** (1971) 31 and **9** (1971) 561.
[14] J.W. Gadzuk, Surf. Sci. **43** (1974) 44.
[15] B.I. Lundqvist, H. Hjelmberg, and O. Gunnarsson: Adsorbate-Induced Electronic States in Photoemission and the Electronic Properties of Surfaces, Eds. B. Feuerbacher, B. Fitton, and R.F. Willis (Wiley, New York, 1978).
[16] Sara E. Mason, Ilya Grinberg, and Andrew M. Rappe, preprint.
[17] A. Nilsson, L.G.M. Pettersson, B. Hammer, T. Bligaard, C.H. Christensen, J.K. Nørskov, Catal. Lett. **100** (2005) 111.
[18] M. Gajdos, A. Eichler, and J. Hafner. J. Phys. (Cond. Mat.) **16** (2004) 1141.
[19] B. Hammer, O.H. Nielsen, and J.K. Nørskov, Catal. Lett. **46** (1997) 31.
[20] J.T. Yates, J. Vac. Sci. Technol. A **13** (1995) 1359.
[21] M.A. van Hove, R.J. Koestner, P.C. Stair, J.P. Biberian, L.L. Kesmodel, I. Bartos, and G.A. Somorjai, Surf. Sci. **103** (1981) 189.
[22] B. Hammer, Top. Catal. **37** (2006) 3.
[23] A. Roudgar and A. Gross, J. Electronanal. Chem. **548** (2003) 121.
[24] G.A. Somorjai and B.E. Bent, Prog. Colloid Polym. Sci. **70** (1985) 38.
[25] C.R. Henry, C. Chapon, C. Goyhenex, and R. Monot, Surf. Sci. **272** (1992) 283.
[26] R. van Hardeveld and A. van Montfoort, Surf. Sci. **4** (1966) 396.
[27] C.E. Tripa, T.S. Zubkov, J.T. Yates, M. Mavrikakis, and J.K. Nørskov, J. Chem. Phys. **111** (1999) 8651.
[28] G. Mills, M.S. Gordon, and H. Metiu, J. Chem. Phys. **118** (2003) 4198.
[29] M. Mavrikakis, B. Hammer, and J.K. Nørskov, Phys. Lett. **81** (1998) 2819.
[30] J. Greeley, M. Mavrikakis, Nature Materials **3** (2004) 810.
[31] N.M. Markovic, P.N. Ross, Surf. Sci. Rep. **45** (2002) 121; T. Toda, H. Igarashi, H. Uchida, and M. Watanabe, J. Electrochem. Soc. **146** (1999) 3750.
[32] J.L. Zhang, M.B. Vukmirovic, K. Sasaki, A.U. Nilekar, M. Mavrikakis, and R.R. Adzic. J. Am. Chem. Soc. **127** (2005) 12480.
[33] J.R. Kitchin, J.K. Nørskov, M.A. Barteau, and J.C. Chen J. Chem.Phys. **120** (2004) 10240.
[34] O.M. Løvvik and R.A. Olsen. J. Chem. Phys. **118** (2003) 3268.
[35] A. Roudgar and A. Gross, Phys. Rev. B **67** (2003) 33409.
[36] J.S. Filhol, D. Simon, and P. Sautet, J. Am. Chem. Soc. **126** (2004) 3228.
[37] J. Meier, J. Schiøtz, P. Liu, J.K. Nørskov, and U. Stimming, Chem. Phys. Lett. **390** (2004) 440.
[38] L.A. Kibler, A.M. El-Aziz, R. Hoyer, and D.M. Kolb, Angew. Chem. Int. Ed. **44** (2005) 2080.
[39] R.J. Behm, Acta Phys. Pol. **93** (1998) 259.
[40] J.C. Davies, B.E. Hayden, and D.J. Pegg, Electrochim. Acta **44** (1998) 1181.
[41] G. Hoogers and D. Thompsett, CATTECH **3** (1999) 106.
[42] H. Igarashi, T. Fujino, Y. Zhu, H. Uchida, and M. Watanabe, Phys. Chem. Chem. Phys. **3** (2001) 306.

[43] P. Strasser, Q. Fan, M. Devenney, H.W. Weinberg, P. Liu, and J.K. Nørskov, Phys. Chem B **107** (2003) 11013.
[44] A. Ruban, B. Hammer, P. Stoltze, H.L. Skriver, and J.K. Nørskov, J. Mol. Catal. A **115** (1997) 421.
[45] D.P. Woodruff and T.A. Delchar, Modern Techniques of Surface Science, Cambridge University Press, Cambridge, (1986).
[46] B.S. Mun, C. Lee, V. Stamenkovic, N.M. Markovic, and P.N. Ross, Phys. Rev. B **71**, (2005) 115420.
[47] M. Weinert and R. E. Watson, Phys. Rev. B **51** (1995) 17168.
[48] D. Hennig, M.V. Ganduglia-Pirovano, and M. Scheffler, Phys. Rev. B **53** (1996) 10344.
[49] J.A. Rodriguez and D.W. Goodman, Science **257**, 897 (1992).
[50] B. Hammer, Y. Morikawa, and J.K. Nørskov Phys. Rev. Lett. **76** (1996) 2141.
[51] T. Toda, H. Igarashi, H. Uchida, M. Watanabe, J. Electrochem. Soc. **146** (1999) 3750.
[52] J.W.A. Sachtler and G.A. Somorjai, J. Catal. **81** (1983) 77.
[53] C.J.H. Jacobsen, S. Dahl, B.S. Clausen, S. Bahn, A. Logadottir, and J.K. Nørskov, J. Am. Chem. Soc. **123** (2001) 8404.
[54] P. Liu and J.K. Nørskov, Phys. Chem. Chem. Phys. **3** (2001) 3814.
[55] M.P. Andersson, T. Bligaard, A. Kustov, K.E. Larsen, J. Greeley, T. Johannessen, C.H. Christensen, and J.K. Nørskov, J. Catal. **239** (2006) 501.
[56] T. Bligaard, J.K. Nørskov, S. Dahl, J. Matthiesen, C.H. Christensen, and J.S. Sehested, J. Catal. **224** (2004) 206.
[57] F. Abild-Pedersen, J.P. Greeley, and J.K. Nørskov, Catal. Lett. **105** (2005) 9.
[58] K. Honkala, A. Hellman, I.N. Remediakis, A. Logadottir, A. Carlsson, S. Dahl, C.H. Christensen, and J.K. Nørskov, Science **307** (2005) 555.
[59] F. Besenbacher, I. Chorkendorff, B.S. Clausen, B. Hammer, A. Molenbroek, J.K. Nørskov, and I. Stensgaard, Science **279** (1998) 1913.
[60] F. Abild-Pedersen, O. Lytken, J. Engbæk, G. Nielsen, I. Chorkendorff, and J.K. Nørskov, Surf. Sci. **590** (2005) 127.
[61] B. Hammer, Phys. Rev. B **63** (2001) 205423.
[62] S. Dahl, A. Logadottir, R.C. Egeberg, J.H. Larsen, I. Chorkendorff, E. Törnqvist, and J.K. Nørskov, Phys. Rev. Lett. **83**, (1999) 1814.
[63] B. Hammer, Phys. Rev. Lett. **83** (1999) 3681.
[64] H.S. Taylor, Proc. R. Soc. London Ser. A **108** (1925) 105.
[65] A.T. Gwathmey and R.E. Cunningham, Adv. Catal. **10** (1958) 57.
[66] J.T. Yates Jr., J. Vac. Sci. Technol. A **13** (1995) 1359–1367.
[67] T. Zambelli, J. Wintterlinn, J. Trost, and G. Ertl, Science **273** (1996) 1688–1690.
[68] H. Dietrich, P. Geng, K. Jacobi, and G. Ertl, J. Chem. Phys. **104** (1996) 375.
[69] S. Lehwald and H. Ibach, J. Phys. Chem. B **107** (2003) 3808–3812.
[70] R.T. Vang, K. Honkala, S. Dahl, E.K. Vestergaard, J. Schnadt, E. Lægsgaard, B.S. Clausen, J.K. Nørskov, F. Besenbacher, Nature Materials **4** (2005) 160.
[71] N. Brønsted, Chem. Rev. **5** (1928) 231–338.
[72] M.G. Evans and M. Polanyi, Trans. Faraday Soc. **34** (1938) 11.
[73] M. Boudart in 'Handbook of Heterogeneous Catalysis' (Eds. G. Ertl, H. Knözinger, and J. Weitkamp), p. 1., Wiley-VCH, Weinheim (1997).
[74] A. Logadottir, T.H. Rod, J.K. Nørskov, B. Hammer, S. Dahl, and C.J.H. Jacobsen, J. Catal. **197** (2001) 229.
[75] V. Pallassana and M. Neurock, J. Catal. **191** (2000) 301.
[76] F. Bozso, G. Ertl, M. Grunze, and M. Weiss, J. Catal. **49** (1977) 18.
[77] D. Heskett, A. Baddorf, and E.W. Plummer, Surf. Sci. **195** (1988) 94.

[78] R. Alcalá, M. Mavrikakis, and J.A. Dumesic, J. Catal. **218** (2003) 178–190.
[79] R.D. Cortright, R.R. Davda, and J.A. Dumesic, Nature **418** (2002) 964.
[80] J.K. Nørskov, T. Bligaard, A. Logadottir, S. Bahn, L.B. Hansen, M. Bollinger, H. Bengaard, B. Hammer, Z. Sljivancanin, M. Mavrikakis, Y. Xu, S. Dahl, and C.J.H. Jacobsen, J. Catal. **209** (2002) 275.
[81] Z.-P. Liu and P. Hu. J. Am. Chem. Soc. **125** (2003) 1958–1967.
[82] I. M. Ciobîcã and R.A. van Santen, J. Phys. Chem. **107** (2003) 3808–3812.
[83] M. Mavrikakis, M. Bäumer, H.J. Freund, and J.K. Nørskov, Catal. Lett. **81**, 153 (2002).
[84] T. Zubkov, G.A. Morgan Jr., J.T. Yates Jr., O. Kühlert, M. Lisowski, R. Schillinger, D. Fick, and H.J. Jänsch, Surf. Sci. **526** (2003) 57–71.
[85] P. Gambardella, Z. Sljivancanin, B. Hammer, M. Blanc, K. Kuhnke, and K. Kern, Phys. Rev. Lett. **87** (2001) 056103.
[86] A. Michaelides, Z.-P. Liu, C.J. Zhang, A. Alavi, D.A. King, P. Hu, J. Am. Chem. Soc. **125** (2003) 3704.
[87] J. Greeley and M. Mavrikakis, Nature Materials **3** (2004) 810.
[88] H. Eyring, J. Chem. Phys. **3** (1935) 107.
[89] E. Wigner, Trans. Faraday. Soc. **34** (1938) 29.
[90] H. Eyring, Trans. Faraday. Soc. **34** (1938) 41.
[91] A. Voter, Phys. Rev. B **34** (1986) 6819.
[92] M. Neurock and E. W. Hansen, Comp. and Chem. Eng. **22** (1998) 1045.
[93] S. Ovesson, B.I. Lundqvist, W.F. Schneider, and A. Bogicevic, Phys. Rev. B **71** (2005) 115406.
[94] K. Reuter, D. Frenkel, and M. Scheffler, Phys. Rev. Lett. **93** (2004) 116105.
[95] M. Boudart and G. Djéga-Mariadassou, Kinetics of Heterogeneous Catalytic Reactions. Princeton University Press, Princeton, New Jersey (1984).
[96] J.A. Dumesic, J. Catal. **185** (1999) 496.
[97] R.A. van Santen and J.W. Niemantsverdriet, Chemical Kinetics and Catalysis. Plenum Press (1995).
[98] A. Wierzbicki and H.J. Kreuzer, Surf. Sci. **257** (1991) 417.
[99] J.A. Dumesic, D.F. Rudd, L.M. Aparicio, J.E. Rekoske, and A.A. Treviño. The Microkinetics of Heterogeneous Catalysis. Am. Chem. Soc. (1993).
[100] J. Horiuti, Bull. Chem. Soc. Jpn. **13** (1938) 210.
[101] G.H. Jóhannesson and H. Jónsson, J. Chem. Phys. **115** (2001) 9644.
[102] J.C. Keck, J. Chem. Phys. **32** (1960) 1035.
[103] J.C. Keck, Adv. Chem. Phys. **13** (1967) 85.
[104] E.A. Carter, G. Ciccotti, J.T. Hynes, and R. Kapral, Chem. Phys. Lett. **156** (1989) 472.
[105] M. Sprik and G. Ciccotti, J. Chem. Phys. **109** (1998) 7737.
[106] J.C. Keck, Disc. Faraday Soc. **33** (1962) 173.
[107] T. Yamamoto, J Chem Phys. **33** (1960) 281.
[108] C.H. Bennett, Exact defect calculations in model substances, in Algorithms for Chemical Computation, edited by A.S. Nowick and J.J. Burton, ACS Symposium Series No. 46 63 (1977).
[109] D. Chandler, J. Chem. Phys. **68** (1978) 2959.
[110] R.A. Olsen, G.J. Kroes, G. Henkelman, A. Arnaldsson, and H. Jonsson, J. Chem. Phys. **121** (2004) 9776.
[111] H. Jónsson, G. Mills, and K.W. Jacobsen, in Classic and Quantum Dynamics in Condensed Phase Simulations, edited by B.J. Berne, G. Ciccotti, and D.F. Coker. World Scientific, Singapore (1998).
[112] G. Henkelman and H. Jónsson, J. Chem. Phys. **113** (2000) 9978.
[113] G. Henkelman, B.P. Uberuaga, and H. Jónsson, J. Chem. Phys. **113** (2000) 9901.

[114] P. Maragakis, S.A. Andreev, Y. Brumer, D.R. Reichman, E. Kaxiras, J. Chem. Phys. **117** (2002) 4651.
[115] J.W. Chu, B.L. Trout, and B.R. Brooks, J. Chem. Phys. **119** (2003) 12708.
[116] P.W. Atkins and R.S. Friedman, Molecular Quantum Mechanics 3rd ed., Section 10.13. Oxford University Press, Oxford (1997).
[117] T. Bligaard, K. Honkala, A. Logadottir, J.K. Nørskov, S. Dahl, and C.J.H. Jacobsen, J. Phys. Chem. B **107** (2003) 9325.
[118] R. van Harrevelt, K. Honkala, J.K. Nørskov, and U. Manthe, J. Chem. Phys. **123** (2005) 234702.
[119] M.C. Evans and M. Polanyi, Trans. Faraday Soc. **33** (1937) 448.
[120] S. Kandoi, J. Greeley, M.A. Sanchez-Castillo, S.T. Evans, A.A. Gokhale, J.A. Dumesic, and M. Mavrikakis, Top. Catal. **37** (2006) 17.
[121] S. Linic and M.A. Barteau, J. Catal. **214** (2003) 200.
[122] T. de Donder, L'Affinité. Gauthier-Villers (1927).
[123] J.A. Dumesic, J. Catal. **204** (2001) 525.
[124] P.W. Atkins, Physical Chemistry, 6th ed. Oxford University Press, Oxford (1998).
[125] K.W. Kolansinski, Surface Science. Wiley, (2002).
[126] A. Eichler, J. Hafner, A. Gross, and M. Scheffler. Phys. Rev. B **59** (1999) 13297.
[127] A. Eichler, F. Mittendorfer, and J. Hafner. Phys. Rev. B **62** (2000) 4744.
[128] C.J.H. Jacobsen, S. Dahl, A. Boisen, B.S. Clausen, H. Topsøe, A. Logadottir, and J.K. Nørskov, J. Catal. **205** (2002) 382.
[129] T. Bligaard, 'A Le Chatelier-like Principle in Heterogeneous Catalysis', Submitted.
[130] P. Stoltze and J.K. Nørskov, Phys. Rev. Lett. **55** (1985) 2502.
[131] S. Dahl, J. Sehested, C.J.H. Jacobsen, E. Törnqvist, and I. Chorkendorff. J. Catal. **192** (2000) 391.
[132] P. Sabatier, Hydrogénations et déshydrogénations par catalyse. Ber. Deutsch. Chem. Gesellshaft **44** (1911) 1984.
[133] A.A. Balandin, Adv. Catal. **19** (1969) 1.
[134] H. Topsøe, M. Boudart, J.K.Nørskov, Ed., Frontiers in Catalysis: Ammonia Synthesis and Beyond. *Top. Catal.* **1** (1994).
[135] M.A. Vannice, *J. Catal.* **50** (1977) 228.
[136] M. Muhler, in Handbook of Heterogeneous Catalysis, Eds. G. Ertl, Knözinger, H., Weitkamp, J. (Wiley-VCH, Weinheim, 1997), vol. 5, pp. 2274.
[137] B.E. Nieuwenhuys, *Surf. Rev. Lett.* **50** (1996) 1869.
[138] P.J. Cong, R.D. Doolen, Q. Fan, D.M. Giaquinta, S.H. Guan, E.W. McFarland, D.M. Poojary, K. Self, H.W. Turner, and W.H. Weinberg, Ang. Chem. Int. Ed. **38** (1999) 484.
[139] N. Schumacher, A. Boisen, S. Dahl, A.A. Gokhale, S. Kandoi, L.C. Grabow, J.A. Dumesic, M. Mavrikakis, and I. Chorkendorff, J. Catal. **229** (2005) 265.
[140] M.P. Andersson, T. Bligaard, C.H. Christensen, and J.K. Nørskov, To be submitted.
[141] P. Sabatier and J.B. Senderens, C.R. Acad. Sci. Paris **134** (1902) 514.
[142] D.W. Goodman, R.D. Kelly, T.E. Madey, and J.T. Yates, J. Catal. **63** (1980) 226.
[143] K. Yaccato, R. Carhart, A. Hagemeyer, A. Lesik, P. Strasser, A.F. Volpe, H. Turner, H. Weinberg, R.K. Grasselli, and C. Brooks, *App. Catal. A: Gen.* **296** (2005) 30.

Chapter 5

Semiconductor Surface Chemistry

Stacey F. Bent

Department of Chemical Engineering, Stanford University, Stanford, CA 94305, USA

1. Inroduction

Semiconductor materials, including the group IV elements Si and Ge, play a dominant role in much of modern technology. Most significantly, semiconductor materials form the foundation for integrated circuits (ICs), which make up the over $200 billion semiconductor industry. These integrated circuits, in turn, fuel the nearly $1 trillion worldwide electronics market. Semiconductor-based devices are now pervasive in modern life, residing in our cars, our cell phones, our kitchen appliances, our computers, and countless other places. Other non-IC uses abound as well, including in solar cells, microelectromechanical systems, chemical sensors, and infrared detectors. Given the huge market for semiconductors, understanding the properties of the surfaces of semiconductor materials is an important area of study. Moreover, the significance of such study has grown in the past few years as semiconductor devices have become increasingly smaller, leading to very high surface-to-volume ratios in most modern devices. Since the advent of the integrated circuit by Jack Kilby in 1959, the number of transistors on an integrated circuit has been increasing at a rate of approximately two times every 18–24 months. These trends follow a prediction known as Moore's Law. This scaling has been achieved by shrinking the dimensions of the transistors. Whereas in 1992 the minimum feature size was 0.5 μm, in a transistor made in 2006 this value is well under 100 nm. At such a small scale, surface or interfacial properties can dominate the behavior of the device.

The study of semiconductor surface structure and surface chemistry was actually begun several decades ago together with the advent of surface analytical techniques. Many of the earliest surface science studies examined semiconductors. However, for many years the volume of literature on semiconductor surface structure and chemistry was far smaller than that of metals, which were subjected to extensive study motivated largely by catalysis [1–3]. With the rapid proliferation of semiconductor-based technologies, studies of semiconductor surface chemistry have also grown, and current literature contains more than a thousand publications each year on the structure, adsorption, and reaction of single-crystalline semiconductor surfaces.

Most of these studies have focused on silicon. For more than 40 years, the semiconductor of choice for transistors and integrated circuits has been silicon. Silicon's dominance in the microelectronics industry is due in part to several important properties. Silicon can be produced in single-crystalline form at very high purity; it forms an excellent oxide at its surface; and, its electronic properties can be tuned dramatically by substituting only a small fraction of silicon atoms in the lattice with another element in the process of doping. The silicon oxide (SiO_2) has traditionally been used for the dielectric in metal oxide semiconductor field effect transistors (MOSFET). The superb interface that forms naturally between SiO_2 and Si has been critical to the operation of the transistor, as the interface between the dielectric (SiO_2) and the underlying semiconductor (Si) greatly influences the electrical properties of the device.

Recent trends in the semiconductor industry, however, are causing a shift away from SiO_2 as the dielectric material. At the nanometer-scale dielectric thicknesses used today, leakage currents caused by electron tunneling through the dielectric can be significant. To prevent this leakage, new dielectric materials are being developed with much larger dielectric constants than SiO_2; these materials can elicit the same electrical response in the channel of the transistor with thicker dielectric films, greatly reducing the leakage current. Consequently, SiO_2 dielectrics are being phased out of the transistor structure.

The elimination of SiO_2 from some of the core structures of an integrated circuit, particularly its elimination as a fundamental part of the transistor, is leading to a re-evaluation of the Si platform in which transistors are built, and the use of other semiconductor materials for device fabrication is being explored. In fact, it has long been known that other semiconductors (e.g., germanium and GaAs) have better inherent electrical properties than does Si [4]. Such considerations motivate studies of the surface chemistry of these other semiconductors. Consequently, much recent work has focused on Ge. Far fewer studies exist of the surface chemistry of compound semiconductors such as GaAs, InP, and GaN, providing a rich area for future study.

Several excellent reviews are available concerning both surface structure of semiconductors and surface chemistry of semiconductors, including Refs. [5–23]. Here, a comprehensive review is not attempted and the reader is referred instead to those references. The focus of this chapter is primarily on the surface chemistry of silicon and germanium, as these are the two most heavily studied systems. We strive to provide insight into the chemical reactivity of these two surfaces, and hence select some specific systems that are most highly illustrative of important chemical principles. One of the key themes of this chapter is the covalent nature of the semiconductor surface and its influence on the reactivity. The covalent nature of the semiconductor surface permits its reactivity to be described within a molecular framework. This covalency leads to a reactivity that is often quite distinct from that of metals. On metals, bonding may not be strongly site-specific. This characteristic is markedly different from that of a surface such as silicon, where bonding is both localized and directional.

To a significant degree, the reactivity of semiconductor surfaces can be understood by drawing analogies to known molecular systems. In this chapter, we concentrate particularly on adsorption of organic molecules to the surface, part of the field known as organic functionalization of semiconductors. The mechanisms and products of adsorption of organic molecules onto semiconductor surfaces can in many cases be described by analogy to fundamental reactions in organic chemistry. In addition, new applications may be enabled by organic functionalization. Tethering organic materials to inorganic semiconductors allows the incorporation of new optical, electronic, mechanical, and chemical properties. This field is growing based on expected applications in a number of new areas, including chemical and biological sensors, molecular electronics, microelectromechanical (MEMS) devices, area-selective deposition, and passivation of microelectronic devices.

We begin by reviewing the surface structure of the most commonly studied crystal faces of these materials. We will then briefly discuss the oxidation of both silicon and germanium, then review the passivation of silicon and germanium surfaces. In addition, we explore the reactivity of Si, Ge, and to a lesser extent GaAs and InP, toward various organic molecules that can be used to functionalize the surface and impart new properties to the semiconductor.

2. Structure of semiconductor surfaces

It has been known since the 1950s that the surfaces of semiconductors have a tendency to undergo reconstructions compared to the bulk atomic positions [24]. Duke [5,25] has set out some general principles describing reconstructions of tetrahedrally bonded semiconductor surfaces. The first principle states that atoms rearrange (reconstruct) to satisfy the valence of the broken or 'dangling' bonds which result from truncation of the bulk geometry (or to convert these dangling bonds into non-bonding orbitals). This is very clearly seen, e.g., in the formation of dimers on the Si(100)–2×1 reconstructed surface, as will be discussed below. The second principle states that in many cases, surfaces can lower their energy with atomic relaxations that generate insulating or semiconducting surface energy bands (i.e., an energy gap) instead of metallic states [5,25]. This energy consideration leads, e.g., to the tilting of the silicon dimers on Si(100)–2 × 1 out of the plane of the surface. The third principle distinguishes between activated and non-activated reconstructions. In other words, the lowest energy structure which is kinetically accessible under the preparation conditions will be observed [5,25]. The realization of this principle is seen in the multiple reconstructions that can be formed under different preparation conditions for the GaAs(001) surface, for example. Finally, compound semiconductors follow a fourth principle. In such surfaces, charge transfer occurs between anions and cations so that the surface is 'autocompensated' and leads to a charge neutral surface [5,6,25].

In the following section, we provide a brief review of the structures of the major semiconductor surfaces for which the adsorption and reaction chemistry will be covered in this chapter. This includes the (100) and (111) crystal faces of silicon and germanium. Chapter 1 of this book also provides a brief overview of the structure of the silicon surface. The surface structures of compound semiconductors, including GaAs and InP, can be quite complex and are not covered here. A number of reviews describe the structure of these surfaces much more extensively [5,6,25–29], and the reader is referred to those references for more detail.

2.1. Silicon surface structure

We will begin with a discussion of the structure of the silicon bulk and surface, since its structure will lay the foundation for germanium, as well. Silicon is a covalent solid that crystallizes into a diamond cubic lattice structure. A unit cell of this structure is illustrated in Figure 5.1, with the [001], [010], and [100] directions designated. Like carbon, its Group IV homolog, silicon atoms hybridize into a tetrahedral bonding configuration. In the bulk solid, the diamond cubic lattice allows silicon atoms to achieve this tetrahedral configuration. At the surface of the material, however, the bulk is truncated, so the stable bulk tetrahedral bonding is disturbed.

The reactivity of the silicon surface is controlled in part by the unsatisfied bonding orbitals, or so-called dangling bonds, that remain upon truncating the bulk. At the surface, silicon atoms are bonded to fewer than four bulk silicon atoms, leaving dangling bonds. As described in Duke's first principle of semiconductor surface reconstruction, the atoms will readjust, i.e., undergo a reconstruction, to minimize the total free energy of the system and eliminate the dangling bonds. The energy minimization is a trade-off between energy gained by forming new local bonds (to eliminate the dangling bonds) with energy lost because of bond strain that results from its new configuration.

The crystallographic face of silicon that is most important industrially is the Si(100) surface, followed by the Si(111) surface. Both the (100) and (111) surfaces undergo extensive reconstructions, i.e., their surface atomic geometry differs significantly from that of the bulk. Moreover, the two surfaces have markedly different

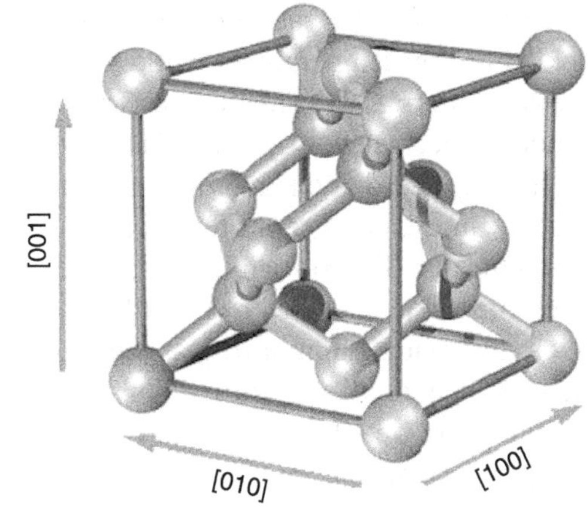

Figure 5.1. Illustration of a unit cell of the diamond cubic lattice. The arrows designate the [001], [010], and [100] directions. Both silicon and germanium crystallize into the diamond cubic lattice structure, where each atom is bonded to four neighboring atoms in a tetrahedral geometry. Figure reproduced from Ref. [30] with permission.

surface structures [5,25,31]. Under suitable vacuum preparation conditions, which usually include argon ion sputtering and/or high temperature annealing, the Si(100) surface reconstructs to form silicon dimers, thereby reducing the number of dangling bonds per surface Si atom from two to one [5]. These dimers are arranged in rows separated by trenches, with the row direction changing by 90° at an atomic step. Figure 5.2 illustrates the key features of the surface, particularly the dimer rows and the trenches which separate each dimer row. This rearrangement, known as a (100)–2 × 1 reconstruction, minimizes the surface energy because the stability gained by creating a chemical bond outweighs that lost from strain. Due to solid state electronic effects (Duke's second principle), the surface dimers tilt (or buckle) out of the surface plane, as shown in Figure 5.2 [5].

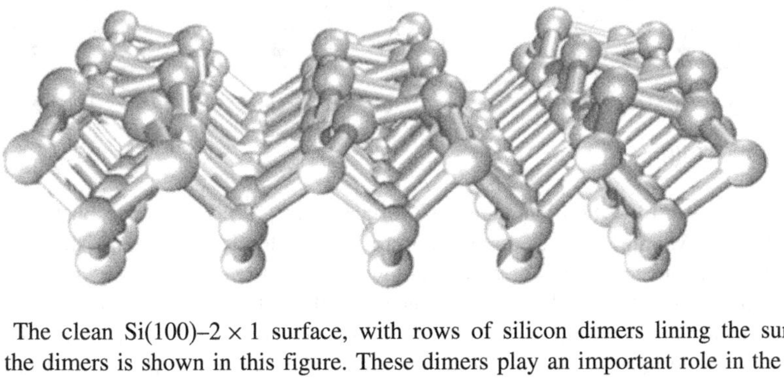

Figure 5.2. The clean Si(100)–2 × 1 surface, with rows of silicon dimers lining the surface. The buckling of the dimers is shown in this figure. These dimers play an important role in the chemistry of organic molecules at this surface.

Scanning tunneling microscopy (STM) measurements show that the silicon dimers are asymmetrically buckled at low temperatures while measurements at room temperature display symmetric dimers [32]. This suggests that silicon surface dimers dynamically tilt when enough thermal energy is available. Molecular dynamics simulations agree with this finding and demonstrate that silicon dimers should dynamically tilt on the picosecond timescale at room temperature [33]. As a result, while the room temperature structure of Si(100) can be described as (2×1), at temperatures below 200 K, the reconstructed surface has a $c(4 \times 2)$ structure [5], in which the buckling of the dimers alternates along a row as well as across the trench. The differences between the (2×1) and $c(4 \times 2)$ structures are illustrated in Figure 5.3.

It is useful to consider in more detail the bonding of the dimers on the Si(100)–2×1 surface, because they play such an important role in the reactivity of this surface toward organic molecules. The Si(100)–2×1 dimers are attached by something akin to a double bond, which consists of both a σ and a π bond [5,34]. Although the strength of this weak π bond interaction is still under debate, reported values are approximately 5–10 kcal/mol [35–41] far less than the 64 kcal/mol of traditional alkenes [42]. Nonetheless, the partial π bond of the surface dimer mimics that of an alkene from classic organic chemistry. Furthermore, because the dimers tilt, charge transfer occurs between the dimer atoms. The recessed atom, known as the 'down atom,' donates significant electronic charge to the protruding or 'up atom'. The effect is shown in Figure 5.4 for a Si(100)–2×1 dimer cluster model, in which the highest occupied molecular orbital is drawn. Consequently, the down atom is

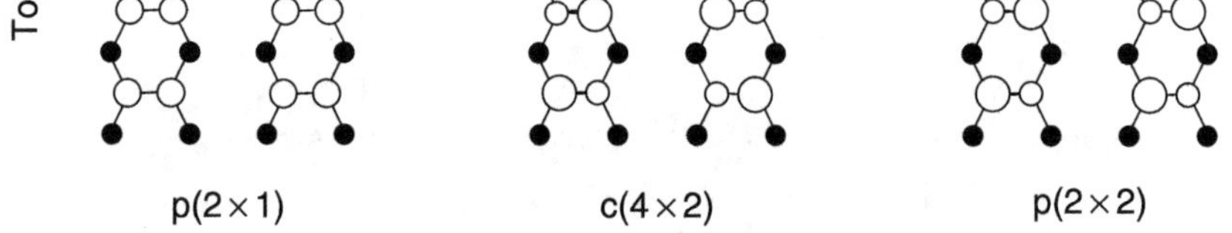

Figure 5.3. Models of the Si(100) and Ge(100) surface: (Left) (2×1) dimer reconstruction involving symmetric dimers; (Middle) $c(4 \times 2)$ dimer reconstruction with buckled dimers; These two structures are observed for silicon at room temperature and lower temperature, respectively. For germanium, the structure at (Right), the $p(2 \times 2)$ dimer reconstruction with buckled dimers, is also observed at lower temperatures. In the top view model, the open circles represent the top layer atoms, with the larger and smaller circles designating the up and down atoms of the dimer, respectively. The filled circles represent the next layer of atoms.

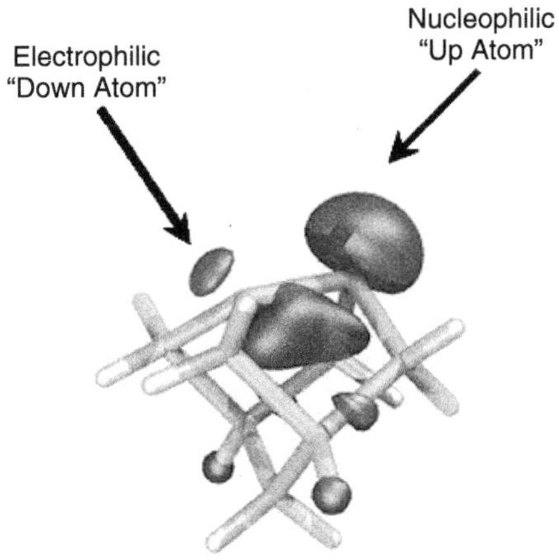

Figure 5.4. The highest occupied molecular orbital of a Si_9H_{12} dimer cluster. The top two silicon atoms comprise the surface dimer, and the remaining seven Si atoms contain three subsurface layers which are hydrogen terminated to preserve the sp^3 hybridization of the bulk diamond cubic lattice. The up atom is nucleophilic and the down atom is electrophilic.

electrophilic and the up atom is nucleophilic [43]. Both the π-bonded and the electrophilic/nucleophilic nature of the silicon dimer influence the types of reaction products formed at the surface.

Si(111), on the other hand, reconstructs into a complex (7 × 7) structure that contains 49 surface atoms in the new unit cell [5]. This reconstruction is known as the dimer-adatom-stacking fault (DAS) model, and it was first proposed by Takayanagi et al. [44] based on electron diffraction data [44]. Soon thereafter, the structure was confirmed by STM, as one of the earliest surfaces imaged by the newly-invented technique [45]. The Si(111)–7 × 7 surface is displayed schematically in Figure 5.5. The reconstructed surface contains 19 dangling bonds per unit cell, reduced from the 49 dangling bonds that would be present at the unreconstructed Si(111) surface. These dangling bonds reside on three types of silicon surface sites: adatoms, rest atoms, and the corner hole. The adatoms account for 12 dangling bonds per unit cell, the rest atoms for 6, and the corner-hole sites for 1. In addition, there is a stacking fault over half of the unit cell [6] which can be observed as a small height asymmetry in the STM image.

The different surface atoms on the Si(111)–7 × 7 surface have chemically distinct properties. Because of the electron redistribution, each rest atom and corner hole is negatively charged, while each of the adatoms has a partial positive charge [19,47]. Consequently, the rest atoms and corner hole can be considered nucleophilic, and the adatoms electrophilic. An adjacent rest atom-adatom pair hence forms a dipolar or diradical entity; although the closest rest atom-adatom spacing on Si(111)–7 × 7 (4.5 Å) is much larger than the dimer distance on Si(100)–2 × 1 (2.4 Å) [48], this pair may be expected to show some similarities in reactivity to the dimer on the

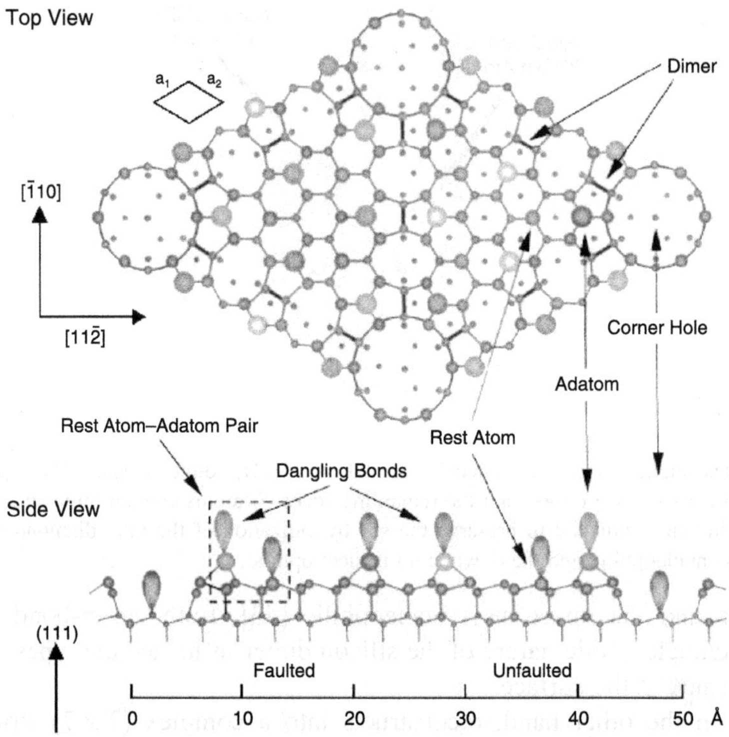

Figure 5.5. Dimer-adatom-stacking fault (DAS) model of the Si(111)–7 × 7 surface unit cell by Takayanagi et al. [44]. Figure reused with permission from Ref. [46], Y. Cao, Journal of Chemical Physics, 115, 3287 (2001). Copyright 2001, American Institute of Physics.

Si(100)–2 × 1 surface. Indeed, similarities in adsorption products between the two surfaces have been reported, as will be seen later in this chapter.

Even after reconstruction, both the Si(100)–2 × 1 and Si(111)–7 × 7 surfaces are highly reactive. Both surfaces are typically cleaned and prepared in an ultrahigh vacuum environment ($p = 10^{-10}$ torr). The resulting surfaces are not stable in air, instead quickly oxidizing to form an SiO_2 layer. Hence, experiments on the clean Si(100)–2 × 1 and Si(111)–7 × 7 surfaces must be performed in vacuum, and reagents for adsorption and reaction are introduced in vapor form into the vacuum chamber. On the other hand, silicon surfaces can be rendered relatively stable in air (i.e., relatively resistant to oxidation) by coating the surface with hydrogen or other terminating groups. These processes are discussed in later sections.

2.2. *Germanium surface structure*

In its crystalline state, germanium, similar to silicon, is a covalent solid that crystallizes into a diamond cubic lattice structure. Like for Si, both the (100) and (111) surfaces of germanium undergo reconstructions when prepared under vacuum conditions in which their surface atomic geometry differs significantly from that of the bulk to minimize their surface energy [5].

Analogous to silicon, the Ge(100) surface reconstructs to form germanium dimers, reducing the number of dangling bonds per surface Ge atom from two to one. The nature of the bonding of the dimers on the Ge(100) surface strongly influences the reactivity of this surface toward gas-phase reactants. Locally, the surface structure of reconstructed Ge(100) is similar to that of Si(100)–2 × 1; e.g., both exhibit dimer rows with similar geometrical spacings (the Ge lattice constant is only 4 % larger than that of Si). The homologous Si(100)–2 × 1 dimers were described above as being somewhat akin to a double bond, i.e., containing both a σ bond and a π bond [5], but with the π bond being significantly weaker than a C=C π bond in alkenes. For Ge, the π bond is even weaker, estimated between 2 and 8 kcal/mol [16]. Unlike a true double bond, the dimer is buckled away from a symmetric configuration even at room temperature. Like for Si, the resulting buckled dimer exhibits charge separation: the up Ge atom is more nucleophilic, and the down Ge atom is more electrophilic [49]. Structurally, the longer-range interactions between dimers lead to various higher-order surface reconstructions [50]. These are illustrated in Figure 5.3. Symmetric dimers lead to a 2 × 1 structure (referred to as p(2 × 1)). However, depending on the relative ordering of the buckled dimers, Ge(100)–c(4 × 2) or Ge(100)–p(2 × 2) dimer reconstructions are formed. All three of these reconstructions can be observed locally at the Ge(100) surfaces by scanning-tunneling microscopy (STM) [50].

Although Si(100) and Ge(100) undergo similar dimer reconstructions, the Ge(111) surface reconstructions differ from those of Si(111). As described above, Si(111) reconstructs into a (7 × 7) structure that contains 49 surface atoms in the new unit cell. Ge(111) is found in various reconstructed forms depending on surface preparation, but the most common reconstruction under vacuum is Ge(111)–c(2 × 8) [51–53]. This structure involves charge transfer from adatoms to restatoms [5]. On the other hand, most of the passivation and functionalization studies reviewed here lead to the Ge(111)–1 × 1 surface structure. This structure, in which the surface Ge atoms retain their bulk positions, can be achieved by hydrogen, chlorine, or alkyl termination of the surface (discussed below). The structure is analogous to that for H-terminated Si(111).

3. Surface oxidation

3.1. Silicon

Silicon possesses an excellent oxide. Silicon dioxide (SiO_2) is a remarkably stable passivating layer, acts as a good electrical insulator, and forms an excellent interface with Si. It also serves as a superb barrier against diffusion of dopants and other impurities, can resist a number of chemical and thermal processes, and adheres well to silicon [54,55]. In addition, silicon dioxide can be formed on silicon by a variety of thermal or chemical oxidation processes. Due to its importance in the microelectronics industry, there is a vast body of knowledge available on the formation of the silicon oxide, and the properties of the oxide and its interface with silicon.

Silicon will spontaneously oxidize in air to form a thin (~10 nm) native oxide at its surface. It can also be oxidized thermally or chemically, a property, which is extensively exploited in the IC fabrication process. Thermal oxidation usually takes place in a furnace at temperatures between 900–1200 °C in either H_2O or O_2 ambient, forming SiO_2 [54,55]. Thick silicon dioxide films can be formed by thermal oxidation, although the oxidation process slows with film thickness. The thermal oxidation mechanism is known as the Deal-Grove model [56]; in this model, oxidant (O_2 or H_2O) must diffuse through the SiO_2 film to reach the SiO_2/Si interface, followed by reaction at the interface. For thicker silicon dioxide films, the thermal oxidation enters the so-called parabolic regime in which the process is diffusion limited. UHV mechanistic studies of oxidation of silicon have shown that room temperature oxygen adsorption followed by a high-temperature anneal results in growth of SiO_2 from suboxides [57,58]. The suboxide (SiO) itself is thermally unstable at elevated temperatures.

The other prevalent method of oxidizing silicon uses wet chemical treatment, often in a sequence of steps called the RCA process [59]. In this process, silicon is both cleaned and oxidized by a process involving treatment with an alkaline mixture of ammonium hydroxide and hydrogen peroxide (called standard clean 1, or SC1), followed by treatment with an acidic HCl and hydrogen peroxide mixture (SC2). The process leads to what is called a chemical oxide. This, like the thermal oxide, is SiO_2, but the wet oxidation process typically leads to a more hydroxylated SiO_2 surface.

Due to the historical importance of the initial stages of silicon oxidation to microelectronics fabrication, there has been a great deal of interest in the reaction of the water oxidant on the Si(100)–2 × 1 surface. A number of studies have shown that water adsorbs in a dissociated state consisting of OH(a) and H(a) species adsorbed on the Si surface dimer at room temperature [60–69]. More recent studies have closely investigated the mechanism of water oxidation. A series of density functional theory calculations (DFT) calculations by Konecny and Doren indicated that water first molecularly adsorbs through one of its lone pairs in a weakly bound precursor state, then transfers a proton to form OH(a) and H(a) species on the surface dimer [43]. The pathway to proton transfer is found to be unactivated with respect to the entrance channel, which suggests that OH(a) and H(a) are the dominant surface species at room temperature, in agreement with the previous experimental work [60–69].

Chabal and coworkers investigated this system in detail with several infrared (IR) spectroscopy techniques including multiple internal reflection (MIR), external reflection, and transmission that allowed high resolution access to the low energy spectral region and enabled identification of Si−H bending modes, Si−O stretching modes, and other modes that were invisible in previous studies [70–73]. Interestingly, they found that upon heating to temperatures near 330 °C, the oxygen of the surface-bound hydroxyl group can insert into the silicon dimer bond to form a suboxide (Si−O−Si) structure. DFT calculations indicated that the suboxide structure is approximately 50 kcal/mol more stable than the dissociated state, but the pathway consists of multiple activation barriers in excess of 40 kcal/mol [70,74], explaining why oxygen insertion is only detected at elevated temperatures. Oxygen insertion is a key step in ultimately oxidizing the silicon to form SiO_2.

3.2. Germanium

In contrast to silicon, the instability of the germanium oxides has prevented the widespread use of germanium in integrated circuits. Native germanium oxide (GeO_2) is water soluble and forms a poor interface with Ge, which results in facile removal of the oxide layer and a high density of electronic defects [75]. These properties render GeO_2 unsuitable for use in transistor applications.

Another difference between silicon and germanium oxides is that whereas thermal oxide growth is readily achieved for silicon, this is not the case for germanium. Oxidation of Ge by O_2 results in growth of the Ge+I, +II, and +III oxidation states, but little or no growth of the +IV state (GeO_2). Upon a high-temperature anneal, all states, including GeO_2, disappear at the expense of the +II (GeO) state; this behavior differs from that of Si, where the +III and +IV states appear upon annealing [76,77]. GeO desorbs completely from the surface at 450 °C, and thus annealing above this temperature results in the oxide-free surface [76].

Wet chemical oxidation of germanium is an important step in the sample cleaning process. Similar to the process of silicon cleaning, alternating cycles of wet oxide formation and etching are used to remove contaminants from the germanium surface. The most common oxidants used are H_2O_2, HNO_3, and H_2O, in a variety of concentrations. Although germanium surface cleaning is not covered in detail here, several recent studies have addressed the issue [78–87]. In a study by Prabhakaran and Ogino [76], aqueous H_2O_2 treatment resulted in the formation of the suboxide GeO_x with no formation of GeO_2 for both the Ge(100) and Ge(111) surfaces. This is expected because GeO_2 is readily soluble in water [88] and would dissolve upon formation in the aqueous-oxidant solution or subsequent deionized water rinse. Because the various oxides are unfavorable for device characteristics, the prevention of oxide growth is an important part of germanium processing.

4. Passivation of semiconductor surfaces

4.1. Silicon passivation

The passivation of silicon, motivated by the centrality of this semiconductor to the microelectronics industry, has been well studied. In addition to excellent passivation allowed by the silicon oxide, silicon can also be passivated with silicon nitride (Si_3N_4), other dielectrics, metal layers, and hydrogen. Here we focus only on hydride termination, since in addition to acting as a passivating layer for the underlying silicon, the hydride groups provide a versatile starting point for subsequent attachment chemistry.

4.1.1. Hydride termination of silicon

A number of recipes have been developed to terminate the silicon surface with Si–H (silicon hydride) groups. An excellent overview of this process is given by Weldon et al. [21]. The most common methods involve exposure to dilute, aqueous hydrofluoric acid solutions, and these wet etch procedures are extensively used at scales ranging from bench-top research laboratories to large-scale microfabrication facilities. Hydrofluoric acid (or, commonly, an NH_4F/HF buffered solution) etches SiO_2; following removal of the oxide, etching leads to hydrogen termination of the silicon surface. The Si–F bond is much stronger than the Si–H bond (140 vs. 80 kcal/mol), and initially it was thought that the surface was F-terminated [21]. However, many studies have since shown that the surface is actually H-terminated. This result is explained mechanistically by recognizing that the Si–F bond is highly polarized and susceptible to further HF attack, which etches the silicon as SiF_4 and leaves hydrogen at the surface [89].

Not only does terminating a silicon surface with hydrogen atoms render it passivated toward many reactions, but it also changes the surface structure. Because the reconstructed surface contains some weak Si–Si bonds, replacement of these with strong Si–H bonds can relax the surface reconstructions. One well-studied surface is that of hydride-terminated Si(111). Due to the strong bonding and passivating nature of hydrogen at the interface, Si(111) that is hydride-terminated does not reconstruct, revealing instead a nearly bulk-like periodicity of the surface silicon atoms [90]. An STM image of hydride-terminated Si(111) surface is shown in Figure 5.6. It is clear that a very high degree of order is achievable at the H–Si(111) surface. Furthermore, with the right etching conditions, the surface exhibits in the infrared a single strong, narrow Si–H stretching mode originating from silicon monohydride (SiH) species at the terraces, again indicating a high degree of perfection [91]. HF etching of the Si(100) surface, while also generating hydride-termination, does not produce

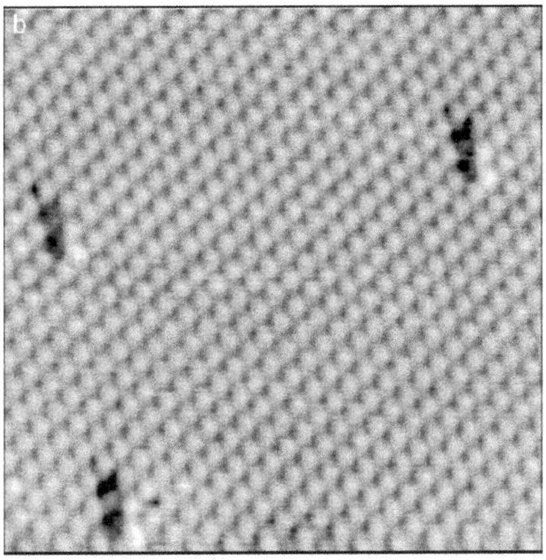

Figure 5.6. STM image of a hydrogen-terminated Si(111) surface. The image shows a 10 nm × 10 nm region of the surface, from Ref. [20]. Reproduced by permission of the Royal Society of Chemistry.

such a highly ordered surface and contains mono-, di-, and trihydride species (SiH, SiH_2, and SiH_3) [91].

Silicon can also be terminated by hydrogen in vacuum. The most commom method in vacuum is by adsorption of atomic hydrogen on a cleaned, ordered Si surface. Molecular H_2 has a very low sticking probability at room temperature (less than 10^{-6}) on the silicon surface, so pre-cracking of the hydrogen is required [23]. Hydrogen-terminated silicon can also be generated by reaction of silane or disilane with the clean silicon surface. Three hydrides can be formed – the monohydride (SiH), dihydride (SiH_2), and trihydride (SiH_3) structures – depending on the surface temperature and the coverage.

4.2. Germanium passivation

Because of the poor oxide, passivation of germanium surfaces is required for practical use of this semiconductor in devices. Although an ideally passivated surface would resist oxidation and degradation perfectly, such complete resistance is not possible in practice. For this discussion, we consider passivated surfaces as those that strengthen resistance to oxidation in both ambient air and aqueous solution. Three different surface terminating layers are reviewed: sulfide-, chloride-, and hydride-terminated germanium. To date, sulfide termination creates the most ideal passivating layer, whereas both chloride and hydride termination add limited stability sufficient to alter the surface reactivity in a way that allows for further reaction.

4.2.1. Sulfide passivation of germanium

Whereas germanium oxide does not sufficiently passivate the germanium surface, sulfide termination leads to a well-passivated interface. As a congener of oxygen, sulfur is bivalent and thus is expected to make two bonds with the surface. Because each of the surface atoms on the Ge(100)–2 × 1 surface has two dangling bonds, sulfur may occupy the germanium lattice sites at the surface, with each Ge surface atom bonded to two sulfur atoms. In this way, each of the sulfur atoms can eliminate two dangling bonds while maintaining the bulk germanium structure at the surface, leading to a 1 × 1 surface structure.

Such a 1 × 1 structure was demonstrated by Weser et al. [92,93] in the formation of a S/Ge(100)–1 × 1 surface by chemisorption of elemental sulfur on the Ge(100)–2 × 1 surface under ultrahigh vacuum (UHV) conditions. The sulfur atoms sit in a bridge-like configuration [92]. In this bonding arrangement the surface valences are saturated at 1-monolayer (ML) coverage. It is worth noting that similar results for the Si(100)–2 × 1 surface could not be obtained by the same authors [94]. Theoretical calculations by Krüger and Pollmann [95] confirmed the bridge-bonded adsorption as the lowest energy conformation for the S/Ge(100)–2 × 1 surface, with the overlayer of sulfur atoms positioned in the Ge lattice sites. Sulfur passivation has also been achieved by H_2S adsorption, instead of elemental sulfur [96,97], on the Ge(111) surface. However, repeated steps of adsorption of H_2S followed by annealing to desorb hydrogen were found necessary to achieve a saturation coverage of 0.5 ML, with one S atom adsorbed per two Ge surface atoms [97,98]. The same method applied to Ge(100)–2 × 1 also led to S atoms in bridge sites, but at a lower saturation coverage than that for elemental sulfur adsorption [99,100]. Despite the lower saturation coverage than that observed for elemental sulfur adsorption, this sulfur overlayer surface showed enhanced stability relative to the clean Ge(100)–2 × 1 surface, resisting contamination for days in UHV conditions.

In addition to deposition in vacuum, sulfur can also be deposited onto hydrogen-terminated germanium surfaces from aqueous solution [101–103]. The first aqueous sulfidation was carried out by Anderson et al. [101] using $(NH_4)_2S$. Although others have repeated this approach, there is disagreement in the results, with different sulfur coverages (ranging from 1 to 3 ML) and final structures (ranging from well-defined ordered overlayers to amorphous thin layers) observed [101,102]. A possible cause for the amorphous overlayer is a high step density from crystal miscut [102]; this finding is supported by studies on the same aqueous sulfidation treatment of germanium nanowires, which resulted in a 5-nm thick GeS_x surface coating [103]. Even though sulfur overlayers obtained by the aqueous treatment differed, all cases resulted in a self-limiting overlayer that was resistant to oxidation in air for days [101–104], which makes aqueous sulfide termination a promising technique for passivating germanium surfaces.

4.2.2. Chloride passivation of germanium

Germanium surface passivation by chloride termination inhibits oxide formation and maintains a well-ordered surface. The chloride-terminated surface can also be used as a reactive precursor for wet organic functionalization. For example, Cullen et al. [105] first demonstrated the reaction of a chloride-terminated Ge(111) surface with ethyl Grignard as a means of ethylation for use in surface stabilization. The chlorination was performed by a mixture of Cl_2 and HCl gas with N_2 above atmospheric pressures [105]. Although this resulted in approximately a one-to-one ratio of adsorbed chlorine atoms with Ge surface atoms, the high pressures resulted in severe etching of the substrate [105].

Citrin et al. [106] later investigated the adsorption of Cl_2 on the Ge(111)–2×8 surface by synchrotron X-ray photoemission and found that Cl binds in a one-to-one ratio with Ge surface atoms in the atop site. This configuration allows for saturation of the Cl valence and the single dangling bond per Ge(111) surface atom, maintaining the bulk structure at the surface resulting in a Cl/Ge(111)–1×1 LEED pattern [106]. Theoretical calculations by Bachelet and Schlüter [107] confirm that adsorption in the atop site is the lowest energy configuration.

Cl adsorption on the Ge(100) surface changes the LEED pattern from c(4×2) to (2×1), indicating a monochloride termination at saturation [108]. In forming the monochloride phase, the asymmetric dimers change to symmetric dimers. This monochloride phase results from the addition of a chlorine atom to one dangling bond on each surface atom, breaking the weak π interaction between dimer atoms. Upon monochloride adsorption, the 2×1 reconstruction persists, as the σ-like bond character of the dimer remains. At higher adsorption pressures, a dichloride ($GeCl_2$) phase is formed on the surface with a 1×1 symmetry [109]. Upon heating, desorption of $GeCl_2$ resulted in atomic layer etching [109].

Lu [110] prepared the first solution phase Cl-terminated surface by immersing a Ge(111) sample in dilute HCl. This resulted in a well-ordered, atop adsorption, similar to the surface formed by Cl adsorption in vacuum [110]. The chloride-terminated surface product is thermodynamically favored over the hydride-terminated surface product, as the Ge–Cl bond and Ge–H bond strengths are 103 and 77 kcal/mol, respectively [88]. The chloride-terminated surface demonstrates passivation in its stability against oxidation on the scale of hours in ambient air [104,111].

4.2.3. Hydride termination of germanium

Hydride surface termination has the capability for ideal surface passivation, with each hydrogen atom bonding to a single surface-dangling bond. On silicon, hydride termination has been well researched and shown to provide many advantages, including aqueous stability and limited air stability [13]. The hydride-terminated surface is also of interest as it can be used as a precursor for wet chemical reactions.

Hydride-terminated germanium can be obtained in vacuum by adsorption of atomic hydrogen on the germanium surface. On the Ge(100)–2 × 1 reconstructed surface, saturation is seen with the monohydride phase [112–115], which is analogous to the monochloride phase discussed above. Upon higher exposures of hydrogen, a dihydride phase is formed, again similar to the chlorine adsorption, but reports on this are mixed. Whereas Appelbaum et al. [112,113] and Chabal [112,113] were both unable to observe the dihydride phase using LEED, UV photoelectron spectroscopy (UPS), and IR spectroscopy, Papagno et al. [114,115] and Maeng et al. [114,115] observed small amounts of the dihydride – using high-resolution electron energy loss spectroscopy (HREELS) and STM – at exposures an order of magnitude greater than those for monohydride saturation. Results of hydrogen desorption studies from Ge(100) using TPD and laser-induced desorption vary as well [116–118], from the observation of no dihydride [116] to a clearly distinguishable dihydride phase [117]. This behavior differs from that of the Si(100)–(2 × 1) surface, where a clear dihydride phase is formed upon increasing hydrogen exposure from monohydride saturation [119]. On the Ge(111) surface, the hydrogen caps the Ge surface atoms, adsorbing in a one-to-one ratio in the atop position, with a product similar to the chloride-terminated Ge(111) surface. Upon adsorption, the surface changes from a c(2 × 8) reconstruction to a 1 × 1 surface [120].

Wet chemical methods for hydrogen termination of Ge evolved from the surface cleaning processes. In the cleaning process, the oxide is repeatedly formed and etched away. During the etch step, use of HF (the most common etchant) produces a hydrogen-terminated surface for both the Ge(100) and Ge(111) surfaces [79]. The nature of the hydride termination has been directly studied on Ge(100). Choi and Buriak [121] argue that, unlike the HCl-treated chloride-terminated surface, the hydride-terminated surface is a kinetic product because the Ge−F bond (116 kcal/mol) is thermodynamically favored over the Ge−H bond (77 kcal/mol). The same surface reactivity is observed for the Si(100) surface, which also forms the kinetic hydride-terminated product when exposed to solutions of HF [13]. IR data from hydride-terminated Ge(100) show a broad Ge−H peak arising from mono-, di-, and trihydride termination; this peak has an optimal HF exposure time, after which the peak intensity degrades [121]. After HF etch, the surface is not atomically flat, and the surface roughness of 1 nm likely stabilizes the formation of the dihydride and trihydride products that are observed. This surface containing multiple hydrides with nanometer roughness is also obtained for Si(100) under the same treatment conditions [13]. A high-quality monohydride surface can be obtained on Si(111) upon treatment with 40% NH_4F aqueous solution [122–124], but the analogous treatment of Ge(111) has not been reported.

Despite the range of hydrides present, hydride termination by HF etching stabilizes the surface against oxidation and maintains surface ordering for further wet chemistry. Hydride-terminated germanium shows no oxidation after exposure to ambient atmosphere for at least 1 h [79,104,121] and little oxidation after 1 week [79]. Although the surface has multiple hydride phases, subsequent hydrogermylation and alkanethiol reactions with the surface yield densely packed organic monolayers (discussed below) [121,125,126]. When compared with chloride termination, the wet hydride termination seems less desirable, because it lacks a well-ordered, atomically flat surface. However, even with some amount of disorder, the hydride-terminated surface is sufficiently passivated to withstand oxidation in ambient on a practical timescale and can be used as a reactive precursor for subsequent surface reactions.

5. Reactions at passivated semiconductor surfaces

5.1. Organic functionalization of semiconductor surface

There is a significant and actively growing body of work examining the attachment of organic groups to both clean [10–19] and passivated [20] semiconductor surfaces, particularly for the case of silicon. The field is commonly referred to as organic functionalization of semiconductors. A broad range of chemistries has been used to accomplish the functionalization, and the results have provided fundamental insight into the surface reactivity of these semiconductors. Moreover, the functionalization step can change the properties of the surface, and the organic-terminated surfaces are increasingly being formed for a variety of applications, including electrical devices, biosensing, electrochemistry, and nanopatterning.

Organic functionalization reactions have been carried out both in vacuum and in solution. The vacuum studies typically use the clean, reconstructed (100) or (111) crystal faces of the semiconductor, and the reactants are dosed in the gas phase. Because the semiconductor surfaces are readily oxidized and otherwise contaminated in air or solution, the usual approach for solution-based functionalization is to first passivate the semiconductor (e.g., with hydrogen or halogens) through solution processing, then carry out a reaction which replaces the passivating layer with the organic molecules.

The literature on organic functionalization is too large to allow for a full review of the field within this chapter. Instead, we focus on the historical development of the functionalization reactions and highlight some recent developments that illustrate the utility of these approaches.

5.2. Reaction with passivated silicon (Si–H and Si–Cl)

5.2.1. Hydrosilylation

The first report of a densely-packed organic monolayer bound directly to silicon through Si–C bonds was based upon the hydrosilylation reaction on the H:Si(111) surface [127,128]. In this groundbreaking work, Chidsey and coworkers demonstrated that long-chain 1-alkenes could be attached to hydrogen-terminated silicon surfaces at 100°C in the presence of a radical initiator [127,128]. The reaction consists of an unsaturated C−C bond added across a Si−H bond. Linford et al. proposed that the surface attachment mechanism was analogous to the hydrosilylation reaction in solution phase chemistry. In the proposed surface mechanism (illustrated in Figure 5.7), the radical initiator starts the reaction by abstracting H from a surface Si−H bond, producing a silicon dangling bond (i.e., a silicon radical). The silicon dangling bond then reacts with the alkene molecule, forming a new Si−C bond at the surface and leading to the attachment of the organic molecule, which itself is left with a carbon radical as shown in Figure 5.7. This surface bound organic radical then abstracts a hydrogen atom either from an unreacted alkene molecule or from a neighboring Si−H group on the surface. The net reaction sequence produces a stable, closed-shell organic group bound directly to the Si(111) surface. Furthermore, if the adsorbate abstracts a hydrogen atom from a neighboring Si−H group, a new silicon dangling bond is produced, which can begin the reaction anew. This process is known as a radical chain reaction.

Studies by Linford et al. [127,128] showed a remarkable stability and robustness to the organic monolayers formed by this process. A number of diagnostic techniques were used to characterize the monolayers, including vibrational spectroscopy, X-ray spectroscopy and reflectivity, ellipsometry and wetting measurements. These showed that the monolayers were tightly packed, with a density close to 90% that of crystalline hydrocarbons. More importantly, the organic layers formed by hydrosilylation were stable in a number of environments, including boiling water, organic solvents, acids, and air. The silicon substrate was found to be protected by the monolayer against oxidation after many weeks of exposure to air [127,128].

After the reports of the radical-initiated hydrosilylation process, a number of different approaches for functionalization based on this reaction were developed [128–136]. It has now been demonstrated that a large variety of monolayers can

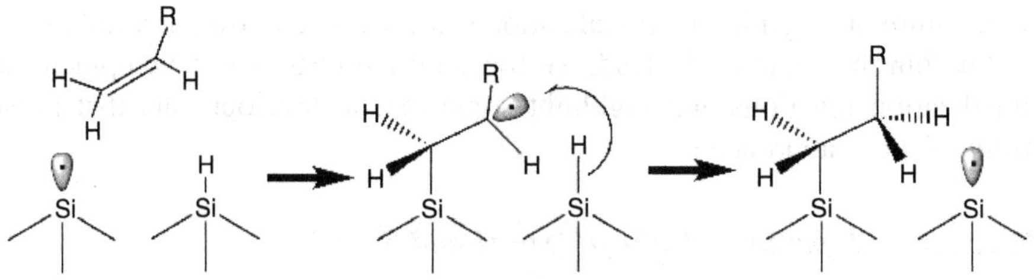

Figure 5.7. Proposed mechanism for surface hydrosilylation. The initial loss of silicon hydride generates a silicon dangling bond. Reaction between the silicon and an alkene molecule leads to an attached alkyl radical, which may abstract a hydrogen atom from a neighboring silicon.

be formed from alkenyl and alkynyl reactions with the hydride-terminated surfaces of both Si(111) and Si(100) [13], as well as from reaction of aldehydes and other unsaturated molecules [135]. It has also been shown that the reaction can proceed not just with a radical initiator, but also with other methods for generating the initial silicon radical (dangling bond). The use of heat (>150°C) works by thermally breaking the Si−H bond, although the monolayers produced by this method were of a lower quality [128]. Another method involves use of acidic molecules (called Lewis acids) to catalyze the reaction [130,131].

The proposed radical-chain reaction mechanism was confirmed in hydrosilylation experiments performed in vacuum. In one report, Cicero et al. prepared hydride-terminated Si(111) surfaces using standard wet-etch methods, then introduced the sample into an ultrahigh vacuum chamber [137]. In vacuum, the tip of a scanning tunneling microscope was used to generate isolated silicon dangling bond sites surrounded by silicon hydride groups [138,139]. This surface was then exposed to styrene, a molecule which consists of an alkene group attached to a benzene ring. The STM image revealed islands of adsorbed styrene on the silicon surface in areas surrounding the locations of the initial dangling bonds [137]. In other words, one dangling bond on the surface led to the deposition of many styrene molecules. This result supports the presence of a radical-chain reaction during which hydrogen atoms are abstracted from neighboring silicon atoms. The observation of bunched-up islands of styrene rather than one-dimensional lines is attributed to a 'random walk' propagation of the reaction. Note that this system provides a demonstration of the localization of the reaction and limited mobility of the organic molecules at silicon surfaces. If the styrene molecules were mobile, the bunching effect would not be observed.

A recent theoretical study by Takeuchi et al. [140] has examined the mechanism for the reaction of both alkenes and alkynes with the H-terminated silicon surface using periodic DFT calculations, and the results are in good agreement with the proposed radical-based mechanism [137]. In particular, the calculations show that the reaction occurs through a carbon-based radical intermediate which must be sufficiently stabilized to proceed by abstraction of a surface hydrogen (as in the case of styrene); if the intermediate is not stable enough, it will preferentially desorb (as in the case of ethylene). The calculations also show that reaction with terminal alkynes should proceed faster and lead to more stable products than with terminal alkenes [140].

Due to the relative ease of carrying out the reaction and the versatility of the process, the hydrosilylation reaction has been used in a number of interesting extensions and applications. Here several of them are highlighted. In one report, Lopinski and coworkers used the same concept of the radical-initiated hydrosilylation reaction on the Si(100)–2 $\times$ 1 surface to induce self-directed growth of molecular 'wires' on the surface [141]. On the Si(100)–2 $\times$ 1 surface, the radical chain reaction propagates primarily along the direction of the dimer row, leading to 'lines' of organic adsorbates. Molecular assemblies as long as 130 Å (corresponding to 34 styrene molecules) were observed. Lopinski et al. suggested that by using different reactant alkenes and seed conditions at the surface, this reaction may be used to engineer more complex molecular nanostructures on silicon. Recently, Wolkow and coworkers [142] have used the styrene nanostructures on Si(100)–2 × 1 to demonstrate regulation in the conductivity of single styrene molecules by a nearby fixed point charge. In their studies, a dangling bond present near a styrene line was found to influence the tunneling current (measured via the STM tip) in the styrene molecules as a function of distance from the dangling bond due to the electrostatic field.

More recently, Hossain, Kato and Kawai have found a molecular system for the hydrosilylation reaction on the hydride-terminated Si(100)–2 × 1 surface that adds molecular lines *across* the dimer rows [143,144]. They have shown that allyl mercaptan ($CH_2{=}CH{-}CH_2{-}SH$) reacts, like styrene, by a hydrosilylation reaction, but that the reaction propagates across the trench to the neighboring dimer row. Consequently, by using both styrene and allyl mercaptan in controlled experiments, they could fabricate perpendicularly connected lines of allyl mercaptan and styrene, as shown in Figure 5.8. In Figure 5.8, the brightest line along the dimer row is from styrene, while the less bright lines across the dimer rows shows the allyl mercaptan chain reaction product at the surface. The authors speculate that in the allyl mercaptan molecule, the carbon-centered radical formed upon reaction at the surface tautomerizes to a thiyl radical. Once the radical is on the sulfur atom, the distance between the anchoring site and the S radical is large enough to favor abstraction of H on the next dimer row. Other molecules, such as propylene or allyl methyl sulfide

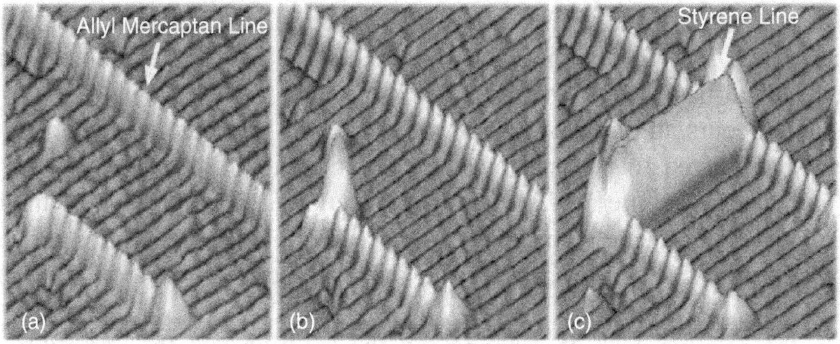

Figure 5.8. STM images of the Si(100)–(2 × 1)–H surface after hydrosilylation reaction using both styrene and allyl mercaptan, from Hossain, Kato and Kawai. The image shown in (a) contains two allyl mercaptan lines. Image (b) shows the system after a new dangling bond is generated using the STM tip. After the surface is exposed to styrene molecules, a styrene line forms which joins the two allyl mercaptan lines (c). The allyl mercaptan lines form across the Si dimer rows, while the styrene line forms along the dimer row. Figure reproduced with permission from Ref. [143]. Copyright 2005 American Chemical Society.

($CH_2{=}CH{-}CH_2{-}S{-}CH_3$), in contrast to allyl mercaptan, did not show any line growth, suggesting that the presence of the –SH group in the mercaptan is important for stabilizing the intermediate radical [143,144].

Linford and coworkers have shown that the attachment of alkenes to H-terminated silicon surfaces can also be initiated by direct mechanical scribing, in a process termed chemomechanical functionalization [145–147]. The reaction of 1-alkenes (as well as 1-alkynes) leads to attachment of the molecule to the surface through two new Si−C bonds. The proposed mechanism is the mechanical cleavage of Si−H and Si−Si bonds, leading to silicon radicals that then react with the reactive liquid. Interestingly, Linford and coworkers have also extended this work to show that chemomechanical functionalization can be carried out not only on H-terminated Si, but also on silicon covered with oxide, and have shown that the process works with a variety of halides, alcohols, and epoxides in both the liquid and gas phase [146].

Another attractive extension of the hydrosilylation reaction involves photoinitiation. Chidsey and coworkers showed that shining ultraviolet light on the H:Si(111) surface while exposing the surface to a 1-alkene leads to the formation of organic monolayers [148,149]. Others have shown that the hydrosilylation reaction also can occur with visible light activation [132,150]. The light-induced reaction opens the possibility for direct photopatterning of the organic at the surface. For example, Buriak and coworkers have used the photoinitiated hydrosilylation reaction to produce patterns of alkyl termination on porous silicon [151]. Porous silicon is generated by specialized etching of crystalline silicon, and contains a complex array of pores at the surface; like crystalline silicon, it can be hydride terminated [129].

Buriak et al. demonstrated that the hydrosilylation reaction of alkenes could be induced by visible white light on photoluminescent samples of hydride-terminated porous silicon [129,151]. Using a shadow mask, regions that were exposed to white light in the presence of alkene molecules became covered with organic groups, whereas regions that were kept dark remained terminated with hydrogen. Subsequently, the porous silicon was etched in basic solution. Areas of the substrate that were functionalized remained intact because the organic layer acted to protect the underlying silicon, but areas that were hydride-terminated were destroyed during the etch process. This simple example serves to demonstrate the potential of organic functionalization in manipulating the properties of a silicon substrate. Adding the ability to pattern these attached organics, as shown in the work by Buriak and coworkers [151], only expands the possibilities for future device applications.

In a noteworthy application, Hamers and coworkers recently extended the radical-based chemistry developed for the silicon surfaces to functionalize gallium nitride surfaces for eventual use in biosensing [152]. In those experiments, a GaN(0001) surface was first terminated by hydrogen using a hydrogen plasma, then exposed to alkenes under irradiation with 254 nm light through a shadow mask. The reaction was carried out with bifunctional alkenes containing an alkene group at one end of the molecule and a protected amine at the other. The alkenes were successfully attached to the GaN in the irradiated regions. Subsequently, the amine was deprotected, reacted with a heterofunctional crosslinker, then reacted with thiol-modified DNA, leading ultimately to a DNA-functionalized GaN substrate [152]. The process is illustrated in Figure 5.9. The work demonstrated a reliable route for functionalizing GaN with biomolecular layers. Similar work has also been demonstrated by the same group for diamond surfaces [153–155].

Since the hydrosilylation reaction with alkenes can lead to well-passivated, alkyl-terminated silicon surfaces, this reaction can also be used to protect the surface against subsequent film deposition steps. This use was applied recently by Chen et al. [156,157] in the development of an area-selective atomic layer deposition (ALD) process. In this work, the authors utilized the inherent chemical selectivity of the hydrosilylation process toward H-terminated silicon over oxide-coated silicon. Under thermal conditions, 1-alkenes do not react with the oxide of silicon. Consequently, Chen et al. reacted 1-alkenes under thermal conditions with a pre-patterned SiO_2/Si substrate, in which the silicon regions were hydrogen-terminated by dipping in HF. The process sequence is illustrated in Figure 5.10(a). No reaction occurred on the silicon dioxide, while alkyl monolayers were attached to the silicon regions. This substrate, now spatially patterned with both alkyl-terminated silicon and SiO_2 regions, was loaded into a reactor for deposition of Pt by ALD. The alkyl monolayer served as a resist against Pt ALD, leading to deposition of the Pt only on the bare SiO_2 regions of the substrate. Figure 5.10(b) shows an SEM image of the initial patterned SiO_2/Si substrate together with an elemental analysis (by Auger electron spectroscopy) of Pt showing excellent selectivity toward the deposition of Pt using the resist formed by hydrosilylation [156].

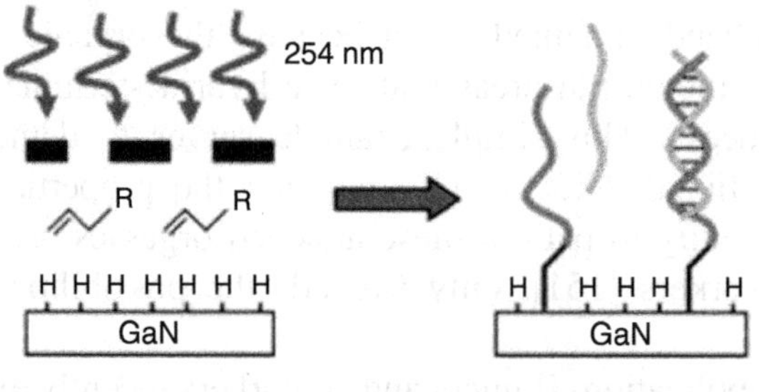

Figure 5.9. Functionalization of GaN using the UV-photoinduced reaction with alkenes. Using a shadow mask for the irradiation allows for patterned functionalization. In this study, the alkyl terminated layer was later functionalized with DNA. Figure reproduced with permission from Ref. [152]. Copyright 2006 American Chemical Society.

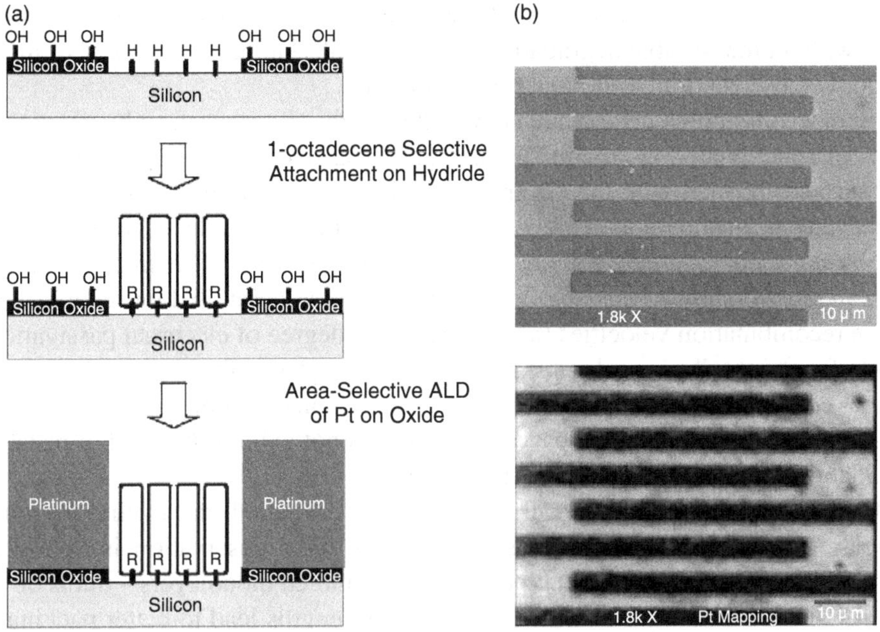

Figure 5.10. (a) Schematic illustration of the area-selective ALD scheme using the hydrosilylation reaction to selectively passivate the silicon surface, leaving only the silicon oxide regions free for Pt deposition. (b) Top: SEM image of a test structure; the brighter part in this image shows the thermal-oxide region, the darker part is the Si−H region. Bottom: AES elemental Pt mapping on the test structure after area-selective ALD leading to Pt deposited only on the oxide region [156] .

5.2.2. Grignard reactions on silicon

In addition to reaction at the hydride-terminated silicon surface, silicon can also be functionalized by reaction at a halide-terminated surface. In 1996, Bansal et al. first reported a two-step alkylation process involving halogenation of a H-terminated silicon to produce Cl−Si(111), followed by reaction with an alkyl-Grignard reagent (RMgX, where R is the alkyl group such as $-CH_3$ or $-C_{10}H_{21}$, and X is Br or Cl) [158]. Organolithium reagents were also shown to work with this scheme.

Soon thereafter, studies showed that organic attachment could be achieved by direct reaction between the Grignard reagent and the hydrogen-terminated silicon surface, without the need for the intermediate halogenation step. This was demonstrated with reaction of alkyl Grignard or aryl lithium compounds both on H-terminated porous silicon [159,160] and on the single-crystalline Si(111) surface [161]. Investigation by Boukherroub et al. [161] revealed that covalently bonded alkyl monolayers formed by direct reaction between alkylmagnesium bromide and the H−Si(111) surface led to surfaces with similar properties to that formed in the hydrosilylation reactions described in the previous section. Specifically, these samples were resistant to a variety of chemical treatments and could be stored for weeks without measurable degradation [161]. The alkyl monolayers could ultimately be removed by reaction with potassium hydroxide.

A number of studies have since shown the Grignard approach to be an excellent means of functionalizing silicon surfaces [159,162–167]. In particular, the derivatized surfaces formed with the Grignard reaction on the chloride-terminated silicon surface have been shown to achieve among the best chemical and electrical passivation observed. For example, a study of electrical properties and chemical stability by Webb and Lewis showed that alkylated Si(111) surfaces exhibited low charge carrier surface recombination velocities (a measure of the degree of electrical passivation of the interface) as well as good resistance to oxidation [165]. The Grignard-passivated surfaces could resist oxidation in air for at least 50 hours. Although the H-terminated surface also provides electrical passivation, it does not exhibit the level of resistance to oxidation found for alkyl-terminated surfaces.

Interestingly, of the different alkyl chain lengths studied, the methyl-terminated surfaces were found to be the most passivating. This observation is opposite to results for the two most common types of self-assembled monolayers – thiols on gold and alkylsilanes on SiO_2 – in which *longer* chain lengths lead to better packing and better passivation [168–170]. Webb et al. have explained this observation by a size argument: the methyl group is the only alkyl group that can terminate every silicon atop atom site on the Si(111) surface; the other alkyl chains have too large a van der Waals radius to fit within the Si–Si atomic spacing of 3.8 Å [165]. STM studies have confirmed a high degree of order on the CH_3-terminated Si(111) surfaces using the two-step chlorination/alkylation procedure [167,171]. An STM image from Ref. [167] is shown in Figure 5.11; the three-fold symmetry of each methyl group can be observed in the image. Yu et al. [171] have shown by analyzing their STM images that a commensurate 1×1 overlayer is formed by the methyl groups at the surface. Furthermore, within this overlayer, the methyl groups are rotated by 7° relative to the underlying Si−Si bonds [171], i.e., very nearly in the 'eclipsed' form. The orientation is attributed to competition between repulsions between hydrogen atoms on adjacent methyl group and methyl interactions with the underlying Si−Si bonds.

5.3. Reaction with passivated germanium (Ge−H and Ge−Cl)

The wet chemical functionalization of Ge surfaces has not been as well studied as that of Si surfaces [13,20], despite the fact that the first report on Ge functionalization was made by more than 40 years ago [105]. Three methods of wet chemical functionalization have been reported on Ge surfaces to date: (*a*) Grignard reactions on chloride-terminated surfaces, (*b*) hydrogermylation reactions on

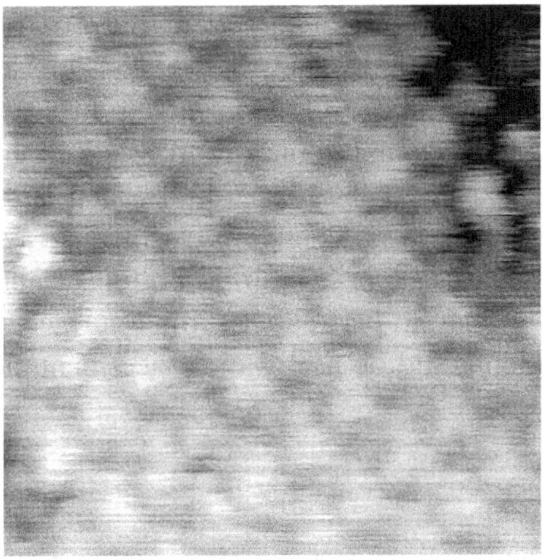

Figure 5.11. High resolution STM image of the methylated Si(111) surface (from Yamada et al.). The size is 4 × 4 nm, sample bias −1.16 V, 7.12 nA. The triangular shapes results from the three-fold symmetry of the methyl groups. Figure reused with permission from Ref. [167], Taro Yamada, Journal of Chemical Physics, 121, 10660 (2004). Copyright 2004, American Institute of Physics.

hydride-terminated surfaces, and (*c*) alkanethiol reactions on hydride-terminated surfaces [105,111,121,125,126].

5.3.1. Grignard reactions on germanium

The original work on wet chemical functionalization of germanium was reported as early as 1962 by Cullen et al. (researchers in RCA labs) [105]. They exposed ethyl Grignard to a chloride-terminated Ge(111) surface and used radiotracer studies with tritiated ethyl groups to show that ethylation of the surface occurs via formation of Ge−C bonds in a one-to-one ratio with surface atoms. Because the chlorination procedure used by the authors can lead to formation of mono-, di-, and trichloride species, there may be some uncertainty as to the Ge/C ratio. However, these surfaces did demonstrate hydrophobic behavior and stability in both atmosphere and aqueous solution.

Following the surface Grignard reactions, another 35 years passed before further research in the field of wet functionalization of Ge was pursued. These more recent studies have used dilute HCl to prepare the chloride-terminated Ge(111) surface, avoiding the production of di- and tri-chloride species and leading instead to an atomically flat, monochloride Ge surface [110]. Surface alkylation by Grignard reagents can be achieved at temperatures ranging from 60 °C to 80 °C, using different lengths of alkane chains ranging from ethyl to octadecyl, as well as phenyl and alkenyl functionalized ethyl groups [111]. Reaction time to achieve a well-ordered monolayer increased with chain length, from 6 hours (ethyl) to 7 days (octadecyl). He et al. [111] reported that reaction of chloride-terminated Ge(111) with alkyllithium reagents did not result in formation of a well-ordered monolayer, in contrast to Si(111), for which Grignard and alkyllithium reagents both yield well-ordered monolayers [158]. The Grignard-alkylated surfaces show promise in increasing surface stability, with no change in surface X-ray photoelectron spectra and IR spectra after 5 days in air or 30 min in boiling water [111]. Attachment of these organic monolayers may impart either further surface stability or specifically enhanced reactivity, which could be useful in device manufacturing.

5.3.2. Hydrogermylation

Of the wet chemical functionalization methods for silicon, hydrosilylation has been the most extensively studied. A large variety of monolayers have been formed from alkenyl and alkynyl reactions with the hydride-terminated surfaces of both Si(111) and Si(100) [13], as described in Section 5.2.1. The corresponding reaction on germanium, hydrogermylation, has potential for direct attachment of carbon to germanium surfaces. Choi and Buriak [121] first demonstrated hydrogermylation, using the hydride-terminated Ge(100) surface. Hydrogermylation has been achieved using the following three approaches: Lewis acid mediation, UV light, and thermal activation. The hydrogermylation reaction is generally faster than the Grignard attachment described above. For example, thermally induced hydrogermylation using 1-dodecyne led to well-ordered monolayers in 2 h at 220 °C [121]. Contact angle measurements of the thermally induced 1-hexadecene product on Ge were comparable to values reported for the hydrosilylation reaction on Si.

Similar to the surface alkylated via Grignard reagents, the surfaces alkylated by hydrogermylation showed enhanced surface stability. However, the stability depended on the hydrogermylation route. There was no change in the IR alkyl peak intensities for the thermally induced monolayer upon immersion in boiling water for 20 min followed by immersion in boiling chloroform for 20 min, but similar treatment of the Lewis acid-mediated monolayer results in a 20% IR decrease for boiling water and a 30% IR decrease for boiling chloroform [121]. Thus, whereas the thermally induced hydrogermylation route provides well-ordered monolayers, the Lewis acid-mediated route appears to be insufficient for uniform organic functionalization of germanium.

Interestingly, hydrogermylation chemistry has been applied to Ge nanowires, providing an organic passivating layer on the nanowire [103]. Hydride-terminated Ge nanowires can be fabricated by etching Ge nanowires in HF. Unlike the surfaces of bulk Ge [13,121], H-terminated Ge nanowires are only stable for a few minutes in dry air [103]. However, the organic-terminated nanowires exhibit much better stability.

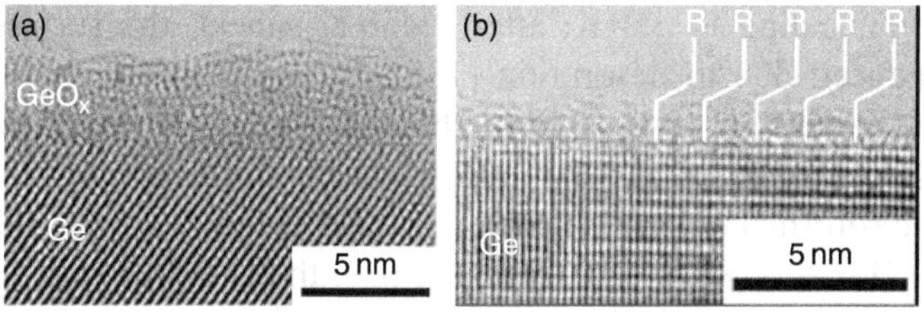

Figure 5.12. High resolution TEM images of Ge nanowires from Korgel and coworkers. Image (a) shows an untreated Ge nanowire with a native germanium suboxide layer. Image (b) shows a Ge nanowire with a covalently bonded hexyl monolayer attached using the hydrogermylation reaction. An abrupt surface is observed. Figure adapted with permission from Ref. [103]. Copyright 2004 American Chemical Society.

In work by Hanrath and Korgel, H-terminated Ge nanowires were exposed to hexene in a supercritical fluid reactor at 220 °C, and the resulting hexyl-terminated nanowires appear resistant to oxidation in either air or water. The TEM image reveals an abrupt interface for the alkyl-terminated Ge nanowire (Figure 5.12(b)) compared to the nanowires removed from the reactor without termination (Figure 5.12(a)) [103].

5.3.3. Alkanethiol reactions on germanium

In addition to Grignard chemistry and hydrogermylation, for which analogs also exist for silicon, there is another type of functionalization chemistry that has been carried out on germanium surfaces: that of alkanethiol attachment to H-terminated Ge. Alkanethiols are well known for their use in self-assembled monolayers (SAMs) on gold surfaces.

Unlike either the Grignard or hydrogermylation reactions, which create a Ge–C bond in the organic attachment to the surface, the alkanethiol reaction creates a Ge–S–C bond configuration upon functionalization. Various 1-alkanethiols react with the hydride-terminated Ge(111) surface at room temperature, with a well-ordered monolayer resulting from 1 day of exposure to a solution of alkanethiol in 2-propanols [125]. Kosuri et al. [126] speculate that the reaction proceeds with the evolution of H_2; however, the mechanism is not well understood. X-ray photoelectron spectroscopy (XPS) measurements indicate that an alkanethiolate monolayer is formed at the Ge surface, with no GeO_x observed after short exposures to air. Vibrational measurements show alkyl modes similar to alkanethiolated Au and alkylated Si, but the Ge–S stretch is not observed [125]. Contact angle measurements match those of the Grignard-alkylated Ge(111) surface, corresponding to a well-ordered alkane monolayer [125]. On the other hand, the surface is less stable than the corresponding monolayers containing direct Ge–C bonds formed by Grignard or hydrogermylation. For example, although no change in the surface coverage is observed upon annealing to 350 K, after a 450 K anneal, the surface coverage is significantly reduced due to desorption [125]. In addition, although a 12-h exposure to ambient results in no change in the contact angles, after 24 h a decrease is observed [125]. Decreases are also observed upon immersion into boiling water and boiling chloroform [125]. This lower surface stability likely results from the presence of the Ge−S bond, which is weaker than the Ge−C bond. Han et al. [125] suggest that the weaker bond allows water to attack at the Ge interface, leading to oxidation and subsequent desorption of the monolayer upon oxide dissolution. Thus, although alkanethiol monolayers present an alternate method for wet chemical organic functionalization, the decreased surface stability may lead to the choice of direct Ge−C bond formation created by both the Grignard and hydrogermylation reactions in applications where stability is important.

5.4. Reaction with compound semiconductors

Compound semiconductors provide capabilities in optoelectronics that cannot be obtained from silicon or germanium. One of the challenges of compound semiconductors is the presence of surface defects that introduce states within the bandgap, leading to surface charge carrier recombination and a degradation of optoelectronic performance [172,173]. Consequently, passivation of these surfaces is important to the ultimate success of such devices. It has been shown by a number of studies that compound semiconductors such as GaAs can be partially passivated with sulfur from inorganic sulfides (e.g., $(NH_4)_2S$ or Na_2S) [173,174]. That work motivated the exploration of organic thiols for passivation of compound semiconductors.

The first reported organic self-assembled monolayer on any semiconductor surface was actually that of thiols on GaAs [172]. Since then, organothiols have been studied at several oxide-free compound semiconductor surfaces, including GaAs, InP, and InAs. The results indicate a significant improvement of the semiconductor electrical properties, and as a result thiols have attracted increasing attention as a means of electrically passivating the surface.

Here we describe some of the results. In each of these studies, the compound semiconductor was first etched in either acid or base to remove the oxide. The specific surface groups following the etch are not well understood. However, Pluchery et al. have followed the acid etching of InP by in situ infrared spectroscopy [175] and observed the removal of the oxide. Unlike Si, for which an acid (HF) etch leaves the surface hydrogen-terminated and temporarily passivated, acid etching of InP does not produce a chemically passivated surface. Presumably, the surface is left unprotected, and quickly oxidizes if not passivated by another process. Similar results showing reduction or removal of the oxide are seen for GaAs [174,176,177].

Exposing etched GaAs to a range of organothiols leads to the formation of reasonably stable and ordered SAMs [172,174,176–182]. The mechanism in solution is not well understood, but a study under vacuum conditions has shown that ethanethiol reacts on clean GaAs(100) at room temperature via S−H bond scission [183]. One of the first studies, carried out by Allara and coworkers [172], indicated that conformationally ordered SAMs could be formed by reacting octadecanethiol on freshly etched GaAs(100). Although no improvement in photoluminescence intensity (corresponding to a decrease in surface recombination) was observed following octadecanethiol passivation in that study, Lewis and coworkers [174,178] reported a large increase in photoluminescence intensity for a variety of organothiols on GaAs(100). Other studies have shown that thiol-modification of GaAs(100) produces a surface which is more stable with respect to degradation than inorganic treatments or the etched surface [179,181]. Although a number of inconsistent results have been reported in the literature, a recent study has determined that if the reaction conditions are tightly controlled, a nearly vertical octadecanethiol SAM (with a chain tilt angle of 14° and thickness of nearly 25Å) which is stable to 100 °C can be reproducibly obtained [181]. Interestingly, the formation of the alkanethiol SAMs on GaAs is reportedly 2 orders of magnitude slower than that on Au(111) [181].

Organothiol SAMs also form on InP. An X-ray photoelectron spectroscopic study of thiol SAMs on InP found that the sulfur atom of the thiol binds to the In atoms, and not the P atoms, of the InP surface [184]. In that study, the authors also found that a C_{12} alkanethiol chain formed a monolayer 14 Å thick, with the chains tilting at 44° from the surface normal. This was compared with the tilt angle of alkanethiols on gold, for which an angle of ~30° on Au(111) and 5° on Au(100) are seen. In another study, the C_{18} alkanethiol SAM on InP(100) was found to be be stable up to 100°C in vacuum and able to resist acidic solution. Overall, alkanethiols were found to be relatively effective in protecting the surface from oxidation [185]. InP passivated by alkanethiols also exhibits improved electrical properties. For example, in a study by Schvartzman et al. [173], an improvement in the steady-state photoluminescence (a parameter that is sensitive to surface states) of over one order of magnitude was obtained upon organothiol SAM formation. Measurement of diodes also showed improvement for the SAM-coated InP, leading the authors to conclude that the number of surface dangling bonds is reduced upon formation of the SAM. Alkanethiol self-assembled monolayers were also found to passivate the InAs(100) surface against oxidation [186].

6. Adsorption of organic molecules under vacuum conditions

Studying the adsorption of organic molecules with the semiconductor surfaces under vacuum conditions allows the reactivity of the reconstructed surfaces to be investigated. In the case of (100)–2 × 1 reconstructed surfaces of silicon and germanium, such studies permit investigation into the reactive nature of the intriguing Si and Ge dimers. Recall that the dimers can be described in different ways—either as a weakly double-bonded entity, or as a nucleophile/electrophile pair. Over the past decade, a number of studies in vacuum have revealed reactivity of the dimers toward organic species that can be interpreted using one or both of these descriptions. Specifically, many of the reactions of organic species at the Si(100)–2 × 1 or Ge(100)–2 × 1 surfaces can be categorized into one of two types of reactions—cycloadditions and nucleophilic/electrophilic reactions. Interestingly, similar behavior has been observed on the Si(111)–7 × 7 surface as well. For example, in some systems the Si(111)–7 × 7 acts as an electron acceptor [47], and in others, the adatoms and/or rest atoms participate in cycloaddition reactions [19]. In other words, the silicon or germanium surfaces to some extent mimic organic reagents.

In the following section, a description of the semiconductor surface and its reactions will be developed with reference to the standard framework of organic chemistry. Rather than give a comprehensive survey of the literature, we have selected examples that illustrate key features of the reactivity. Our discussion focuses on the (100)–2 × 1 reconstructed surface of Si and Ge, with some comparison made to the (111)–7 × 7 face of silicon.

6.1. Silicon surface chemistry

6.1.1. Cycloaddition reaction on Si(100)–2 × 1

Because of the presence of both σ and π bonds in the Si dimer, analogies can be drawn between the Si dimer and the double bonds formed by its Group IV congener, carbon. Carbon–carbon double bonds (alkenes) constitute a very well studied functional group in organic chemistry, and a number of reactions are known which can create bonds with an alkene group. One synthetically useful class of these reactions is 'cycloaddition' reactions, which fall under the more general category of pericyclic reactions. Cycloadditions are widely used in organic synthesis as a means of forming new carbon-carbon bonds and new carbon rings because of their versatility and high steroselectivity [187–189]. Cycloadditions are reactions in which two π bonded molecules come together to make a new cyclic molecule with the formation of two new σ bonds. Two important cycloaddition reactions are the [2+2] cycloaddition and [4+2] cycloaddition, also known as the Diels-Alder reaction. In a [2+2] cycloaddition, a pair of π electrons from each alkene reacts to form a four-membered ring. This reaction only proceeds after ultraviolet excitation, even at considerably elevated temperatures [190]. In contrast to a [2+2] cycloaddition, a diene can readily react with an alkene (the dieneophile) near room temperature in a [4+2] cycloaddition reaction to form a six-membered ring [187].

The cycloaddition reactions in organic chemistry are subject to the Woodward–Hoffman selection rules [191]. These selection rules stem from an analysis based on frontier orbital theory, which examines the symmetries of the highest occupied and lowest unoccupied molecular orbitals (HOMO and LUMO, respectively) of the reacting molecules. The resulting Woodward-Hoffman selection rules are widely used for predicting how readily an organic reaction will occur. In cycloaddition reactions, the frontier π orbitals must overlap in phase for the reaction to be symmetry allowed. In organic chemistry, [2+2] cycloadditions are found to be 'symmetry forbidden'. That is, the Woodward-Hoffman symmetry rules dictate that this reaction should not occur without significant energy activation, and consequently in organic chemistry this reaction is largely limited to synthesis involving photochemical activation. The [4+2] reaction, in contrast, is 'allowed' by symmetry considerations, and Diels-Alder reactions are commonly used in organic synthesis as a means of forming new C−C bonds and ring structures.

If the analogy that is drawn between the Si=Si dimer on the Si(100)–2 × 1 surface and an alkene group is reasonable, then certain parallels might be expected to exist between cycloaddition reactions in organic chemistry and reactions that occur between alkenes or dienes and the silicon surface. In other words, cycloaddition products should be observed on the Si(100)–2 × 1 surface. Indeed, this prediction has been borne out in a number of studies of cycloaddition reactions on Si(100)–2 × 1 [14], as well as on the related surfaces of Ge(100)–2 × 1 (see Section 6.2.1) and C(100)–2 × 1 [192–195]. On the other hand, because the double-bonded description is only an approximation, deviations from the simple picture are expected. A number of studies have shown that the behavior differs from that of a double bond, and the asymmetric character of the dimer will be seen to play an important role. For example, departures from the symmetry selection rules developed for organic reactions are observed at the surface. Several review articles address cycloaddition and related chemistry at the Si(100)–2 × 1 surface; the reader is referred to Refs. [10–18] for additional detail.

6.1.1.1. [2+2] Cycloaddition reaction on silicon

The first examples of what can be categorized as [2+2] type cycloaddition product formed by reaction between an alkene and a silicon surface were reported in the late 1980s. Alkenes such as ethylene, as well as the related alkyne molecule acetylene, were reacted with the clean Si(100)–2 × 1 surface in vacuum [196–213]. The adsorption of these unsaturated C_2 molecules (ethylene and acetylene) on Si(100)–2 × 1 is also discussed in Chapter 1. The alkenes were found to chemisorb at room temperature, forming stable species that 'bridge-bonded' across the silicon dimers on the surface. The reaction proceeded by formation of two new σ bonds between Si and C atoms, hence the bonding was referred to as 'di-sigma' bonding. In addition, it was shown that while the π bonds of the alkene and of the Si−Si dimer are broken, the σ bonds remain intact [205,206,209]. This reaction product is analogous to a [2+2] cycloaddition product in which ethylene reacts with the silicon dimer (Si=Si) on the Si(100)–2 × 1 surface to form a four-membered $Si_2(CH_2)_2$ ring with the surface, as shown in Figure 5.13(a).

Early investigators recognized that although the formation of a [2+2] C=C cycloaddition product is symmetry forbidden in traditional organic chemistry [191], the sticking probability of alkenes on Si(100)–2 × 1 is near unity at room temperature [11]. Liu et al. used infrared spectroscopy to investigate the mechanism of this reaction on Si(100)–2 × 1 by probing its stereoselectivity [198]. By using *cis*- and *trans*-1,2-dideuteroethylene on Si(100)–2 × 1 and comparing the splitting of ν_s(−C−H) and ν_{as}(−C−H) stretching modes to those calculated theoretically, they showed that the reaction is stereoselective but does not proceed through the traditional symmetry-allowed mechanism. A subsequent STM study by Lopinski et al. found that there is a small, but measurable, percentage (~2–3%) of alkenes that have isomerized during the adsorption to form a non-stereoselective [2+2] C=C cycloaddition product [214]. This result confirmed that [2+2] cycloaddition reactions, unlike their solution phase analogs, are not concerted and likely occur in a stepwise fashion on the Si(100)–2 × 1 surface. It has been proposed that the tilting of the dimers on Si(100)–2 × 1 allows the alkene to adsorb through a low-symmetry, lower energy pathway in which the molecule approaches from the side of the dimer [10,198,207,208]. Recently, Yoshinobu and coworkers reported direct observation of this precursor state to the [2+2] product using high-resolution electron energy loss spectroscopy (HREELS). At low temperature, they were able

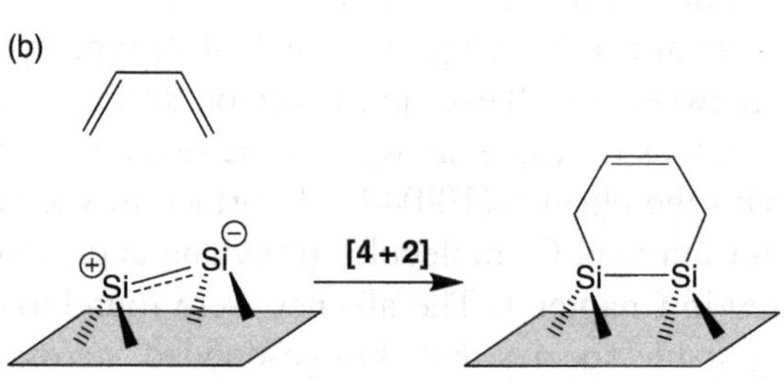

Figure 5.13. Cycloaddition products at the silicon dimer of the Si(100)–2 × 1 surface. (a) shows the [2+2] cycloaddition product formed in the reaction with ethylene, and (b) shows the [4+2], or Diels-Alder, cycloaddition product formed in the reaction with 1,3-butadiene.

to isolate and spectroscopically detect a precursor in which the ethylene forms a π-complex with the down silicon atom of the dimer [215].

The result of the [2+2]-like reaction is a tightly bonded organic group, directly attached to the silicon surface through strong Si−C bonds (~82 kcal/mol each [40]), as illustrated in Figure 5.13(a). Furthermore, the reaction occurs readily for many alkenes, and larger alkenes have yielded similar results. Consequently, [2+2] cycloaddition has been utilized in a number of other surface functionalization reactions on silicon. Here we highlight a few interesting examples.

One example is the work by Hamers and coworkers. Hamers et al. showed with IR spectroscopy that cyclic alkenes, such as cyclopentene [216] and 1,5-cyclooctadiene [217], also bond to the Si(100)–2 × 1 surface to give a [2+2] C=C cycloaddition product. Furthermore, they demonstrated with STM that cyclopentene and 1,5-cyclooctadiene form well-ordered monolayers, as shown in Figure 5.14 [218]. The cyclopentene experiments by Hamers and coworkers were performed on a vicinal Si surface that was intentionally cut slightly (4°) off-axis

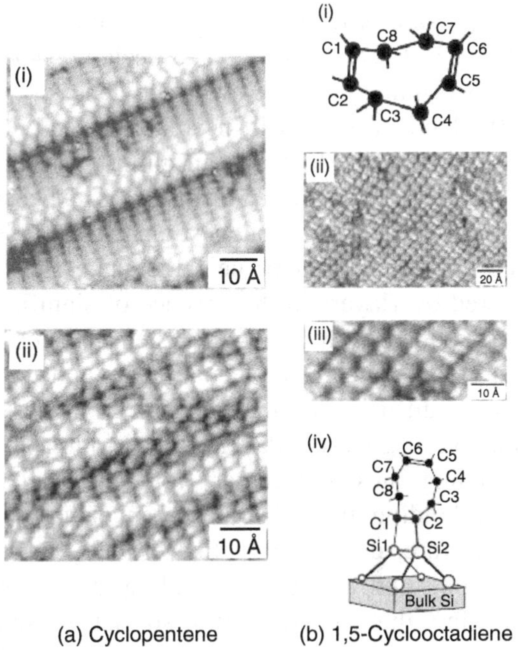

Figure 5.14. STM images of the well-ordered monolayers formed from the reaction of (a) cyclopentene and (b) 1,5-cyclooctadiene with Si(100)–2 × 1. The images in (a) are collected on a vicinal silicon surface, as described in the text. (i) and (ii) are before and after saturation exposure to cyclopentene, respectively. The images in (b) include: (i) the 1,5-cyclooctadiene molecule, (ii) and (iii) STM images of saturation coverge of 1,5-cyclooctadiene, and (iv) a model of the molecule bonded across a Si-Si dimer. Both molecules form well-ordered monolayers on Si(100)–2 × 1. Figure adapted from Ref. [218] with kind permission of Springer Science and Business Media.

from the (100) crystal face. Typical Si crystals cut close to the (100) plane contain a distribution of steps separating terraces which differ in height by one atom, leading to a surface in which the orientation of the dimers is rotated by 90° on adjacent terraces. In contrast, the intentional miscut has the effect of producing a surface containing terraces separated by steps that are mostly two atoms high, leading to retention of the dimer orientation from terrace to terrace [219,220]. The orientation of the dimer rows on the terraces can be seen in the top STM image of Figure 5.14(a) of the clean 4° miscut Si(100) surface [220]. On the flat terrace regions, the Si=Si dimers, imaged as oblong white features, all maintain the same orientation with respect to the step. After adsorption of cyclopentene (bottom image of Figure 5.14(a), long-range order is clearly observed in the new cyclopentene overlayer. A high degree of molecular order is also observed after adsorption of 1,5-cyclooctadiene on Si(100)–2×1, as seen in Figure 5.14(b). Long-range order on a length scale of millimeters was confirmed by measuring the retention of orientational anisotropy across the length of the silicon sample using polarized vibrational spectroscopy [219]. These studies provide support for the use of cycloaddition chemistry at the Si(100)–2×1 surface as a method to prepare ordered monolayer organic films.

The success in forming ordered overlayers is especially significant in light of the following characteristics of this system. As with ethylene, cyclopentene was found to react at the Si(100)–2×1 surface with near unity reactivity, that is, each gas molecule that collides with the surface has an almost perfect chance of sticking and bonding at the surface. Furthermore, diffusion is expected to be very slow in these systems at room temperature, suggesting that once the molecules have bonded to the surface, they remain at or near the same site on the surface [11]. The formation of ordered overlayers in the absence of significant surface mobility is in contrast to the behavior of other growth systems. In most common methods of semiconductor film growth, such as molecular beam epitaxy (MBE) or chemical vapor deposition (CVD), diffusion of the depositing species at the surface is a requisite for the formation of good overlayers [55]. Motion of atoms or molecules to find the optimal binding sites is necessary for conformal film coverage. This is usually achieved by holding the substrate at an elevated temperature. These cyclopentene films, therefore, do not grow like ordinary metal or semiconductor crystals. Rather, the strong and directional bonding at the interface is sufficient to direct the molecules into the necessary sites, so that the semiconductor surface acts as a template for the placement of the organic molecules. The effect is one in which the structure of the underlying surface is propagated into ordered overlayers of organic molecules.

Ordered overlayers are not achieved in all systems, however, and it appears that certain criteria need to be met to form well-ordered films. Although cyclopentene and 1,5-cyclooctadiene could be used to produce ordered monolayer films through [2 + 2] attachment [221], a number of other alkenes investigated did not adsorb with a high degree of order. For example, cyclohexadiene and cyclooctatetraene while reacting readily with the Si(100)–2 × 1 surface, did not form ordered films on Si(100)–2 × 1 [217,222]. In fact, it is only a minority of cycloaddition systems for which well-ordered monolayers have been reported. This lower degree of order in most cycloaddition systems may result from several different reasons. First, the final (saturation) coverage of molecules at the surface may be too low. Second, the molecules may bind in several different configurations, prohibiting the degree of order necessary for close packing [222]. This may occur because there are multiple reaction pathways possible, e.g., a multifunctional reactant may react at different parts of the molecule. In this regard, having a single type of functional group or a single means of bonding to the surface is advantageous. Even for monofunctional reactants, however, the molecule may bind at a variety of surface sites (e.g., intradimer vs. interdimer), as will be discussed below. Finally, even with uniform, selective formation of one type of product, there may be disorder in the configurations of individual adsorbates. Rigidity of the organic molecule, generally provided by a ring structure, was found to help in the formation of ordered layers [222].

In addition to forming densely-packed organic layers, the [2 + 2] cycloaddition reaction has been used to generate synthetic chiral surfaces, as demonstrated by Lopinski et al. in the adsorption of 1S(+)-3-carene on the Si(100)–2 × 1 surface [223]. Carene is a chiral, two-ringed molecule containing a C=C double bond and a three-carbon-atom ring (cyclopropane). Carene is thought to adsorb into an initial [2 + 2] cycloaddition product via the alkene group, then undergo a second reaction at the cyclopropane group to eventually form an adduct that bridges across two dimers. The carene molecule contains a bulky dimethyl group that geometrically hinders adsorption at one face of the molecule. This steric hindrance forces the [2+2] reaction to take place with only one facial orientation of the molecule. Lopinski et al. verified that the reaction was enantiospecific, i.e., produced only one of the possible chiral products, using scanning tunneling microscopy. The resulting carene adsorbate contains chiral centers and produces a chiral surface [223]. This result suggests that cycloaddition may be useful for creating chiral, organically modified surfaces for possible applications in molecular recognition and chemical sensing.

Alkynes have also been shown to form the [2 + 2] cycloaddition product. Acetylene (H−C≡C−H), the simplest alkyne, forms an interesting adsorption case, because the specific adsorption geometries of acetylene on Si(100)–2 × 1 have been debated [11,201,207,210,224–236]. Acetylene was first found experimentally to form a [2 + 2] C≡C cycloaddition product that exhibits a cyclobutene-like surface structure on Si(100)–2 × 1 [210,227]. Later STM measurements revealed that at least two different surface products were present [228,231,233], and identified a product that is oriented perpendicularly to the dimer row. From these images, it was argued that in addition to an intradimer [2 + 2] C≡C cycloaddition geometry, acetylene also forms a surface adduct that bridges two dimers along a row. Several theoretical investigations of acetylene on Si(100)–2 × 1 have been performed and these studies do find other stable structures, including a bridge-bonded state, although they disagree somewhat on its identity and relative energetics [207, 224–226, 229, 230, 234–236].

6.1.1.2. [4+2] Cycloadditions, or Diels-Alder reactions, on silicon

The other common type of cycloaddition reaction in organic chemistry is the [4+2] or Diels-Alder cycloaddition. In 1997, a series of papers were published in which the Diels-Alder reaction product (see Figure 5.13b) was first predicted and measured on the Si(100)–2 × 1 surface. Konecny and Doren performed a theoretical study in which they predicted that Diels-Alder reactions should occur across silicon dimers on a Si(100)–2 × 1 surface, just as they do across C=C double bonds in a molecule [237,238]. Their prediction was based on ab initio calculations of the reaction of a model conjugated diene (1,3-cyclohexadiene) with a silicon surface. In the calculations, the silicon surface was modeled by a silicon cluster containing nine silicon atoms and featuring the silicon dimer. This silicon cluster and related larger clusters have been shown to provide reasonable models of the silicon surface while remaining computationally tractable. We note that many theoretical studies using similar cluster models have been performed for both Si and Ge over the past several years, particularly in the area of organic functionalization. Doren et al. followed the reaction of 1,3-cyclohexadiene with the silicon cluster, and examined two possible reaction pathways: a [4+2] reaction or a [2+2] reaction [237,238]. Figure 5.15 illustrates the two possible reaction products for a related molecule, 1,3-butadiene. The calculations showed that there is a substantial thermodynamic driving force (54 kcal/mol) to form the [4+2] cycloaddition product, significantly more energy than is gained in the competing [2+2] reaction (39 kcal/mol). The difference was attributed to lower ring strain in the 6-membered ring product of the [4+2] cycloaddition compared to the 4-membered ring formed by the [2+2]

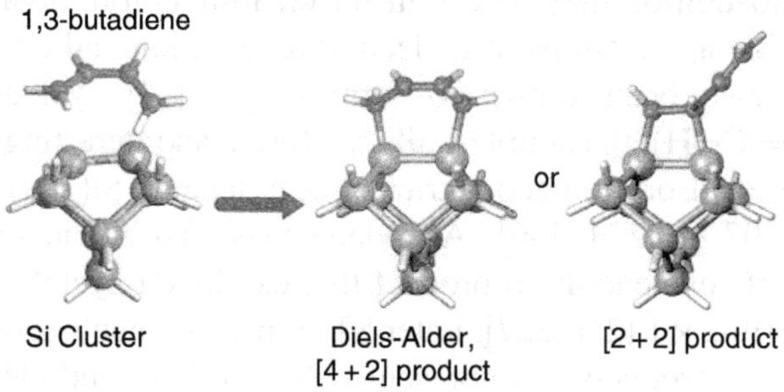

Figure 5.15. Density functional theory study of the reaction of 1,3-butadiene with the Si(100)–2 × 1 surface, modeled using a nine silicon atom cluster and examining two possible products: the product of the Diels-Alder cycloaddition, and the product of the [2+2] cycloaddition [237,238].

reaction. In addition, the calculations predicted that the reaction should occur with little or no energetic barrier across a symmetric silicon dimer, consistent with the Woodward-Hoffman rules for a [4+2] reaction [237,238].

Studies by Teplyakov et al. provided the experimental evidence for the formation of the Diels-Alder reaction product at the Si(100)–2 × 1 surface [239,240]. A combination of surface-sensitive techniques was applied to make the assignment, including surface infrared (vibrational) spectroscopy, thermal desorption studies, and synchrotron-based X-ray absorption spectroscopy. Vibrational spectroscopy in particular provides a molecular 'fingerprint' and is useful in identifying bonding and structure in the adsorbed molecules. An analysis of the vibrational spectra of adsorbed butadiene on Si(100)–2 × 1 in which several isotopic forms of butadiene (i.e., some of the H atoms were substituted with D atoms) were compared showed that the majority of butadiene molecules formed the Diels-Alder reaction product at the surface. Very good agreement was also found between the experimental vibrational spectra obtained by Teplyakov et al. [239,240] and frequencies calculated for the Diels-Alder surface adduct by Konecny and Doren [237,238].

A subsequent study using STM suggested that although the majority (80%) of the product formed by adsorption of 2,3-dimethyl-1,3-butadiene on Si(100)–2 × 1 was the [4+2] cycloaddition adduct, the remaining 20% formed a [2+2] C=C cycloaddition product [222]. However, theoretical analysis led Doren [241] to propose that the [2+2] C=C cycloadduct observed in that study is actually a [4+2] cycloadduct that bridges two adjacent dimers, i.e., an interdimer product.

Competition between various products is likely to remain an important issue when considering functional groups such as dienes for which more than one reaction pathway is possible (e.g., [2+2] vs. [4+2] cycloaddition). Although some generalizations can be reached from a thorough understanding of the reactivity of different functional groups, ultimately the product distribution will depend on the specific molecule. For example, 1,3-cyclohexadiene [242–247] and cyclopentadiene [248], both cyclic dienes, have been studied by a variety of techniques. 1,3-cyclohexadiene was found to exhibit features characteristic of both [4+2] and [2+2] cycloaddition products, while cyclopentadiene primarily formed a [4+2] cycloaddition product. On the basis of the calculated binding energies, the expected ratio of [2+2] to [4+2] cycloaddition products is approximately 10^{-10}. Therefore, the presence of a quantifiable amount of [2+2] side product suggests that these reactions are governed by the difference in activation barriers (i.e., kinetics) and not relative product stabilities. Further, the competition is not just between these two types of intradimer cycloadditions. An STM study of 1,3-cyclohexadiene by Teague and Boland [243] has identified six states at the surface, including an interdimer [4+2] product in addition to the intradimer [2+2] and [4+2] cycloaddition products. They propose, in agreement with the above statement, that kinetics favor the interdimer [4+2] product over the intradimer [4+2] product even though it is thermodynamically less stable, and attribute the difference to reduced ring strain. Subsequent theoretical studies of this system support the importance of kinetics in this system [245–247].

Given the observation of interdimer [4+2] products, in which the diene reacts with silicon atoms from two different dimers, it is clear that the cycloaddition product can form at pairs of surface silicon atoms other than the double-bond-like Si=Si dimer. Recent experiments on the Si(111)–7 × 7 surface show, interestingly, that this surface can also serve as a substrate for formation of a product analogous to the Diels-Alder product. One of the first molecules for which a cycloaddition-like product was reported on Si(111)–7 × 7 is thiophene, a sulfur-containing, five-membered, aromatic ring. Thiophene adsorption has been investigated by a variety of techniques, including HREELS, STM, and synchrotron-based photoemission [46,48]. Based on these methods, the product was assigned to a 2,5-dihydrothiophene-like species which is di-sigma bonded to an adjacent rest atom-adatom pair of silicon atoms, i.e., the Diels-Alder-like product. Note that the rest atom-adatom separation on Si(111)–7 × 7 is 4.5 Å, significantly larger than the 2.4 Å separation between dimer atoms on the Si(100)–2 × 1 surface [48]. Interestingly, valence band photoemission spectra show that thiophene adsorption leads to the same moiety on both Si(100)–2 × 1 and Si(111)–7 × 7, as well as on Ge(100)–2 × 1. As commented by Rousseau et al., the observation of this product on the Si(111)–7 × 7 surface, which lacks a π-bond, suggests a different reaction chemistry than the standard Diels-Alder mechanism [48].

Butadiene has also been reported to form a [4+2] cycloaddition-like product at the Si(111)–7 × 7 surface. Theoretically, the reaction has been predicted to occur between a rest atom-adatom pair via a step-wise, diradical path pathway that is unactivated [19,249]. Recent STM studies of butadiene adsorbed on Si(111)–7 × 7 have observed the formation of either a [4+2]-like or a [2+2]-like product with the Si adatom-rest atom pair, but were not able to definitely assign the product as one or the other [250]. Interestingly, the STM images also reveal that about a third of the cycloaddition product forms between two Si adatoms. Because both adatoms are positively charged, the stepwise reaction does not seem likely, and the authors suggest that this reaction occurs in a concerted fashion [250].

A recent theoretical study has nicely addressed the question of mechanism on the silicon surface. Minary and Tuckerman carried out an ab initio molecular dynamics (MD) study of the [4+2] cycloaddition reaction on Si(100)–2 × 1 [251]. Because the previously reported ab initio DFT models were 'static', these were not able to address in detail the mechanisms by which the [4+2] product was formed. The results of the MD study indicate that rather than being concerted, the dominant mechanism is a stepwise zwitterionic process in which an initial nucleophilic attack of one of the C=C bonds by the down atom of the dimer leads to a carbocation. This carbocation exists for up to 1–2 ps, stabilized by resonance, and depending on which positively charged carbon atom reacts with which Si surface atom, can form either inter- and interdimer [4 + 2] adducts as well as the [2 + 2] product [251]. By extension, a similar mechanism may apply on Si(111)–7 × 7, giving rise to analogous products.

Both the [2 + 2] and [4 + 2] products of reaction between a diene and the Si(100)–2 × 1 (or Si(111)–7 × 7) surface have a number of features that are desirable for surface functionalization. Both reactions produce organic molecules that are strongly bound to the silicon surface through two covalent Si−C bonds. The high strength of these bonds means that the organic layer at the surface is typically stable to temperatures in excess of 100 °C to 200 °C [242]. In addition, both of the cycloaddition reaction products between the diene and the surface retain one double bond in the organic layer. This remaining alkene group can possibly be used as a starting point for further derivatization of the surface. In this way, there is the potential to perform controlled, layer-by-layer synthesis of organic films on the Si surface.

6.1.2. Heterocycloadditions

In addition to alkenes and dienes, heteronuclear unsaturated compounds can also undergo cycloadditions. Several research groups have studied related compounds that contain, for example, carbonyl (C═O), nitrile (C≡N), azo (N═N), or nitro (NO_2) functional groups, to provide insight into pericyclic reactions of functional groups with polarized charge distributions. It was found that such unsaturated functional groups do often undergo cycloaddition chemistry at the Si(100)–2 × 1 surface. Interestingly, these heteroatom functional groups typically form the cycloaddition products by traversing a reaction pathway involving a dative-bonded precursor state. Dative bonds will be discussed in the following section. For example, in a theoretical investigation of a series of carbonyl-containing compounds, Barriocanal et al. identified a barrierless pathway that passes through a dative-bonded precursor state for the [2 + 2] C═O cycloaddition product of glyoxal [252]. Armstrong et al., using a combination of TPD, XPS, and HREELS, reported that the majority surface adducts for acetone, acetaldehyde, and biacetyl at low temperature are [2 + 2] C═O cycloaddition products, forming a 4-membered Si-C-O-Si ring structure [253,254].

Hetero-Diels-Alder products (i.e., Diels-Alder products involving an atom other than carbon) have also been observed for a number of systems at the Si(100)–2 × 1 surface. Some examples include the reaction of unsaturated ketones RC═C−CR′═O (e.g., ethylvinylketone) [255], 2-propenenitrile [256–260], and dicarbonyls O═C−R−C═O [261]. For a review specifically of heteroatom chemistry at the silicon surface, the reader is referred to Ref. [262]. Here we use the dicarbonyl example to provide an illustration of this class of surface chemistry. Hamers and coworkers have used the [4 + 2] heteroatom cycloaddition reaction of a dicarbonyl to attach π-conjugated molecules to the Si(100)–2 × 1 surface [261]. As pointed out in a theoretical paper by Barriocanal and Doren [252], when dicarbonyl compounds undergo [4 + 2] cycloadditions, the resulting 6-membered ring stands normal to the surface, in contrast to the nearly 40° tilt in the butadiene [4 + 2] cycloaddition product. Taking advantage of this geometry, Hamers and coworkers reacted 9,10-phenanthrenequinone with Si(100)–2 × 1. They found that the molecule reacted selectively to form the heteroatomic Diels-Alder product, leading to the attachment of aromatic rings oriented perpendicular to the surface, as shown in Figure 5.16. They suggest that such monolayers may exhibit anisotropy in their optical and electrical properties, and may be useful for fabricating π-stacked molecular overlayers [261]. We note that although 9,10-phenanthrenequinone reacted selectively, competition between the [2 + 2] and [4 + 2] products, as discussed for the diene [4 + 2] reaction, may exist in hetero-[4 + 2] chemistry as well.

6.1.3. Nucleophilic/electrophilic reactions

In addition to forming dimers with double-bond-like character that can participate in cycloaddition-type reactions, the reconstruction of the (100) semiconductor surface is also responsible for the presence of electrophilic and nucleophilic sites. The bond of a tilted dimer deviates from the plane of the surface, and the resulting structure of each dimer consists of an 'up' dimer atom protruding from the surface and a 'down' dimer atom recessed on the surface. The electron density at the up atom of the dimer is higher than that at the down atom, leading to nucleophilic and electrophilic behavior, respectively [49].

Nucleophilic/electrophilic reactions are frequently observed in solution phase organic chemistry and have thus been subjected to detailed study [42]. In the majority

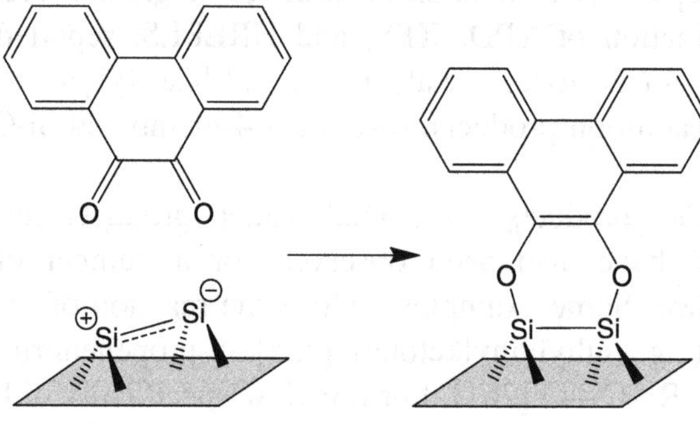

Figure 5.16. Attachment of a π-conjugated molecule to the Si(100)–2 × 1 surface using a hetero-Diels-Alder cycloaddition. Hamers and coworkers reacted 9,10-phenanthrenequinone, a di-carbonyl, with the surface, leading to aromatic rings oriented perpendicular to the surface [261].

of these reactions, each participating molecule shares one of its electrons to form a new covalent bond. Dative bonding, also known as coordinate covalent bonding, occurs when one molecule donates both of the electrons needed to form a covalent bond. The molecule that donates electronic charge is known as the nucleophile and the molecule that accepts the charge is the electrophile. Since the down atom of a silicon dimer is electron deficient, it is an electrophile and can react with an impinging nucleophile to form a dative bond [49]. A number of studies have now shown that dative bonding is a phenomenon common to many organic reactions on Si(100)–2×1, either as a precursor state for formation of more thermodynamically stable products, or as the final product itself. In this section, several examples of dative bonding at the silicon surface will be presented.

The first example is that of ammonia, which has a lone pair on the nitrogen (making it a good Lewis base) and exhibits characteristic dative bonding behavior. The reaction of ammonia with silicon has been well studied due to its importance as a nitriding agent for forming Si_3N_4 in microelectronic devices [263–266]. Studies have shown that ammonia dissociatively adsorbs on Si(100)–2×1 into NH_2(a) and H(a) at room temperature, preserving the (100)–2×1 reconstruction [267–272]. At temperatures below 120 K, ammonia coexists in both the molecularly adsorbed and dissociated forms, suggesting that a barrier is located between these two states that not all molecules can traverse [270,273]. Musgrave and coworkers [274–276] investigated the bonding of ammonia on Si(100)–2×1 with density functional theory using a cluster approximation, and dative bonding was proposed as the mechanism of molecular adsorption. Their calculations also revealed that the donation of electronic charge from the ammonia lone pair to the surface delocalizes across adjacent surface dimers in the same dimer row, and the authors concluded that surface-mediated electron transfer plays a significant role in the adsorption of nucleophiles on the Si(100)–2×1 surface. A combined theoretical and spectroscopic study of ammonia adsorption on Si(100)–2×1 by Queeney et al [277] showed evidence for these non-local effects, indicating that because of changes in the electronic structure of the surface near adsorbed ammonia molecules, further adsorption of ammonia is directed to the opposite side of the neighboring dimer along a row. The result is predicted to be the formation of a two-dimensional ordered pattern in which ammonia dissociatively bonds in a zig-zag arrangement along the dimer row.

Several groups have investigated the bonding of organic amines on Si(100)–2×1 in an effort to understand the kinetic and thermodynamic factors controlling reactions on these surfaces [47,49,278–286]. On Si(100)–2×1, the most important factor influencing the final surface product was found to be the presence of an N−H bond in the reacting molecule. Primary and secondary amines, both of which contain at least one N−H bond, underwent N−H dissociation (referred to as proton transfer) while tertiary amines did not. Tertiary amines were found to form a stable, dative-bonded product at the surface, as illustrated in Figure 5.17. Non-aromatic

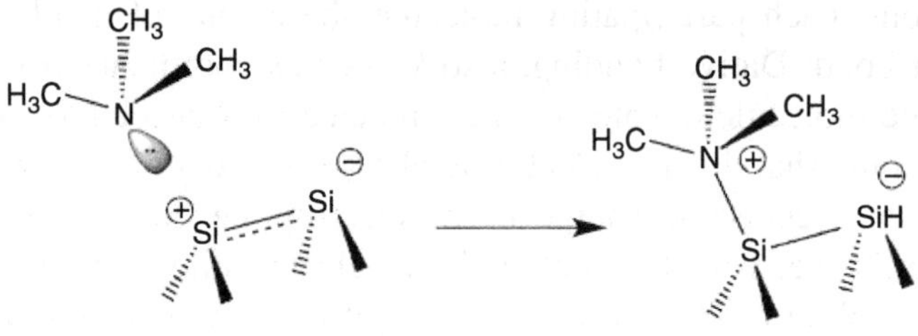

Figure 5.17. Illustration of the dative bonded product formed when trimethylamine adsorbs on Si(100)–2 × 1.

cyclic amines [216,287,290] were found to have a reactivity similar to that for noncyclic amines.

The methylamines have proved to be an important system for developing a deeper understanding of the nucleophilic and electrophilic behavior of the semiconductor surface dimers. Mui et al. investigated the bonding of methylamines on Si(100)–2 × 1 with a combined experimental and theoretical approach [49,279]. Their calculations showed that although the surface reactions of methylamine and dimethylamine involving N−C bond cleavage are thermodynamically more favorable than N−H dissociation reactions, the activation energies for N−C bond cleavage are significantly higher than those for N−H dissociation. Figure 5.18 shows energetics for both the N−H and $N-CH_3$ bond dissociation pathways for dimethylamine calculated at a Si_9H_{12} dimer silicon cluster. Initial adsorption of dimethylamine results in a stable, molecularly adsorbed state with an adsorption energy of nearly 25 kcal/mol. This molecularly adsorbed precursor state involves a dative bond between the nitrogen lone-pair and the electrophilic down-atom of the buckled Si dimer [49,275,276,279]. From the dative-bonded state, the calculations show that the barrier for the N−H bond cleavage pathway is below the entrance channel, and the formation of surface $N(CH_3)_2$(a) and H(a) species is exothermic by over 50 kcal/mol. In contrast, although the pathway for N−C bond cleavage is more exothermic (−65 kcal/mol) than that for N−H bond cleavage, the barrier for $N-CH_3$ dissociation is nearly 20 kcal/mol above the entrance channel energy. Hence, this process is activated, and N−C bond cleavage would not be expected to occur at room temperature. Similar theoretical results were obtained by Cao et al. [47].

In order to understand why the activation energies differ between the two pathways, Mui et al. examined the transition state geometries [279]. They found that as electron density is donated from the amine lone pair to the down silicon atom upon adsorption into the precursor state, the up Si atom in the dimer becomes electron rich. At this stage, the dative bonded precursor can be described as a quaternary ammonium ion. The N−H dissociation pathway can thus be interpreted as the transfer of a proton from the ammonium ion to the electron-rich up Si atom through a Lewis acid-base reaction. In the transition state for this proton transfer, the N−H and Si−H

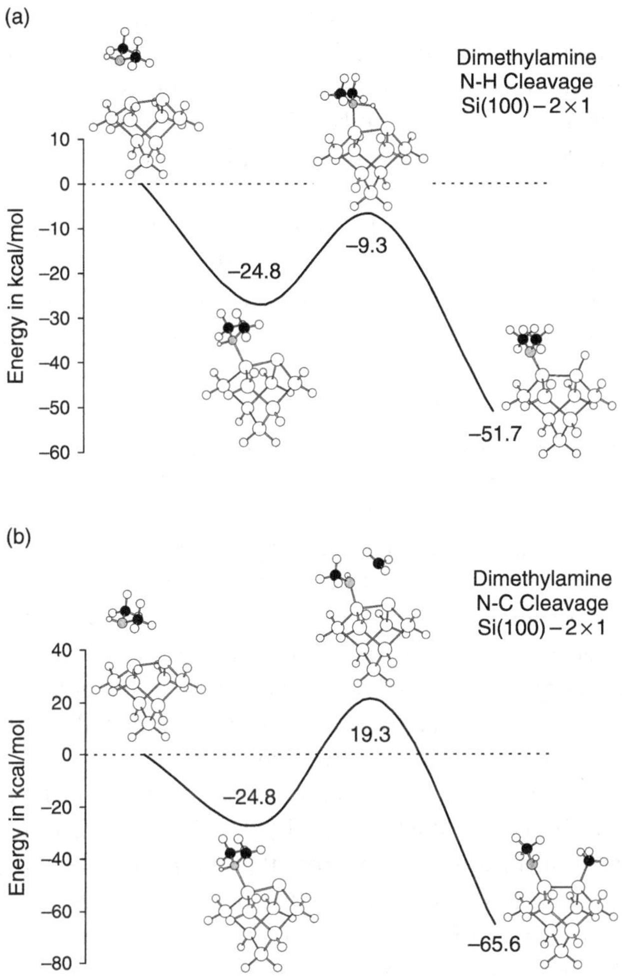

Figure 5.18. Calculated reaction path and optimized structures using DFT for (a) N−H dissociation and (b) $N-CH_3$ dissociation of dimethylamine on the one-dimer silicon cluster. Figure reused with permission from Collin Mui, Journal of Chemical Physics, 114, 10170 (2001). Copyright 2001, American Institute of Physics [279].

bond lengths were found to be only ~25% longer than those in the molecularly adsorbed state and the H dissociation state, respectively, leading to a relatively low activation barrier. In contrast, the $N-CH_3$ dissociation pathway was described as a nucleophilic substitution reaction, where the electron-rich up Si atom of the dimer acts as a nucleophile, and the amine molecule attached to the down Si atom is the leaving group. Due to steric hindrance of the methyl hydrogens, the N−C bond has to be stretched significantly in order to accommodate front side nucleophilic attack, leading to a relatively high activation barrier [279]. The results were consistent with organonitrogen chemistry, where for quaternary ammonium ions, abstraction of a proton from tetravalent nitrogen by a base (in this case the electron-rich 'up' Si atom) is typically kinetically easy compared to N−C cleavage [291]. Moreover, the results indicated that the selectivity of N−H cleavage over N−CH_3 dissociation for amines on the Si(100)–2 × 1 surface is kinetically controlled, and not the result of thermodynamic control.

Direct experimental evidence for the presence of a dative bonded state has been found using a variety of techniques, including infrared spectroscopy, X-ray photoelectron spectroscopy, and STM. The infrared studies take advantage of vibrational modes specific to the presence of a lone pair. In a molecule such as trimethylamine, strong ν(C−H) stretching modes known as Bohlmann bands are observed below 2800 cm^{-1}. These modes originate from the interaction of the C−H σ-orbitals located *trans* periplanar to the lone pair of the nitrogen. If the lone pair is involved in bonding, as in a dative bond, these modes should disappear. Mui et al. observed a large attenuation of the Bohlmann bands in the IR spectrum upon adsorption of trimethylamine on Si(100)–2 × 1, indicating that the majority surface species is adsorbed molecularly in a stable dative-bonded state at room temperature [279]. X-ray photoelectron spectroscopy (XPS) data from Cao and Hamers revealed two N(1s) photoelectron peaks present near 402.2 and 398.9 eV at room temperature. The high binding energy peak, which comprised 85 % of the area, was characteristic of an atom in an extremely electron deficient environment [47,292]. The authors assigned this feature to dative-bonded trimethylamine.

Interestingly, Cao and Hamers found in the same set of studies that trimethylamine also forms a dative-bonded adduct on the Si(111)–7 × 7 surface, reacting similarly to the Si(100)–2 × 1 surface [47]. They note that the positively charged adatoms on Si(111)–7 × 7 act as Lewis acids, and are the most likely site for the nucleophilic amine to bond. The main difference observed between the two surfaces is that the coverage of trimethylamine molecules on Si(111)–7 × 7 is only about half the coverage on Si(100)–2 × 1.

Yoshinobu and coworkers performed STM studies on the trimethylamine/Si(100)–c(4 × 2) system. Recall from Section 2 that the c(4 × 2) notation refers to a dimer-reconstructed surface in which the dimer buckling alternates row to row, as shown in Figure 5.3 [286]. They found that, in agreement with the IR and XPS results, trimethylamine adsorbs via a dative bond at the surface. Further, they concluded that the reaction is highly selective, with the TMA adsorbing selectively on the down atom of the silicon dimer. This result is consistent with the chemical description of the down atom as an electrophile, and of the overall reaction as a Lewis acid-base reaction. Moreover, two theoretical studies have addressed whether dissociatively-adsorbed methylamine exhibits 2-dimensional ordering as was inferred for the case of ammonia [283,285]. The studies both predict that ordering will occur, but disagree on the type of two-dimensional arrangement that will form. One study concludes that methylamine dissociation will reverse the dimer tilt, leading to dissociated methylamine species aligned along one side of the dimer row [283], while the other study concludes that a zigzag structure similar to that for ammonia will be favored energetically over the aligned structure [285]. In either case, it appears that this and related systems may well exhibit long-range ordering of adsorbates.

Evidence for the reverse process, donation of electron density from the nucleophilic dimer atom to an electron-deficient molecule, also exists. Konecny and Doren theoretically found that borane (BH_3) will dissociatively adsorb on Si(100)–2×1 [293]. While much of the reaction is barrierless, they note an interaction between the boron atom and the nucleophilic atom of the Si dimer during the dissociation process. Cao and Hamers have demonstrated experimentally that the electron density of the nucleophilic dimer atom can be donated to the empty orbital of boron trifluoride (BF_3) [278]. XPS on a clean Si(100)–2×1 surface at 190 K indicates that BF_3 dissociates into BF_2(a) and F(a) species. However, when BF_3 is exposed on a Si(100)–2×1 surface previously covered with a saturation dose of trimethylamine, little B–F dissociation occurs, as evidenced by the photoelectron spectrum. They conclude that BF_3 molecularly adsorbs to the nucleophilic dimer atom and DFT calculations indicate that the most energetically favorable product is a surface-mediated donor-acceptor complex (trimethylamine–Si–Si–BF_3) as shown in Figure 5.19.

One goal of organic functionalization of semiconductor surfaces is to achieve selective deposition of a monolayer that still contains sufficient functionality to allow attachment of subsequent layers. In other words, a *reactive* rather than *passivating* monolayer is desired. Among possible applications, the initial monolayer can then be used to grow a multilayer of controlled thickness at the surface through a process called 'molecular layer deposition' (see Section 6.2.4), or used to bind biologically active species to the surface (as for the GaN example discussed in Section 5.2.1). In pursuit of such reactive monolayers, a number of studies have investigated the adsorption of multifunctional molecules, i.e., those containing two or more reactive

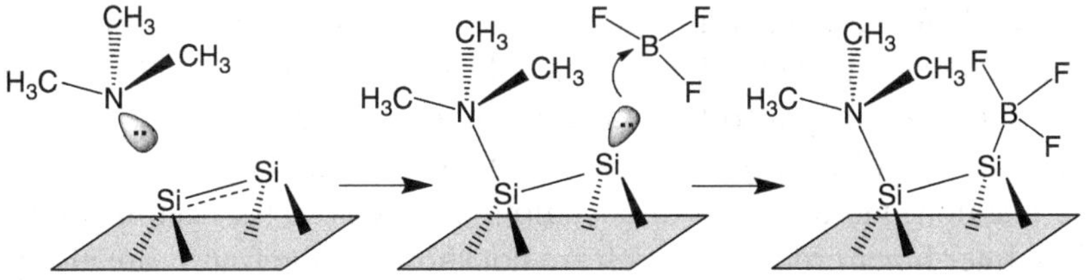

Figure 5.19. Illustration of the formation of a donor–acceptor complex by the sequential adsorption of trimethylamine and boron trifluoride. The Lewis-basic trimethylamine forms a dative bond with the electrophilic Si dimer atom, and the Lewis-acidic boron trifluoride bonds to the nucleophilic Si dimer atom [278].

moieties, at semiconductor surfaces. In addition to enabling multilayer deposition, multifunctional molecules allow fundamental study of the competition between various functional groups to be carried out.

Many studies have shown that competition between different available reaction pathways is observed even for relatively simple multifunctional compounds [16]. Consider, for example, the formation of at least six different [4+2] and [2+2] cycloaddition products observed for the simple cyclic diene (1,3-cyclohexadiene) reported by Teague and Boland [243]. The competition has been attributed to the low barriers observed for many pathways on Si(100)–2 × 1, which results in most reactions at this surface falling under kinetic control. The following explanation has been put forward [16,294]. In a kinetically controlled regime, the product distribution is controlled by the relative kinetic rates of the various reactions; in contrast, in the thermodynamically controlled regime, the product distribution is controlled by the relative thermodynamic stabilities of the products. Often, the thermodynamically controlled regime is reached at temperatures sufficiently high that every activation barrier can be surmounted. However, it has been shown that annealing of covalently bound organics on Si(100)–2 × 1 usually leads to decomposition of the adsorbates rather than conversion to the more thermodynamically stable product [10,15]. Consequently, there is a relatively narrow window (up to ~200 °C) within which most adsorption systems have been studied, and at these temperatures it is generally the relative kinetics of the competing pathways rather than the thermodynamics of the products that determine the final product distribution on Si(100). We note that due to weaker bonding at the germanium surface, the situation is notably different for Ge, as discussed in Section 6.2.

A study by Wang et al. [255] specifically examined the competition and selectivity of organic reactions at Si(100)–2 × 1 and Ge(100)–2 × 1 using unsaturated ketones as a model system. Figure 5.20 shows the possible reaction pathways for reaction of ethylvinyl ketone (EVK) at the Si(100)–2 × 1 surface. EVK is a simple bifunctional molecule which contains conjugated C=C and C=O double bonds. It is clear from Figure 5.20 that a variety of products can possibly form at the surface. These include [2+2] cycloaddition reaction products across either the C=C or C=O double bonds. Additionally, EVK may undergo a [4+2] hetero-Diels-Alder reaction (or a related [4+2] trans cycloaddition) with the surface through its conjugated C=C and C=O bonds. The molecule may also form a dative bond with the surface via donation of the lone-pair charge of its oxygen, or undergo an ene reaction (a group-transfer type pericyclic reaction) via R−C−H dissociation, as observed for acetone on Ge(100)–2 × 1 [294]. Interdimer products, which are not shown in the figure, are also possible.

Ab initio theoretical cluster calculations showed that all of the products, with the exception of the O-dative bonded product, have sufficient binding energies (from ~40–60 kcal/mol) to be thermodynamically stable at room temperature within the timescale of the experiments [255]. Furthermore, most of the activation barriers

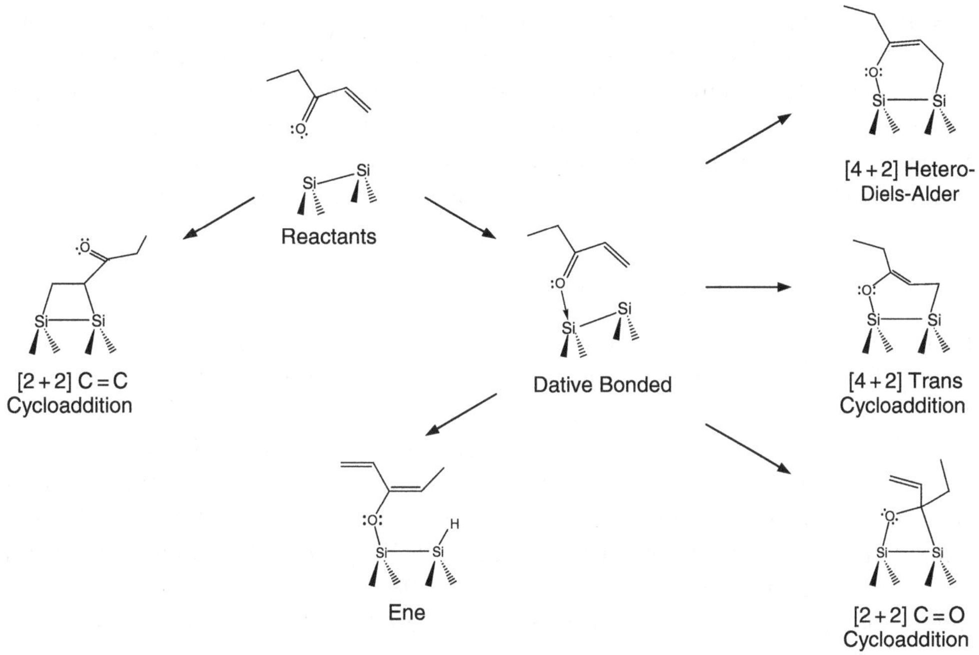

Figure 5.20. Possible reaction pathways for ethylvinylketone on Si(100)–2 × 1. Reproduced with permission from Ref. [255]. Copyright 2002 American Chemical Society.

resided below the energy of the entrance channel. Even the [2 + 2] C=C cycloaddition reaction, which was activated, had a barrier of only ~5 kcal/mol. Consistent with expectation, experimental measurements using infrared spectroscopy showed formation of a mix of products, including the [4 + 2] cycloaddition adducts as well as the ene product and possibly [2 + 2] C=O cycloaddition products [255].

A recent study by Hossain et al. [295] found that the adsorption products formed by reaction of the bifunctional molecule 1-dimethylamine-2-propyne (DMAP) could be switched by changing the temperature. DMAP ($N(CH_3)_2CH_2C{\equiv}CH$) can adsorb either by a dative bond through the nitrogen atom, or by a [2 + 2] cycloaddition reaction through the C≡C triple bond. A combination of electron energy loss spectroscopy (EELS) and STM showed that at low temperature (65–90 K) the molecule reacted selectively through the dative bond, whereas upon annealing to room temperature only the reaction through the C≡C triple bond was observed. Hence, the surface could be selectively functionalized by careful choice of temperature.

6.2. Germanium surface chemistry

The existing literature on reactions at germanium surfaces under vacuum conditions is not as extensive as that of silicon. However, germanium surface chemistry is an actively growing area of research due to the rising interest in alternative semiconductors to replace silicon. In this section, we discuss several molecular systems that provide particular insight into the germanium surface reactivity. These exemplary systems cover the two main classes of dry functionalization reactions: cycloaddition and dative bonding-mediated chemistry. As for silicon, these two classes of reactions dominate the interactions of organic molecules with the germanium surface.

6.2.1. Cycloaddition reactions on Ge(100)–2 × 1

Because of the similarity between the structures of the Ge(100)–2 × 1 and Si(100)–2 × 1 surfaces, cycloaddition products like those observed on Si(100)–2 × 1 are also expected to form at the Ge surface. Indeed, studying butadiene, Teplyakov et al. [240] showed that a similar Diels-Alder product formed on the surface of Ge(100)–2 × 1. Studies of alkenes have also revealed the formation of [2 + 2] cycloaddition products on germanium. For example, cyclopentene has been shown to form the [2 + 2] cycloaddition product on both surfaces [224,296,297]. In further studies of several other dienes and alkenes (including ethylene [298–303], acetylene [304–306], and cyclohexadiene [307]), cycloaddition products were found for Ge(100)–2 × 1 similar to those observed for Si(100)–2 × 1.

However, despite the similarity in reactivity between Si and Ge toward cycloaddition products, important differences between the two surfaces have become evident as the number of cases examined has grown. One such difference is the degree of selectivity that can be achieved between possible reaction products, as will be discussed in Section 6.2.2. Another difference is in the reversibility of the adsorption reaction. For example, upon heating of the germanium surface, the butadiene adsorption product reversibly desorbs, providing evidence for a so-called retro-Diels-Alder reaction [240]. This reversible evolution on Ge is different from the behavior on Si, where the butadiene primarily decomposes instead of desorbing [308,309]. Using DFT, Mui et al. [40] showed that the binding energy of the Diels-Alder product was lower on Ge than on Si; the discrepancy was attributed to differences in the Si−C vs. Ge−C bond strengths; e.g., the Ge−C bonds were 7–9 kcal/mol weaker than the Si−C bonds [40]. This lower binding energy allows desorption to compete effectively with decomposition on the Ge surface, whereas on silicon decomposition dominates at higher temperatures. The increased propensity for molecular desorption over decomposition on Ge compared with Si has been observed in a number of cycloaddition systems. For example, in the adsorption of ethylene and acetylene, both molecules desorb cleanly from Ge(100)–2 × 1, whereas decomposition is reported for the analogous reaction on Si(100)–2 × 1 [298,299,301].

In order to more deeply explore the cycloaddition process on Ge(100)–2 × 1, we will examine the case of ethylene, the smallest alkene. Rather than exhibit straightforward adsorption behavior, as might be expected for such a simple molecule,

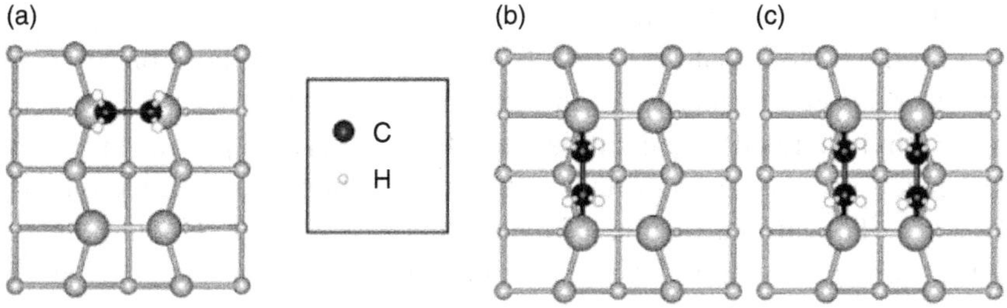

Figure 5.21. Possible bonding configurations of C_2H_4 chemisorbed on the Ge(100) surface; (a) on-top, (b) single end-bridge, and (c) paired end-bridge configurations. Figure reproduced with permission from Ref. [301]. Copyright 2004 American Chemical Society.

ethylene undergoes complex adsorption chemistry. The simplest adsorption product for ethylene on Ge(100)–2×1 is a $[2+2]$ cycloaddition product across two germanium atoms in a single dimer, as shown in Figure 5.21(a). The first experimental studies on ethylene/Ge(100)–2×1, carried out by Lal et al. [298] using IR spectroscopy and TPD, demonstrated vibrational spectra consistent with a $[2+2]$ cycloaddition product in which the double bond of ethylene is broken and two σ-bonds to surface Ge atoms are formed. However, TPD measurements of this system by the same group showed two clear molecular desorption peaks, indicating that not just one, but at least two major molecular adsorption states were formed. Furthermore, both the TPD and IR measurements showed time-dependent behavior, in which changes in the adsorption states occurred at the surface over time [298].

A subsequent study using synchrotron-based angle-resolved UPS and TPD confirmed the molecular adsorption of ethylene on Ge via di-σ bonding. While that study attributed one of the two states to adsorption at minority step sites rather than a second major adsorption species [299], a recent STM investigation of ethylene on Ge(100)–2×1 confirmed the presence of two distinct bonding geometries stable at room temperature at the surface [301], consistent with the earlier TPD study. The STM images show two principal features that are interpreted as the $[2+2]$ product across a single Ge dimer (Figure 5.21(a)) as well as a structure that bridges two neighboring Ge dimers within the same dimer row (called an end-bridge structure), shown in Figure 5.21(b). The first structure is an intradimer adduct, whereas the second is an interdimer adduct.

Interestingly, the authors found that the interdimer product never formed as a single end-bridge structure; instead, the ethylene molecules in this configuration always paired up as shown in Figure 5.21(c) [301]. Thermodynamically, the paired end-bridge configuration is expected to be more favorable than two single end bridges because placing a second adsorbed ethylene across two dimers that have already lost the π-bond interaction from the first adsorption event costs less energy than breaking two new dimer π-bonds at a new location. The STM data suggest that the paired end-bridge structure is more stable than the on-top $[2+2]$ cycloaddition product and that the higher temperature desorption peak in TPD arises from this interdimer product. Note that these results are different from those on Si(100)–2 × 1, where ethylene is seen primarily as an intradimer product.

From the ethylene results, and similar results on acetylene [306], it is evident that interdimer reactions play an important role in the chemistry of organic molecules on Ge(100)–2 × 1. The simple picture of reaction across a single Ge–Ge dimer, while capturing a number of important reaction pathways, is incomplete. Even small C_2 molecules such as ethylene and acetylene can bridge across dimers along a dimer row. Other molecules are found to bridge across the wider trench. Furthermore, these studies indicate that multiple reaction products can form even for simple systems.

Larger unsaturated hydrocarbons have also been studied on Ge(100)–2 × 1. Examples include cyclopentene [224,297], cyclohexene [297], cyclohexadiene [307], and butadiene [40,240]. In addition, the reactivity of the aromatic compounds benzene [310] and styrene [311,312] toward Ge have also been investigated. The alkenes (cyclopentene and cyclohexene) behave similarly to ethylene in that they form a di-σ-bonded product at the surface. As on Si(100)–2 × 1, cyclopentene forms a [2+2] product across the Ge dimer. However, STM shows that, although there are regions of partial ordering at the surface, the degree of ordering is not as high as that on Si(100)–2 × 1. The coverage at saturation is approximately two cyclopentene molecules per three germanium dimers [296,297].

In addition, the sticking probability of cyclopentene on germanium is significantly lower than that on silicon. TPD measurements as a function of exposure estimate the sticking probability of cyclopentene on Ge(100)–2 × 1 as 0.1, a factor of 10 less than that on silicon [297]. According to Lee et al., the lower sticking probability may arise from the slightly larger Ge=Ge dimer bond length (0.1–0.2 Å longer than Si=Si [40]), which leads to a larger activation barrier for adsorption. This explanation assumes a mechanism whereby the [2 + 2] cycloaddition product is not formed in a concerted reaction, but rather occurs in a two-step process in which the cyclopentene π-bond interacts with the electrophilic down Ge atom before moving across the dimer to form the final four-membered ring product [297]. In the case of Ge, the longer dimer bond length may make the reaction more highly activated. In a recent theoretical investigation of the cyclopentene/Ge(100)–2 × 1 system [224], the reaction pathway through this 'three-atom' intermediate state was studied. Cho and Kleinman [224] found that the reaction going through the precursor state was more highly activated for the case of Ge than for Si. Similar reaction pathways are expected to apply to the adsorption of other alkenes on Ge(100)–2 × 1.

6.2.2. Heterocycloadditions

The adsorption of several heteronuclear unsaturated organics, including ketones (RC=O), nitriles (RC≡N), and isocyanates (RN=C=O), has been studied on the germanium surface. These compounds have been found to undergo a range of pericyclic reactions on Ge(100)–2 × 1, including group transfer, hetero-Diels-Alder cycloaddition, and [2+2] cycloaddition. Most of these products have also been observed on Si(100)–2 × 1. Although only a few molecules in this category have been studied on germanium, some key themes have already emerged. For example, as also observed for silicon, the presence of a heteroatom such as O or N drives the reaction pathway through a dative-bonded precursor state. However, the products of adsorption of the heteronuclear compounds formed on Ge(100)–2 × 1 often differ significantly from those formed on Si(100)–2 × 1. Furthermore, the reaction product distribution can be more selective on Ge than on Si. The higher selectivity is attributed to the prevalence of thermodynamic control in reactions on Ge, whereas on silicon, most reactions are under kinetic control. The increased thermodynamic control on Ge arises from the weaker bonds that organics form with Ge compared with Si, allowing for reversibility and interconversion between products. Typically, reactions on the Si(100)–2 × 1 surface have lower activation barriers and are more exothermic than those on Ge(100)–2 × 1 [313].

Of the heteronuclear molecules, we will focus on ketones because they provide an excellent illustration of the differences between the reactivity on Ge and Si. Ketones that have been studied on Ge(100)–2 × 1 include acetone and a series of vinylketones. It was found through a combination of IR spectroscopy and DFT that acetone undergoes attachment at the carbonyl oxygen with loss of an α hydrogen at room temperature on the Ge(100)-2 × 1 surface to form an enol-like adduct [294]. The resulting adduct can also be characterized as the product of an 'ene' reaction. This 'enolization' reaction was not seen on the Si(100) surface in which a [2+2] addition across the carbonyl bond is observed instead [254]. To understand the difference between the reactivity on silicon and germanium, DFT calculations were used to map out the potential energy curves for the reaction. The DFT calculations showed that on both Si(100)–2 × 1 and Ge(100)−2 × 1 surfaces, the [2+2] C=O addition reaction is kinetically favored, whereas the enolization reaction is thermodynamically favored [294]. Thus, the formation of the enol product at the expense of the [2+2] C=O product indicated that at room temperature, the reaction of acetone on the Ge(100)–2 × 1 surface was under thermodynamic control, in which the kinetically favored product was so weakly bound that it was thermodynamically unstable on the timescale of the experiment. In contrast, on Si(100)–2 × 1, the kinetically favored product was sufficiently stable and hence was observed as the majority product. The strategic use of thermodynamically controlled reactions on the Ge(100)–2 × 1 surface to create new organic-semiconductor interfaces that are kinetically uncompetitive on the Si(100)–2 × 1 surface was proposed as a new approach for the selective formation of organic-functionalized semiconductor surfaces [294].

In the related system of ethylvinylketone (EVK) on Ge(100)–2 × 1, the energetics on germanium provide for greater selectivity toward reaction products than those on silicon [255]. As described in Section 6.1.4, EVK can, in principle, form a number of reaction products on either surface, including the [2 + 2] C=C cycloaddition product, the [2 + 2] C=O cycloaddition product, the ene reaction product (as was seen for acetone), or the [4 + 2] cycloaddition products. On silicon, experiments showed evidence for the hetero-Diels-Alder product as well as the ene and [2 + 2] C=O cycloaddition products. However, on germanium, the [4 + 2] hetero-Diels-Alder product forms selectively [255]. This increased selectivity supports the suggestion that Ge may be a superior material for cases where selective attachment of organic compounds onto semiconductor surfaces is required.

6.2.3. Nucleophilic/electrophilic reactions

The participation of the germanium dimers in nucleophilic/electrophilic or Lewis acid/base reactions has been the subject of several investigations on the Ge(100)–2×1 surface [16,49,255,288,294,313–318]. As for the case of silicon, adsorption of amines has provided an excellent system for probing such reactions. Amines contain nitrogen lone pair electrons that can interact with the electrophilic down atom of a tilted Ge dimer to form a dative bond via a Lewis acid/base interaction (illustrated for trimethylamine at the Si(100)–2×1 surface in Figure 5.17). In the dative bond, the lone pair electrons on nitrogen donate charge to the Ge down atom [49].

In the case of primary and secondary amines, the complex has the possibility of undergoing N−H dissociation from the adsorbed state; this pathway was observed on Si(100)–2×1 at room temperature for both mono- and dimethylamine [279]. However, for Ge(100)–2×1, Mui et al. [49] found that N−H cleavage did not occur for any of the methylamines: methylamine, dimethylamine, and trimethylamine all remained datively bonded at the surface. Theoretical calculations by the same authors showed that there was a larger activation barrier for the N−H cleavage reaction on Ge compared with that on Si. Recall that in the studies on Si(100)–2×1, the N−H dissociation pathway was interpreted as the transfer of a proton from the ammonium ion to the electron-rich up Si atom through a Lewis acid-base reaction [279]. The difference in chemical reactivities of the Ge(100)–2×1 and Si(100)–2×1 surfaces toward N−H cleavage, therefore, was interpreted as a decrease of proton affinity down a group in the periodic table, with Ge having a lower proton affinity than Si [49].

A related study of ethylenediamine at the Ge(100)–2×1 surface provides further insight into the role of dative bonding in amine compounds. Using a combination of infrared spectroscopy and density functional theory [319], it was found that unlike simple alkylamines, which adsorb on Ge(100)–2×1 through a simple dative bond, ethylenediamine formed different surface-bound products depending on the coverage. At higher coverage (near saturation), adsorption took place predominantly through dative bonding, as observed for the methylamine series. However, at low coverage, the dominant reaction was dual N−H dissociation from both amine groups of ethylenediamine, leading to a species tethered to the Ge surface through both nitrogen atoms. The presence of N−H dissociation at low coverage is an interesting deviation from the related methylamine studies, in which dissociation was not observed. This observation was explained by a shift in the energetics of the N−H dissociation reaction when both amine groups are datively bonded to neighboring dimers at the surface, yielding a lower dissociation activation barrier. Consequently, the dual dative bonded state acts as a precursor toward subsequent N−H dissociation at both nitrogens. At higher coverage, however, there is a decrease in the number of adjacent dimers available, which may shift the reaction product to that of a species dative-bonded through only one nitrogen. The barrier for dissociation from the single dative bonded state is too high [319].

Pyridine, a six-membered cyclic aromatic amine, has also been studied on Ge(100)–2 × 1 both theoretically [315,316] and experimentally by STM [314]. It adsorbs selectively through a Ge−N dative bond on the surface. Theoretical calculations showed that the dative-bonded adduct is more stable than other possible reaction products (e.g., cycloaddition products) on Ge [315,316]. Furthermore, STM images show formation of a highly ordered monolayer at the surface with a coverage of 0.25 ML. The pyridine overlayer forms a c(4 × 2) structure in which the molecules bind to the down atoms of every other dimer to minimize repulsive interactions between pyridine molecules.

Whereas pyridine adsorbs selectively through a nitrogen dative bond to the Ge(100)–2 × 1 surface, pyrrole forms a mix of dissociative reaction products at the surface. In the adsorption of pyrrole, which is a five-membered cyclic aromatic amine, the main reaction product was an N−H dissociation product [288]. The difference between the reactivity of pyrrole and the alkylamines was attributed to the fact that in pyrrole, the nitrogen 'lone pair' is delocalized over the ring and comprises part of the aromatic π system. Consequently, the adsorption energy of the N dative-bonded state is weakened by loss of the resonance energy. For example, on Ge(100)–2 × 1, binding energy of the dative-bonded state was calculated to be ~1 kcal/mol [288]. Interestingly, Wang et al. [288] noted that N−H dissociation would likely occur through an alternative pathway involving, not the weak N dative bond, but a dative bond at a ring carbon. This C dative-bonded state is stabilized by the aromaticity of pyrrole.

A number of other studies have now shown that dative bonding is a phenomenon common to many organic reactions on Ge(100)–2 × 1, as it is for Si(100)–2 × 1. In some cases, e.g., with the methylamines and pyridine, the dative-bonded state is the final surface species. This dative-bonded state can be quite stable. For example, the nitrogen dative bonds formed via exposure of methylamines to Ge(100)–2 × 1 have binding energies near 25 kcal/mol [49]. The STM study of pyridine on Ge(100)–2 × 1 revealed that 90% of the dative-bonded surface adducts remain after one day at room temperature under UHV [314]. Oxygen can also form a dative bond. For oxygen dative bonds formed by the interaction of acetone with the germanium surface, the binding energy was calculated as approximately 12 kcal/mol, which, although too low for the dative bond to be stable at room temperature, was high enough to observe the dative-bonded state at 115 K [294]. Furthermore a recent study of amides (RCONR′R″) has shown that tertiary amides adsorb through an oxygen dative bond at room temperature on Ge(100)–2 × 1, and assigned a binding energy of approximately 18 kcal/mol to this product. At this strength of binding, the dative-bonded product, though observable, desorbed over the course of minutes during room temperature experiments [320].

In other cases, the dative-bonded state acts as a precursor for formation of more thermodynamically stable products. Whether the reaction progresses toward other products depends on the stability of the product and the size of any activation barrier between the dative-bonded precursor state and the other product(s). Cases where the dative-bonded state is a precursor to the final product include the heterocycloadditions (discussed in Section 6.2.2) in addition to pyrrole. Another case is that of secondary amides. A new study has shown that in the reaction of N-methylformamide, the oxygen-dative-bonded state is a precursor to N−H dissociation, via a cyclic species shown in Figure 5.22. At room temperature, a mixture of the dative-bonded product and the N−H dissociation product are observed. However, the product distribution can be tuned using thermal control: at low temperatures (240 K), only the dative bonded species was observed, whereas upon annealing to higher temperatures (450 K), only the N−H dissociation product was found [321].

6.2.4. Multiple-layer reactions

One of the ultimate goals in organic functionalization is to be able to control deposition of multiple organic layers so that precisely tailored surfaces may be

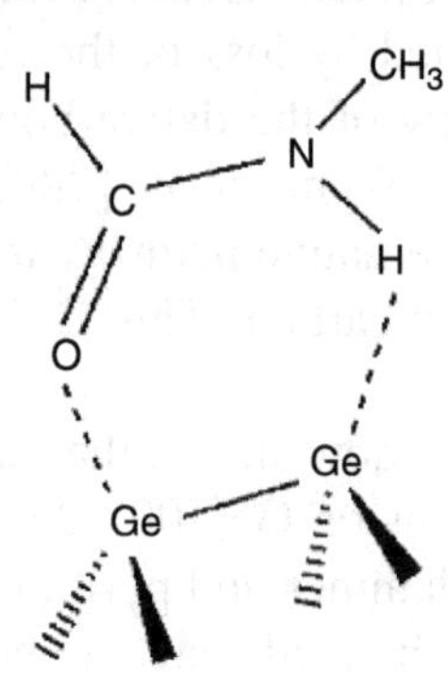

Figure 5.22. Structure of the O-dative bonded adduct formed by adsorption of N-methylformamide on Ge(100)–2 × 1, showing the presence of a Ge−H interaction, which stabilizes the dative bonded state [321].

prepared with a variety of useful functionalities. Such molecular-level control is currently achievable in solution with well-known chemistries including Langmuir–Blodgett film deposition [322,323] and the creation of self-assembled monolayers (SAMs) [322,324]. In contrast, vacuum-based growth methods are in their infancy. Some initial work on Si(100) demonstrated the growth of an ultrathin organic film on Si(100)–2 × 1 via sequential imide coupling reactions with thermal curing at intermediate steps [325,326]. Building upon the work carried out on bifunctional organic reactants, a multiple-layer organic functionalization system was recently demonstrated on Ge(100)–2 × 1 using a combination of a diamine and a diisocyanate to form urea linkages at room temperature [318]. The idealized multiple layer film is illustrated in Figure 5.23.

The study was carried out using initial attachment of ethylenediamine (ED) at the Ge(100)–2 × 1 surface. As previously discussed, ED reacts to form an

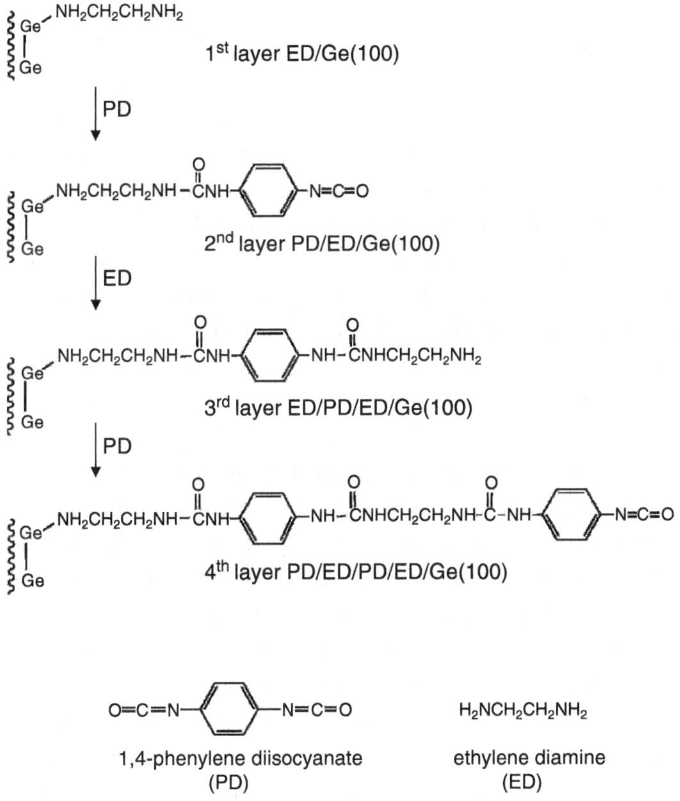

Figure 5.23. Schematic illustration of the synthetic route for the layer-by-layer growth of polyurea films on the Ge(100)–2 × 1 surface. Reproduced with permission from Ref. [318]. Copyright 2005 American Chemical Society.

N–H-dissociated product at low coverage and a dative-bonded product at high coverage [319]. Both products leave NH_x groups at the surface. These groups, in turn, can react with an isocyanate moiety in phenylene diisocyanate (PD) to form a urea bond. Kim et al. [318] studied the incremental changes in measured IR vibrational spectra following alternating exposure of PD and ED to the ED-functionalized surface. The IR spectra showed consecutive formation of urea linkages, as detected by three new IR bands characteristic of amide bonds and a urea linkage [318]. The spectra confirm the reaction between phenylene diisocyanate and ethylenediamine to form urea bonds, and the polyurea coupling reaction was demonstrated for up to four molecular layers at room temperature. The work established for the first time that the binary exposure of two bifunctional precursors at room temperature could be used to form an ultrathin film on Ge(100) under vacuum conditions. Such a molecular layer deposition approach holds promise for controllable functionalization of Ge surfaces.

6.3. Summary of concepts in organic functionalization

The study of organic functionalization of semiconductors, in which organic molecules are adsorbed onto the surface in vacuum, is providing new insight into the reactivity of the reconstructed semiconductor surfaces. Several common themes have emerged in our understanding of the reactivity of both silicon and germanium based on these studies. One motif is the prevalence of cycloaddition and dative bonding among the products and reaction pathways of many molecules studied. The observation of these products fits with the description of the Si or Ge dimers on the Si(100)–2×1 or Ge(100)–2×1 surface as containing both weak double bonds and significant nucleophilic and electrophilic character. Interestingly, despite the large differences in structure, similar products have been reported on Si(111)–7×7, suggesting similarities in the chemical properties at that surface.

Furthermore, the organic functionalization studies have indicated that multiple reaction products can form even for simple systems. Kinetic and thermodynamic influences must be considered in any analysis of the product distribution. Moreover, the studies have revealed differences in the dominance of kinetic vs. thermodynamic control between the silicon and germanium surfaces. The dissimilarity primarily stems from the fact that adsorbate bonds are usually weaker on Ge than on Si. This difference in energetics leads to observable differences in the degree of selectivity that can be achieved on the two surfaces. Another important motif is the significance of interdimer bonding in the products. Many molecules, even as small as ethylene, have been observed to form products that bridge across two dimers. Consequently, each analysis of adsorption products should include consideration of interdimer as well as intradimer species.

Finally, the level of understanding of the behavior of silicon and germanium surfaces toward organic reagents has increased substantially over the past decade. This understanding, in turn, will enable the next generation of studies, in which organic functionalization can be achieved selectively and controllably for a variety of new applications.

Acknowledgments

The author expresses appreciation for the many fruitful interactions with students and colleagues that contributed greatly to this review. This Chapter builds upon previous review articles written by the author and two very talented coauthors, Dr. Michael Filler and Paul Loscutoff. SFB is indebted to both of them for their contributions. The National Science Foundation (Grants CHE 0245260 and CHE 0615087) is gratefully acknowledged for financial support that made this review possible.

References

[1] Goodman, D. W. Correlations between surface science models and 'Real-World' catalysts. Journal of Physical Chemistry 100, 13090 (1996).
[2] Bent, B. E. Mimicking aspects of heterogeneous catalysis: Generating, isolating, and reacting proposed surface intermediates on single crystals in vacuum. Chemical Reviews 96, 1361 (1996).
[3] Ma, Z. and Zaera, F. Organic chemistry on solid surfaces. Surface Science Reports 61, 229 (2006).
[4] Meuris M., 2003. High k strides reopen door to germanium. EE Times 8/22/2003.
[5] Duke, C. B. Semiconductor surface reconstruction: The structural chemistry of two-dimensional surface compounds. Chemical Reviews 96, 1237–1259 (1996).
[6] Kubby, J. A. and Boland, J. J. Scanning tunneling microscopy of semiconductor surfaces. Surface Science Report 26, 61 (1996).
[7] Boland, J. J. Scanning tunnelling microscopy of the interaction of hydrogen with silicon surfaces. Advances in Physics 42, 129 (1993).
[8] Engel, T. The interaction of molecular and atomic oxygen with Si(100) and Si(111). Surface Science Reports 18, 91 (1993).
[9] Oura, K., Lifshits, V. G., Saranin, A. A., Zotov, A. V. and Katayama, M. Hydrogen interaction with clean and modified silicon surfaces. Surface Science Reports 35, 1 (1999).
[10] Hamers, R. J., Coulter, S. K., Ellison, M. D., Hovis, J. S., Padowitz, D. F., Schwartz, M. P., Greenlief, C. M. and Russell, J. N. Cycloaddition chemistry of organic molecules with semiconductor surfaces. Accounts of Chemical Research 33, 617 (2000).
[11] Wolkow, R. A. Controlled molecular adsorption on silicon: Laying a foundation for molecular devices. Annual Review of Physical Chemistry 50, 413–441 (1999).
[12] Hamers, R. J., Hovis, J. S., Coulter, S. K., Ellison, M. D. and Padowitz, D. F. Ultrathin organic layers on silicon surfaces. Japanese Journal of Applied Physics, Part 1 39, 4366 (2000).

[13] Buriak, J. M. Organometallic chemistry on silicon and germanium surfaces. Chemical Reviews 102, 1271–1308 (2002).

[14] Bent, S. F. Organic functionalization of group IV semiconductor surfaces: principles, examples, applications, and prospects. Surface Science 500, 879–903 (2002).

[15] Bent, S. F. Attaching organic layers to semiconductor surfaces. Journal of Physical Chemistry B 106, 2830–2842 (2002).

[16] Filler, M. A. and Bent, S. F. The surface as molecular reagent: organic chemistry at the semiconductor interface. Progress in Surface Science 73, 1–56 (2003).

[17] Loscutoff, P. W. and Bent, S. F. Reactivity of the germanium surface: Chemical passivation and functionalization. Annual Review of Physical Chemistry 57, 467 (2006).

[18] Yoshinobu, J. Physical properties and chemical reactivity of the buckled dimer on Si(100). Progress in Surface Science 77, 37 (2004).

[19] Tao, F. and Xu, G. Q. Attachment chemistry of organic molecules on Si(111)–7×7. Accounts of Chemical Research 37, 882–893 (2004).

[20] Wayner, D. D. M. and Wolkow, R. A. Organic modification of hydrogen terminated silicon surfaces. Journal of the Chemical Society-Perkin Transactions 2, 23 (2002).

[21] Weldon, M. K., Queeney, K. T., Eng, J., Raghavachari, K. and Chabal, Y. J. The surface science of semiconductor processing: gate oxides in the ever-shrinking transistor. Surface Science 500, 859 (2002).

[22] Hines, M. A. In search of perfection: understanding the highly defect-selective chemistry of anisotropic etching. Annual Review of Physical Chemistry 54, 29 (2003).

[23] Brenig, W. and Hilf, M. F. Reaction dynamics of H_2 and D_2 on Si(100) and Si(111). Journal of Physics: Condensed Matter 13, R61 (2001).

[24] Schlier, R. E. and Farnsworth, H. E. Structure and adsorption characteristics of clean surfaces of germanium and silicon. Journal of Chemical Physics 30, 917 (1959).

[25] Duke, C. B. Surface structures of tetrahedrally coordinated semiconductors: principles, practice, and universality. Applied Surface Science 65–66, 543–552 (1993).

[26] Duke, C. B. Structure and bonding of tetrahedrally coordinated compound semiconductor cleavage faces. Journal of Vacuum Science & Technology A 10, 2032 (1992).

[27] Schmidt, W. G. III-V compound semiconductor (001) surfaces. Applied Physics A: Materials Science and Processing 75, 89 (2002).

[28] Xue, Q. K., Hashizume, T. and Sakurai, T. Scanning tunneling microscopy of III-V compound semiconductor (001) surfaces. Progress in Surface Science 56, 1 (1997).

[29] LaBella, V. P., Krause, M. R., Ding, Z. and Thibado, P. M. Arsenic-rich GaAs(0 0 1) surface structure. Surface Science Reports 60, 1 (2005).

[30] Mui, C. Growth and Functionalization of Electronic Materials. PhD Thesis, Stanford University, 2003.

[31] Waltenburg, H. N. and Yates, J. T. Surface chemistry of silicon. Chemical Reviews 95, 1589 (1995).

[32] Wolkow, R. A. Direct observation of an increase in buckled dimers on Si(001) at low temperature. Physical Review Letters 68, 2636 (1992).

[33] Weakliem, P. C. and Carter, E. A. Constant temperature molecular dynamics simulations of Si(100) and Ge(100): equilibrium structure and short-time behavior. Journal of Chemical Physics 96, 3240 (1992).

[34] Hamers, R. J., Tromp, R. M. and Demuth, J. E. Electronic and geometric structure of Si(111)–(7×7) and Si(001) surfaces. Surface Science 181, 346 (1987).

[35] D'Evelyn, M. P., Yang, Y. L. and Sutcu, L. F. Pi-bonded dimers, preferential pairing, and first-order desorption kinetics of hydrogen on Si(100)–(2×1). Journal of Chemical Physics 96, 852 (1992).

[36] Wu, C. J. and Carter, E. A. Adsorption of hydrogen atoms on the Si(100)–2 × 1 surface: implications for the H_2 desorption mechanism. Chemical Physics Letters 185, 172 (1991).
[37] D'Evelyn, M. P., Cohen, S. M., Rouchouze, E. and Yang, Y. L. Surface π bonding and the near-first-order desorption kinetics of hydrogen from Ge(100)2 × 1. Journal of Chemical Physics 98, 3560–3 (1993).
[38] Flowers, M. C., Jonathan, N. B. H., Yong, L. and Morris, A. Temperature programmed desorption of molecular hydrogen from a Si(100)–2 × 1 surface: theory and experiment. Journal of Chemical Physics 99, 7038 (1993).
[39] Hofer, U., Leping, L. and Heinz, T. F. Desorption of hydrogen from Si(100)2 × 1 at low coverages: the influence of pi -bonded dimers on the kinetics. Physical Review B 45, 9485 (1992).
[40] Mui, C., Bent, S. F. and Musgrave, C. B. A theoretical study of the structure and thermochemistry of 1,3-butadiene on the Ge/Si(100)–2 × 1 surface. Journal of Physical Chemistry A 104, 2457–2462 (2000).
[41] Nachtigall, P., Jordan, K. D. and Sosa, C. Ab initio calculation of the energy of recombinative hydrogen desorption from the monohydride phase of Si(100). Journal of Physical Chemistry 97, 11666 (1993).
[42] Carey, F. A. and Sundberg, R. J. Advanced Organic Chemistry (Plenum Publishing Corp., 2001).
[43] Konecny, R. and Doren, D. J. Adsorption of water on Si(100)–(2 × 1): A study with density functional theory. Journal of Chemical Physics 106, 2426–2435 (1997).
[44] Takayanagi, K. T., Y.; Takahashi, S.; Takahashi, M. Surface Science 164, 367 (1985).
[45] Tromp, R. M. H., R. J.; Demuth, J. E. Physical Review B 34, 1388 (1986).
[46] Cao, Y., Yong, K. S., Wang, Z. H., Deng, J. F., Lai, Y. H. and Xu, G. Q. Cycloaddition chemistry of thiophene on the silicon (111)–7 × 7 surface. Journal of Chemical Physics 115, 3287 (2001).
[47] Cao, X. P. and Hamers, R. J. Silicon surfaces as electron acceptors: Dative bonding of amines with Si(001) and Si(111) surfaces. Journal of the American Chemical Society 123, 10988 (2001).
[48] Rousseau, G. B. D., Dhanak, V. and Kadodwala, M. Photoemission studies of the surface reactivity of thiophene on Si(100)–(2 × 1), Si(111)–(7 × 7) and Ge(100)–(2 × 1). Surface Science 494, 251–264 (2001).
[49] Mui, C., Han, J. H., Wang, G. T., Musgrave, C. B. and Bent, S. F. Proton transfer reactions on semiconductor surfaces. Journal of the American Chemical Society 124, 4027–4038 (2002).
[50] Zandvliet, H. J. W. The Ge(001) surface. Physics Reports 388, 1–40 (2003).
[51] Becker, R. S., Golovchenko, J. A. and Swartzentruber, B. S. Tunneling images of germanium surface reconstructions and phase boundaries. Physical Review Letters 54, 2678–80 (1985).
[52] Mercer, J. L. and Chou, M. Y. Energetics of the Si(111) and Ge(111) surfaces and the effect of strain. Physical Review B 48, 5374–85 (1993).
[53] Abe, M., Sugimoto, Y. and Morita, S. Imaging the restatom of the Ge(111)–c(2 × 8) surface with noncontact atomic force microscopy at room temperature. Nanotechnology 16, S68–S72 (2005).
[54] Plummer, J. D., Griffin, P. B. and Deal, M. D. Silicon VLSI Technology: Fundamentals, Practice, and Modeling (Prentice Hall, 2000).
[55] Campbell, S. A. The Science and Engineering of Microelectronic Fabrication (Oxford, New York, 2001).
[56] Deal, B. E. and Grove, A. S. General relationship for the thermal oxidation of silicon. Journal of Applied Physics 36, 3770 (1965).

[57] Hollinger, G. and Himpsel, F. J. Oxygen chemisorption and oxide formation on Si (111) and Si (100) surfaces. Journal of Vacuum Science & Technology A 1, 640-645 (1982).
[58] Hollinger, G. and Himpsel, F. J. Probing the transition layer at the SiO_2-Si interface using core level photoemission. Applied Physics Letters 44, 93–5 (1984).
[59] Kern, W. and Puotinen, D. A. Cleaning solutions based on hydrogen peroxide for use in silicon semiconductor technology. RCA Review 31, 187–206 (1970).
[60] Ibach, H., Wagner, H. and Bruchmann, D. Dissociative chemisorption of H_2O on Si(100) and Si(111): a vibrational study. Solid State Communications 42, 457–459 (1982).
[61] Chabal, Y. J. Hydride formation on the Si(100)–H_2O surface. Physical Review B 29, 3677–3680 (1984).
[62] Chabal, Y. J. and Christman, S. B. Evidence of dissociation of water on the Si(100)2 × 1 surface. Physical Review B 29, 6974–6976 (1984).
[63] Stucki, F., Anderson, J., Lapeyre, G. J. and Farrell, H. H. Multiple vibrational excitations of H_2O and D_2O on Si(100)(2 × 1): a HREELS study. Surface Science 143, 84–92 (1984).
[64] Gao, Q., Dohnalek, Z., Cheng, C. C., Choyke, W. J. and Yates, J. T. Direct images of isotropic and anisotropic vibrations in the Cl-Si and H-O-Si chemisorption bonds on Si(100). Surface Science 312, 261–270 (1994).
[65] Johnson, A. L., Walczak, M. M. and Madey, T. E. ESDIAD of 1st-row protic hydrides adsorbed on Si(100): structure and reactivity. Langmuir 4, 277–282 (1988).
[66] Larsson, C. U. S., Johnson, A. L., Flodstrom, A. and Madey, T. E. Adsorption of H_2O on planar and stepped Si(100): structural aspects. Journal of Vacuum Science & Technology A 5, 842–846 (1987).
[67] Schulze, R. K. and Evans, J. F. Room-temperature water-adsorption on the Si(100) surface examined by UPS; XPS; and static SIMS. Applied Surface Science 81, 449–463 (1994).
[68] Zhou, X. L., Flores, C. R. and White, J. M. Adsorption and decomposition of water on Si(100): a TPD and SSIMS study. Applied Surface Science 62, 223-237 (1992).
[69] Flowers, M. C., Jonathan, N. B. H., Morris, A. and Wright, S. The adsorption and reactions of water on Si(100)–2 × 1 and Si(111)-7 × 7 surfaces. Surface Science 351, 87–102 (1996).
[70] Stefanov, B. B., Gurevich, A. B., Weldon, M. K., Raghavachari, K. and Chabal, Y. J. Silicon epoxide: unexpected intermediate during silicon-oxide formation. Physical Review Letters 81, 3908–3911 (1998).
[71] Weldon, M. K., Queeney, K. T., Chabal, Y. J., Stefanov, B. B. and Raghavachari, K. Mechanistic studies of silicon oxidation. Journal of Vacuum Science & Technology B 17, 1795–1802 (1999).
[72] Weldon, M. K., Queeney, K. T., Gurevich, A. B., Stefanov, B. B., Chabal, Y. J. and Raghavachari, K. Si-H bending modes as a probe of local chemical structure: Thermal and chemical routes to decomposition of H_2O on Si(100)–(2 × 1). Journal of Chemical Physics 113, 2440–2446 (2000).
[73] Weldon, M. K., Stefanov, B. B., Raghavachari, K. and Chabal, Y. J. Initial H_2O-induced oxidation of Si(100)–(2 × 1). Physical Review Letters 79, 2851–2854 (1997).
[74] Stefanov, B. B. and Raghavachari, K. Pathways For initial water-induced oxidation of Si(100). Applied Physics Letters 73, 824–826 (1998).
[75] Kingston, R. H. Review of germanium surface phenomena. Journal of Applied Physics 27, 101–114 (1956).
[76] Prabhakaran, K. and Ogino, T. Oxidation of Ge(100) and Ge(111) surfaces: an UPS and XPS study. Surface Science 325, 263–71 (1995).
[77] Schmeisser, D., Schnell, R. D., Bogen, A., Himpsel, F. J., Rieger, D., Landgren, G. and Morar, J. F. Surface oxidation states of germanium. Surface Science 172, 455–65 (1986).

[78] Cho, J. W. and Nemanich, R. J. Surface electronic states of low-temperature H-plasma-exposed Ge(100). Physical Review B 46, 12421-6 (1992).

[79] Deegan, T. and Hughes, G. An X-ray photoelectron spectroscopy study of the HF etching of native oxides on Ge(111) and Ge(100) surfaces. Applied Surface Science 123/124, 66–70 (1998).

[80] Hovis, J. S., Hamers, R. J. and Greenlief, C. M. Preparation of clean and atomically flat germanium(001) surfaces. Surface Science 440, L815–19 (1999).

[81] Kim, J., Saraswat, K. and Nishi, Y. Study of germanium surface in wet chemical solutions for surface cleaning applications, ECS Transactions 1, 214–219 (2005).

[82] Ma, Q., Moldovan, N., Mancini, D. C. and Rosenberg, R. A. Synchrotron-radiation-induced wet etching of germanium. Applied Physics Letters 81, 1741–1743 (2002).

[83] Okumura, H., Akane, T. and Matsumoto, S. Carbon contamination free Ge(100) surface cleaning for MBE. Applied Surface Science 125, 125–8 (1998).

[84] Prabhakaran, K., Ogino, T., Hull, R., Bean, J. C. and Peticolas, L. J. An efficient method for cleaning Ge(100) surface. Surface Science 316, L1031–3 (1994).

[85] Zhang, X. J., Xue, G., Agarwal, A., Tsu, R., Hasan, M. A., Greene, J. E. and Rockett, A. Thermal desorption of ultraviolet-ozone oxidized Ge(001) for substrate cleaning. Journal of Vacuum Science & Technology A 11, 2553–61 (1993).

[86] Zhang, X. J., Xue, G., Agarwal, A., Tsu, R., Hasan, M. A., Greene, J. E. and Rockett, A. Thermal desorption of ultraviolet-ozone oxidized Ge(001) for substrate cleaning. Journal of Vacuum Science & Technology A 11, 2553–61 (1993).

[87] Chan, L. H., Altman, E. I. and Liang, Y. Development of procedures for obtaining clean, low-defect-density Ge(100) surfaces. Journal of Vacuum Science & Technology A 19, 976–81 (2001).

[88] Weast, R. C. (ed.) CRC Handbook of Chemistry and Physics (CRC Press, Boca Raton, 1983).

[89] Ubara, H., Imura, T. and Hiraki, A. Formation of Si-H bonds on the surface of microcrystalline silicon covered with SiO_x by HF treatment. Solid State Communications 50, 673 (1984).

[90] Higashi, G. S., Chabal, Y. J., Trucks, G. W. and Raghavachari, K. Ideal hydrogen termination of the Si(111) surface. Applied Physics Letters 56, 656 (1990).

[91] Dumas, P., Chabal, Y. J. and Jakob, P. Morphology of hydrogen-terminated Si(111) and Si(100) surfaces upon etching in HF and buffered-HF solutions. Surface Science 269/270, 867 (1992).

[92] Weser, T., Bogen, A., Konrad, B., Schnell, R. D., Schug, C. A. and Steinmann, W. Photoemission surface core-level study of sulfur adsorption on Ge(100). Physical Review B 35, 8184–8 (1987).

[93] Weser, T., Bogen, A., Konrad, B., Schnell, R. D., Schug, C. A., Moritz, W. and Steinmann, W. Chemisorption of sulfur on Ge(100). Surface Science 201, 245–56 (1988).

[94] Weser, T., Bogen, A., Konrad, B., Schnell, R. D., Schug, C. A. and Steinmann, W. in 18th International Conference on the Physics of Semiconductors, 11–15 Aug. 1986, Stockholm, Sweden (ed. Engstrom, O.) p. 97–100 vol.1 (Singapore: World Scientific, 1987).

[95] Kruger, P. and Pollmann, J. 1st-principles theory of sulfur adsorption on semi-infinite Ge(001). Physical Review Letters 64, 1808–1811 (1990).

[96] Boonstra, A. H. and Van Ruler, J. Adsorption of various gases on clean and oxidized Ge surfaces. Surface Science 4, 141–149 (1966).

[97] Van Bommel, A. J. and Meyer, F. LEED measurement of H_2S and H_2Se adsorption on germanium (111). Surface Science 6, 391–394 (1967).

[98] Robey, S. W., Bahr, C. C., Hussain, Z., Barton, J. J., Leung, K. T., Ji-ren, L., Schach von Wittenau, A. E. and Shirley, D. A. Surface structure of (2×2)S/Ge(111) determined by angle-resolved photoemission fine structure. Physical Review B 35, 5657–65 (1987).

[99] Newstead, K., Robinson, A. W., Daddato, S., Patchett, A., Prince, N. P., McGrath, R., Whittle, R., Dudzik, E. and McGovern, I. T. Adsorbate-induced de-reconstruction in the interaction of H_2S with Ge(001) 2×1. Journal of Physics: Condensed Matter 4, 8441–8446 (1992).
[100] Nelen, L. M., Fuller, K. and Greenlief, C. M. Adsorption and decomposition of H_2S on the Ge(100) surface. Applied Surface Science 150, 65–72 (1999).
[101] Anderson, G. W., Hanf, M. C., Norton, P. R., Lu, Z. H. and Graham, M. J. The S-passivation of Ge(100)–(1×1). Applied Physics Letters 66, 1123–5 (1995).
[102] Lyman, P. F., Sakata, O., Marasco, D. L., Lee, T. L., Breneman, K. D., Keane, D. T. and Bedzyk, M. J. Structure of a passivated Ge surface prepared from aqueous solution. Surface Science 462, L594–8 (2000).
[103] Hanrath, T. and Korgel, B. A. Chemical surface passivation of Ge nanowires. Journal of the American Chemical Society 126, 15466–15472 (2004).
[104] Bodlaki, D., Yamamoto, H., Waldeck, D. H. and Borguet, E. Ambient stability of chemically passivated germanium interfaces. Surface Science 543, 63–74 (2003).
[105] Cullen, G. W., Amick, J. A. and Gerlich, D. The stabilization of germanium surfaces by ethylation, I. Chemical treatment. Journal of the Electrochemical Society 109, 124–127 (1962).
[106] Citrin, P. H., Rowe, J. E. and Eisenberger, P. Direct structural study of Cl on Si(111) and Ge(111) surfaces: new conclusions. Physical Review B 28, 2299–301 (1983).
[107] Bachelet, G. B. and Schluter, M. Structural determination of Cl chemisorption on Si(111) and Ge(111) by total-energy minimization. Physical Review B 28, 2302–4 (1983).
[108] Schnell, R. D., Himpsel, F. J., Bogen, A., Rieger, D. and Steinmann, W. Surface core-level shifts for clean and halogen-covered Ge(100) and Ge(111). Physical Review B 32, 8052–6 (1985).
[109] Ikeda, K., Imai, S. and Matsumura, M. Atomic layer etching of germanium. Applied Surface Science 112, 87–91 (1997).
[110] Lu, Z. H. Air-stable Cl-terminated Ge(111). Applied Physics Letters 68, 520–2 (1996).
[111] He, J. L., Lu, Z. H., Mitchell, S. A. and Wayner, D. D. M. Self-assembly of alkyl monolayers on Ge(111). Journal of the American Chemical Society 120, 2660–2661 (1998).
[112] Appelbaum, J. A., Baraff, G. A., Hamann, D. R., Hagstrum, H. D. and Sakurai, T. Hydrogen chemisorption on 100 (2×1) surfaces of Si and Ge. Surface Science 70, 654–673 (1978).
[113] Chabal, Y. J. High-resolution infrared spectroscopy of adsorbates on semiconductor surfaces: hydrogen on Si(100) and Ge(100). Surface Science 168, 594–608 (1986).
[114] Maeng, J. Y., Lee, J. Y., Cho, Y. E., Kim, S. and Jo, S. K. Surface dihydrides on Ge(100): A scanning tunneling microscopy study. Applied Physics Letters 81, 3555–3557 (2002).
[115] Papagno, L., Shen, X. Y., Anderson, J., Spagnolo, G. S. and Lapeyre, G. J. Hydrogen adsorption on Ge(100) studied by high-resolution energy-loss spectroscopy. Physical Review B 34, 7188–91 (1986).
[116] Lewis, L. B., Segall, J. and Janda, K. C. Recombinative desorption of hydrogen from the Ge(100)–(2×1) surface: a laser-induced desorption study. Journal of Chemical Physics 102, 7222–8 (1995).
[117] Shimokawa, S., Namiki, A., Gamo, M. N. and Ando, T. Temperature dependence of atomic hydrogen-induced surface processes on Ge(100): Thermal desorption, abstraction, and collision-induced desorption. Journal of Chemical Physics 113, 6916–25 (2000).
[118] Cohen, S. M., Hukka, T. I., Yang, Y. L. and D'Evelyn, M. P. Hydrogen-halogen chemistry on semiconductor surfaces. Thin Solid Films 225, 155–9 (1993).
[119] Oura, K., Lifshits, V. G., Saranin, A. A., Zotov, A. V. and Katayama, M. Hydrogen interaction with clean and modified silicon surfaces. Surface Science Reports 35, 1–69 (1999).
[120] Bringans, R. D. and Hochst, H. Angular resolved photoemission measurements on clean and hydrogen covered Ge(111) surfaces. Applications of Surface Science 11/12, 368–74 (1982).

[121] Choi, K. and Buriak, J. M. Hydrogermylation of alkenes and alkynes on hydride-terminated Ge(100) surfaces. Langmuir 16, 7737–7741 (2000).

[122] Higashi, G. S., Becker, R. S., Chabal, Y. J. and Becker, A. J. Comparison of Si(111) surfaces prepared using aqueous-solutions of NH_4F versus HF. Applied Physics Letters 58, 1656–1658 (1991).

[123] Higashi, G. S., Chabal, Y. J., Trucks, G. W. and Raghavachari, K. Ideal hydrogen termination of the Si(111) surface. Applied Physics Letters 56, 656–8 (1990).

[124] Newton, T. A., Boiani, J. A. and Hines, M. A. The correlation between surface morphology and spectral lineshape: a re-examination of the H−Si(111) stretch vibration. Surface Science 430, 67–79 (1999).

[125] Han, S. M., Ashurst, W. R., Carraro, C. and Maboudian, R. Formation of alkanethiol monolayer on Ge(111). Journal of the American Chemical Society 123, 2422–2425 (2001).

[126] Kosuri, M. R., Cone, R., Li, Q. M., Han, S. M., Bunker, B. C. and Mayer, T. M. Adsorption kinetics of 1-alkanethiols on hydrogenated Ge(111). Langmuir 20, 835–840 (2004).

[127] Linford, M. R. and Chidsey, C. E. D. Alkyl monolayers covalently bonded to silicon surfaces. Journal of the American Chemical Society 115, 12631 (1993).

[128] Linford, M. R., Fenter, P., Eisenberger, P. M. and Chidsey, C. E. D. Alkyl monolayers on silicon prepared from 1-alkenes and hydrogen-terminated silicon. Journal of the American Chemical Society 117, 3145 (1995).

[129] Buriak, J. M. Organometallic chemistry on silicon surfaces: formation of functional monolayers bound through Si–C bonds. Chemical Communications, 1051 (1999).

[130] Buriak, J. M. and Allen, M. J. Lewis acid mediated functionalization of porous silicon with substituted alkenes and alkynes. Journal of the American Chemical Society 120, 1339 (1998).

[131] Buriak, J. M., Stewart, M. P., Geders, T. W., Allen, M. J., Choi, H. C., Smith, J., Raftery, D. and Canham, L. T. Lewis acid mediated hydrosilylation on porous silicon surfaces. Journal of the American Chemical Society 121, 11491 (1999).

[132] de Smet, L. C. P. M., Pukin, A. V., Sun, Q. Y., Eves, B. J., Lopinski, G. P., Visser, G. M., Zuilhof, H. and Sudholter, E. J. R. Visible-light attachment of Si-C linked functionalized organic monolayers on silicon surfaces. Applied Surface Science 252, 24 (2005).

[133] Boukherroub, R. Chemical reactivity of hydrogen-terminated crystalline silicon surfaces. Current Opinion in Solid State & Materials Science 9, 66 (2005).

[134] Liu, Y., Yamazaki, S., Yamabe, S. and Nakato, Y. A mild and efficient Si (111) surface modification via hydrosilylation of activated alkynes. Journal of Materials Chemistry 15, 4906 (2005).

[135] Effenberger, F., Gotz, G., Bidlingmaier, B. and Wezstein, M. Photoactivated preparation and patterning of self-assembled monolayers with 1-alkenes and aldehydes on silicon hydride surfaces. Angewandte Chemie-International Edition 37, 2462 (1998).

[136] Quayum, M. E., Kondo, T., Nihonyanagi, S., Miyamoto, D. and Uosaki, K. Formation of organic monolayer on a hydrogen terminated Si(111) surface via silicon-carbon bond monitored by ATR FT-IR and SFG spectroscopy: Effect of orientational order on the reaction rate. Chemistry Letters, 208 (2002).

[137] Cicero, R. L., Chidsey, C. E. D., Lopinski, G. P., Wayner, D. D. M. and Wolkow, R. A. Olefin additions on H-Si(111): Evidence for a surface chain reaction initiated at isolated dangling bonds. Langmuir 18, 305 (2002).

[138] Avouris, P., Walkup, R. E., Rossi, A. R., Akpati, H. C., Nordlander, P., Shen, T. C., Abeln, G. C. and Lyding, J. W. Breaking individual chemical bonds via STM-induced excitations. Surface Science 363, 368 (1996).

[139] Shen, T. C., Wang, C., Abeln, G. C., Tucker, J. R., Lyding, J. W., Avouris, P. and Walkup, R. E. Atomic-Scale desorption through electronic and vibrational excitation mechanisms. Science 268, 1590 (1995).
[140] Takeuchi, N., Kanai, Y. and Selloni, A. Surface reaction of alkynes and alkenes with H-Si(111): A density functional theory study. Journal of the American Chemical Society 126, 15890 (2004).
[141] Lopinski, G. P., Wayner, D. D. M. and Wolkow, R. A. Self-directed growth of molecular nanostructures on silicon. Nature 406, 48 (2000).
[142] Piva, P. G., DiLabio, G. A., Pitters, J. L., Zikovsky, J., Rezeq, M., Dogel, S., Hofer, W. A. and Wolkow, R. A. Field regulation of single-molecule conductivity by a charged surface atom. Nature 435, 658 (2005).
[143] Hossain, M. Z., Kato, H. S. and Kawai, M. Fabrication of interconnected 1D molecular lines along and across the dimer rows on the Si(100)–(2 × 1)–H surface through the radical chain reaction. Journal of Physical Chemistry B 109, 23129 (2005).
[144] Hossain, Z., Kato, H. S. and Kawai, M. Controlled fabrication of 1D molecular lines across the dimer rows on the Si(100)–(2 × 1)–H surface through the radical chain reaction. Journal of the American Chemical Society 127, 15030 (2005).
[145] Niederhauser, T. L., Jiang, G. L., Lua, Y. Y., Dorff, M. J., Woolley, A. T., Asplund, M. C., Berges, D. A. and Linford, M. R. A new method of preparing monolayers on silicon and patterning silicon surfaces by scribing in the presence of reactive species. Langmuir 17, 5889 (2001).
[146] Yang, L., Lua, Y. Y., Lee, M. V. and Linford, M. R. Chemomechanical functionalization and patterning of silicon. Accounts of Chemical Research 38, 933 (2005).
[147] Wacaser, B. A., Maughan, M. J., Mowat, I. A., Niederhauser, T. L., Linford, M. R. and Davis, R. C. Chemomechanical surface patterning and functionalization of silicon surfaces using an atomic force microscope. Applied Physics Letters 82, 808 (2003).
[148] Terry, J., Linford, M. R., Wigren, C., Cao, R. Y., Pianetta, P. and Chidsey, C. E. D. Determination of the bonding of alkyl monolayers to the Si(111) surface using chemical-shift, scanned-energy photoelectron diffraction. Applied Physics Letters 71, 1056 (1997).
[149] Terry, J., Linford, M. R., Wigren, C., Cao, R. Y., Pianetta, P. and Chidsey, C. E. D. Alkyl-terminated Si(111) surfaces: A high-resolution, core level photoelectron spectroscopy study. Journal of Applied Physics 85, 213 (1999).
[150] Sun, Q. Y., de Smet, L. C. P. M., van Lagen, B., Giesbers, M., Thune, P. C., van Engelenburg, J., de Wolf, F. A., Zuilhof, H. and Sudholter, E. J. R. Covalently attached monolayers on crystalline hydrogen-terminated silicon: Extremely mild attachment by visible light. Journal of the American Chemical Society 127, 2514 (2005).
[151] Stewart, M. P. and Buriak, J. M. Photopatterned hydrosilylation on porous silicon. Angewandte Chemie-International Edition 37, 3257 (1998).
[152] Kim, H., Colavita, P. E., Metz, K. M., Nichols, B. M., Sun, B., Uhlrich, J., Wang, X. Y., Kuech, T. F. and Hamers, R. J. Photochemical functionalization of gallium nitride thin films with molecular and biomolecular layers. Langmuir 22, 8121 (2006).
[153] Knickerbocker, T., Strother, T., Schwartz, M. P., Russell, J. N., Butler, J., Smith, L. M. and Hamers, R. J. DNA-modified diamond surfaces. Langmuir 19, 1938 (2003).
[154] Hamers, R. J., Butler, J. E., Lasseter, T., Nichols, B. M., Russell, J. N., Tse, K. Y. and Yang, W. S. Molecular and biomolecular monolayers on diamond as an interface to biology. Diamond and Related Materials 14, 661 (2005).
[155] Nichols, B. M., Butler, J. E., Russell, J. N. and Hamers, R. J. Photochemical functionalization of hydrogen-terminated diamond surfaces: A structural and mechanistic study. Journal of Physical Chemistry B 109, 20938 (2005).

[156] Chen, R. and Bent, S. F. Chemistry for positive pattern transfer using area-selective atomic layer deposition. Advanced Materials 18, 1086 (2006).

[157] Chen, R. and Bent, S. F. Highly stable monolayer resists for atomic layer deposition on germanium and silicon. Chemistry of Materials 18, 3733 (2006).

[158] Bansal, A., Li, X. L., Lauermann, I., Lewis, N. S., Yi, S. I. and Weinberg, W. H. Alkylation of Si surfaces using a two-step halogenation Grignard route. Journal of the American Chemical Society 118, 7225–7226 (1996).

[159] Kim, N. Y. and Laibinis, P. E. Derivatization of porous silicon by Grignard reagents at room temperature. Journal of the American Chemical Society 120, 4516 (1998).

[160] Song, J. H. and Sailor, M. J. Functionalization of nanocrystalline porous silicon surfaces with aryllithium reagents: Formation of silicon-carbon bonds by cleavage of silicon-silicon bonds. Journal of the American Chemical Society 120, 2376 (1998).

[161] Boukherroub, R., Morin, S., Bensebaa, F. and Wayner, D. D. M. New synthetic routes to alkyl monolayers on the Si(111) surface. Langmuir 15, 3831 (1999).

[162] Bansal, A., Li, X. L., Lauermann, I., Lewis, N. S., Yi, S. I. and Weinberg, W. H. Alkylation of Si surfaces using a two-step halogenation Grignard route. Journal of the American Chemical Society 118, 7225 (1996).

[163] Nemanick, E. J., Hurley, P. T., Brunschwig, B. S. and Lewis, N. S. Chemical and electrical passivation of silicon (111) surfaces through functionalization with sterically hindered alkyl groups. Journal of Physical Chemistry B 110, 14800 (2006).

[164] Nemanick, E. J., Hurley, P. T., Webb, L. J., Knapp, D. W., Michalak, D. J., Brunschwig, B. S. and Lewis, N. S. Chemical and electrical passivation of single-crystal silicon(100) surfaces Through a two-step chlorination/alkylation process. Journal of Physical Chemistry B 110, 14770 (2006).

[165] Webb, L. J. and Lewis, N. S. Comparison of the electrical properties and chemical stability of crystalline silicon(111) surfaces alkylated using Grignard reagents or Olefins with Lewis acid catalysts. Journal of Physical Chemistry B 107, 5404 (2003).

[166] Yamada, T., Shirasaka, K., Noto, M., Kato, H. S. and Kawai, M. Adsorption of unsaturated hydrocarbon moieties on H:Si(111) by Grignard reaction. Journal of Physical Chemistry B 110, 7357 (2006).

[167] Yamada, T., Kawai, M., Wawro, A., Suto, S. and Kasuya, A. HREELS, STM, and STS study of CH_3-terminated Si(111)–(1x1) surface. Journal of Chemical Physics 121, 10660 (2004).

[168] Ulman, A. Formation and structure of self-assembled monolayers. Chemical Reviews 96, 1533 (1996).

[169] Xia, Y. N. and Whitesides, G. M. Soft lithography. Angewandte Chemie-International Edition 37, 551 (1998).

[170] Schreiber, F. Structure and growth of self-assembling monolayers. Progress in Surface Science 65, 151 (2000).

[171] Yu, H., Webb, L. J., Ries, R. S., Solares, S. D., Goddard Iii, W. A., Heath, J. R. and Lewis, N. S. Low-temperature STM images of methyl-terminated Si(111) surfaces. Journal of Physical Chemistry B 109, 671 (2005).

[172] Nakagawa, O. S., Ashok, S., Sheen, C. W., Martensson, J. and Allara, D. L. GaAs interfaces with octadecyl thiol self-assembled monolayer: structural and electrical properties. Japanese Journal of Applied Physics, Part 1 30, 3759 (1991).

[173] Schvartzman, M., Sidorov, V., Ritter, D. and Paz, Y. Passivation of InP surfaces of electronic devices by organothiolated self-asse.mbled monolayers. Journal of Vacuum Science & Technology B 21, 148 (2003).

[174] Lunt, S. R., Ryba, G. N., Santangelo, P. G. and Lewis, N. S. Chemical studies of the passivation of GaAs surface recombination using sulfides and thiols. Journal of Applied Physics 70, 7449 (1991).

[175] Pluchery, O., Chabal, Y. J. and Opila, R. L. Wet chemical cleaning of InP surfaces investigated by in situ and ex situ infrared spectroscopy. Journal of Applied Physics 94, 2707 (2003).

[176] Shaporenko, A., Adlkofer, K., Johansson, L. S. O., Tanaka, M. and Zharnikov, M. Functionalization of GaAs surfaces with aromatic self-assembled monolayers: A synchrotron-based spectroscopic study. Langmuir 19, 4992 (2003).

[177] Jun, Y., Zhu, X. Y. and Hsu, J. W. P. Formation of alkanethiol and alkanedithiol monolayers on GaAs(001). Langmuir 22, 3627 (2006).

[178] Lunt, S. R., Santangelo, P. G. and Lewis, N. S. Passivation of GaAs surface recombination with organic thiols. Journal of Vacuum Science & Technology B 9, 2333 (1991).

[179] Dorsten, J. F., Maslar, J. E. and Bohn, P. W. Near-surface electronic structure in GaAs (100) modified with self-assembled monolayers of octadecylthiol. Applied Physics Letters 66, 1755 (1995).

[180] Adlkofer, K., Eck, W., Grunze, M. and Tanaka, M. Surface engineering of gallium arsenide with 4-mercaptobiphenyl monolayers. Journal of Physical Chemistry B 107, 587 (2003).

[181] McGuiness, C., Blasini, D. R., Uppilil, S., Shaporenko, A., Zharnikov, M., Smilgies, D. M. and Allara, D. L. Structure and assembly of ordered alkanethiol monolayers on GaAs (001). Abstracts of Papers of the American Chemical Society 230, U1194 (2005).

[182] Ding, X., Moumanis, K., Dubowski, J. J., Tay, L. and Rowell, N. L. Fourier-transform infrared and photoluminescence spectroscopies of self-assembled monolayers of long-chain thiols on (001) GaAs. Journal of Applied Physics 99 (2006).

[183] Singh, N. K. and Doran, D. C. Decomposition reactions of ethanethiol on GaAs(100). Surface Science 422, 50 (1999).

[184] Yamamoto, H., Butera, R. A., Gu, Y. and Waldeck, D. H. Characterization of the surface to thiol bonding in self-assembled monolayer films of $C_{12}H_{25}SH$ on InP(100) by angle-resolved X-ray photoelectron spectroscopy. Langmuir 15, 8640 (1999).

[185] Lim, H., Carraro, C., Maboudian, R., Pruessner, M. W. and Ghodssi, R. Chemical and thermal stability of alkanethiol and sulfur passivated InP(100). Langmuir 20, 743 (2004).

[186] Tanzer, T. A., Bohn, P. W., Roshchin, I. V., Greene, L. H. and Klem, J. F. Near-surface electronic structure on InAs(100) modified with self-assembled monolayers of alkanethiols. Applied Physics Letters 75, 2794 (1999).

[187] Wassermann, A. Diels-Alder Reactions: Organic Background and Physico-Chemical Aspects (Elsevier, New York, 1965).

[188] Carruthers, W. Cycloaddition Reactions in Organic Synthesis (Pergamon Press, New York, 1990).

[189] Gill, G. B. and Willis, M. R. Pericyclic Reactions (Chapman and Hall, London, 1974).

[190] Fleming, I. Pericyclic Reactions (Oxford University Press, Oxford, 1999).

[191] Woodward, R. B. and Hoffmann, R. The Conservation of Orbital Symmetry (Academic Press, New York, 1970).

[192] Hossain, M. Z., Aruga, T., Takagi, N., Tsuno, T., Fujimori, N., Ando, T. and Nishijima, M. Diels-Alder reaction on the clean diamond (100) 2×1 surface. Japanese Journal of Applied Physics Part 2-Letters 38, L1496 (1999).

[193] Hovis, J. S., Coulter, S. K., Hamers, R. J., D'Evelyn, M. P., Russell, J. N. and Butler, J. E. Cycloaddition chemistry at surfaces: Reaction of alkenes with the diamond(001)-2×1 surface. Journal of the American Chemical Society 122, 732 (2000).

[194] Wang, G. T., Bent, S. F., Russell, J. N., Butler, J. E. and D'Evelyn, M. P. Functionalization of diamond(100) by Diels-Alder chemistry. Journal of the American Chemical Society 122, 744 (2000).

[195] Fitzgerald, D. R. and Doren, D. J. Functionalization of diamond(100) by cycloaddition of butadiene: First-principles theory. Journal of the American Chemical Society 122, 12334 (2000).

[196] Clemen, L., Wallace, R. M., Taylor, P. A., Dresser, M. J., Choyke, W. J., Weinberg, W. H. and Yates, J. T. Adsorption and thermal behavior of ethylene on Si(100)–(2 × 1). Surface Science 268, 205 (1992).

[197] Huang, C., Widdra, W. and Weinberg, W. H. Adsorption of Ethylene On the Si(100)–(2 × 1) Surface. Surface Science 315, L953–L958 (1994).

[198] Liu, H. B. and Hamers, R. J. Stereoselectivity in molecule-surface reactions: Adsorption of ethylene on the silicon(001) surface. Journal of the American Chemical Society 119, 7593 (1997).

[199] Mayne, A. J., Avery, A. R., Knall, J., Jones, T. S., Briggs, G. A. D. and Weinberg, W. H. An STM study of the chemisorption of C_2H_4 on Si(100)(2 × 1). Surf. Sci. 284, 247–256 (1993).

[200] Sorescu, D. C. and Jordan, K. D. Theoretical study of the adsorption of acetylene on the Si(001) surface. Journal of Physical Chemistry B 104, 8259 (2000).

[201] Widdra, W., Huang, C., Yi, S. I. and Weinberg, W. H. Coadsorption of hydrogen with ethylene and acetylene on Si(100)–(2 × 1). Journal of Chemical Physics 105, 5605 (1996).

[202] Xu, S. H., Keeffe, M., Yang, Y., Chen, C., Yu, M., Lapeyre, G. J., Rotenberg, E., Denlinger, J. and Yates, J. T. Photoelectron diffraction imaging for C_2H_2 and C_2H_4 chemisorbed on Si(100) reveals a new bonding configuration. Physical Review Letters 84, 939 (2000).

[203] Bozack, M. J., Choyke, W. J., Muehlhoff, L. and Yates, J. T. Reaction chemistry at the Si (100) surface-control through active-site manipulation. Journal of Applied Physics 60, 3750 (1986).

[204] Cheng, C. C., Wallace, R. M., Taylor, P. A., Choyke, W. J. and Yates, J. T. Direct determination of absolute monolayer coverages of chemisorbed C_2H_2 and C_2H_4 on Si(100). Journal of Applied Physics 67, 3693 (1990).

[205] Craig, B. I. and Smith, P. V. Structures of small hydrocarbons adsorbed on Si(001) and Si terminated beta-SiC(001). Surface Science 276, 174 (1992).

[206] Fisher, A. J., Blochl, P. E. and Briggs, G. A. D. Hydrocarbon adsorption on Si(001): when does the Si dimer bond break? Surface Science 374, 298 (1997).

[207] Liu, Q. and Hoffmann, R. The bare and acetylene chemisorbed Si(001) surface, and the mechanism of acetylene chemisorption. Journal of the American Chemical Society 117, 4082 (1995).

[208] Lopinski, G. P., Moffatt, D. J., Wayner, D. D. M. and Wolkow, R. A. How stereoselective are alkene addition reactions on Si(100)? Journal of the American Chemical Society 122, 3548 (2000).

[209] Meng, B. Q., Maroudas, D. and Weinberg, W. H. Structure of chemisorbed acetylene on the Si(001)–(2 × 1) surface and the effect of coadsorbed atomic hydrogen. Chemical Physics Letters 278, 97 (1997).

[210] Nishijima, M., Yoshinobu, J., Tsuda, H. and Onchi, M. The adsorption and thermal decomposition of acetylene on Si(100) and vicinal Si(100) 9 degrees. Surface Science 192, 383 (1987).

[211] Pan, W., Zhu, T. H. and Yang, W. T. First-principles study of the structural and electronic properties of ethylene adsorption on Si(100)–(2 × 1) surface. Journal of Chemical Physics 107, 3981 (1997).

[212] Taylor, P. A., Wallace, R. M., Cheng, C. C., Weinberg, W. H., Dresser, M. J., Choyke, W. J. and Yates, J. T. Adsorption and decomposition of acetylene on Si(100)–2 × 1. Journal of the American Chemical Society 114, 6754 (1992).

[213] Yoshinobu, J., Tsuda, H., Onchi, M. and Nishijima, M. The adsorbed states of ethylene on Si(100)c(4 × 2), Si(100)(2 × 1), and vicinal Si(100) 9 degrees Electron-energy loss spectroscopy and low-energy electron-diffraction studies. Journal of Chemical Physics 87, 7332 (1987).
[214] Lopinski, G. P., Moffatt, D. J., Wayner, D. D. and Wolkow, R. A. Determination of the absolute chirality of individual adsorbed molecules using the scanning tunnelling microscope. Nature 392, 909 (1998).
[215] Nagao, M., Umeyama, H., Mukai, K., Yamashita, Y., Yoshinobu, J., Akagi, K. and Tsuneyuki, S. Precursor mediated cycloaddition reaction of ethylene to the Si(100)c(4 × 2) surface. Journal of the American Chemical Society 126, 9922 (2004).
[216] Hovis, J. S., Lee, S., Liu, H. B. and Hamers, R. J. Controlled formation of organic layers on semiconductor surfaces. Journal of Vacuum Science & Technology B 15, 1153–1158 (1997).
[217] Hovis, J. S. and Hamers, R. J. Structure and bonding of ordered organic monolayers of 1,3,5,7-cyclooctatetraene on the Si(001) surface: Surface cycloaddition chemistry of an antiaromatic molecule. Journal of Physical Chemistry B 102, 687 (1998).
[218] Hovis, J. S., Liu, H. and Hamers, R. J. Scanning-tunneling-microscopy of cyclic unsaturated organic-molecules on Si(001). Applied Physics A-Materials Science & Processing 66, S553–S557 (1998).
[219] Hamers, R. J., Hovis, J. S., Lee, S., Liu, H. B. and Shan, J. Formation of ordered, anisotropic organic monolayers on the Si(001) surface. Journal of Physical Chemistry B Materials 101, 1489 (1997).
[220] Hovis, J. S., Liu, H. and Hamers, R. J. Cycloaddition chemistry and formation of ordered organic monolayers on silicon (001) surfaces. Surface Science 404, 1–7 (1998).
[221] Hovis, J. S. and Hamers, R. J. Structure and bonding of ordered organic monolayers of 1,5-cyclooctadiene on the silicon(001) surface. Journal of Physical Chemistry B 101, 9581 (1997).
[222] Hovis, J. S., Liu, H. B. and Hamers, R. J. Cycloaddition chemistry of 1,3-dienes on the silicon(001) surface: Competition between [4 + 2] and [2 + 2] reactions. Journal of Physical Chemistry B 102, 6873–6879 (1998).
[223] Lopinski, G. P., Moffatt, D. J., Wayner, D. D. M., Zgierski, M. Z. and Wolkow, R. A. Asymmetric induction at a silicon surface. Journal of the American Chemical Society 121, 4532 (1999).
[224] Cho, J. H. and Kleinman, L. First-principles study of the adsorption and reaction of cyclopentene on Ge(001). Physical Review B 67, 115314–5314 (2003).
[225] Dyson, A. J. and Smith, P. V. A Molecular-Dynamics Study of the Chemisorption of C_2H_2 and CH_3 On the Si(001)–(2 × 1) Surface. Surface Science 375, 45–54 (1997).
[226] Hofer, W. A., Fisher, A. J. and Wolkow, R. A. Adsorption sites and STM images of C_2H_2 on Si(100): A first principles study. Surface Science 475, 83–88 (2001).
[227] Huang, C., Widdra, W., Wang, X. S. and Weinberg, W. H. Adsorption of acetylene on the Si(100)–(2 × 1) surface. Journal of Vacuum Science & Technology A 11, 2250–2254 (1993).
[228] Li, L., Tindall, C., Takaoka, O., Hasegawa, Y. and Sakurai, T. STM study of C_2H_2 adsorption on Si(001). Physical Review B 56, 4648–4655 (1997).
[229] Lu, X. and Lin, M. C. Bonding configurations of acetylene adsorbed on the Si(100)–2 × 1 surface predicted by density functional cluster model calculations. Physical Chemistry Chemical Physics 2, 4213–4217 (2000).
[230] Miotto, R., Ferraz, A. C. and Srivastava, G. P. Acetylene adsorption on the Si(001) surface. Physical Review B 65, 075401 (2002).
[231] Kim, W., Kim, H., Lee, G., Chung, J., You, S. Y., Hong, Y. K. and Koo, J. Y. Acetylene molecules on the Si(0 0 1) surface: Room-temperature adsorption and structural modification upon annealing. Surface Science 514, 376 (2002).

[232] Kim, W., Kim, H., Lee, G., Young-Kyu, H., Kidong, L., Chanyong, H., Dal-Hyun, K. and Ja-Yong, K. Initial adsorption configurations of acetylene molecules on the Si(001) surface. Physical Review B 64, 193313 (2001).
[233] Mezhenny, S., Lyubinetsky, I., Choyke, W. J., Wolkow, R. A. and Yates, J. T. Multiple bonding structures of C_2H_2 chemisorbed on Si(100). Chemical Physics Letters 344, 7 (2001).
[234] Rintelman, J. M. and Gordon, M. S. Adsorption of acetylene on Si(100)–(2 × 1). Journal of Physical Chemistry B 108, 7820 (2004).
[235] Cho, J. H. and Kleinman, L. Adsorption kinetics of acetylene and ethylene on Si(001). Physical Review B 69, 075303 (2004).
[236] Silvestrelli, P. L., Pulci, O., Palummo, M., Del Sole, R. and Ancilotto, F. First-principles study of acetylene adsorption on Si(100): The end-bridge structure. Physical Review B 68, 235306 (2003).
[237] Konecny, R. and Doren, D. J. Cycloaddition reactions of unsaturated hydrocarbons on the Si(100)–(2 × 1) surface: theoretical predictions. Surface Science 417, 169 (1998).
[238] Konecny, R. and Doren, D. J. Theoretical prediction of a facile Diels-Alder reaction on the Si(100)–2 × 1 surface. Journal of the American Chemical Society 119, 11098–11099 (1997).
[239] Teplyakov, A. V., Kong, M. J. and Bent, S. F. Vibrational spectroscopic studies of Diels-Alder reactions with the Si(100)–2 × 1 surface as a dienophile. Journal of the American Chemical Society 119, 11100 (1997).
[240] Teplyakov, A. V., Lal, P., Noah, Y. A. and Bent, S. F. Evidence for a retro-Diels-Alder reaction on a single crystalline surface: Butadienes on Ge(100). Journal of the American Chemical Society 120, 7377–7378 (1998).
[241] D. J. Doren, Private Communication.
[242] Kong, M. J., Teplyakov, A. V., Lyubovitsky, J., Jagmohan, J. and Bent, S. F. Interaction of C6 cyclic hydrocarbons with a Si(100)–2 × 1 Surface: adsorption and hydrogenation reactions. J. Phys. Chem. 104, 3000 (2000).
[243] Teague, L. and Boland, J. STM study of multiple bonding configurations and mechanism of 1,3-cyclohexadiene attachment on Si(100)–2 × 1. Journal of Physical Chemistry B 107, 3820–3823 (2003).
[244] Yamashita, Y., Nagao, M., Machida, S., Hamaguchi, K., Yasui, F., Mukai, K. and Yoshinobu, J. High resolution Si 2p photoelectron spectroscopy of unsaturated hydrocarbon molecules adsorbed on Si(100)c(4 × 2): the interface bonding and charge transfer between the molecule and the Si substrate. Journal of Electron Spectroscopy and Related Phenomena 114/116, 389 (2001).
[245] Choi, C. H. and Gordon, M. S. Cycloaddition reactions of dienes on the Si(100)-2×1 surface. International Journal of Modern Physics B 17, 1205 (2003).
[246] Teague, L. C., Chen, D. and Boland, J. J. DFT investigation of product distribution following reaction of 1,3-cyclohexadiene on the Si(100)–2 × 1 surface. Journal of Physical Chemistry B 108, 7827 (2004).
[247] Lee, H. S., Choi, C. H. and Gordon, M. S. Cycloaddition isomerizations of adsorbed 1,3-cyclohexadiene on Si(100)–2 × 1 surface: First neighbor interactions. Journal of the American Chemical Society 127, 8485 (2005).
[248] Wang, G. T., Mui, C., Musgrave, C. B. and Bent, S. F. Cycloaddition of cyclopentadiene and dicyclopentadiene on Si(100)–2 × 1: Comparison of monomer and dimer adsorption. Journal of Physical Chemistry B 103, 6803 (1999).
[249] Lu, X., Wang, X. L., Yuan, Q. H. and Zhang, Q. Diradical mechanisms for the cycloaddition reactions of 1,3-butadiene, benzene, thiophene, ethylene, and acetylene on a Si(111)–7 × 7 surface. Journal of the American Chemical Society 125, 7923 (2003).

[250] Baik, J., Kim, M., Park, C. Y., Kim, Y., Ahn, J. R. and An, K. S. Cycloaddition reaction of 1,3-butadiene with a symmetric Si adatom pair on the Si(111)7 $\times$ 7 surface. Journal of the American Chemical Society 128, 8370 (2006).
[251] Minary, P. and Tuckerman, M. E. Reaction mechanism of cis-1,3-butadiene addition to the Si(100)–2 $\times$ 1 surface. Journal of the American Chemical Society 127, 1110 (2005).
[252] Barriocanal, J. A. and Doren, D. J. Cycloaddition of carbonyl compounds on Si(100): New mechanisms and approaches to selectivity for surface cycloaddition reactions. Journal of the American Chemical Society 123, 7340–7346 (2001).
[253] Armstrong, J. L., Pylant, E. D. and White, J. M. Thermal chemistry of biacetyl on Si(100). Journal of Vacuum Science & Technology A 16, 123–130 (1998).
[254] Armstrong, J. L., White, J. M. and Langell, M. Thermal decomposition reactions of acetaldehyde and acetone on Si(100). Journal of Vacuum Science & Technology A 15, 1146–1154 (1997).
[255] Wang, G. T., Mui, C., Musgrave, C. B. and Bent, S. F. Competition and selectivity of organic reactions on semiconductor surfaces: Reaction of unsaturated ketones on Si(100)–2 $\times$ 1 and Ge(100)–2 $\times$ 1. Journal of the American Chemical Society 124, 8990–9004 (2002).
[256] Tao, F., Sim, W. S., Xu, G. Q. and Qiao, M. H. Selective binding of the cyano group in acrylonitrile adsorption on si(100)–2 $\times$ 1. Journal of the American Chemical Society 123, 9397 (2001).
[257] Choi, C. H. and Gordon, M. S. Cycloaddition reactions of acrylonitrile on the Si(100)–2 $\times$ 1 surface. Journal of the American Chemical Society 124, 6162–6167 (2002).
[258] Schwartz, M. P. and Hamers, R. J. The role of Pi-conjugation in attachment of organic molecules to the Silicon (001) surface. Surface Science 515, 75–86 (2002).
[259] Bournel, F., Gallet, J., Kubsky, S., Dufour, G., Rochet, F., Simeoni, M. and Sirotti, F. Adsorption of acetonitrile and acrylonitrile on Si(001)–2 $\times$ 1 at room temperature studied by synchrotron radiation photoemission and NEXAFS spectroscopies. Surface Science 513, 37–48 (2002).
[260] Filler, M. A., Mui, C., Musgrave, C. B. and Bent, S. F. Competition and selectivity in the reaction of nitriles on Ge(100)–2 $\times$ 1. Journal of the American Chemical Society 125, 4928–4936 (2003).
[261] Fang, L. A., Liu, J. M., Coulter, S., Cao, X. P., Schwartz, M. P., Hacker, C. and Hamers, R. J. Formation of π-conjugated molecular arrays on silicon (0 0 1) surfaces by heteroatomic Diels-Alder chemistry. Surface Science 514, 362 (2002).
[262] Lu, X. and Lin, M. C. Reactions of some [C;N;O]-containing molecules with Si surfaces: experimental and theoretical studies. International Reviews in Physical Chemistry 21, 137–184 (2002).
[263] Hashimoto, A., Kobayashi, M., Kamijoh, T., Takano, H. and Sakuta, M. Properties of PECVD SiO_xN_y films as selective diffusion barrier. Journal of the Electrochemical Society 133, 1464–1467 (1986).
[264] Kooi, E., Vanlierop, J. G. and Appels, J. A. Formation of silicon-nitride at a Si-SiO_2 interface during local oxidation of silicon and during heat-treatment of oxidized silicon in NH_3 gas. Journal of the Electrochemical Society 123, 1117–1120 (1976).
[265] Ma, Y., Yasuda, T. and Lucovsky, G. Fixed and trapped charges at oxide-nitride-oxide heterostructure interfaces formed by remote plasma-enhanced chemical-vapor-deposition. Journal of Vacuum Science & Technology B 11, 1533–1540 (1993).
[266] Osenbach, J. W. Sodium diffusion in plasma-deposited amorphous oxygen-doped silicon-nitride (alpha-SiON: H) films. Journal of Applied Physics 63, 4494–4500 (1988).
[267] Dresser, M. J., Taylor, P. A., Wallace, R. M., Choyke, W. J. and Yates, J. T. The adsorption and decomposition of NH_3 on Si(100): Detection of the NH_2(a) species. Surface Science 218, 75–107 (1989).

[268] Hamers, R. J., Avouris, P. and Bozso, F. A scanning tunneling microscopy study of the reaction of Si(001)–(2 × 1) with NH_3. Journal of Vacuum Science & Technology A 6, 508–511 (1988).
[269] Fujisawa, M., Taguchi, Y., Kuwahara, Y., Onchi, M. and Nishijima, M. Electron-energy-loss spectra of the Si(100)–(2 × 1) surface exposed to NH_3. Physical Review B 39, 12918–12920 (1989).
[270] Hossain, M. Z., Yamashita, Y., Mukai, K. and Yoshinobu, J. Microscopic observation of precursor-mediated adsorption process of NH_3 on Si(100)c(4 × 2) using STM. Physical Review B 68, 235322 (2003).
[271] Chung, O. N., Kim, H., Chung, S. and Koo, J. Y. Multiple adsorption configurations of NH_3 molecules on the Si(001) surface. Physical Review B 73, 033303 (2006).
[272] Lee, J. Y. and Cho, J. H. Two dissociation pathways of water and ammonia on the Si(001) surface. Journal of Physical Chemistry B 110, 18455 (2006).
[273] Bischoff, J. L., Lutz, F., Bolmont, D. and Kubler, L. UPS differentiation between molecular NH_3 and partially dissociated NH_2 fragments adsorbed at low-temperature on Si(001) surfaces. Surface Science 248, L240–L244 (1991).
[274] Widjaja, Y. and Musgrave, C. B. ab initio study of the initial growth mechanism of silicon nitride on Si(100)–(2 × 1) using NH_3. Physical Review B 64, 205303 .
[275] Widjaja, Y. and Musgrave, C. B. A density functional theory study of the nonlocal effects of NH_3 adsorption and dissociation on Si(100)–(2 × 1). Surface Science 469, 9-20 (2000).
[276] Widjaja, Y., Mysinger, M. M. and Musgrave, C. B. ab initio study of adsorption and decomposition of NH_3 on Si(100)–(2 × 1). Journal of Physical Chemistry B 104, 2527–2533 (2000).
[277] Queeney, K. T., Chabal, Y. J. and Raghavachari, K. Role of interdimer interactions in NH_3 dissociation on Si(100)–(2 × 1). Physical Review Letters 86, 1046 (2001).
[278] Cao, X. P. and Hamers, R. J. Formation of a surface-mediated donor-acceptor complex: Coadsorption of trimethylamine and boron trifluoride on the silicon (001) surface. Journal of Physical Chemistry B 106, 1840–1842 (2002).
[279] Mui, C., Wang, G. T., Bent, S. F. and Musgrave, C. B. Reactions of methylamines at the Si(100)–2 × 1 surface. Journal of Chemical Physics 114, 10170–10180 (2001).
[280] Kato, T., Kang, S. Y., Xu, X. and Yamabe, T. Possible dissociative adsorption of CH_3OH and CH_3NH_2 on Si(100)–2 × 1 surface. Journal of Physical Chemistry B 105, 10340–10347 (2001).
[281] Mulcahy, C. P. A., Carman, A. J. and Casey, S. M. The adsorption and thermal decomposition of dimethylamine on Si(100). Surface Science 459, 1 (2000).
[282] Carman, A. J., Zhang, L. H., Liswood, J. L. and Casey, S. M. Methylamine adsorption on and desorption from Si(100). Journal of Physical Chemistry B 107, 5491 (2003).
[283] Cho, J.-H. and Kleinman, L. Self-assembled molecular array in methylamine dissociation on Si(001). Physical Review B 67, 201301 (2003).
[284] Cho, J.-H. and Kleinman, L. Contrasting structural and bonding properties of trimethylamine and dimethylamine adsorbed on Si(001). Physical Review B 68, 245314 (2003).
[285] Wang, Y. and Hwang, G. S. Two-dimensional arrangement of CH_3NH_2 adsorption on Si(001)–2 × 1. Chemical Physics Letters 385, 144 (2004).
[286] Hossain, M. Z., Machida, S.-I., Nagao, M., Yamashita, Y., Mukai, K. and Yoshinobu, J. Highly selective surface Lewis acid-base reaction: Trimethylamine on Si(100)c(4 × 2). Journal of Physical Chemistry B 108, 4737 (2004).
[287] Wang, G. T., Mui, C., Musgrave, C. B. and Bent, S. F. The effect of a methyl-protecting group on the adsorption of pyrrolidine on Si(100)–2 × 1. Journal of Physical Chemistry B 105, 3295–3299 (2001).
[288] Wang, G. T., Mui, C., Tannaci, J. F., Filler, M. A., Musgrave, C. B. and Bent, S. F. Reactions of cyclic aliphatic and aromatic amines on Ge(100)–2 × 1 and Si(100)–2 × 1. Journal of Physical Chemistry B 107, 4982–4996 (2003).

[289] Cao, X., Coulter, S. K., Ellison, M. D., Liu, H., Liu, J. and Hamers, R. J. Bonding of nitrogen-containing organic molecules to the silicon(001) surface: The role of aromaticity. Journal of Physical Chemistry B 105, 3759–3768 (2001).
[290] Liu, H. B. and Hamers, R. J. An X-ray photoelectron-spectroscopy study of the bonding of unsaturated organic-molecules to the Si(001) surface. Surface Science 416, 354–362 (1998).
[291] Bailey, P. D. and Morgan, K. M. Organonitrogen Chemistry (Oxford University Press, Oxford, 1996).
[292] Cao, X. P. and Hamers, R. J. Interactions of alkylamines with the silicon (001) surface. Journal of Vacuum Science & Technology B 20, 1614–1619 (2002).
[293] Konecny, R. and Doren, D. J. Adsorption of BH_3 on Si(100)–(2 × 1). Journal of Physical Chemistry B 101, 10983–10985 (1997).
[294] Wang, G. T., Mui, C., Musgrave, C. B. and Bent, S. F. Example of a thermodynamically controlled reaction on a semiconductor surface: acetone on Ge(100)–2 × 1. Journal of Physical Chemistry B 105, 12559–12565 (2001).
[295] Hossain, M. Z., Yamashita, Y., Mukai, K. and Yoshinobu, J. Selective functionalization of the Si(100) surface by switching the adsorption linkage of a bifunctional organic molecule. Chemical Physics Letters 388, 27 (2004).
[296] Hamers, R. J., Hovis, J. S., Greenlief, C. M. and Padowitz, D. F. Scanning tunneling microscopy of organic molecules and monolayers on silicon and germanium (001) surfaces. Japanese Journal of Applied Physics, Part 1 38, 3879–87 (1999).
[297] Lee, S. W., Hovis, J. S., Coulter, S. K., Hamers, R. J. and Greenlief, C. M. Cycloaddition chemistry on germanium(001) surfaces: The adsorption and reaction of cyclopentene and cyclohexene. Surface Science 462, 6–18 (2000).
[298] Lal, P., Teplyakov, A. V., Noah, Y., Kong, M. J., Wang, G. T. and Bent, S. F. Adsorption of ethylene on the Ge(100)–2 × 1 surface: Coverage and time-dependent behavior. Journal of Chemical Physics 110, 10545–10553 (1999).
[299] Fink, A., Huber, R. and Widdra, W. Ethylene adsorption on Ge(100)–(2 × 1): A combined angle-resolved photoemission and thermal desorption spectroscopy study. Journal of Chemical Physics 115, 2768–2775 (2001).
[300] Miotto, R., Ferraz, A. C. and Srivastava, G. P. Comparative study of the adsorption of C_2H_4 on the Si(001) and Ge(001) surfaces. Surface Science 507/510, 12–17 (2002).
[301] Kim, A., Choi, D. S., Lee, J. Y. and Kim, S. Adsorption and thermal stability of ethylene on Ge(100). Journal of Physical Chemistry B 108, 3256–3261 (2004).
[302] Lu, X., Zhu, M. P. and Wang, X. L. Diradical mechanisms for the cycloaddition chemistry of ethylene on X(100) surfaces (X = C, Si, and Ge). Journal of Physical Chemistry B 108, 7359–7362 (2004).
[303] Lu, X. and Zhu, M. P. Beyond the intradimer [2+2] cycloaddition chemistry of ethylene on Si(100): theoretical evidence on the occurrence of interdimer reaction. Chemical Physics Letters 393, 124–127 (2004).
[304] Kim, A., Maeng, J. Y., Lee, J. Y. and Kim, S. Adsorption configurations and thermal chemistry of acetylene on the Ge(100) surface. Journal of Chemical Physics 117, 10215–10222 (2002).
[305] Miotto, R. and Ferraz, A. C. A theoretical study of C_2H_2 adsorption on the Ge(001) surface. Surface Science 513, 422–30 (2002).
[306] Cho, J. H. and Kleinman, L. Adsorption structure of acetylene on Ge(001): A first-principles study. Journal of Chemical Physics 119, 2820–2824 (2003).
[307] Lee, S. W., Nelen, L. N., Ihm, H., Scoggins, T. and Greenlief, C. M. Reaction of 1,3-cyclohexadiene with the Ge(100) surface. Surface Science 410, L773–L778 (1998).

[308] Teplyakov, A. V., Kong, M. J. and Bent, S. F. Vibrational spectroscopic studies of Diels-Alder reactions with the Si(100)–2 × 1 surface as a dienophile. Journal of the American Chemical Society 119, 11100–11101 (1997).

[309] Teplyakov, A. V., Kong, M. J. and Bent, S. F. Diels-Alder reactions of butadienes with the Si(100)–2 × 1 surface as a dienophile: vibrational spectroscopy, thermal-desorption and near-edge X-ray-absorption fine-structure studies. Journal of Chemical Physics 108, 4599–4606 (1998).

[310] Fink, A., Menzel, D. and Widdra, W. Symmetry and electronic structure of benzene adsorbed on single-domain Ge(100)–(2 × 1) and Ge/Si(100)–(2 × 1). Journal of Physical Chemistry B 105, 3828–3837 (2001).

[311] Zhang, Y. P., Yang, L., Lai, Y. H., Xu, G. Q. and Wang, X. S. Self-assembly of one-dimensional molecular nanostructures on the Ge-covered Si(100) surface. Applied Physics Letters 84, 401–3 (2004).

[312] Hwang, Y. J., Kim, A., Hwang, E. Y. and Kim, S. Chiral attachment of styrene mediated by surface dimers on Ge(100). Journal of the American Chemical Society 127, 5016–5017 (2005).

[313] Mui, C., Filler, M. A., Bent, S. F. and Musgrave, C. B. Reactions of nitriles at semiconductor surfaces. Journal of Physical Chemistry B 107, 12256–12267 (2003).

[314] Cho, Y. E., Maeng, J. Y., Kim, S. and Hong, S. Y. Formation of highly ordered organic monolayers by dative bonding: Pyridine on Ge(100). Journal of the American Chemical Society 125, 7514–7515 (2003).

[315] Kim, H. J. and Cho, J. H. Different adsorption structures of pyridine on Si(001) and Ge(001) surfaces. Journal of Chemical Physics 120, 8222–8225 (2004).

[316] Hong, S., Cho, Y. E., Maeng, J. Y. and Kim, S. Atomic and electronic structure of pyridine on Ge(100). Journal of Physical Chemistry B 108, 15229–15232 (2004).

[317] Lee, J. Y., Jung, S. J., Hong, S. and Kim, S. Double dative bond configuration: Pyrimidine on Ge(100). Journal of Physical Chemistry B 109, 348–351 (2005).

[318] Kim, A., Filler, M. A., Kim, S. and Bent, S. F. Layer-by-layer growth on Ge(100) via spontaneous urea coupling reactions. Journal of the American Chemical Society 127, 6123–6132 (2005).

[319] Kim, A., Filler, M. A., Kim, S. and Bent, S. F. Ethylenediamine on Ge(100)–2 × 1: The role of interdimer interactions. Journal of Physical Chemistry B 109, 19817–19822 (2005).

[320] Keung, A. J., Filler, M. A., Porter, D. W. and Bent, S. F. Tertiary amide chemistry at the Ge(1 0 0)−2 × 1 surface. Surface Science 599, 41 (2005).

[321] Keung, A. J., Filler, M. A. and Bent, S. F. Thermal control of amide product distributions at the Ge(100)–2 × 1 surface. Journal of Physical Chemistry C 111, 411–419 (2007).

[322] Ulman, A. An introduction to ultrathin organic films from Langmuir-Blodgett to self-assembly (Academic Press, Inc., 1991).

[323] Tredgold, R. H. Order in thin organic films (Cambridge University Press, Cambridge, 1994).

[324] Ulman, A. Formation and structure of self-assembled monolayers. Chemical Reviews 96, 1533–1554 (1996).

[325] Bitzer, T. and Richardson, N. V. Demonstration of an imide coupling reaction on a Si(100)–2 × 1 surface by molecular layer deposition. Applied Physics Letters 71, 1890–1892 (1997).

[326] Bitzer, T. and Richardson, N. V. Route for controlled growth of ultrathin polyimide films with Si-C bonding to Si(100)–2 × 1. Applied Surface Science 145, 339–343 (1999).

Chapter 6

Surface Electrochemistry

Peter Strasser[1] and Hirohito Ogasawara[2]

[1]*Department of Chemical and Biomolecular Engineering, Cullen College of Engineering, University of Houston, Houston, TX 77204, USA;* [2]*Stanford Synchrotron Radiation Laboratory, Post Office Box 20450, Stanford, CA 94309, USA*

1. Introduction

Electrochemistry focuses on chemical reactions involving the transfer of electric charge across the electrified interface between electronic and ionic conductors. The combination of surface chemical and electrical phenomena makes electrochemistry a very interdisciplinary subject. Electrochemical reactions are inherently associated with currents and voltages. Controlling and monitoring interfacial currents and voltages during an electrochemical reaction has traditionally been used to classify the reaction and to study its interfacial reaction kinetics. Until about the end of the 1970s, kinetic and mechanistic studies of electrochemical interfaces were restricted to current and voltage measurements. This was because the surface of the typically solid electronic conductor (electrode) is buried in liquid or solid ionic conductor (electrolyte) which rendered direct observation of interfacial processes very difficult. Experimental techniques to characterize bonding at electrochemical interfaces were lacking. At that time, the thermodynamic and kinetic analysis of interfacial currents and voltages was the basis for understanding what controls electrochemical reactions. Although current–voltage measurement continues to be an integral approach in this field, it describes the electrochemical interfaces only phenomenologically and often neglects the role of chemical bonding at the interface.

In the last 40 years, modern surface science techniques have been developed to provide information about solid surfaces and interfaces on the atomic or molecular level [1]. Surface Science studies have revealed adsorption sites, electronic structure of atoms and molecules on surfaces including bond distances. These studies deal with well-defined surfaces in a well-controlled environment, in other words, focus on structurally well-defined single-crystal surfaces, which are clean and without adsorbates on the atomic or molecular level or covered with atoms or molecules with controlled amounts and structure. By understanding the reactions on simple model surfaces, we can extrapolate to reactions on real catalysts that are more complex.

Inspired by these Surface Science studies at the gas-solid interface, the field of electrochemical Surface Science ('Surface Electrochemistry') has developed similar conceptual and experimental approaches to characterize electrochemical surface processes on the molecular level. Single-crystal electrode surfaces inside liquid electrolytes provide electrochemical interfaces of well-controlled structure and composition [2–9]. In addition, novel in situ surface characterization techniques, such as optical spectroscopies, X-ray scattering, and local probe imaging techniques, have become available and helped to understand electrochemical interfaces at the atomic or molecular level [10–18]. Today, Surface electrochemistry represents an important field of research that has recognized the study of chemical bonding at electrochemical interfaces as the basis for an understanding of structure–reactivity relationships and mechanistic reaction pathways.

In this chapter, we will first discuss thermodynamic and kinetic concepts of electrified interfaces and point out some distinct features of electrochemical reaction processes. Subsequently, we will relate these concepts to chemical bonding of adsorbates on electrode surfaces. Finally, a discussion of the surface electrocatalytic mechanism of some important technological electrochemical reactions will highlight the importance of understanding chemical bonding at electrified surfaces.

2. Special features of electrochemical reactions

For electrochemical reactions to occur at an electrode/electrolyte interface, equal amount of electrical charges need to enter and leave the electrolyte solution. This implies that electrochemical reactions always require two (or a multiple of two) interfaces at which charge transfer occurs.

The collision between reacting atoms or molecules is an essential prerequisite for a chemical reaction to occur. If the same reaction is carried out electrochemically, however, the molecules of the reactants never meet. In the electrochemical process, the reactants collide with the electronically conductive electrodes rather than directly with each other. The overall electrochemical Redox reaction is effectively split into two half-cell reactions, an oxidation (electron transfer out of the anode) and a reduction (electron transfer into the cathode).

There is another difference between chemical and electrochemical reactions: while reactive collisions of reactant molecules during chemical reactions are associated with a specific heat profile (exothermic or endothermic), the Gibbs free energy of the corresponding electrochemical reactions is converted directly to electricity without heat reaching the surroundings. This feature of electrochemical reactions has been termed *cold combustion* and is what makes electrochemical energy conversion in fuel cells so energy efficient.

2.1. Electrochemical current and potential

Figure 6.1 shows a schematic view of a simple electrochemical cell. In electrochemical studies, we measure a current flow and a potential difference between two half-cells, which are associated with an electrochemical reaction on the respective electrode surface. The flow of electrons and ions between the two half-cell reactions and the associated potential differences between the electrodes are the most obvious special features of electrochemical processes. The electric currents flowing in the external portion of the closed ionic/electronic circuit are directly proportional to the rate of conversion of reactants (Faraday's law). Hence, measurements of overall electrochemical reaction rates become simple and straightforward. In contrast to chemical reactions, the use of a constant current device in the external circuit allows to force controlled reaction rates onto the reacting system. Similarly, potentiostatic devices can be used to deliberately control the thermodynamic driving force of an electrochemical process. At the quantum level of metal electrodes, this means that changing the applied external potential effectively amounts to controlling the relative energy of the Fermi levels of the two electrodes. In essence, current and voltage are two additional powerful kinetic and thermodynamic control parameters in the realm of electrochemistry.

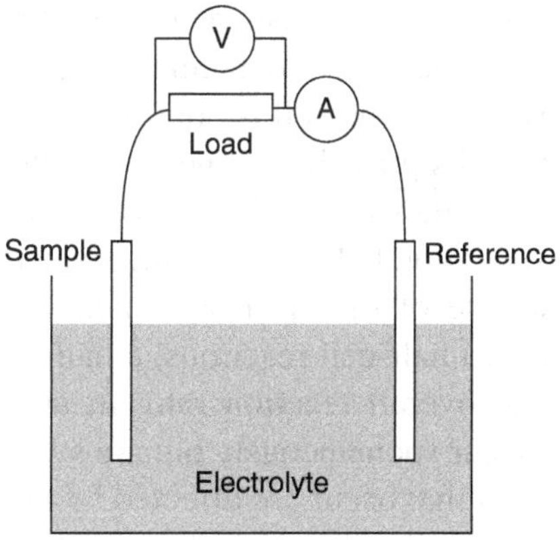

Figure 6.1. Schematic of a simple electrochemical cell.

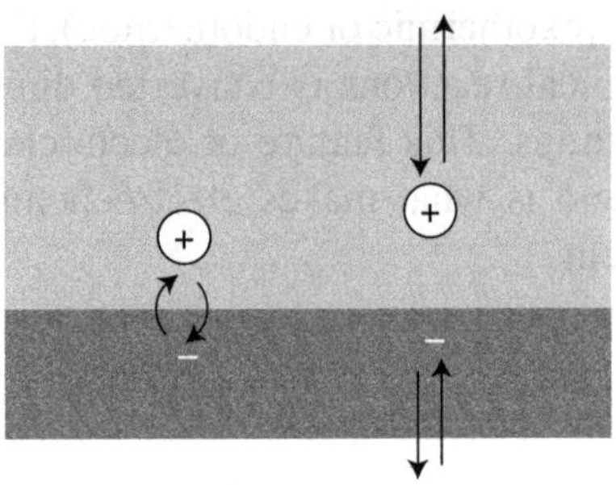

Figure 6.2. Schematic illustration of faradic (left) and capacitive (right) currents.

Two types of current are observed in electrochemical measurements (see Figure 6.2). An electric current corresponds to the reduction or oxidation of chemical species in the cell (i.e., ions, molecules in the electrolyte or atoms or molecules on the electrode surface), which involves the transfer of charge across the interface between a metallic electrode surface and a solution phase. This is called the faradic current. The faradic current accompanies the charge transfer across the interface. A current also flows in the absence of electrochemical reactions: capacitive current. The capacitive current flow is observed when the potential of the electrode is changing, or the distribution of ions is changing near the interface. The interface between the electrode and the electrolyte works like a capacitor to block the faradic current flow. On the other hand, the electrochemical reaction behaves like a switch to trigger the faradic current flow, and it is the barrier to initiate this electrochemical reaction that blocks the faradic current flow. The electrical work is consumed to accumulate or remove charges on the electrode and move ions in the electrolyte near the interface in a process that can be viewed as charging or discharging an electrochemical double-layer capacitor. In other words, the electrode potential is defined by how ions and molecules are distributed near the interface.

In galvanic cells, e.g., batteries, electrochemical reactions occur spontaneously and the faradic current flows when the cell is connected to an external load. Chemical reactions in the two half-cells provide the energy for the galvanic cell operations. The maximum voltage that the galvanic cell can produce is the difference in chemical potential between the oxidation half-cell and the reduction half-cell. The faradic current corresponds to the rate of the electrochemical reaction. Since the faradic current is the same for the two half-cell reactions, a half-cell reaction that is slower than another determines the overall reaction rate. In an electrolytic cell, electrochemical reactions do not occur spontaneously but are switched on by application of an external voltage. Reactions that occur are affected by the external voltage supply. To drive the reaction, the external voltage needs to be larger than the difference in the chemical potential between the oxidation half-cell and the reduction half-cell.

Electric energy is then used to force non-spontaneous electrochemical reactions to occur.

The potential difference between two electrodes is defined as the amount of work done in transporting a charge from one electrode to the other. There is an analogy between the work function Φ in vacuum and the electrochemical potential ψ in the electrochemical cell: the work function Φ is the minimum energy required to remove an electron from a solid, i.e., to take out an electron from the Fermi level in solid materials. The work function Φ is often measured experimentally by photoemission spectroscopy.

Let's consider how an electron is transported between two media in an experimental situation. We can start with the simplest case: a metallic solid placed in vacuum. We set the electrical potential at infinity in vacuum to be zero. The work function Φ is based on measurements of minimum work, in which a unit charge, q, is removed from the inside of the bulk crystal and transferred through the surface to a region outside, but not too far away from the surface (approximately 1 micron from the surface). The potential of unit charge at this distance r is called the Volta potential $\theta(r)$. When the unit charge, an electron for instance, is placed at this distance from the surface, the electrostatic potential produced by the unit charge ($U(r) = q/4\pi\varepsilon_0 r$) is small enough as not to produce an image charge inside the solid; assuming $r = 1$ micron, $U(r)$ becomes less than 1 meV. Thus the Volta potential θ is defined by the electrostatic potential produced by the solid that reflects the charge distribution near the solid surface. By definition, the Volta potential is the potential of the lowest state of an electron in vacuum at the distance r from the surface, i.e., a zero kinetic-energy electron at this point. Since this distance is still small compared with the distance from any other face of the solid surface with a different work function, the work function Φ of different surface crystallographic orientations can be discriminated using this definition.

Consequently, the energy of an electron at the Fermi level (E_F) in a solid is expressed in electron volt as

$$E_F(\text{eV}) = -\psi - \theta(r). \tag{1}$$

This is schematically illustrated in Figure 6.3. Per definition, the Fermi energy is a true bulk property and independent from the specific surface conditions. E_F and $\theta(r)$ are however influenced by external means, such as a connection to a voltage source with respect to 0 V. The work function Φ, however, is not influenced by such external means, but depends on the other hand strongly on the surface conditions.

The work function is influenced by several factors. On a clean surface in vacuum, the coordination number of the surface atoms or molecules is lower and the electron density leaks out from the lattice of positive ion cores to vacuum: electron

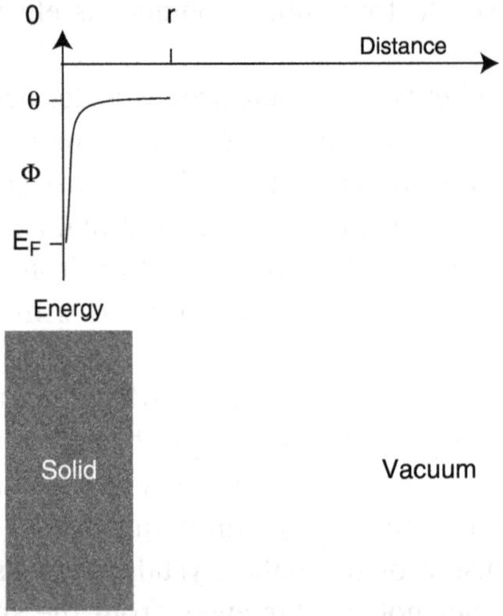

Figure 6.3. Schematic of the Fermi energy, Volta potential and work function.

overspill. To retain overall electrical neutrality of the solid, the electron overspill into the vacuum is balanced by accumulation of counter (positive) charges at the solid surface. As a result, this gives rise to a dipole layer at the surface, which the electron must pass through. The work function is sensitive to structural inhomogeneities of surfaces. The presence of defects affects the degree of electron overspill and changes the work function. Adsorbed atoms and molecules generally have a significant influence on the work function. If the adsorbed molecule has its own static dipole moment, the dipole moment normal to the surface influences the work potential. In the case of chemisorption, charges may furthermore be shifted from the substrate to the adsorbate or vice versa which gives rise to additional dipoles that also influence the electrons. They rearrange the electronic charge within the chemical bond and can add elementary dipoles. Even for physisorption, i.e., molecules that are adsorbed through van der Waals bonding, image charges just underneath the surface are produced by screening producing resulting dipole moments (see Chapter 2).

Next, we can look at the charge transport in electrochemical cells and consider how the electrochemical potential is defined (Figure 6.4). The electrochemical potential in electrochemistry corresponds to the amount of work required to move a charge between two electrodes through the electrolyte solution. Similar to the solids in vacuum the transport of charge can be defined as a two-step process. First, a unit charge is brought close to the interface in an electrolyte. In this region, the unit charge does not change the distribution of electrons and ions at the surface of the electrode. The work required to transport the charge from infinite distance in vacuum to this point defines the Volta potential $\theta(r)$ in the electrolyte solution. For

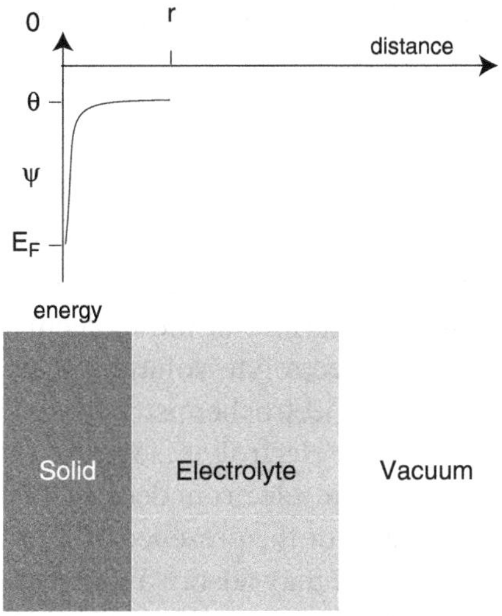

Figure 6.4. Schematic of the Fermi energy, Volta potential and interfacial potential.

an electron, the Volta potential now corresponds to the lowest state of the electron in the electrolyte solution. Similar to the work function measurements for solids in vacuum, the second step of charge transport is related to charge transport across the interface and an interfacial potential, ψ, is then defined as the work corresponding to the transport of charge from this point through the interface to the interior of the solid. Therefore, the measured electric potential difference between sample and reference electrodes corresponds to

$$E_F^{\text{sample}} - E_F^{\text{reference}}(\text{eV}) = -(\psi^{\text{sample}} - (-\psi^{\text{reference}})) - (\theta(r)^{\text{sample}} - \theta(r)^{\text{reference}}) \quad (2)$$

There are similarities and differences between the behavior of the change in work function (1) and that of the interfacial potential (2). Similar to the work function change, the interfacial potential is linked to the conditions of the surface on a molecular scale, i.e., surface structure, adsorbed molecules. Adsorption of ions on the surface can lead to charging, resulting in a change in the interfacial potential. Even when the adsorption energy is small, solvent molecules or ions on the surface influence the interfacial potential. In addition, the presence of ions near the surface can lead to charging of the surface. In this region, the concentration of ions in the electrolyte is different from that in the bulk, and the amount of positive or negative ions can be non-equivalent. In addition, the interfacial potential ψ can also be influenced by external means. If the electrode is connected to a voltage source, an induced electric field attracts counter ions to neutralize the surface charging accompanied by a capacitance current flow. In other words, we can control the interfacial potential ψ by the applied voltage. This is very unique in the electrochemical environment.

2.2. Electrochemical interfaces

The electrochemical current and potential parameters are connected to the properties of the solid electrolyte interface and we will devote this section to the structure of the electrode–electrolyte solution interface at the molecular scale.

The structure of the electrode–electrolyte solution interface has been one of the most intensively studied matters in electrochemistry for over a hundred years. When a metal is immersed in an aqueous electrolyte solution that contains ions, several processes at the interface affect the electron density distribution at the surface. Adsorption of an ion on the surface, or the presence of ions near the surface, induces surface charging. A solvent molecule may adsorb at the surface, and alter the electron density distribution. Surface charging can also be induced by application of an external electric voltage between two electrode surfaces immersed in an electrolyte solution: sample and counter electrodes. The external voltage attracts ions near the electrode surface. Ions near the electrode surface cause an accumulation of counter charges on the surface to neutralize the charges in the electrolyte solution side. As a result, a layer consisting of surface charges and counter ions is formed at the interface: the electrochemical double-layer.

Figure 6.5 depicts the well-accepted *electrochemical double-layer* on a molecular scale. One plate is the metal surface, and the other plate is built up by solvated ions from the electrolyte. Circles with charges denote the ions in the electrolyte. Circles with an arrow represent the polar solvent molecules. The layer of counter ions that

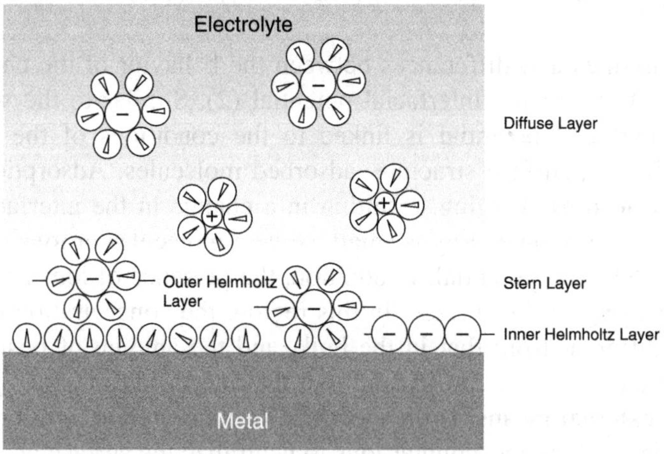

Figure 6.5. Schematic view of electrochemical interface.

shields the charge of the surface is called the Stern layer [19]. The ions in the Stern layer are temporarily bound to the electrode surface and immobile compared to the ions in the bulk. There are two different types of ions in the Stern layer. One has its complete solvation shell and is held in place only by purely electrostatic forces. The other loses a part of the solvation shell and undergoes direct chemical bonding with the metal surface.

Cations and anions with a strong solvation shell retain their solvation shell and thus interact with the electrode surface only through electrostatic forces. Since the interaction is exclusively electrostatic, the amount of these ions at the interface is defined by the electrostatic bias between the sample and the counter electrodes and independent from the chemical properties of the electrode surface: non-specific adsorption. Considering the size effect of their hydration shell, these ions are able to approach the electrode to a distance limited by the size of the solvation shell of the ion. The center of these ions at a distance of closest approach defined by the size of the solvation shell is called the outer Helmholtz layer. The electrode surface and the outer Helmholtz layer have charges of equal magnitude but opposite sign, resulting in the formation of an equivalent of a plate condenser on a scale of a molecular layer. Helmholtz proposed such a plate condenser on such a molecular scale for the first time in the middle of the nineteenth century.

There is a balance between the thermal energy of the ions and the electrostatic force to hold them within the Stern layer. The thermal energy of the ions provides a tendency to drive the ions in the Stern layer to desorb and diffuse into the bulk. The charge accumulated on the surface changes the adsorption and desorption kinetics of ions in the Stern layer. As a result, a part of the counter charge is stored in a diffuse layer, which is called the Gouy–Chapman layer and is more extended and on a length scale larger than a molecular layer [20,21]. The Stern layer is separated from the Gouy-Chapman layer by the outer Helmholtz plane. Per definition, the mass transport of ions in the Gouy-Chapman layer is influenced by the unshielded charge of the electrode surface.

Ions with a weak solvation shell, anions in general, lose a part of or the complete solvation shell in the double layer and form a chemical bond to the metal surface. The adsorption is termed 'specific' since the interaction occurs only for certain ions or molecules and is not related to the charge on the ion. The plane where the center of these ions are located is called the inner Helmholtz layer. In the 'specific' adsorption, ions are chemically bound to the surface and the interaction has a covalent nature. In the case of non-specific adsorption, in which an electrostatic force binds ions to the surface, the coverage of ions is below 0.1 – 0.2 ML due to electrostatic repulsion between the ions. In contrast, the coverage of specifically adsorbed ions exceeds this value, and a close-packed layer of specifically adsorbed ions is often observed. Specifically adsorbed ions are easily observed by STM [22], indicating that the junction between the electrode surface and the inner Helmholtz layer is highly conductive to a DC current, i.e., it does not behave as a plate condenser. Due to the covalent nature of the bonding to the surface, negatively charged ions can adsorb on the sample electrode even when negatively biased with respect to the counter electrode. In this case, the sum of charges on the metal surface and the adsorbed ions defines the total charges of the electrode, which become negative with respect to the reference electrode.

2.3. Models of electrochemical electron transfer kinetics

In studying the activation polarization (*overpotential*, η) of an electrode interface, i.e., the difference between actual electrode potential, E, and the equilibrium potential, E_0, of a given half-cell reaction, and the associated observed current density, j, electrochemists have established an empirical semilogarithmic relation in the form

$$\eta = a + b \log j \tag{3}$$

This relation is termed the Tafel equation. The Tafel equation essentially describes the observed current density at a given electrode potential (potentiostatic mode) or, equivalently, yields the observed electrode potential at a given current density (galvanostatic mode).

Any half-cell reaction occurring at a given electrode surface can be characterized by a set of parameters a and b. Parameter a, which represents the deviation of the electrode potential from its thermodynamic equilibrium potential at unit current density, assumes different values for different electrode materials. The absolute value of parameter b, called the Tafel slope, changes in a much more narrow range from about 60 to 120 mV. The value of b indicates by how much the electrode potential changes for every decade of current density. Parameter a, the intercept of the Tafel line in η versus $\log j$ plots, can be used to compare the intrinsic activity of different electrode materials with respect to a given electrochemical half-cell reaction. Smaller intercepts indicate that a smaller thermodynamic driving force is necessary to achieve a given reaction rate, hence the electrochemical process is associated with a smaller kinetic activation barrier.

Single-electron transferring across electrified interfaces are traditionally characterized as either outer- or inner-sphere processes according to whether or not they are accompanied in a concerted manner with the formation or the breaking of chemical bonds. Outer-sphere reactions therefore solely involve the reorganization of the outer solvent sphere after the electron transfer has occurred. There are only very few truly outer-sphere reactions known to date such as [23]

$$[Ru(NH_3)_6]^{2+} \Leftrightarrow [Ru(NH_3)_6]^{3+} + e^-$$

or

$$Ferrocene \Leftrightarrow Ferrocenium^{+} + e^{-}.$$

Consider a simple outer-sphere half-cell reaction process

$$A + e^{-} \Leftrightarrow B^{-} \tag{4}$$

with the forward and backward reaction being the reduction of A and the oxidation of B, respectively.

The Butler–Volmer (BV) approximation is the simplest approach to model and capture the essential features of the empirical Tafel equation. It considers an electrochemical half-cell reaction as an activated process, with the forward and backward reaction rates following an Arrhenius type law according to

$$r_f \propto e^{-\frac{\Delta G_f^{\neq}}{RT}} \tag{5}$$

and

$$r_b \propto e^{-\frac{\Delta G_b^{\neq}}{RT}} \tag{6}$$

The parameter $\Delta G^{\neq}$ is the Gibbs energy of activation of the forward and backward process. The BV approximation assumes $\Delta G^{\neq}$ to change linearly with changes in the thermodynamic driving force, that is, the overpotential $\eta = (E - E_0)$, according to

$$\frac{\partial \Delta G_f^{\neq}}{\partial (E - E_0)} = \alpha F \tag{7}$$

$$\frac{\partial \Delta G_b^{\neq}}{\partial (E - E_0)} = -(1 - \alpha) F \tag{8}$$

with F denoting the Faraday constant.

The dimensionless parameter α is called the transfer coefficient; it represents the fraction of the thermodynamic driving force that is used in favor of the reduction, while $(1 - \alpha)$ is the fraction favoring the oxidation rate. Integrating equations (7) and (8) yields the linear free-energy relationships

$$\Delta G_f^{\neq} = \Delta G_0^{\neq} + \alpha F(E - E_0) \tag{9}$$

and

$$\Delta G_b^{\neq} = \Delta G_0^{\neq} - (1-\alpha)F(E-E_0) \tag{10}$$

Here, $\Delta G_0^{\neq}$ is the standard Gibbs energy of activation and represents the intrinsic barrier of the process. $\Delta G_0^{\neq}$ becomes the common activation barrier of the forward and backward reactions if the driving force (overpotential $\eta = (E - E_0)$) is zero (that is $E = E_0$).

Combining equations (5)–(10), the forward and backward reaction rates, $r_f(E)$ and $r_b(E)$ can be written as

$$r_f(E) \propto k_0\, e^{-\frac{\alpha F}{RT}(E-E_0)} \text{ and } r_b(E) \propto k_0\, e^{\frac{(1-\alpha)F}{RT}(E-E_0)} \tag{11}$$

Where k_0 denotes the *standard rate constant*. The overall *Butler-Volmer equation* assuming the reactant and product concentrations to be the same, $c_A = c_B = c$, can be formulated as

$$j = j_0\left(-e^{-\frac{\alpha F}{RT}(E-E_0)} + e^{\frac{(1-\alpha)F}{RT}(E-E_0)}\right) \tag{12}$$

with

$$j_0 = nFk_0c \tag{13}$$

j_0 is termed the *exchange current density* of a reaction on a specific electrode material in a specific environment. This quantity represents an intrinsic 'equilibrium' rate of the elementary processes on the electrode surface.

Figure 6.6 illustrates the linear Gibbs energy profile of the reaction process along the reaction coordinate at two values of E. For $E = E_1$ (Figure 6.6(a)) starting at reactant A, a barrier $\Delta G_f^{\neq}$ needs to be overcome to reach the product B. The corresponding change in the Gibbs energy of reaction ΔG_r is indicated. This picture rests on the notion that electron transfer from the electrode to reactant A occurs at the intersection point of the two energy curves where the reaction coordinates and nuclear configurations and energies of A and B are the same.

Changing the electrode potential to $E = E_2 < E_1$ (Figure 6.6(b)) lowers the Gibbs energy curve of the charged species B. As a result of this, the forward activation barrier is now reduced by $\alpha e(E_1 - E_2)$, and hence the reaction rate increases.

The BV formalism is extremely useful in the prediction of electrochemical rates, yet its assumptions of linear free energy relationships are not justified on a molecular basis. Also, it does not account for the influence of the reacting medium, the electrolyte, nor the changes in the molecular structure of the reacting species.

The Marcus-Hush (MH) formalism is an attempt to better account for changes in the outer solvation sphere during an electron transfer process. It incorporates the

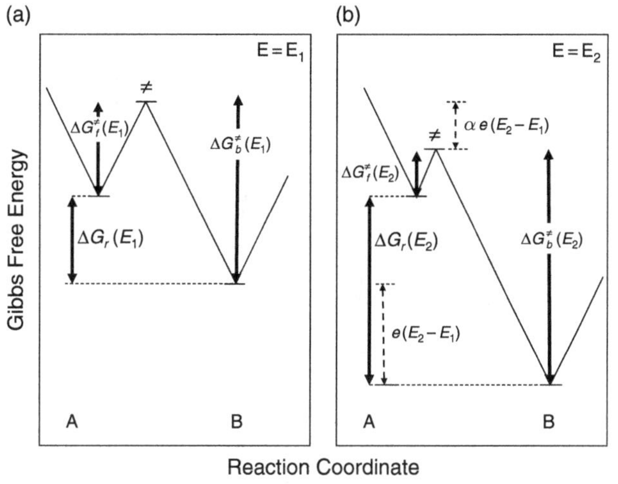

Figure 6.6. Gibbs energy profiles of the outer-sphere one-electron transfer process (4) in the Butler-Volmer formalism at electrode potential E_1 (a) and electrode potential E_2 (b).

electrostatic solvent reorganization energy λ_0^{el} as well as internal vibrational reorganization energy contributions, λ_i, as a combined solvent reorganization parameter $\lambda_t = \lambda_0^{el} + \lambda_i$. The MH formalism considers the dependence of reactant and product energies on vibration coordinates within the harmonic approximation. One of its key results is a quadratic dependence of the Gibbs energies of activation on the overpotentials according to [24]

$$\Delta G_f^{\neq} = \Delta G_0^{\neq}\left(1 + \frac{(E - E_0)F}{4\Delta G_0^{\neq}}\right)^2 \text{ and } \Delta G_b^{\neq} = \Delta G_0^{\neq}\left(1 - \frac{(E - E_0)F}{4\Delta G_0^{\neq}}\right)^2 \quad (14)$$

with the intrinsic barrier

$$\Delta G_0^{\neq} = \lambda_t/4.$$

The quadratic dependence of the Gibbs energy of activation on the driving force implies that the transfer coefficient α is no longer a constant. Instead, it depends linearly on the overpotential. Figure 6.7 illustrates the free energy profile curves along the reaction coordinates for the reaction

$$A + e^- \Leftrightarrow B^- \quad (4)$$

in the MH formalism. The energy curves are now represented as quadratic parabolas. Their curvature at the transition state causes α to change with varying electrode

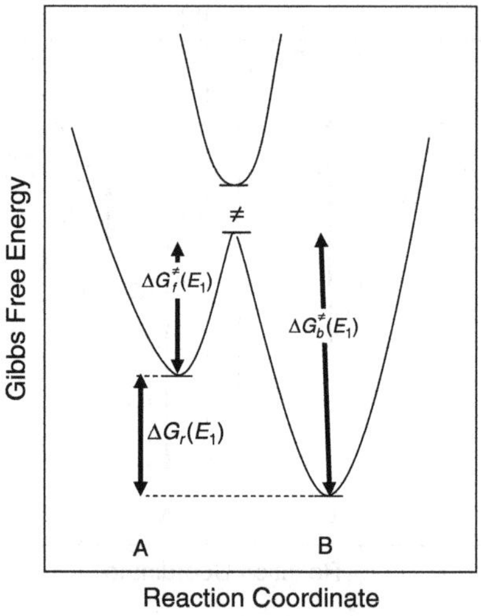

Figure 6.7. Gibbs energy profiles of the outer-sphere one-electron transfer process (4) in the Marcus-Hush formalism.

potential *E*. The figure also indicates the energy gap between the adiabatic reaction surface (bottom) and the non-adiabatic reaction branch.

A great benefit of the MH formalism is the fact that predictions of the electron transfer rates are possible based on molecular structure of the reactants and the reacting medium.

The two models discussed above relate to electron transfer without bond breaking in the inner sphere of a reactant. Varying the electrode potential results in variations in the Gibbs energy of reaction of the electron transfer step and consequently alters the Gibbs energy of activation and therefore the rate of this process. The Gibbs energy required for the activation is due to the difference in molecular coordinates including bond distance, bond angles, solvation shell, and orientation of surrounding dipoles, which are varied by the electrode potential. The electrode potential allows the experimenter to deliberately tune the catalytic qualities of a chosen electrode material.

The concept of electrocatalysis and its relation to chemical surface bonding of reactive intermediates is closely related to that of heterogeneous catalysis. Following the previous section, simple Gibbs energy curves can illustrate the essential ideas of how adsorption of intermediates and their associated Gibbs energy affect the rate of an inner-sphere reaction.

Consider the one-electron transfer from a metal electrode, *M*, to a hydronium ion H_3O^+ in which protons at the electrode surface are discharged and form a metal hydrogen bond, M−H, according to

$$M(e^-) + H_3O^+ \Leftrightarrow M(e^-) \cdots H^+ \cdots H_2O \Leftrightarrow M - H + H_2O \tag{15}$$

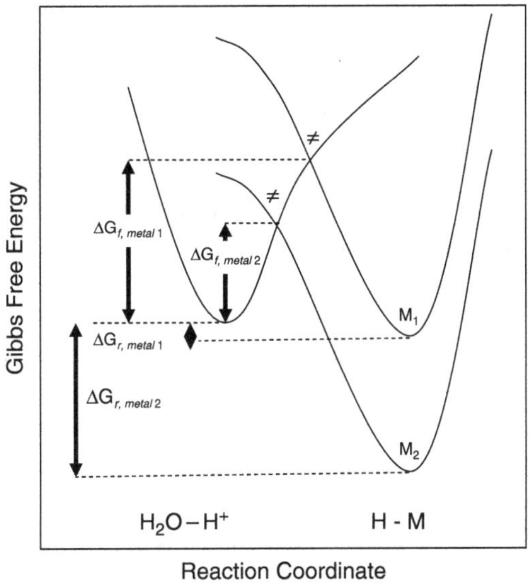

Figure 6.8. Gibbs energy profiles of a proton discharge process resulting in a metal-hydrogen bond formation. The difference in the Gibbs energy of adsorption of hydrogen between metal 1 to metal 2 lowers the activation barrier for the discharge and makes metal 2 the electrocatalytically more favorable (active) electrode material.

Figure 6.8 illustrates the Gibbs energy profile of the forward reaction from left to right on two distinct metal surfaces. The Gibbs energy of adsorption of hydrogen atoms on metal 2 is assumed to be larger than on metal 1. Therefore, in accordance with the previous section, carrying out reaction (15) on metal electrode 2 rather than on metal 1 results in a reduced forward activation barrier associated with a higher electrocatalytic proton discharge rate. Electrode metal 2 is the better electrocatalyst for the proton discharge (hydrogen adsorption) reaction process.

The two models discussed were developed for electrochemical outer-sphere reactions. Most electrochemical reactions, especially those of technological usefulness, however, involve the breaking or formation of chemical bonds during the electron transfer.

The development of a consistent theory for a *dissociative electron transfer* is a recent challenge in the field of theoretical electrocatalysis. Progress in this field of electrochemistry has involved the use of an harmonic Morse curves [25] instead of harmonic approximations. Applying the principles of the theory of the activated complex to adiabatic dissociative electron transfer reactions, the work of Saveant resulted in the following expressions [24] for the Gibbs energy of activation

$$\Delta G^{\neq} = \Delta G_0^{\neq}\left(1 + \frac{(E - E_0)F}{4\Delta G_0^{\neq}}\right)^2 \qquad (16)$$

with

$$\Delta G_0^{\neq} = \frac{D_b + \lambda_0^{el}}{4} \tag{17}$$

and D_b denoting the bond dissociation energy. As with the Marcus-Hush model, the Gibbs energy of activation is a quadratic function of the overpotential, suggesting a potential-dependent transfer coefficient.

3. Electrochemistry at the molecular scale

The electrochemical properties of metals in an electrolyte solution are strongly connected to their surface properties on an atomic or molecular scale such as the structure of the surface and the presence of adsorbates. In this section, we consider how the interfacial potential is linked to the structure of the surface and the adsorption of electrolyte ions and solvent molecules

3.1. Surface structure

At the surface of a solid metal the three-dimensional translation symmetry is broken and the surface atoms have lower coordination number than those in the bulk. At the surface the atoms rearrange themselves in a surface lattice that is denser than that of the ideal truncated lattice and this process is called reconstruction. In vacuum, the clean (110) surface of the late 5d metals shows a missing-row reconstruction with every other <110> row missing. The reconstruction phenomenon is strongly connected to the amount of surface charge [26]. On 5d metals, the missing-row type reconstruction was found to disappear upon atomic halogen adsorption. For 4d and 3d metals, the reconstruction is not observed for the clean surface. On these surfaces, however, the addition of alkali metals induces the missing-row type reconstruction. The adsorption of alkali metals or halogen atoms induces either a negative or positive charge on the surface. The surface can accommodate an excess charge up to a certain threshold. When the change in the charge exceeds a threshold, the surface tends to reconstruct to accommodate the charge. For negatively charged surfaces, low-coordination sites are favored, and the surface undergoes the missing-row type (1×2) or (1×3) reconstruction. On the other hand, for positively charged surfaces the (1×1) structure with high-coordination sites is favored. There is a tendency of electrons at the surface to occupy a large volume to minimize their kinetic energy: the Smoluchowski electron smoothing effect. While the missing-row structure provides a space between the atomic top rows that allows the electrons to occupy a larger volume, the surface with (1×1) structure does not have such a space. Therefore, the (1×1) structure is favored only at lower surface charges.

A consequence of a change in the surface charge is that the work function is affected. A positive charge forces the surface to have a larger work function while a negative charge forces the surface to have a smaller work function. In electrochemical systems, the application of an external potential can change the amount of surface charge. Here the change in the work function corresponds to the change in the electrochemical potential of the electrode. The reconstruction behavior in an electrochemical environment shows a similar tendency to that in vacuum. In general, open surfaces are favored at negative potential, and high-coordination sites are favored at positive potential. The surface of Au(110) shows the missing-row reconstruction at high negative charges at negative potential in a high pH solution [27–29]. As the surface charge decreases by changing from negative to positive potentials, Au(110) undergoes a series of transitions, (1×3) to (1×2) and (1×2) to (1×1), see Figure 6.9. A similar reconstructing behavior is observed for surfaces with crystallographic orientation other than (110). On Au(100) and Pt(100) at negative potentials, a reconstructed hexagonal phase with a close-packed lattice is favored. On the other hand, at positive potentials, a (1×1) phase is observed [30,31]. Although the stability of the surface structure can be tuned by the external potential, the diffusion barrier of surface atoms limits the transition kinetics. The transition is reversible on Au surfaces, but irreversible on Pt surfaces [32–35] where the diffusion barrier of surface atoms is substantially larger than that on Au surfaces [36,37].

3.2. Bonding of ions

It has been known for a long time that the presence of specifically adsorbed anions affects the electrochemical reactivity of a metal electrode in deposition, etching, corrosion and electrocatalysis. For example, blocking of reaction sites through specific adsorption may result in a reduced reaction rate.

In specific adsorption, ions are chemically bound to the electrode surface. What rules the degree of specific adsorption? Specific adsorption of chlorine, bromine, and iodine is often observed, while specific adsorption of fluorine is hardly ever observed. The degree of specific adsorption increases in the order of $F<Cl<Br<I$. Table 6.1 shows the bond energy of halogen to the metal surface in vacuum. The bond strength of these ions increases in the order $I<Br<Cl<F$, which is opposite to the electrochemical system [38–40]. Thus, the degree of specific adsorption is not related to the strength of the bonding to the surface. This implies the importance of considering the solvation of these ions in specific and non-specific adsorption.

Table 6.2 shows the detachment energy of one water molecule from a hydrated halide ion cluster [41]. The strength of the water-halide interactions is reduced as the ionic radius increases in the order of $F<Cl<Br<I$, i.e., in the same order as found for specific adsorption in an electrochemical environment. It is clear that the non-specific adsorption behavior of F^- is due to its strongly bound solvation shell. Due to

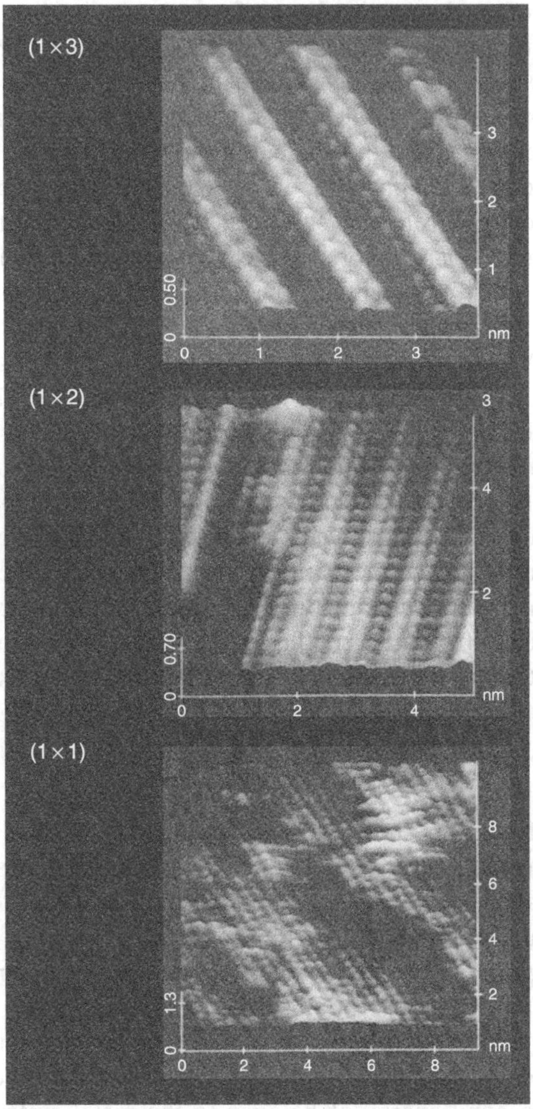

Figure 6.9. Electrochemical STM images showing (1 × 3), (1 × 2) and (1 × 1) Au(110). Taken from Ref. [29].

their weak water-halide interaction, specifically adsorbed layers of I^- and Br^- lose their solvation shell completely. These adsorbed ions form an ordered close-packed layer on the surface. The structure of these specifically adsorbed layers shares a similarity with layers obtained by dosing these species in vacuum. Cl^- has solvation shells of moderate strength, and the specifically adsorbed layer is partially solvated.

Table 6.1
Halide binding energy to Pt [39].

F	−2.31 eV
Cl	−1.73 eV
Br	−1.55 eV
I	−1.39 eV

Table 6.2
Water detachment energy from solvated halide ion clusters [41].

$F(H_2O)_5^- \rightarrow F(H_2O)_4^-$	0.58 eV
$Cl(H_2O)_4^- \rightarrow Cl(H_2O)_3^-$	0.48 eV
$Br(H_2O)_4 \rightarrow Br(H_2O)_3^-$	0.47 eV
$I(H_2O)_3 \rightarrow I(H_2O)_2^-$	0.41 eV

Due to the solvation, the specifically adsorbed layer has a different structure from layers prepared in vacuum in this case [22].

The bonding of ions to metals is dominated by Coulomb attraction since there is a significant difference in electron affinity between the metals and ions. The bonding also involves a redistribution of charge through intermolecular charge transfer (between adsorbed ions and the surface) and intramolecular polarization (in ions and on the surface), which reduces the Pauli repulsion.

3.3. Bonding of water

When ions are non-specifically adsorbed on the electrode surface, there is at least a monolayer of solvation water between the ion and the electrode surface. We should thus also consider the bonding between these water molecules and the electrode surface. In Chapter 2 of the present book, the bonding mechanism of water to metals is discussed in detail. The water molecule bonds to the surface either through oxygen or through a hydrogen atom [42]: O-bonding and H-bonding. Although these interactions are weak in terms of bond energy (approximately 0.5 eV) [43,44], both O-bonding and H-bonding lead to a remarkable redistribution of charge on the metal surface. As we can see in Figure 2.55 of Chapter 2, O-bonding and H-bonding alter the distribution of the surface electronic charge in opposite ways. The electron density between water and metal is decreased upon O-bonding and increased upon H-bonding, respectively. In other words, the bonding scheme of water is strongly connected to the surface charge density.

Studies in vacuum have shown that water adsorbs on metals in either a dissociative or non-dissociative manner [45]. The dissociation of water involves the splitting

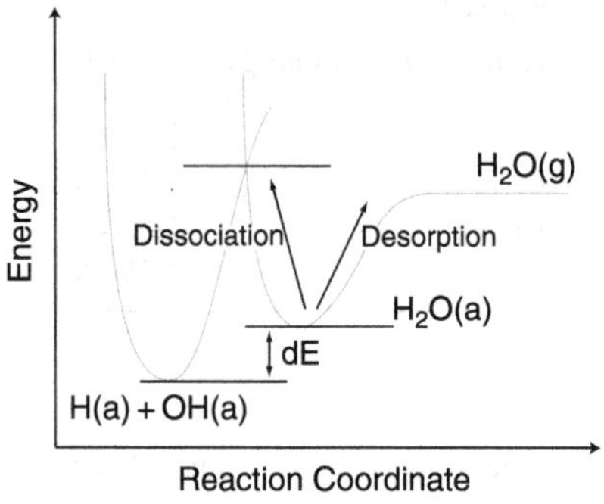

Figure 6.10. Schematic potential energy curve for adsorbed water and its dissociation products.

of one of the internal O–H bonds to produce adsorbed hydroxyl and hydrogen. A schematic potential energy diagram is shown in Figure 6.10. On most metal surfaces, the dissociated phase is energetically stable, but there is a competition between the dissociation and desorption channels. When the activation barrier for the dissociation becomes as large as or larger than that for desorption, the thermal decomposition is hindered kinetically. This activation barrier of dissociation is small on Ru(001), and water dissociates even at 150 K in vacuum. In ambient conditions, just immersion of Ru(001) into pure water at room temperature leads to the formation of the dissociated phase and no electrochemical activation is required [46]. On the other hand, this activation barrier is substantial on Pt(111). Electrochemical activation by applying an external potential is necessary to form the dissociated water layer on Pt(111).

3.4. Experimental aspects of current/voltage properties

The measurement of electrode currents as a function of the applied voltage provides information about the cell reaction. The simple two-electrode setup is suitable for current/voltage measurements in which only a tiny current is passed. However, there is a difficulty in conducting accurate measurements for large samples. If a high current is flowing between two electrodes the resistance of the electrolyte solution cannot be neglected. Moreover, the passage of a high current may also change the interfacial potential of the reference electrode. This issue is resolved using a three-electrode setup as shown in Figure 6.11. An external electronics device (potentiostat or galvanostat) ensures no current flow between the reference electrode and sample electrode. The potential of the reference electrode then becomes constant and we can define the potential of the sample electrode with respect to the known reference electrode. The potentiostat provides a desired electric potential difference between the sample and reference electrodes by forcing a current flow between the sample and counter electrodes. On the other hand, the galvanostat imposes a potential

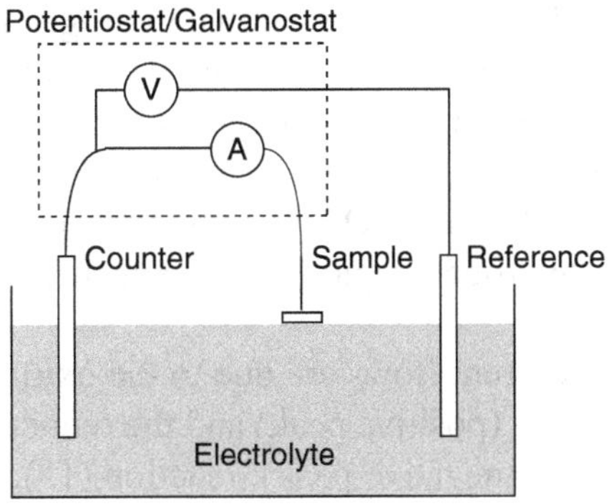

Figure 6.11. Schematic of a three-electrode electrochemical cell.

difference between the sample and reference electrodes to maintain a desired current flow between the sample and counter electrode.

In the electrochemical cell, the sample can be positioned in such a way that only one crystallographic face of the metal is exposed to the electrolyte solution. Figure 6.12 shows the current-potential curve for a single-crystalline Pt electrode in a sulfuric acid solution measured through cyclic voltammetry [47]. Cyclic voltammetry measures the current response to the forward and backward voltage sweep. In Figure 6.12, a triangular wave starting from an initial value (0.05 V) is the applied potential function. Positive and negative current peaks correspond to the oxidation and reduction of adsorbed species, respectively. Both symmetrical and non-symmetrical forward and reverse peaks are seen. The symmetry of the voltammetric peaks occurs since the electrode reaction is now controlled only by electron transfer kinetics and not coupled to either diffusion or mass transport. In other words, both reactants and products are adsorbed on the surface. In Figure 6.12, the

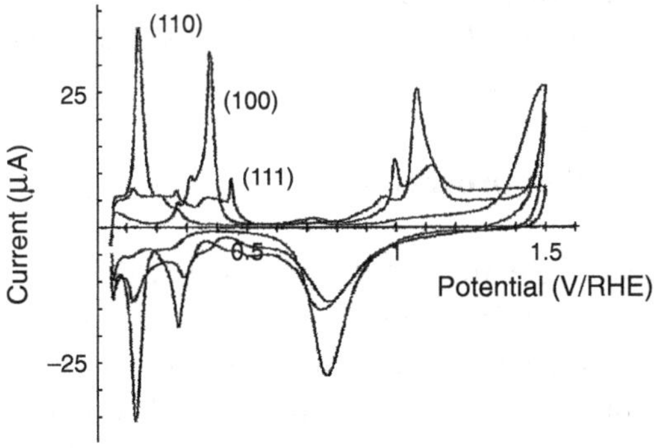

Figure 6.12. Voltammograms of Pt(111), (110) and (100) electrodes in 0.5 M H_2SO_4. From Ref. [47].

applied voltage is measured against the reversible hydrogen reference electrode. The hydrogen evolution reaction occurs if an external potential below 0 V is applied.

The species on the surface varies depending on the applied potential. There are three regions in the cyclic voltammogram. They are 'Hydrogen region (<0.5V)', 'Double-layer region (0.5 – 1V)', and 'Oxygen region (>1V)'. Hydrogen may evolve at a potential below the hydrogen region, and oxygen may evolve at a potential above the oxygen region.

In the hydrogen region, current flows are due to the oxidation of adsorbed atomic hydrogen to proton in solution (positive peak) and the reduction of proton in solution to adsorbed atomic hydrogen (negative peak): reaction (15). The adsorbed hydrogen is considered to occupy the hollow site. However at potential close to 0 V where hydrogen evolves, on-top site adsorbed hydrogen is observed as determined by vibrational spectroscopy [48–50]. There is also a difference in hydrogen evolution kinetics between surface crystallographic orientations. The difference in the catalytic activity can be explained by the stability of on-top hydrogen on these surfaces. No faradic current flow is observed in the double-layer region. In this region, neither oxygen nor hydrogen is adsorbed on the surface. Instead, the surface is covered with solvent or electrolyte. In a sulfuric acid solution, it has been proved that the electrolyte anions are specifically adsorbed on the surface [51–53]. In the oxygen region, the current flows are associated with the oxidation and reduction of the surface giving positive and negative peaks, respectively. This oxidation of Pt occurs below the oxygen evolution potential. Similar to the peaks in the hydrogen region, the features of peaks in the oxygen region strongly depend on the crystal faces of Pt.

The voltammograms are unique for each basal plane and are often used as a diagnostic tool for the identification of primary Pt-surface sites [2–6,54]. The structure-sensitivity of the voltammogram serves as a means to characterize the preferential orientation of a given sample [55]. The voltammogram for platinum nanoparticles obtained by different preparation procedures is shown in Figure 6.13. The difference in voltammetric behavior displays the influence of preparation procedures on the fraction of (100), (110) and (111) atomic sites on the surface of the nanoparticles.

4. Electrocatalytic reaction processes

In this section, a number of electrocatalytic processes will be discussed where surface chemical bonding plays a central role in the reaction mechanism. The selection of reactions is far from complete and not representative of the wide range of technologically important electrocatalytic processes. The selection is biased towards the areas of electrochemical energy conversion and fuel cell electrochemistry, which have been catalyzing a renewed interest in the field of electrochemistry.

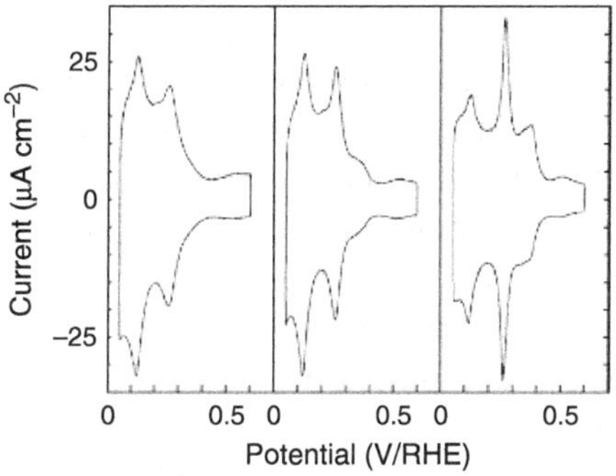

Figure 6.13. Voltammograms of Pt nanoparticles obtained by different preparations in 0.5 M H_2SO_4. Taken from Ref [55].

Unlike classical electroanalytical chemistry, recent work on the selected reactions has been focusing on the 'surface electrochemistry' aspects, that is, the molecular basis of the interfacial processes. Of chief interest is insight in the molecular mechanisms as well as the nature, formation conditions, and reactivity of the surface intermediates. The development and application of in-situ surface-sensitive spectroscopic methods, especially using synchrotron radiation, has been aiding this objective.

In addition to more sophisticated experimental methods, modern theoretical-computational tools, such as state-of-the-art DFT calculations, have greatly improved our understanding of basic electrocatalytic processes. Ever growing computational resources and more and more sophisticated algorithms raise computational methods from the role of a complementary tool to experiments to a viable alternative to experiments.

With a basic mechanism at hand, a rational approach for designing catalysts with desired properties becomes possible. However, despite progress in the direct observation of surface intermediates using high pressure, realistic in situ spectroscopic methods and deeper insight into basic reaction processes, the capability of rationally designing an electrocatalytic surface with a set of desired properties has not yet fully been achieved.

Faced with the experimental and theoretical challenges, experimentalists and theoreticians have very recently started to complement the ongoing 'Surface Electrochemistry' approach by yet another strategy to establish dependable structure-property relationships for a broad range of catalytic surfaces in a reasonable time frame: Combinatorial Materials Discovery and High-Throughput Screening Methods [56–59]. Experimental high-throughput exploration methods initially focus on a reduced set of material properties during the 'primary screening stage' to rapidly eliminate a large portion of the screened parameter spaces. A drastically reduced number of candidates is then tested for a more complete set of properties during the 'secondary screening stage' until a small number of candidates remain with optimized properties. Combinatorial strategies were employed for the discovery of improved heterogeneous [60–62], homogeneous [56,57], as well as electrocatalysts [63–69]. High-Throughput Screening concepts were also adopted for theoretical-computational exploration of large parameter spaces [69,70]. The goal of such high-throughput computational studies is often the rapid qualitative and semi-quantitative prediction of materials or catalyst activity trends as function of materials composition at a given surface geometry. A simple microscopic descriptor with significant importance on the final desired materials property needs to be identified and serves as the computational analogue of the experimental primary screening variable. Particularly powerful for the rapid discovery and mechanistic understanding of new electrocatalysts has proved to be the combination of high-throughput predictive computational and experimental screening [69], followed by conventional electrochemical surface science techniques.

4.1. The electrocatalytic reduction of oxygen

The oxygen/water half-cell reaction has been one of the most challenging electrode systems for decades. Despite enormous research, the detailed reaction mechanism of this complex multi-step process has remained elusive. Also elusive has been an electrode material and surface that significantly reduces the rate-determining kinetic activation barriers, and hence shows improvements in the catalytic activity compared to that of the single-noble-metal electrodes such as Pt or Au.

Oxygen reduction is considered to be one of the most important electrochemical processes due to its central role in electrochemical energy conversion devices and its potential for enhanced efficiencies in industrial Chlorine electrochemistries. Oxygen represents the most convenient and abundant oxidant for conversion of solid, liquid, and gaseous fuels. For decades it has been the focus of interest to the low-temperature fuel cell as well as battery community; first, in the context of Alkaline Fuel Cell research, then Phosphoric Acid Fuel Cells and Zn/Air Batteries, and then in the realm of Polymer Electrolyte Membrane Fuel Cells. The challenges in the study of the oxygen reduction reaction (ORR) continue to be the establishment of a detailed reaction mechanism as well as the identification of a highly active and stable catalyst material. The oxygen reduction reaction is intrinsically slow and very irreversible. As a result of this, the fuel cell cathodes suffer from high activation polarization (see Figure 6.14), which drastically reduces the cell voltage, and hence the efficiency, of a fuel cell. A number of excellent reviews are available on the electroreduction of oxygen [71–78].

4.1.1. Background

The ORR mechanism is very complex and involves a number of different adsorbate intermediates. In addition to its complex reduction mechanism, the relatively high thermodynamic electrode potential of the ORR causes electrocatalytic surfaces

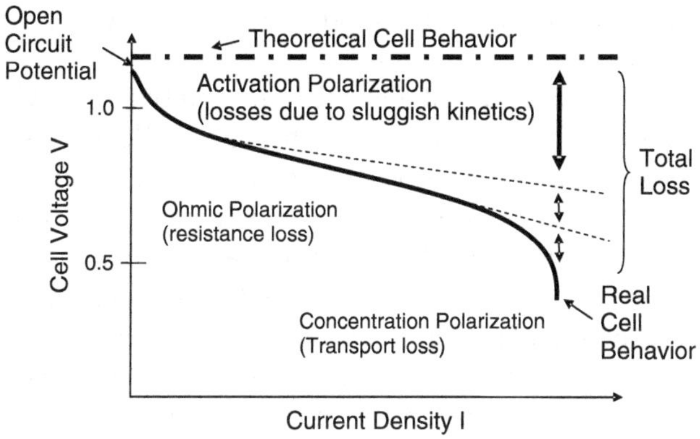

Figure 6.14. Cell Voltage vs. Cell Current profile of a hydrogen – oxygen fuel cell under idealized (dotted-dashed curve) and real conditions. Under real conditions the cell voltage suffers from a severe potential loss (overpotential) mainly due to the activation overpotential associated with the electroreduction process of molecular oxygen at the cathode of the fuel cell. Smaller contributions to the total overpotential losses (resistance loss and mass transport) are indicated.

to undergo oxidative corrosive processes that alter the initial surface structure (roughening) and lead to oxygen-containing surface species that interfere with the oxygen reduction.

The direct four-electron pathway in acid solutions is

$$O_2 + 4H^+ + 4e^- \rightarrow 2H_2O \qquad E_0 = 1.229V \text{ vs. NHE.} \tag{18}$$

The peroxide pathway is

$$O_2 + 2H^+ + 2e^- \rightarrow 2H_2O_2 \qquad E_0 = 0.67V \text{ vs. NHE.} \tag{19}$$

With peroxide undergoing the follow-up reactions

$$H_2O_2 + 2H^+ + 2e^- \rightarrow 2H_2O \qquad E_0 = 1.77V \text{ vs. NHE} \tag{20}$$

$$2H_2O_2 \rightarrow 2H_2O + O_2 \qquad E_0 = 1.77V \text{ vs. NHE} \tag{21}$$

Three principal binding models have been considered for molecular oxygen on electrode surfaces. The Griffith model considers O_2 to interact with a single surface atom by means of a donation bond formed between π-orbitals of O_2 and empty d-orbitals of the surface atom. The Pauling model assumes an end-on adsorption by means of interaction between a σ-bond orbital of O_2 and an empty d-orbital of the electrode atom. Finally, a bridged O_2 adsorption configuration involving two electrode surface atoms was proposed by Yeager, especially for the reaction on Pt.

This model is reasonable under the assumption of dissociative adsorption of oxygen on the electrode surface.

The sluggish reaction kinetics of the ORR in combination with the corrosive environment in which it operates represents the key challenge associated with this reaction. The exchange-current density (see equation (12) and (13)) of the ORR of about 10^{-10} A/cm^2 reflects its highly irreversible character. The experimentally observed open-circuit potentials of this half-cell reaction, that is, the experimentally observed potential at zero current, are considerably less than the expected value of about +1.23 V vs. the potential of a reversible hydrogen electrode. This means that the oxygen electrode is virtually never operating in or even near thermodynamic equilibrium. It is assumed that electrode oxidation processes superimpose on the oxygen reduction reactions and lead to a mixed potential below the expected ORR value. However, despite such attempts to account for the low open-circuit potentials of a resting ORR half-cell, no theory has satisfactorily explained these observations.

4.1.2. Mechanistic pathways

Figure 6.15 illustrates schematically the major ORR reaction pathways. It represents a highly simplified version of a number of previously proposed schemes used to explain and analyze Rotating Ring-Disk Electrode measurements [78].

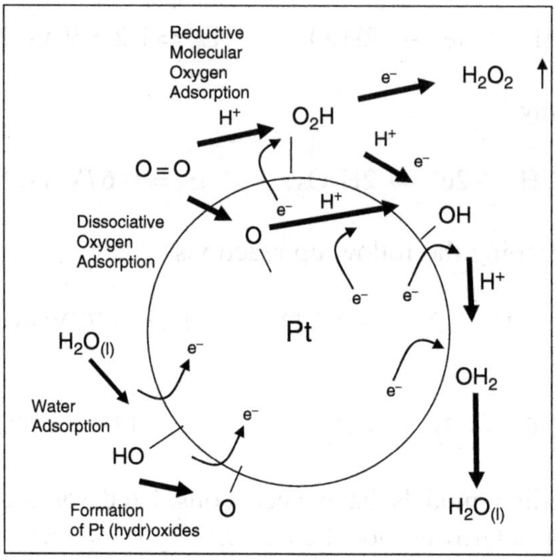

Figure 6.15. Simplified schematic of the most important reaction pathways of the oxygen reduction reaction. The four-electron pathway results in the formation of water. The two-electron pathway forms hydrogen peroxide. Adsorption of molecular oxygen can form atomic oxygen (dissociative pathway) or form a superoxide species (associative pathway). The formation of Pt−OH and Pt−O from water molecules represents the backward reactions of the later portion of the four-electron reduction pathway.

There are two competing views as to the first adsorption step of the oxygen molecule. Following early propositions by Damjanovic [79,80], the rate-determining step of the ORR sequence starts with simultaneous charge and proton transfer to adsorbed oxygen molecules forming an O_2H species adsorbed on the electrode metal atom M:

$$M + O_2 + H^+ + e^- \rightarrow M - O_2H. \quad (22)$$

O–O bond breaking and further reduction occurs in subsequent reaction steps according to

$$M - O_2H + H + +e^- \rightarrow 2M - OH \quad (23)$$

Damjanovic's model also proposed the role of adsorbed surface OH species stemming from the interaction of water with electrode surface atoms M according to

$$M + H_2O \rightarrow M - OH + H^+ + e^-. \quad (24)$$

This reaction was believed to account for the experimentally observed 'low' Tafel slope of about −60 mV per decade of current (unit mV/dec) compared to the normal −20 mV/dec (see definition of Tafel slope in Section 2.3) which is typically indicative of a one-electron transfer process.

Another view was advocated by Yeager [81,82] who proposed dissociative adsorption of molecular oxygen as the initial step, possibly coupled with a simultaneous charge transfer, according to

$$M + {}^1\!/\!_2 O_2 \rightarrow M - O \quad (25)$$

$$M - O + H^+ + e^- \rightarrow M - OH. \quad (26)$$

A missing kinetic isotope effect, as well as adsorption studies in the presence of strongly adsorbing anions, such as chloride, are in favor of Yeager's mechanistic proposition [78].

4.1.3. Electroreduction of oxygen on Pt and Pt alloys

Pt has repeatedly been found to be on the top of activity volcano curves (see Chapter 4) for ORR in acid electrolytes. After a number of inconclusive early studies, Markovic and coworkers demonstrated a clear structural dependence of the ORR activity on the crystallographic faces of Pt single-crystals [83–86]. The activity of the Pt faces decreased in the order Pt(110) > Pt(111) > Pt(100). A four-electron pathway was found on all three surfaces with Tafel slopes of about −120 mV/dec at large current densities. A low Tafel slope of about −60 mV/dec was found at low current densities in accord with the hypothesis of interfering Pt−OH adsorption from activated water molecules. The same order of activity was confirmed in very recent comparative studies of single-crystal Pt and Pt alloy surfaces [87]. Markovic proposed the following kinetic rate law for ORR [86] at low coverage of the reactive intermediates such as O_2H

$$j = nF\,k\,c_{O_2}(1-\Theta_{ad})^x \exp\left[\frac{-\alpha FE}{RT}\right] \exp\left[-\frac{\gamma \Delta G_{ad}}{RT}\right]. \qquad (27)$$

Here, j is the observed current density of the reaction, n is the number of electrons, k is the rate constant, c_{O2} is the concentration of O_2 in the electrolyte, θ_{ad} is the total surface coverage of adsorbed species (both intermediates and adsorbed ions), E is the applied potential, α and γ are symmetry factors and ΔG_{ad} is the Gibbs energy of adsorption of intermediates. F, R, and T have their usual meaning. According to this rate law, the ORR kinetics is either determined by the number of available free surface sites, $(1-\Theta_{ad})^x$, or by the change in Gibbs energy of adsorption of reaction intermediates.

A large number of investigations have been carried out to study the effect of alloying on the ORR activity [87–100]. While Pt alloys have demonstrated activity improvements for ORR all along, pinning down the most active alloying transition metal and stoichiometry has remained difficult. Also the precise mechanism of the Pt alloy activity enhancement is still subject to debate. Mukerjee et al. [101–104] studied the ORR activity and the structural details of a number of Pt alloys using X-ray Absorption Spectroscopy (XAS). The authors demonstrated an increase in d-orbital vacancy for Pt alloys combined with shorter Pt–Pt distances and proposed that these features inhibit the adsorption of OH from water and therefore lessen the blocking of active surface sites.

A similar trend in Pt alloy activities (PtCo > PtNi > Pt) was later found by Paulus et al. [98,105] who studied well-defined and high surface area electrocatalysts using Rotating Disk Electrode (RDE) measurements and high-resolution transmission electron microscope (TEM). Based on similar activation energies for Pt and Pt alloys, the authors concluded that the activity differences between Pt and Pt alloys must stem from the coverage-dependent pre-exponential term that captures differences in the OH coverage of the electrode surfaces. The presence of OH-covered base-metal atoms in the top electrode layer was proposed to cause an inhibition of OH formation on neighboring Pt atoms resulting in a net reduction of the electrode OH coverage (common ion effect). Along these lines, a recent report [106] on Pt alloy monolayer electrocatalyst confirmed that OH-covered transition metal atoms, M-OH, in the top layer of Pt alloys can lead to a reduced OH coverage and consequently to a higher ORR reaction rate (see Section 4.1.5).

More recently, Stamenkovic et al. [95,107] reported on the formation of Pt skins on Pt alloy electrocatalysts after high-temperature annealing. Pt skins were reported to exhibit strongly enhanced ORR activity. It was argued that the electronic properties of the thin Pt layer on top of the alloy alter its adsorption properties in such a way as to reduce the adsorption of OH from water and therefore to provide more surface sites for the ORR process (see Section 5.2 in Chapter 4 for a detailed discussion of skin catalysts, compare also Section 4.1.5 in the present Chapter).

4.1.4. Recent quantum chemical studies of the ORR mechanism

A large number of studies have focused on the chemisorption of molecular oxygen on well-defined Pt and Pt alloys. Chemisorption is the first step in the ORR process, and the structure, bonding type and energy are key elements for understanding trends in ORR activity as function of electrode composition and structure.

Extensive computational studies of atomic oxygen adsorption on transition metals were carried out by Hammer and Nørskov [108]. They developed a simple model to predict the adsorbate bond strength [109–112] emphasizing three surface properties that significantly contribute to surface adsorbate bonding on transition metals: (i) the d-band center energy of the surface atoms (ii) the degree of filling of d-bands, and (iii) the coupling matrix element V between the adsorbate states and the metal d-states (see Chapter 4 for discussion and more details).

The oxygen adsorption properties on Pt skin layers, on Pt surfaces with lattice strain, as well as ordered alloy facets of Pt_3Co and Pt_3Fe were investigated by Xu et al. [113]. The authors revealed a linear relationship between the oxygen adsorption energy and the oxygen dissociation barrier on the transition metals considered. A stronger chemisorption bond of oxygen facilitates the dissociation of the oxygen molecule at the surface. Their study raised the notion that a simple thermochemical calculation can suffice to predict qualitative and semi-quantitative trends in surface reactivities.

Ruban et al. [112,114] stressed the importance of the d-band center energy in controlling the chemisorption energy of reactive intermediates such as oxygen atoms and, therefore, the reaction barriers and ultimately the catalytic activity of transition metal surfaces. Based on Density Functional Theory (DFT) calculations, the authors showed that altering the surroundings of atoms in a transition metal surface by alloying with other metals, the electron density of the surface atoms changes and causes the surface d-bands (as well as their center energy ε_d) to up- or down-shift. As the surface d-bands shift, so do the molecular orbitals (MOs) of surface adsorbates. Given a constant Fermi level energy, this results in a change in occupancy of bonding and antibonding MOs, which, in turn, modifies the chemisorption energy (bond

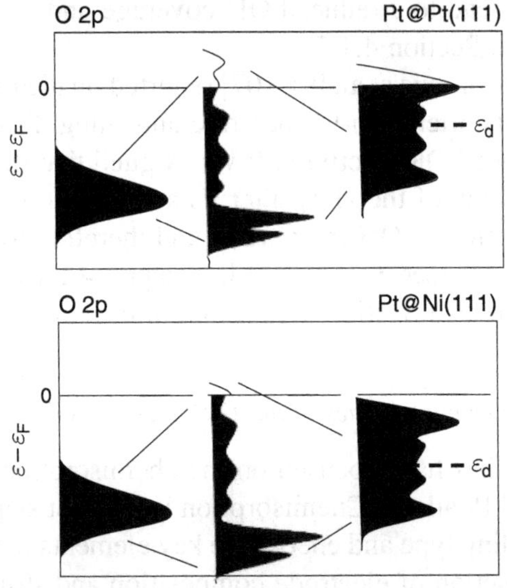

Figure 6.16. Illustration of the d-band model governing surface chemical bonding on transition metal surfaces. As the d-band center of a catalytic surface shifts downward more antibonding orbitals become occupied and the surface bond energy of an adsorbate (here an oxygen atom) decreases. An upward shift in the d-band center predicts strengthening of the surface bond.

strengths) of the adsorbates. Figure 6.16 illustrates the changes in orbital occupancy and bond strength for the Pt−O surface-bond for a Pt monolayer on top of a Pt(111) surface (Pt@Pt(111)) compared to one on top of a Ni(111) surface (Pt@Ni(111)). The d-band center of the latter Pt monolayer shows a more stabilized (down-shifted) d-band center energy resulting in a higher occupancy of antibonding orbitals and a reduced Pt-O bond strength on the Pt@Ni(111) surface (see also Chapter 4).

More recent quantum (electro)chemical calculations in favor of a rate-determining step according to reaction (22) [90,115,116] investigate the adsorption of oxygen on 1-fold and 2-fold sites on platinum surfaces. According to a local reaction-center model, if a dual Pt surface site is present, an OOH species bridging two Pt atoms forms in the rate-determining step, followed by its dissociation. In the potential range of 0.8–0.9 V the proposed model predicted activation energy trends that agreed well with measured activation energies in strong acid electrolytes [90].

Nørskov et al. [117] recently investigated a dissociative (reactions (23) and (24)) as well as an associative ORR mechanism using DFT in a plane-wave pseudo-potential implementation. Calculations were performed on a (3×2) three-layer fcc(111) slab. The authors constructed free-energy diagrams at several electrode potentials as given in Figure 6.17. The theoretical-computational procedure involved an electrode potential, U, which was referenced to that of a standard hydrogen electrode. A simplified electrochemical solvent environment was included in the model by considering a

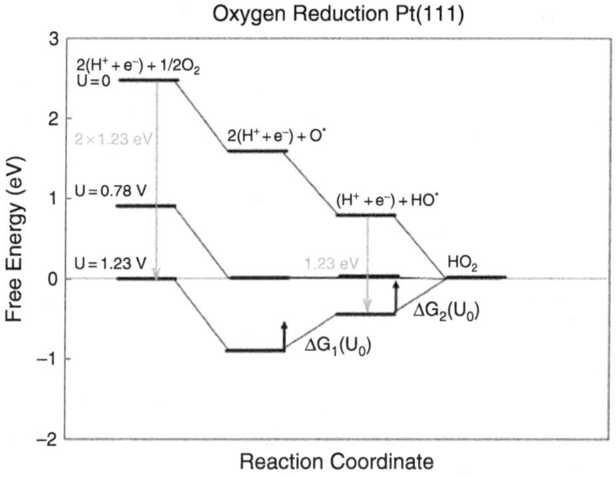

Figure 6.17. Free energy diagram for oxygen reduction over Pt(111) at low surface coverage at zero cell potential (U=0), at the equilibrium potential (zero overpotential) (U=1.23) and at the highest potential where all reaction steps are exothermic (adapted from [117]).

monolayer of water. At low coverages, water was added to fill the surface, while at high O and OH coverages water was added on top of the surface adsorbates. The water layers stabilized OH groups on the surface, while their effect on adsorbed O was negligible. The model further included the effect of a bias on all states involving an extra electron in the electrode. Changing the electrode potential U would shift these states by –eU energy units. The interaction of the electric field with the dipole of the adsorbate was neglected. Finally, the model considered differences in zero-point energies and changes in entropy along the reaction coordinate.

First, free-energy reaction paths were calculated for a dissociative reaction mechanism. Figure 6.17 shows the chemisorption energies of the ORR intermediates along the reaction coordinate at various electrode potentials on Pt(111) for the dissociative mechanism:

$$M + {}^{1}\!/_{2}O_2 \rightarrow M - O \tag{25}$$

$$M - O + H^+ + e^- \rightarrow M - OH \tag{26}$$

$$M - OH + H^+ + e^- \rightarrow M + H_2O \tag{28}$$

where M denotes a free surface site. The electrode potentials U of the ORR electrode may be associated with the cell potential of a single hydrogen/oxygen fuel cell, if one neglects ohmic losses and assumes the anodic hydrogen oxidation reaction to be at equilibrium.

At an electrode potential of U = 0 V, the ORR is running with a high reaction rate. This situation would correspond to a short-circuited fuel cell where all elementary reaction steps are highly exothermic. At an electrode potential of U=1.23 V, where the chemical potential of the electrons is shifted by 1.23 eV, both protonation and electron transfer steps are activated. Process (26) or (28) are therefore rate-limiting under these conditions.

Figure 6.17 also indicates that U = 0.78 V is the highest electrode potential, for which the ORR process is not activated. Conversely, it is the highest potential for which the water dissociation process (reverse of process (28)) is activated. For U > 0.78V water spontaneously dissociates on the Pt surface and blocks surface sites, a process which is likely to be the origin of experimental observations such as changes in Tafel slopes, and a drop in ORR activity [86].

One of the two consecutive reaction steps (26) and (28), associated, respectively, with reaction free energies $\Delta G_1(U)$ and $\Delta G_2(U)$, and its associated activation barrier represents the rate-limiting step of the simplified overall reaction process. Therefore, Nørskov et al. concluded that the origin of the overpotential of the ORR is primarily either the O or the OH surface binding energy, depending on the detailed barrier heights. Using Arrhenius-type expressions of the rate constants and a Sabatier Analysis, the thermochemical data of Figure 6.17 was also converted into microkinetic data and actual electrocatalytic reaction rates. The authors repeated the thermochemical calculations on other metal surfaces and constructed an electrochemical volcano plot (Figure 6.18) with Pt on top of the volcano curve. W, Mo, Fe and other

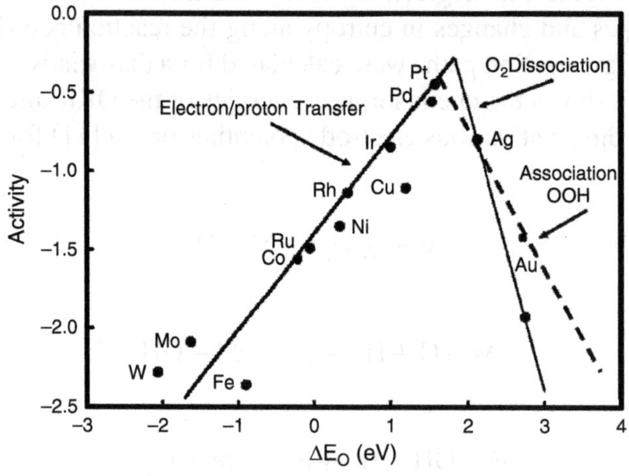

Figure 6.18. Trends in oxygen reduction activity plotted as function of the oxygen binding energy (adapted from Ref. [117]). Solid line: associative mechanism; arrows: rate determining processes; dashed line: associative mechanism.

transition metal surfaces bind oxygen too strongly, resulting in low ORR activities due to the high barriers of protonation/electron transfer. Ag and Au, in contrast, show low activation barriers for the protonation reactions, yet the ORR activity on these metals is limited by the dissociative adsorption of oxygen.

In addition to the dissociative mechanism, Nørskov et al. considered an associative mechanism, where protonation and electron transfer to adsorbed molecular oxygen is assumed to be the first elementary step. Their calculations showed that the oxygen surface coverage strongly determines whether the associative or the dissociative mechanism is energetically favored. At higher oxygen coverage the associative mechanism generally offered lower barriers (higher activities) for the overall process. Also on certain metals, such as Au for instance, the associative mechanism offers an increased reaction rate compared to the dissociative mechanism based on lower activation barriers (see data point for Au on the associative mechanism branch of Figure 6.18).

Based on their quantum chemical investigations of the mechanistic origin of the overpotential of the ORR process (Figure 6.17), Nørskov et al. suggested that gentle shifting of the oxygen chemisorption energy of Pt (1.57 eV) toward more positive values by alloying or any other surface modification would be an effective strategy to increase the overall ORR activity. Based on Nørskov's conclusions, the ideal ORR catalyst surface is such that the Gibbs enthalpy change along the four-electron reaction pathway at potentials near the equilibrium potential $U = 1.23$ V should be close to zero, similar to the situation given on a Pt(111) surface at $U = 0.78$ V (see Figure 6.17).

This theoretical framework accounts for the concept of 'surface poisoning' by adsorbed OH and O at low overpotentials (values of U close to 1.23 V): high OH and O coverages result from the reduced activation barriers associated with the backward reactions of steps (26) and (28) according to

$$M + H_2O \rightarrow M - OH + H^+ + e^- \qquad (24)$$

$$M - OH \rightarrow M - O + H^+ + e^-. \qquad (29)$$

The low activation barriers of these processes are due to the relatively strong Pt−O and Pt−OH chemisorption energy on Pt. In the presence of molecular oxygen, Pt−O and Pt−OH surface species are formed in competition with the water activation process (24) and the dissociative oxygen adsorption (29). Because the Pt−O and Pt−OH chemisorption energies are high and, therefore, slow down the rate of processes (26) and (28), the oxygen-containing intermediates accumulate on the surface. The observed overall ORR rate is low.

The dependence of the overall ORR reaction rate on the surface adsorption energy of Pt−OH and Pt−O species, and consequently on their rate of formation, can experimentally be observed in cyclic voltammetry (CV) and potential sweep voltammetry measurements of Pt alloy electrocatalysts. CV in deaerated solutions reveals the onset potentials of the activation of water that results in oxygenated surface species (processes (24) and (29)).

The solid curve in Figure 6.19 displays the cyclic voltammetric response of a carbon-supported high surface area Pt nanoparticle electrocatalyst in perchloric acid electrolyte under de-aerated conditions. The hydrogen adsorption range between 0.05 and 0.4 V is clearly discernible. On the anodic scan, the formation of Pt−OH and Pt−O surface species from water (processes (24) and (29)) is associated with the peak at about 0.82 V (see inset). The dashed curve in the same figure represents the voltammetric response of a $Pt_{50}Co_{50}$ alloy nanoparticle electrocatalyst. Here, the peak potential of the Pt−OH and Pt−O formation is shifted toward higher electrode potentials (around 0.88 V) suggesting a slower rate of formation and, indirectly, a reduced surface binding energy. This experimental observation is in accordance with the theoretical model predictions that alloying lowers the binding energy of the surface Pt−O bond. Figure 6.20 compares the electrocatalytic activity on the two surfaces: sweeping in the anodic direction, the $Pt_{50}Co_{50}$ alloy electrocatalyst

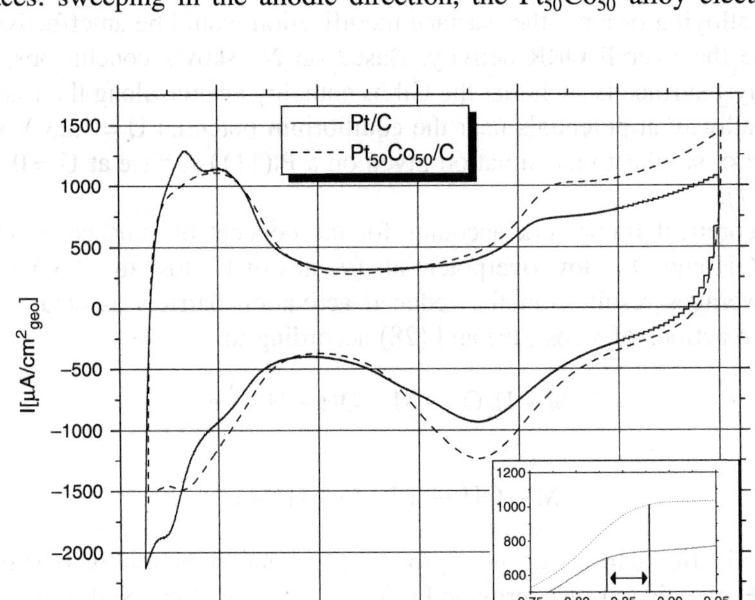

Figure 6.19. Experimental cyclic voltammograms of carbon-supported high surface area nanoparticle electrocatalysts in deaerated perchloric acid electrolyte. Solid curve: pure Pt; dashed curve: $Pt_{50}Co_{50}$ alloy electrocatalyst. Inset: blow up of the peak potential region of Pt−OH and Pt−O formation. Scan rate: 100 mV/s. Potentials are referenced with respect to the reversible hydrogen electrode potential (RHE).

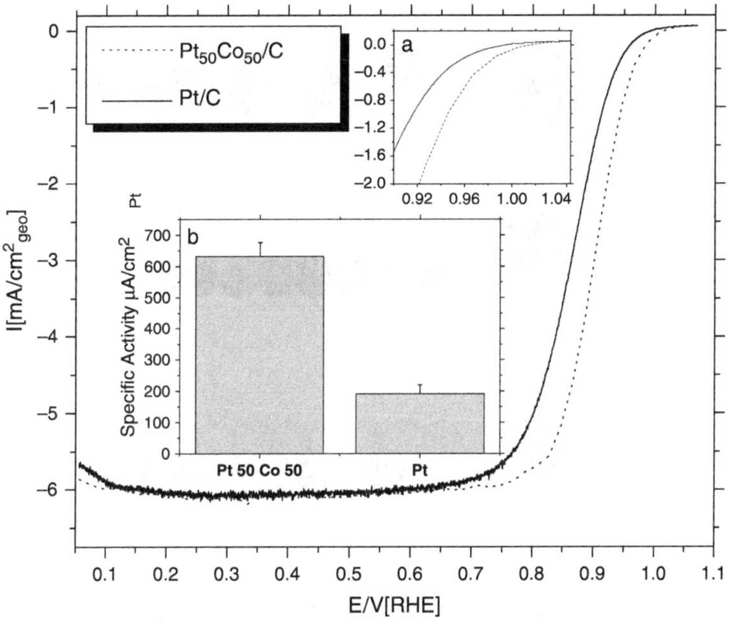

Figure 6.20. Experimental linear sweep voltammogram of carbon-supported high surface area nanoparticle electrocatalyst in oxygen-saturated perchloric acid electrolyte (room temperature). Solid curve: pure Pt; dashed curve: $Pt_{50}Co_{50}$ alloy electrocatalyst. Inset a: blow up of the kinetically controlled ORR regime. Inset b: comparison of the specific (Pt surface area normalized) current density of the Pt and the Pt alloy catalyst for ORR at 0.9 V.

exhibits a decrease in the (absolute) ORR rate from the diffusion-limited current plateau (about−6 mA/cm^2) at a clearly more positive electrode potential compared to the pure Pt electrocatalyst. The shift in the ORR curve in Figure 6.20 (see inset a for details in the high potential range) is about the same as the shift in the peak potential of the oxygen species formation. The delayed accumulation of oxygenated surface species on the electrode surface has been proposed to be associated with the threefold activity increase in the Pt surface-area based (specific) activity at 0.9 V (see inset b in Figure 6.20).

4.1.5. State-of-the-art ORR electrocatalyst concepts

There are a few distinct structural concepts for high-performance Pt alloy ORR electrocatalysts that are currently attracting much attention because they hold the promise of significant activity improvements compared to pure Pt catalysts. As a result of this, these electrocatalysts potentially offer the prospect to impact the future of Polymer Electrolyte Membrane fuel cell catalyst technology.

Figure 6.21 illustrates four different structural catalyst concepts. Structure A represents the conventional bulk alloy surface (single-crystal or polycrystalline) or

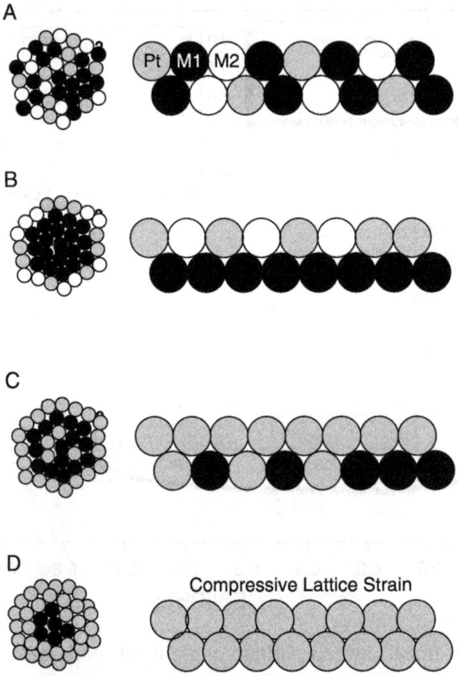

Figure 6.21. Schematics of currently pursued Pt-based electrocatalyst concepts for the ORR. (A) Pt bulk alloys (B) Pt alloy monolayer catalyst concepts (C) Pt skin catalyst concept (D) De-alloyed Pt core-shell catalyst concept.

bulk alloy nanoparticle electrocatalyst. This class of electrocatalysts is characterized by a fairly uniform alloy composition near the surface and inside the catalyst. Catalyst activity improvements using this structural type depend strongly on the nature of the alloying metal, the crystallographic face, the atomic ordering, and the stoichiometry of the catalyst considered. Activity improvements up to a factor of about 3× compared to pure Pt have been reported. It should be noted that smooth single-crystal or polycrystalline alloy surfaces generally exhibit a much higher intrinsic, i.e., Pt surface area normalized, ORR activity compared to alloy nanoparticles of identical surface and bulk composition. This observation has been explained by a particle size effect for Pt and Pt alloys, which is assumed to cause rough surfaces, such as those on small particles, to be less active for the ORR, because they contain a larger portion of low-coordinated surface atoms. The latter are suspected to adsorb oxygenated surface species stronger than higher-coordinated surface atoms. ORR activity improvement factors over a Pt standard, therefore, always refer to a Pt catalyst of a comparable format.

Catalyst Concept B (Figure 6.21(B)) is characterized by a monolayer of Pt or of a Pt alloy sitting on top of a Pt-free nanoparticle (for instance, Ru or Pd) [118–123].

The Pt alloy monolayer nanoparticle catalysts (e.g., Pt–Re layer on Pd cores) showed a clearly improved specific (Pt surface normalized) ORR activity; their Pt mass-based electrocatalytic activity, however, exceeded that of pure Pt catalysts by an impressive factor of $18\times-20\times$. Their noble metal (Pt, Re, and Pd) mass-based activity improvement was still about a factor $4\times$. The Tafel slope in the 800–950 mV/RHE range suggested that the surface accumulation of Pt–OH species is delayed on the Pt monolayer catalyst. The enormous increase in Pt mass-based activity is obviously due to the small amount of Pt metal inside the Pt monolayer.

Computational studies indicated that Pt monolayers on non-Pt substrates exhibit distinct oxygen adsorption and reduction characteristics. In particular, the Pt−OH and Pt−O binding energies were predicted to decrease compared to a pure Pt surface. In the light of the previous discussion in Section 4.1.4 of the origin of the overpotential in the ORR reaction, the Pt−O reduction process becomes activated at a higher electrode potential compared to pure Pt.

Pt alloy monolayer catalysts exhibited even more active ORR behavior compared to Pt monolayer catalysts. To understand this phenomenon computational DFT studies were carried out. The hypothesis to be tested was that, for instance, Ru metal atoms in the Pt−Ru monolayer are OH-covered and could inhibit the adsorption of additional OH on neighboring surface sites (adsorbate–adsorbate repulsion effect). A very similar hypothesis was put forward about three years earlier by Paulus et al. [105] who postulated that Co surface atoms might exhibit a so-called 'common-ion' effect, that is, they could repel like species from neighboring sites. A combined computational-experimental study finally confirmed this hypothesis [123]: If oxophilic atoms such as Ru or Os were incorporated into the Pt monolayer catalysts, the formation of adjacent surface OH was delayed, if not inhibited. Oxophobic atoms, such as Au, displayed the opposite effect, would not inhibit Pt−OH formation, and were found to be detrimental to the overall ORR activity.

While the stability of the monolayer Pt alloy catalyst concept was initially unclear and therefore threatened to make the monolayer catalyst concept a questionable longer term solution, a very recent discovery seems to lend support to the claim that Pt monolayer catalyst could be made into stable catalyst structures: Zhang et al. [94] reported the stabilizing effect of Au clusters when deposited on top of Pt catalysts. The presence of Au clusters resulted in a stable ORR and surface area profile of the catalysts over the course of about 30,000 potential cycles. X-ray absorption studies provided evidence that the presence of the Au clusters modified the Pt oxidation potentials in such a way as to shift the Pt surface oxidation towards higher electrode potentials.

Catalyst Structure C in Figure 6.21 is commonly referred to as the 'Pt-skin' catalysts [87,95,107,124,125]. The term 'Pt skin catalysts' will here be used to refer to a monolayer of pure Pt sitting on a Pt-depleted Pt alloy core; in contrast, a 'Pt monolayer catalyst' was referred above to as a monolayer of pure Pt on top of a Pt-free substrate. Past the second or third layer, the composition of the Pt skin particles or surfaces are close to the bulk value deep inside the single-crystal or particle. Similar to the Pt monolayer catalysts, the presence of a Pt monolayer on a Pt-depleted substrate was found to decrease the chemisorption energy of atomic oxygen (compare the discussion on Pt skin catalysts in Chapter 4, Section 5.2). In the framework of reference [117], as discussed in Section 4.1.4, a gently reduced Pt–O bond energy is beneficial for the overall ORR reaction kinetics. Figure 6.22 compares the predicted and experimentally observed ORR activities and oxygen binding energies of a number of 'Pt-skin' alloy nanoparticles. Pt_3Ni, Pt_3Co, Pt_3Ti, and Pt_3Fe nanoparticles were all found to exhibit a much reduced Pt−O binding energy compared to pure Pt catalysts. The reduction in Pt−O bond energy exceeded the amount necessary to generate a shift close to the maximum of the volcano curve (Figure 6.22).

Pt skin catalysts are prepared by high-temperature annealing and are therefore expected to be thermodynamically stable structures: Thermochemical studies of the metal segregation energies of various metal alloys [114] suggest that Pt-rich alloys prefer to segregate Pt atoms to the surface and form Pt skins.

A particular 'Pt-skin' single-crystal surface, the Pt_3Ni (111) face, was reported to exhibit an extraordinary ORR activity after annealing and formation of the Pt skin structure [87]. This facet exceeded the activity of the Pt(111) single-crystal surface by a factor of 10×, while it was found to be about 90-fold more active than a state-of-the-art high surface area carbon-supported Pt electrocatalyst. The enhancement

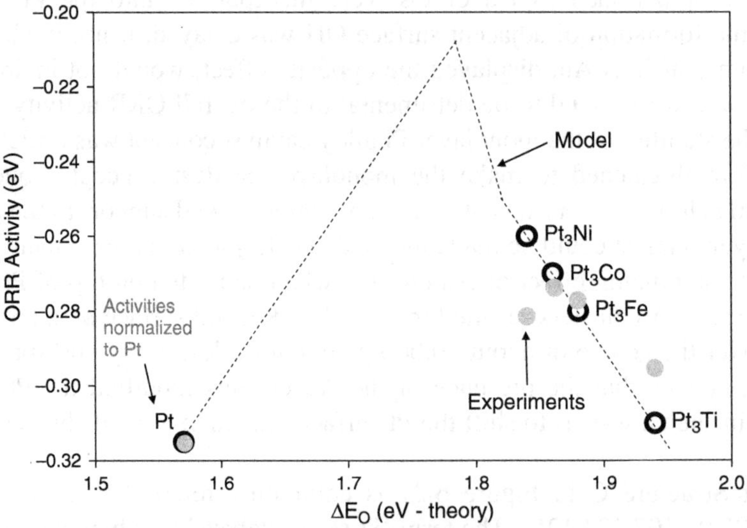

Figure 6.22. Trends in the ORR activity of Pt skin alloy electrocatalysts as function of oxygen binding energy (adapted from Ref. [125])

was attributed to the unusual electronic structure of the Pt_3Ni (111) surface which, in decreasing the Pt−O bond energy, realizes an increase in the availability of more active surface sites for the ORR.

Catalyst Structure D in Fig. 6.21 represents a structurally distinct electrocatalyst compared to catalysts A–C. Unlike A–C, structure D contains no base-metal near the top layer of the smooth or high surface area catalyst. Catalysts like that in Figure 6.21D were prepared [126–129] by rapid electrochemical de-alloying (preferred leaching of the base-metal) of base-metal-rich precursor Pt alloys. The electrocatalytic ORR activities of the catalyst materials obtained after de-alloying exceeded pure Pt nanoparticle catalysts by a factor of 4–6×.

Preliminary experimental and computational characterization indicates that the de-alloyed electrocatalyst particles consist of an essentially pure Pt shell of several atoms diameter sitting on top of a base-metal-rich, and therefore lattice-contracted, particle core. The Pt shell atoms were found to exhibit a smaller than usual Pt−Pt interatomic distance, and are therefore believed to be in a state of compressive strain. DFT computational results have indicated that compressively strained electrocatalytic Pt(111) surfaces indeed reduce their Pt–O binding energy, and lead to a higher ORR activity. More experimental data are necessary to fully understand the origin of the high ORR activity enhancements observed using this kind of electrocatalysts and to thoroughly assess their structural stability in electrochemical environments.

4.2. The electrochemical oxidation of small organic molecules

This section addresses the role of chemical surface bonding in the electrochemical oxidation of carbon monoxide, CO, formic acid, and methanol as examples of the electrocatalytic oxidation of small organics into CO_2 and water. The (electro)oxidation of these small C1 organic molecules, in particular CO, is one of the most thoroughly researched reactions to date. Especially formic acid and methanol [130,131] have attracted much interest due to their usefulness as fuels in Polymer Electrolyte Membrane direct liquid fuel cells [132] where liquid carbonaceous fuels are fed directly to the anode catalyst and are electrocatalytically oxidized in the anodic half-cell reaction to CO_2 and water according to

$$H_3C-OH+H_2O \rightarrow CO_2+6H^+ +6e^- \tag{30}$$

for methanol and according to

$$HCOOH \rightarrow CO_2+2H^+ +2e^- \tag{31}$$

for formic acid. A fuel cell employing the CO oxidation as the anode reaction according to

$$CO+H_2O \rightarrow CO_2+2H^+ +2e^- \tag{32}$$

is theoretically feasible, yet, the high surface bond-strength of CO on many metal surfaces makes a 'CO fuel cell' impractical with respect to its kinetics; due to the high electrode potentials (overpotentials) necessary to remove adsorbed CO by process (32) in combination with overpotentials at the cathode, the resulting cell voltages would be very small. A reaction scheme similar to (30) and (31) can be formulated also for formaldehyde as the anode fuel, but this reaction will not be addressed here. Note that the cathode reaction of a direct liquid fuel cell is the reduction of oxygen described in Section 4.1. (see processes (22),(23),(28) and (25), (26), (28)).

Reaction schemes (30) through (32) are simplified non-elementary steps and do not account for the complexity of the reaction pathways observed in the electrooxidations of oxygenated C1 fuels. Generally, initial adsorption of the fuel on the catalyst surface involves abstraction of hydrogen atoms, which almost instantly lose their electron and form a proton. C1 molecules with carbon being in a lower oxidation state, say methanol, may form formaldehyde and formic acid species as intermediates on the reaction pathway toward CO_2. The mechanistic pathways of the electrooxidations of oxygenated C1 molecules are often discussed using the generalized 'dual-pathway' reaction scheme shown in Figure 6.23. First, the fuel molecule adsorbs and bonds to the surface via the carbon atom after abstraction of a hydrogen atom, thereby forming a reactive intermediate (reaction 1 in Figure 6.23). This intermediate may then be further electrooxidized directly to CO_2 (reaction 2 in Figure 6.23), possibly involving an adsorbed water molecule; such a 'direct oxidation path' mechanism has been clearly established for formic acid [133–136] using the Differentially Pumped Electrochemical Mass Spectrometry method (DEMS).

The reactive intermediate may also decompose into a 'surface poison', that is, a strongly bonded intermediate of little reactivity (reaction 5 in Figure 6.23). The most strongly bonded species, which results from stepwise H abstraction from oxygenated C1 fuels, is typically adsorbed CO, as indicated in the Figure 6.23. In addition, CO may form in parallel from the C1 molecule without involvement of reactive intermediates (reaction 3 in Figure 6.23). To remove CO from the surface, water molecules are needed, to oxidize CO to CO_2 (reaction 4 in Figure 6.23).

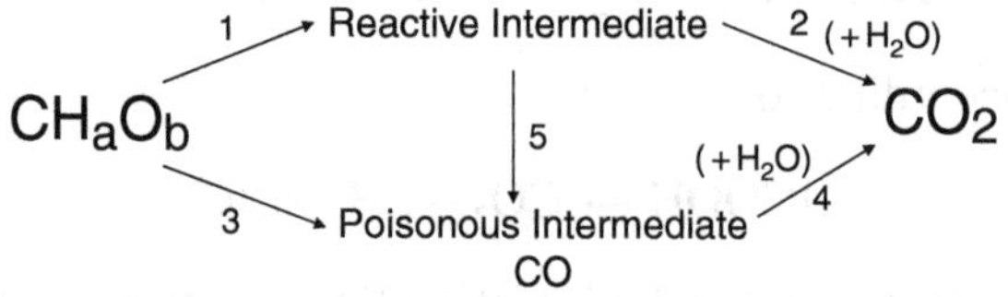

Figure 6.23. Schematic reaction pathways for the electrooxidations of small oxygenated C_1 molecules. The scheme comprises the 'dual path' mechanism proposed for formic acid oxidation (steps 1,2,3,4) as well as the single path mechanism with CO as intermediate (steps 1,5,4).

The multi-step mechanistic pathways of the electrooxidation of C1 fuels, involving a multiplicity of distinct C1 surface intermediates, in combination with co-adsorbed oxygenated species, such as OH or O species (reaction (24)), generate a level of mechanistic complexity which results in experimental parameter ranges associated with non-stationary, complex kinetic regimes, such as multiple stationary states, periodic oscillations, and complex non-periodic oscillations [137–149]. Figure 6.24 displays experimental oscillatory regimes during the potentiostatic formic acid oxidation on a Pt(100) single-crystal surface. Mechanistic analysis has revealed [141–144], [150,151] that the periodic combination of CO surface poisoning at low electrode potentials, rapid OH surface poisoning at high electrode potentials, and the faradaic current due to formic acid oxidation at intermediate electrode potentials, combined with the action of the potentio/galvanostat generates a positive feedback cycle dampened by a negative feedback cycle resulting in sustained current and potential oscillations.

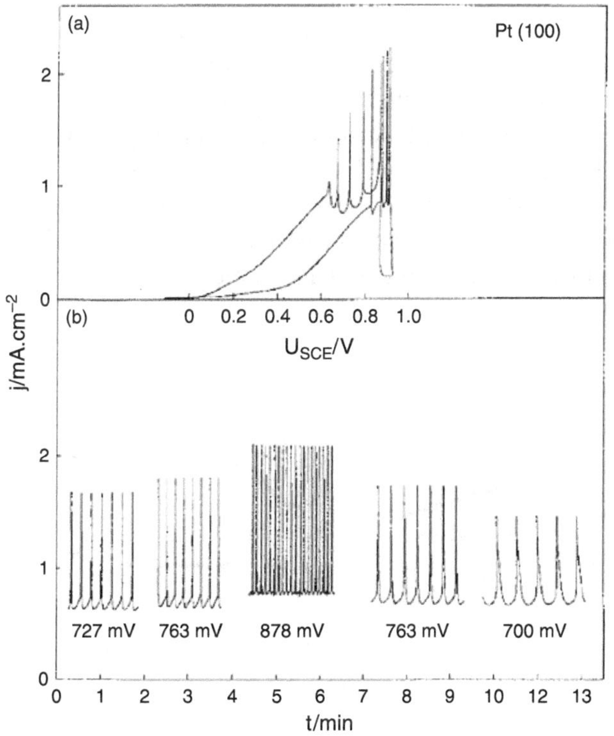

Figure 6.24. (a) Cyclic voltammogram of electro-oxidation of formic acid on a Pt(100) single-crystal showing current spikes indicating sustained current oscillations; and (b) current oscillations measured at various fixed potentials; all potentials are with reference to the Standard Calomel Reference Electrode (SCE) (adapted from [140]).

4.2.1. The electrooxidation of carbon monoxide

The electrooxidation of CO to CO_2 is, similar to its electroless counterpart in gas-phase catalysis, one of the most widely studied electrochemical reaction processes [131,152]. It is generally assumed that the electrooxidation of adsorbed CO proceeds primarily via a Langmuir–Hinshelwood type mechanism involving either adsorbed water molecules or, at higher electrode potentials, adsorbed surface hydroxides (see Figure 6.25) according to

$$CO_{ads} + H_2O_{ads} \rightarrow CO_2 + 2H^+ + 2e^- \tag{32}$$

and

$$CO_{ads} + OH_{ads} \rightarrow CO_2 + H^+ + e^-. \tag{33}$$

A fully CO_{ads} covered Pt surface exhibits surface coverages approaching 1 CO molecule/Pt atom with only a slight dependence on the specific crystal face considered. Because of this, anodic stripping of a monolayer of adsorbed CO_{ads}, that is, the anodic oxidation of CO_{ads} (reaction (32)) during a slow anodic potential sweep, can be used as a means to estimate the electrochemical surface area of Pt surfaces. Figure 6.26 shows the anodic CO stripping voltammetry of a fully CO-covered Pt and Ru surface (solid curves). The integral under the stripping peak corresponds to the charge transferred across the electrochemical interface according to reaction (32). The stripping charge can be converted into moles of adsorbed CO and, assuming 1 CO per electrode surface atom and a suitable electrode surface geometry (100),

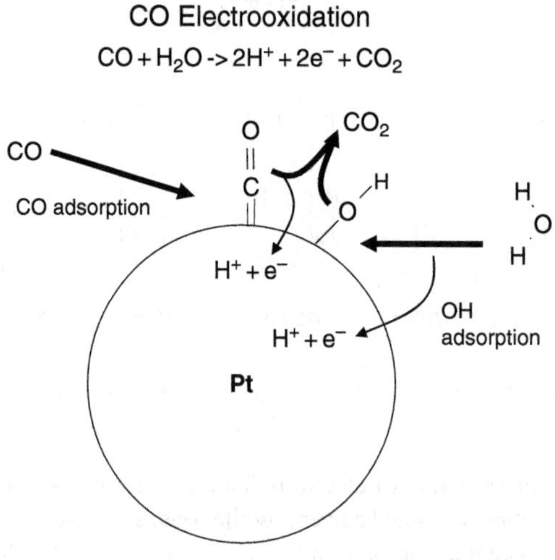

Figure 6.25. Schematic of the mechanism of the CO electrooxidation on a Pt or Pt alloy electrocatalyst.

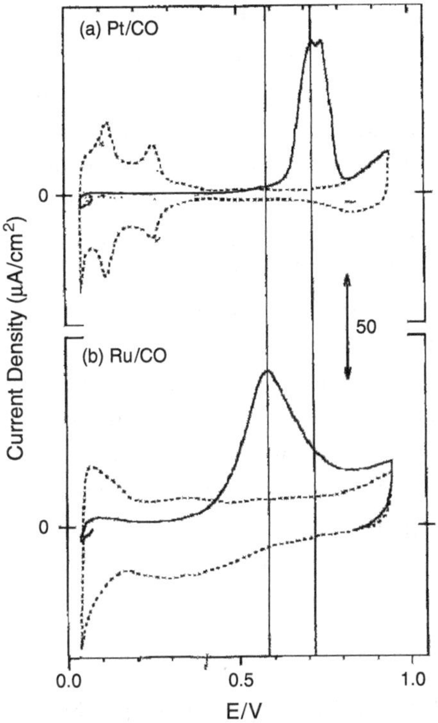

Figure 6.26. CO stripping voltammetry of UHV sputter-cleaned electrodes in 0.5 M H_2SO_4 on (a) Pt and (b) Ru. Solid curves represent the stripping of CO in the first positive-going sweep; dotted lines represent the voltammetric profiles in the absence of CO (adapted from Ref. [153]).

(110), (111) or a mix thereof, subsequently can be converted into the real area of the electrode surface.

Another aspect of the comparative study of the CO stripping voltammetry in Figure 6.26 is relevant to the discussion of surface electrocatalysis of C1 molecules. Figure 6.26(b) clearly shows that the onset of the stripping peak on a Ru surface is shifted toward lower electrode potentials compared to the Pt surface. Similarly, the stripping peak potential on Ru is shifted about 0.15 V negative compared to Pt. It was concluded that reactions (32) and (33) occur earlier, that is, at lower electrode potentials, on Ru than on Pt surfaces. This observation is accounted for by the fact that the adsorption of water and oxygen-containing species onto Ru surfaces commences at potentials as low as 0.2 V/RHE, much more negative than on Pt. It follows that a hypothetical 'designer surface' consisting of alternating metal atoms A and B with metal A capable of adsorbing CO and metal B capable of supplying water or oxygenated species at low potentials would constitute an ideal electrocatalytic interface for CO removal. If each metal atom A, the nucleation site of CO adsorption, has an adjacent metal atom B, the nucleation site of water or OH, the overall CO oxidation rate should be maximized. In fact, this designer surface can be realized by a $Pt_{50}Ru_{50}$ alloy surface that exhibits its CO stripping peak at the lowest electrode potentials among all studied binary Pt−Ru alloy stoichiometries [153–155]. The enhancement mechanism of a Pt−Ru surface has come to be known as a 'bifunctional' mechanism [156], since Pt and Ru surface atoms have their distinct roles in the surface catalytic process. The bifunctional mechanism is a special case of an 'ensemble effect' in electrocatalysis, that is, a modification of the overall catalytic reaction rate based on a specific configuration, often clustering, of one or more types of surface atoms.

Under conditions where reactions (32) and (33) are rate-determining, and given the fact that CO adsorbs strongly on both Pt and Ru, the continuous electrooxidation of dissolved CO gas on Pt−Ru alloys is characterized by negative reaction orders with respect to the CO partial pressure. Pt atoms are CO covered. CO and water compete for Ru sites which may, at sufficiently high CO partial pressures, also become completely CO covered suppressing the nucleation of water species. As a result of this, the overall reaction rates drops with increasing CO partial pressures (negative reaction order).

A 'purely bifunctional' mechanism was assumed to be operative in the CO oxidation on Pt_3Sn surfaces as well as on other Pt alloys with oxophilic transition metals [157–159]. Here, the oxophilic Sn surface atoms are believed to provide nucleation sites for water and, following stepwise hydrogen abstraction, for its subsequent oxygenated surface products OH and O. CO oxidation on Sn atoms is unlikely [131,160] such that no competition for Sn sites occurs between water and CO molecules. All Pt atoms are covered with CO.

Designing catalytically active metal-alloy surfaces for CO electro-oxidation has gained significance for fuel cell technologies employing hydrogen-rich, but CO contaminated anode feeds. Trace amounts of CO (100–10,000 ppm CO depending on the gas pretreatment), left over from the incomplete conversion of carbonaceous fuels into H_2 and CO_2, poison the anode catalyst surface and prevent Pt surface atoms from adsorbing and oxidizing molecular hydrogen. A schematic of the individual surface reactions of this combined process is shown in Figure 6.27. The blockage of surface sites reduces the effective H_2 electrooxidation current densities at the anode resulting in a lower cell voltage and cell power. Hence, feeding fuel cells with H_2/CO mixtures requires fuel cell anode electrocatalysts with a high CO tolerance; that is, catalysts that oxidize CO_{ads} at very low electrode potentials where the surface is almost fully covered with CO. Despite extensive research into active CO oxidation catalysts involving all single-metals and a huge number of binary alloy combinations, $Pt_{50}Ru_{50}$ continues to be the most preferred anode catalyst for reformate-fed, low temperature fuel cells.

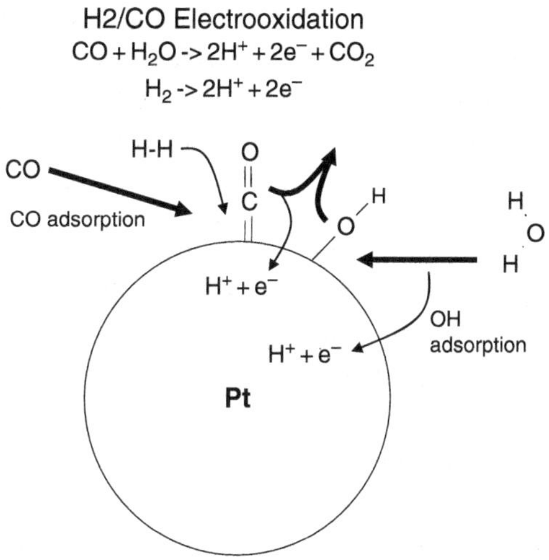

Figure 6.27. Schematic of the CO electro-oxidation on a Pt or Pt alloy electrocatalyst in the presence of hydrogen. CO adsorption blocks the dissociative adsorption and oxidation of hydrogen.

In recent years, the search for a Pt alloy fuel cell anode electrocatalyst superior to $Pt_{50}Ru_{50}$ has remained a scientific priority. Alloying additional metals with Pt and Ru has been a widely used approach toward more active CO electrooxidation activity [64,88,161–169]. Because a rigorous and reliable theoretical description of electrocatalytic activity on multi-metal surfaces was lacking, much of the early multi-metal alloy work was mainly empirical. More recently, systematic exploration of large ternary and quaternary compositional space became possible by the advent of combinatorial and high-throughput screening methods [59,63,64,66,67,164,170,171].

Rational design strategies of improved CO-tolerant electrocatalysts have focused on the chemical bonding of the reactants and intermediates. Based on the bifunctional mechanism two design strategies toward more active multi-metal surfaces compared to $Pt_{50}Ru_{50}$ are obvious: First, a third alloying metal M in a ternary Pt−Ru−M alloy may increase the oxophilicity of the surface by forming strong M−OH surface bonds. This translates into water activation at lower electrode potentials which, in turn, leads to CO removal at lower potentials. Alternatively, given the high CO coverage at low electrode potentials and the competition between CO and water for Ru sites, the atomic proximity of a third alloying metal M may reduce the Pt−CO and/or Ru−CO bond strength. This results in a reduced CO coverage which allows for water adsorption and CO removal at lower electrode potentials. While the former strategy associates the role of the third metal M with the bifunctional effect, the latter strategy focuses on the electronic modification of Pt and/or Ru surface atoms caused by the metal M (ligand effect). The question arises as to the mechanistic role of the individual alloy metals in the overall surface catalysis.

Liu and co-workers [160,172] reported a detailed computational study on the CO and CO/H_2 electrooxidation reactions, in which they clarified the mechanistic role of the alloy metals in the promotion of electrocatalytic activity. First, they presented comparative DFT calculations of adsorption energies of CO, H_2, and OH on surface slabs representing Pt(111), a Ru(0001), a Pt monolayer on Ru (0001), $Pt_{50}Ru_{50}$, as well as Pt_3Sn. In accordance with the d-band model, upward shifts of the d-band center resulted in stronger surface bonds for CO and H_2 (see Figure 6.28). In their formalism, they also considered the effect of the electric field at the reactive interface on the adsorption and reaction energies of intermediates. In combining their thermo-chemical data with a simple kinetic model, Liu et al. succeeded in separating the catalytic activity into the promoting effect associated with the bifunctional mechanism of Ru and into the effect associated with the electronic modification of Pt by Ru (ligand effect). The authors found that the role of Ru in the CO oxidation of Pt−Ru surfaces in the absence of hydrogen is, in fact, the bifunctional effect; this means that Ru effectively is responsible for the supply of water adsorption sites. In contrast, the authors [161] concluded that in the presence of hydrogen (H_2/CO oxidation according to Figure 6.28) the mechanistic role of

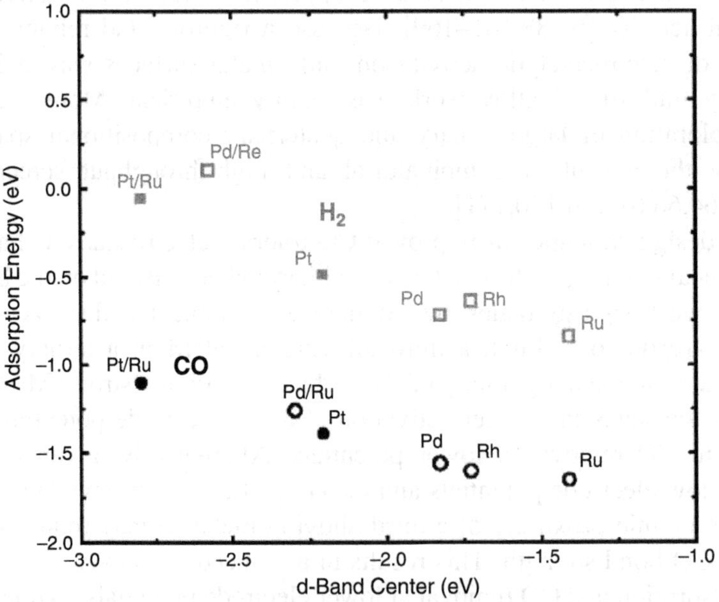

Figure 6.28. Calculated adsorption energies for CO and H_2 on different metals and overlayers plotted as a function of the d-band center (adapted from [160]).

alloyed Ru consists predominantly in the modification of the electronic properties of Pt atoms (ligand effect), in particular the modification of the Pt−CO bond strength. Ru atoms in the surface or underneath a Pt monolayer weaken the Pt−CO bond and therefore lower the CO surface coverage, which allows for more hydrogen to be adsorbed and oxidized per unit time given that hydrogen still adsorbs on the alloy surface. Figure 6.29 illustrates how significantly the electrode potential at a given overall catalytic activity depends on the CO and H_2 chemisorption energies of alloy surfaces. A reduction in the CO binding energy reduces the predicted electrochemical overpotential ($U-U^{Pt}$).

A more recent DFT study [69] extended the predictive modeling of CO tolerance to ternary alloy systems of Pt, Ru and a third metal M. Figure 6.30 shows the predicted trends in CO binding energy (including the hydrogen adsorption energy) that were previously shown to correlate well with the H_2/CO electro-oxidation activity. The surface binding energy on pure Pt ('pure Pt' in Figure 6.30) is taken to be zero. The CO bond strength decreases in the order Pt > Pt−Ru ('Pt' in Figure 6.30) > Pt−Ru−Pd > Pt−Ru−Ni ~ Pt−Ru−Sn > Pt−Ru−Co ('Pd', 'Ni', 'Sn', and 'Co', respectively, in Figure 6.30). This study underlined the critical role of the ligand effect in the promotion of the H_2/CO electrooxidation. The study also suggested specific candidates for the design of more CO-tolerant ternary Pt alloys.

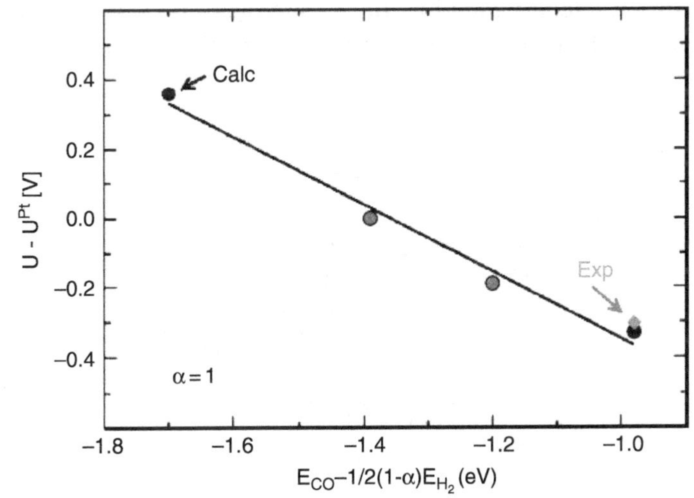

Figure 6.29. Plots of the calculated (circles) and measured (diamonds) overpotentials as a function of the calculated CO adsorption energy, E_{CO}, at 1 atm H_2 and 250 ppm CO at 60°C. The potentials are shown relative to the overpotential on the Pt electrode and are extracted at a current density of 1 mA/cm^2 (adapted from [160]).

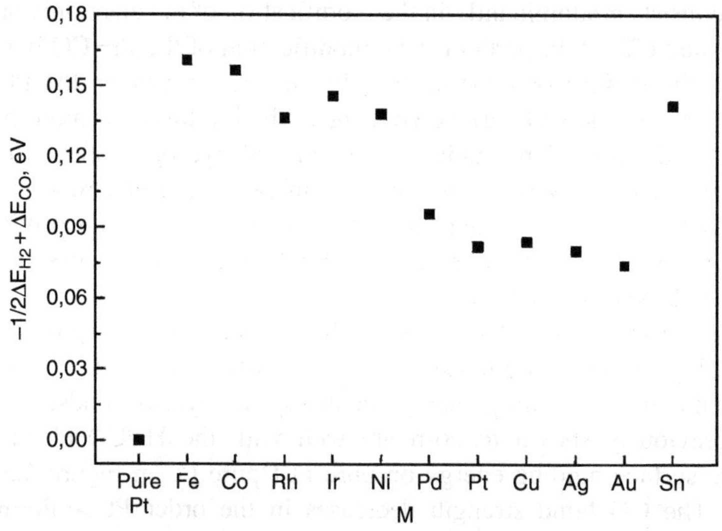

Figure 6.30. The surface activities of various Pt–Ru–M ternary alloys. The additional alloy atoms, M, on the surface are listed along the horizontal axis. 'Pure Pt' denotes monometal Pt; 'Pt' denotes the binary alloy Pt/Ru (adapted from [69]).

4.2.2. *The electrooxidation of formic acid and methanol*

4.2.2.1. *Formic acid electrooxidation*

The mechanism of electrooxidation of formic acid is a well established dual pathway (Figure 6.31) involving a reactive intermediate pathway to CO_2 as well

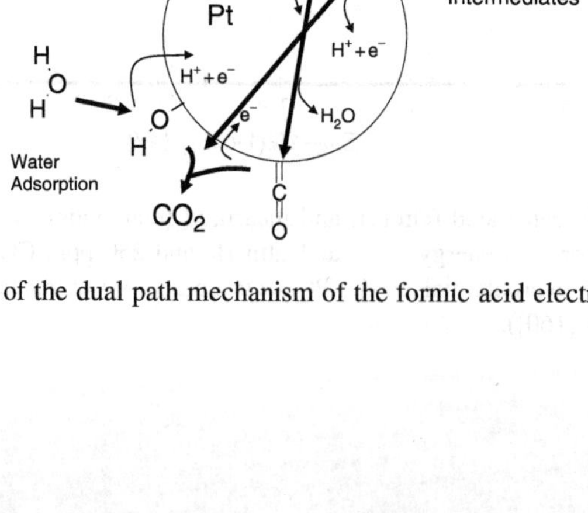

Figure 6.31. Schematic of the dual path mechanism of the formic acid electrooxidation on a Pt or Pt alloy electrocatalyst.

as a CO-poisoned pathway. Figure 6.31 schematically summarizes the current state of knowledge on this reaction. Formic acid adsorbs on Pt forming a Pt−C bonded intermediate which directly reacts to CO_2 via some reactive intermediate RI according to

$$HCOOH \rightarrow RI \rightarrow CO_2 + 2H^+ + 2e^- \tag{34}$$

In parallel, formic acid is known to form surface-bonded CO, either via an adsorbed state (see Figure 6.31) or via an electroless process according to

$$HCOOH \rightarrow CO + H_2O \tag{35}$$

A large body of work has been published and recently reviewed on spectroscopic evidence of the adsorption and transformation of formic acid on Pt and Pt alloy surfaces according to reactions (34) and (35) [130,131]. Pt was found to interact weakly up to a potential of 0.2 V/RHE. Above this electrode potential, both reaction pathways were found to be present. Reaction (34) was established as the dominant process with reaction (35) being the site-blocking side reaction. On Ru, however, formic acid reacted distinctly differently. Here formic acid decomposition into surface CO_{ads} was found to be very pronounced even at very low electrode potentials.

Bulk Pt alloys for the electrooxidation of formic acid have been less frequently studied compared to underpotential deposition (upd) modified Pt surfaces. The $Pt_{50}Ru_{50}$ surface was again found to be one of the most active Pt–Ru surfaces. Underpotentially deposited metals, such as Bi, Se, Sb, were studied as reaction modifiers for Pt surfaces and provided significant electrocatalytic activity increases. Electronic factors (ligand effects) rather than bifunctional effects were held responsible for these activity modifications, because the metal coverages that caused the activity gains were extremely small.

In summary, CO poisoning is held responsible for slow formic acid oxidation rates. Activity improvements should therefore correlate well with the CO tolerance of the catalytic surface. However, to date, no systematic study has been presented comparing the catalytic activity of different binary and ternary alloy surfaces under similar conditions for testing the model predictions of Figure 6.31.

4.2.2.2. Methanol electrooxidation

The electrooxidation of methanol has attracted tremendous attention over the last decades due to its potential use as the anode reaction in direct methanol fuel cells (DMFCs). A large body of literature exists and has been periodically reviewed [130,131,156], [173–199]. Unlike for formic acid, a generally accepted consensus on the specific mechanistic pathways of methanol electrooxidation is still elusive.

Spectroscopic evidence, in particular DEMS [130], strongly suggests that adsorbed CO is not just forming in a parallel poisoning pathway, but that it constitutes a mechanistic intermediate, at least at low to medium electrode potentials of 0.2–0.6 V / RHE. Figure 6.32 provides a schematic illustration of the *series pathway* with CO as intermediate. Methanol adsorbs on Pt forming some H-containing surface intermediate; H atoms are abstracted in a sequence of steps, until absorbed CO is formed. Removal of CO requires adsorption of water to yield water-based oxygenated surface species.

Methanol electrooxidation rates on Pt surfaces are generally much smaller at electrode potentials between 0.05 – 0.4 V / RHE compared to those of formic acid. This can be accounted for by the absence of adsorbed water on Pt which is a requirement for CO_2 formation. Methanol oxidation exhibits a surface structure sensitivity with the catalytic rates decreasing in the order Pt(110) > Pt(100) >> Pt(111). The methanol oxidation rate on Ru at room temperatures is virtually zero, since methanol does not adsorb on Ru under these conditions. However, as pointed out earlier, Ru is more oxophilic than Pt and promotes the adsorption of water molecules at lower electrode potentials. In fact, Pt−Ru alloy catalysts promote the electrooxidation of methanol significantly. A $Pt_{90}Ru_{10}$ alloy was found to be most active [66,154,197,200] at room temperatures where methanol adsorption on Ru is absent. Under these conditions, the high activity of the $Pt_{90}Ru_{10}$ surface was accounted for by invoking an ensemble effect of three adjacent Pt atoms next to one Ru atom. Assuming that methanol adsorption requires three adjacent Pt atoms

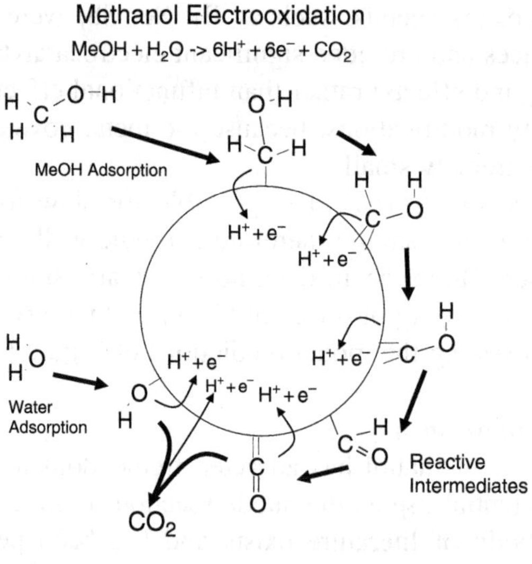

Figure 6.32. Schematic of the series mechanism of the methanol electrooxidation on a Pt or Pt alloy electrocatalyst with CO being formed as an intermediate, not in a parallel poisoning pathway.

with one adjacent Ru atom in order to provide the surface OH needed to form CO_2, Gasteiger et al. showed that the maximum concentration of active site ensembles occurs near 10 at. % Ru. The $Pt_{50}Ru_{50}$ stoichiometry is the most active alloy surface at elevated temperatures, because methanol starts adsorbing on Ru under these conditions and the ensemble effect disappears. At Ru concentrations of about 50 at.% the alloy surface is nearly free from CO indicating that the rate-limiting step has now shifted to the methanol adsorption.

Numerous exploratory studies for improved multi-metal alloy catalysts have been reported using conventional catalyst development [156,173,183,199], [201–207] as well as combinatorial and high-throughput screening techniques [164,165,171], [208–211]. Figure 6.33 displays the summary of an experimental combinatorial screening study involving a 64-member Pt alloy electrocatalyst library [69]. The reactivity of ternary Pt alloy electrocatalysts was found to increase in the order Pt < Pt−Ru < Pt−Ru−Ni < Pt−Ru−Co. It is interesting that the activity trends among Pt, PtRu, PtRuNi and PtRuCo obtained from the experimental screening are consistent with the trends obtained in the computational study of the CO tolerance of ternary Pt alloys in Figure 6.33. These results suggest that, assuming the CO removal to be rate-determining, the CO binding energy may serve as a reasonable predictor for improved catalytic activity: In the case of methanol oxidation, weaker CO bond energies may lead to more active CO_{ads} species that are oxidized

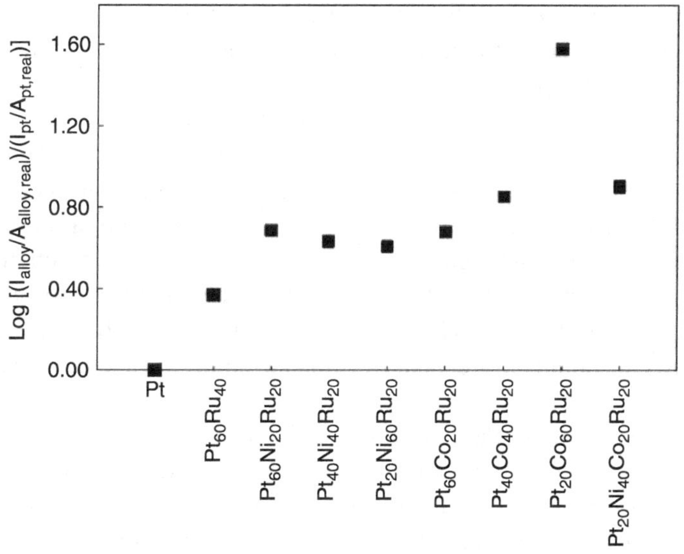

Figure 6.33. Trends in Pt surface-area normalized electrochemical activity of various ternary alloy compositions with respect to the electrooxidation of methanol. Activity gains are seen in the order Pt, PtRu, PtRuNi, and PtRuCo (adapted from [69]).

at lower electrode potentials. Similar to the H_2/CO electrooxidation, modifications of the electronic properties of Pt surface atoms (ligand effects) apparently outweigh bifunctional effects in the development of more active methanol oxidation catalysts.

5. Summary and outlook

In this chapter, we have described modern electrochemical surface science and catalysis and shown fundamental correlations between the structure, bonding, composition and electrocatalytic reactivity at the interface.

From a description of the geometric structure of electrified interfaces we moved to a description of models for electrochemical electron transfer across an electrode interface. The science of atomic scale electrochemistry was presented with an emphasis on the bonding of water molecules and anions on electrode surfaces. Subsequently, we presented an in-depth description of the role of surface bonding in a number of important electrocatalytic processes for energy conversion. We have attempted to illustrate how closely surface bonding and catalytic activity are related.

In electrochemistry, potential and current measured by electroanalytical methods provide kinetic and potential energy pictures of electrochemical reactions. Measured current and potential are strongly connected to the molecular scale properties of the electrode surface, solvent molecules and ions. Currents and potentials represent how molecules and atoms are distributed near the interface, how they are bonded on the electrode surface, and how they are solvated in the electrolyte solution. The electrochemical properties are also sensitive to the atomic arrangements of the electrode surface: crystallographic orientations and defects.

In the future, we need to develop ways to apply novel and more sophisticated surface science techniques, such as the soft X-ray spectroscopies outlined in Chapter 2, to electrified interfaces under realistic conditions. For example, atom-specific probing of the occupied and unoccupied electronic states of electrochemical surface intermediates during the electrocatalytic reactions described in the present chapter has not been achieved to date. Modern synchrotron-based soft X-ray methods will aid these efforts to obtain more insight into the bonding of electrochemical intermediates.

Early in the last century, Paul Sabatier[1] pointed out 'A most important property of an excellent catalyst is that it has an ability to bind many molecules but not too strongly'. This Sabatier's principle is also the principle for how an excellent catalyst for electrochemical reactions works. In electrochemical terms, an active

[1] Paul Sabatier, French chemist. The Nobel Prize winner in Chemistry in 1912 for discovering metal particle catalysts for hydrogenation processes.

electrode catalyst, which exhibits a low overpotential for a reaction, binds reaction intermediates not too strongly but not too weakly. The ultimate goal of surface electrochemistry remains the design of an excellent catalyst by understanding and controlling the nature of the chemical bond formation between reaction intermediates and the surface of electrocatalysts.

References

[1] C.B. Duke (Ed.), Surface Science: The First Thirty Years. Elsevier, North-Holland, Amsterdam, 1994.
[2] A.T. Hubbard, R.N. Ishikawa, I. Katekaru, J. Electroanal. Chem. 86 (1978) 271.
[3] K. Yamamoto, D.M. Kolb, R. Kotz, G. Lehmpful, J. Electroanal. Chem. 135 (1978) 159.
[4] P.N. Ross, J. Electroanal. Chem. 126 (1979) 67.
[5] A.S. Homa, E.B. Yeager, B.D. Cahan, J. Electroanal. Chem. 100 (1983) 181.
[6] J. Clavilier, J. Electroanal. Chem. 107 (1980) 205.
[7] D.M. Kolb, Z. Phys. Chem. Neue Folge 154 (1987) 179.
[8] A.T. Hubbard, Chem. Rev. 88 (1988) 633.
[9] M.P. Soriaga, Chem. Rev. 90 (1990) 771.
[10] K. Itaya, S. Sugawara, K. Sashikata, N. Furuya, J. Vac. Sci. Technol. A 8 (1990) 515.
[11] A.J. Bard, H.D. Abruña, C.E. Chidsey, L.R. Faulkner, S.W. Feldberg, K. Itaya, M. Majda, O. Melroy, R.W. Murray, M.D. Porter, M.P. Soriaga, H.S. White, J. Phys. Chem. 97 (1993) 7147.
[12] M.J. Weaver, X. Gao, Annu. Rev. Phys. Chem. 44 (1993) 459.
[13] A. Bewick, K. Kunimatsu, B.S. Pons, Surf. Sci. 101 (1980) 131.
[14] F. Kitamura, M. Takeda, M. Takahashi, M. Ito, Chem.Phys.Lett. 142 (1987) 318.
[15] T. Iwashita, F.C. Nart, Progess in Surface Science 55 (1997) 271.
[16] K. Ashley, S. Pons, Chem. Rev. 88 (1988) 673.
[17] M.G. Samant, M.F. Toney, G.L. Borges, L. Blum, O.R. Melroy, J. Phys. Chem. 92 (1988) 220.
[18] C.A. Lucas, N.M. MarkovicThe Encyclopaedia of Electrochemistry. Willey-VCH, 2003.
[19] O. Stern, Zeitschrift für Electrochemie 30 (1924) 508.
[20] G. Gouy, Journ. de Phys. 9 (1910) 457.
[21] D.L. Chapman, Phil. Mag. 25 (1913) 475.
[22] O.M. Magnussen, Chem. Rev. 102 (2002) 679.
[23] R.J.D. Miller, G.L. McLendon, A.J. Noszik, W. Schmickler, F. Willig: Surface Electron Transfer Processes, VCH, New York, 1995.
[24] J.-M. Saveant: Elements of Molecular and Biomolecular Electrochemistry – An Electrochemical Approach to Electron Transfer Chemistry, Wiley, New York, 2006.
[25] J.-M. Saveant, J. Am. Chem. Soc. 114 (1992) 10595.
[26] A.Y. Lozovoi, A. Alavi, Phys. Rev. B 68 (2003) 245416.
[27] B.M. Ocko, G. Helgesen, B. Schart, J. Wang, A. Hamelin, Phys. Rev. Lett. 23 (1992) 3350.
[28] O.M. Magnussen, J. Wiechers, R.J. Behm, Surf. Sci. 289 (1993) 139.
[29] X. Gao, A. Hamelin, M.J. Weaver, Phys. Rev. B 44 (1991) 10983.
[30] D.M. Kolb, G. Lehmpfuhl, M.S. Zei, J. Electroanal. Chem. 179 (1984) 289.
[31] A. Hamelin, L. Stoicoviciu, G.J. Edens, X. Gao, M.J. Weaver, J. Electroanal. Chem. 365 (1994) 47.
[32] K. Wu, M.S. Zei, Surf. Sci. 415 (1998) 212.

[33] M.S. Zei, G. Ertl, Surf. Sci. 442 (1999) 19.
[34] C.A. Lucas, N.M. Markovic, P.N. Ross, Phys. Rev. Lett. 77 (1996) 4922.
[35] R. Michaelis, D.M. Kolb, J. Electroanal. Chem. 328 (1992) 341.
[36] C.L. Kellog, Surf. Sci. Rep. 21 (1994) 1.
[37] C.J. Liu, J.M. Cohen, J.B. Adams, A.F. Voter, Surf. Sci. 253 (1991) 334.
[38] P.A. Dowben, Critical Reviews in Solid State and Materials Sciences 13 (1987) 191.
[39] M.T. Koper, R.A.v. Santen, Surf. Sci. 422 (1999) 118.
[40] A. Nigani, F. Illas, J. Phys. Chem. B 110 (2006) 11894.
[41] M. Arshadi, R. Yamdagni, P. Kebarle, J. Phys. Chem. 74 (1990) 1475.
[42] H. Ogasawara, B. Brena, D. Nordlund, M. Nyberg, A. Pelmenschikov, L.G.M. Pettersson, A. Nilsson, Phys. Rev. Lett. 89 (2002) 276102.
[43] P.A. Theil, T.E. Madey, Surf. Sci. Reports 7 (1987) 221.
[44] M.A. Henderson, Surf. Sci. Rep. 46 (2002) 1.
[45] K. Andersson, A. Nikitin, L.G.M. Pettersson, A. Nilsson, H. Ogasawara, Phys. Rev. Lett. 93 (2004) 196101.
[46] E.Y. Cao, D.A. Stern, J.Y. Gui, A.T. Hubbard, J. Electroanal. Chem. 354 (1993) 71.
[47] S. Motoo, N. Furuya, J. Electroanal. Chem. 167 (1984) 309.
[48] R.J. Nichols, A. Bewick, J. Electroanal. Chem. 243 (1988) 445.
[49] H. Ogasawara, M. Ito, Chem. Phys. Lett. 221 (1994) 213.
[50] A. Peremans, A. Tajeddine, Phys. Rev. Lett. 73 (1994) 3010.
[51] P.W. Faguy, N. Markovic, R.R. Adzic, C.A. Fiero, E.B. Yeager, J. Electroanal. Chem. 289 (1990) 245.
[52] H. Ogasawara, Y. Sawatari, J. Inukai, M. Ito, J. Electroanal. Chem. 358 (1993) 337.
[53] M.E. Gamboa-Aldeco, E. Herrereo, P.S. Zelenay, A. Wieckowski, J. Electroanal. Chem. 348 (1993) 452.
[54] J. Clavilier, D. Armand, S.G. Sun, M. Petit, J. Electroanal. Chem. 205 (1986) 267.
[55] J. Solla-Gullon, F.J. Vidal-Iglesias, P. Rodriguez, E. Herrero, J.M. Feliu, J. Clavilier, A. Aldaz, J. Phys. Chem. B 108 (2004) 13573.
[56] B. Archibald, O. Bruemmer, M. Devenney, A. Gorer, B. Jandeleit, T. Uno, W.H. Weinberg, T. Weskamp, in K.C. Nicolaou, R. Hangko, W. Hartwig (Ed.), Handbook of Combinatorial Chemistry - Drugs, Catalyst, Materials. Wiley, New York, 2002.
[57] B. Jandeleit, D.J. Schaefer, T.S. Powers, H.W. Turner, W.H. Weinberg, Angew. Chem., Int. Ed. Engl. 38 (1999) 2495.
[58] X.-D. Xiang, X. Sun, G. Briceno, Y. Lou, K.-A. Wang, H. Chang, W.G. Wallace-Freedman, S.-W. Chen, P.G. Schultz, Science 268 (1995) 1738.
[59] A. Hagemeyer, P. Strasser, A.F. Volpe (Ed.), High-Throughput Screening in Chemical Catalysis – Technologies, Strategies and Applications. Wiley VCH, Weinheim, 2004.
[60] P. Cong, R.D. Doolen, Q. Fan, D.M. Giaquinta, S. Guan, E.W. McFarland, D.M. Poojary, Angew. Chem. Int. Ed. 38 (1999) 484.
[61] S. Senkan, Angew. Chem. Int. Ed. 40 (2001) 312.
[62] Y.M. Liu, P.J. Cong, R.D. Doolen, S.H. Guan, V. Markov, L. Woo, S. Zeyss, U. Dingerdissen, Applied Catalysis A: General 254 (2003) 59.
[63] E. Reddington, A. Sapienza, B. Gurau, R. Viswanathan, S. Sarangapani, E.S. Smotkin, T.E. Mallouk, Science 280 (1998) 1735.
[64] P. Strasser, Q. Fan, M. Devenney, W.H. Weinberg, Proceedings Volume of the AIChE Fall National Meeting (2003).
[65] P. Strasser, S. Gorer, M. Devenney, in O. Yamamoto (Ed.), Proceedings Volume of the international symposium on fuel cells for vehicles-41st Battery Symposium. The Electrochemical Society of Japan, Nagoya, 2000, p. 153.

[66] P. Strasser, S. Gorer, Q. Fan, K. Chondroudis, K. Cendak, D. Giaquinta, M. Devenney, in A. Hagemeyer, P. Strasser, A.F. Volpe (Ed.), High-Throughput Screening in Chemical Catalysis. Wiley VCH, Weinheim, 2004, p. 271.
[67] E.S. Smotkin, R.R. Diaz-Morales, Ann. Rev. of Mater. Res. 33 (2003) 557.
[68] E.S. Smotkin, J. Jiang, A. Nayar, R. Liu, Applied Surface Science 252 (2006) 2573.
[69] P. Strasser, Q. Fan, M. Devenney, W.H. Weinberg, P. Liu, J.K. Nørskov, J. Phys. Chem. B 107 (2003) 11013.
[70] J. Greeley, T. Jaramillo, J. Bonde, I. Chorkendorff, J.K. Nørskov, Nature Materials 5 (2006) 909.
[71] K. Kinoshita: Electrochemical Oxygen Technology, Wiley, New York, 1992.
[72] A.J. Appleby, J.O.M. Bockris, E. Yeager, S.U.M. Khan, R.E. White, Plenum Press 7 (1983) 173.
[73] A.J. Appleby, J. Electroanal. Chem 357 (1993) 117.
[74] A. Damjanovic, in O.J. Murphy, S. Srinivasan (Ed.), Electrochemistry in Transition. Plenum Press, New York, 1992, p. 107.
[75] M.R. Tarasevic, A. Sadkowski, E. Yeager, in B. Conway, J.O.M. Bockris, E. Yeager, S.U.M. Khan, R.E. White (Ed.), Comprehensive Treatise of Electrochemistry. Plenum Press, New York, 1983, p. 301.
[76] Z. Shi, J. Zhang, Z.-S. Liu, H. Wang, D.P. Wildinson, Electrochimica Acta 51 (2006) 1905.
[77] J. Lipkowski, P.N. Ross: Electrocatalysis, John Wiley and Sons Canada, Ltd., Canada, 1998.
[78] R. Adzic, in J. Lipkowski, P.N. Ross (Ed.), Electrocatalysis. Wiley, New York, 1998, p. 197.
[79] A. Damjanovic, V. Brusic, J.O.M. Bockris, J. Phys. Chem. 71 (1967) 2471.
[80] A. Damjanovic, M.A. Genshaw, J.O.M. Bockris, J. Phys. Chem. 45 (1964) 4057.
[81] E. Yeager, M. Razaq, D. Gervasio, A. Razaq, D. Tryk, Structural Effects in Electrocatalysis and Oxygen Electrochemistry 11 (1992) 440.
[82] E. Yeager, M. Razaq, D. Gervasio, A. Razaq, D. Tryk, J. Serb. Chem. Soc. 57 (1992) 819.
[83] N.M. Markovic, H.A. Gasteiger, P.N. Ross, The Journal of Physical Chemistry 100 (1996) 6715.
[84] N.M. Markovic, H.A. Gasteiger, J. Philip N. Ross, J. Phys Chem 99 (1995) 3411.
[85] N.M. Markovic, P.N. Ross, CATTECH 4 (2000) 110.
[86] N.M. Markovic, P.N. Ross, Surf. Sci. Reports 45 (2002) 117.
[87] V.R. Stamenkovic, B. Fowler, B.S. Mun, G. Wang, P.N. Ross, C.A. Lucas, N.M. Markovic, Science 315 (2007) 493.
[88] P. Stonehart, Phys. Chem 94 (1990) 913.
[89] P. Stonehart, Appl. Electrochem. 22 (1992) 995.
[90] A.B. Anderson, J. Roque, S. Mukerjee, V.S. Murthi, N.M. Markovic, V. Stamenkovic, J. Phys. Chem. B 109 (2005) 1198.
[91] M. Neegrat, A.K. Shukla, K.S. Gandhi, J. of Applied electrochemistry 31 (2001) 373.
[92] S. Mukerjee, S. Srinivasan, M.P. Soriaga, J. McBreen, J. Electrochem. Soc. 142 (1995) 1409.
[93] B.C. Beard, P.N. Ross, J. Electrochem. Soc. 137 (1990) 3368.
[94] J. Zhang, K. Sasaki, E. Sutter, R.R. Adzic, Science 315 (2007) 220.
[95] V. Stamenkovic, B.S. Moon, K.J. Mayerhofer, P.N. Ross, N. Markovic, J. Rossmeisl, J. Greeley, J.K. Nørskov, Angew. Chem. Int. Ed. 45 (2006) 2897.
[96] E. Antolini, J.R.C. Salgado, M.J. Giz, E.R. Gonzalez, Int. J. Hydrogen Energy 30 (2005) 1213.
[97] D. Thompsett, in W. Vielstich, A. Lamm, H.A. Gasteiger (Ed.), Handbook of Fuel Cells - Fundamentals, Technology and Applications. Wiley, New York, 2003, p. 467 (chapter 37).
[98] U.A. Paulus, A. Wokaun, G.G. Scherer, T.J. Schmidt, V. Stamenkovic, N.M. Markovic, P.N. Ross, Electrochim. Acta 47 (2002) 3787.
[99] N.M. Markovic, T.J. Schmidt, V. Stamenkovic, P.N. Ross, Fuel Cells 1 (2001) 105.
[100] E. Antolini, J.R.C. Salgado, E.R. Gonzalez, J. Power Sources 160 (2006) 957.
[101] S. Mukerjee, J. McBreen, S. SrinivasanProc. Electrochem. Soc. (Oxygen Electrochemistry), 1996, p. 38.

[102] S. Mukerjee, S. Srinivasan, in W. Vielstich, A. Lamm, H. Gasteiger (Ed.), Handbook of fuel cells – Fundamentals, Technology and Applications Wiley, New York, 2003, p. 502 (chapter 34).
[103] S. Mukerjee, S. Srinivasan, M.P. Soriaga, J. McBreen, J. Phys. Chem. 99 (1995) 4577.
[104] S. Mukerjee, S. Supramaniam, M.P. Soriaga, J. Electrochem. Soc. 142 (1995) 1409.
[105] U.A. Paulus, A. Wokaun, G.G. Scherer, T.J. Schmidt, V. Stamenkovic, V. Radmilovic, N.M. Markovic, P.N. Ross, J. Phys. Chem. B 106 (2002) 4181.
[106] J. Zhang, M.B. Vukmirovic, K. Sasaki, A.U. Nilekar, M. Mavrikakis, R.R. Adzic, J. Am. Chem. Soc. 127 (2005) 12480.
[107] V. Stamenkovic, T.J. Schmidt, P.N. Ross, N.M. Markovic, J. Phys. Chem. B 106 (2002) 11970.
[108] B. Hammer, J.K. Nørskov, Adv. Catal. 45 (2000) 71.
[109] J. Greeley, J.K. Nørskov, M. Mavrikakis, Annu. Rev. Phys. Chem. 53 (2002) 319.
[110] B. Hammer, L.B. Hansen, J.K. Nørskov, Phys. Rev. B 59 (1999) 7413.
[111] B. Hammer, J.K. Nørskov, in R.M. Lambert, G. Pacchioni (Ed.), Chemisorption and Reactivity on Supported Clusters and Thin Films. Kluwer Academic Publishers, Netherlands, 1997, p. 285.
[112] A. Ruban, B. Hammer, P. Stoltze, H.L. Skriver, J.K. Nørskov, J. Mol. Catal. A: Chem 115 (1997) 421.
[113] Y. Xu, A.V. Ruban, M. Mavrikakis, J. Am. Chem. Soc. 126 (2004) 4717.
[114] A.V. Ruban, H.L. Skriver, J.K. Nørskov, Physical Review B 59 (1999) 15990.
[115] T.V. Albu, A.B. Anderson, J. Electrochem. Soc. 147 (2000) 4229.
[116] R.A. Sidik, A.B. Anderson, J. Electroanal. Chem 528 (2002) 69.
[117] J.K. Nørskov, J. Rossmeisl, A. Logadottir, L. Lindqvist, J.R. Kitchin, T. Bligaard, H. Jonsson, J. Phys. Chem. B 108 (2004) 17886.
[118] M.B. Vukmiovic, J. Zhang, K. Sasaki, A.U. Nilekar, F. Uribe, M. Mavrikakis, R.R. Adzic, Electrochim. Acta 52 (2007) 2257.
[119] K. Sasaki, J.X. Wang, M. Balasubramanian, J. McBreen, F. Uribe, R.R. Adzic, Electrochim. Acta 49 (2004) 3873.
[120] J. Zhang, Y. Mo, M.B. Vukmirovic, R. Klie, K. Sasaki, R.R. Adzic, J. Phys. Chem. B 108 (2004) 10955.
[121] S.R. Brankovic, J. McBreen, R.R. Adzic, J. Electroanal. Chem. 503 (2001) 99.
[122] S.R. Brankovic, J.X. Wang, R.R. Adžic, Electrochem. and Solid-State Letters 4 (2001) A217.
[123] J.L. Zhang, M.B. Vukmirovic, K. Sasaki, A.U. Nilekar, M. Mavrikakis, R.R. Adzic, J. Am. Chem. Soc. 127 (2005) 12480.
[124] V. Stamenkovic, T.J. Schmidt, P.N. Ross, N.M. Markovic, J. Electroanal. Chem 554 (2003) 191.
[125] V.R. Stamenkovic, B.S. Mun, K.J.J. Mayrhofer, P.N. Ross, N.M. Markovic, J. Am. Chem. Soc. 128 (2006) 8702.
[126] P. Strasser, Lattice-strained Pt Nanoparticle Catalysts for the electroreduction of oxygen at PEMFC cathodes, ACS fall meeting, San Francisco, 2006.
[127] P. Strasser, Pt alloy nanoparticle catalysts with high lattice strain for the electroreduction of molecular oxygen, ECS fall meeting, Cancun, Mexico, 2006.
[128] S. Koh, J. Leisch, M.F. Toney, P. Strasser, Probing size and composition distribution dynamics of alloy nanoparticles in fuel cell electrodes using anomalous Small Angle X-ray Scattering (ASAXS), ECS fall meeting, Cancun, Mexico, 2006.
[129] S. Koh, M.F. Toney, P. Strasser, Lattice-strained Pt shell Nanoparticle Catalysts for the Electroreduction of Oxygen at PEMFC cathodes, AIChE fall meeting, San Francisco, 2006.
[130] T.D. Jarvi, E.M. Stuve, in J. Lipkowski, P.N. Ross (Ed.), Electrocatalysis. Wiley-VCH, New York, 1998, p. 376.
[131] P.N. Ross, in J. Lipkowski, P.N. Ross (Ed.), Electrocatalysis. Wiley, New York, 1998, p. 43.

[132] W. Vielstich, A. Lamm, H. Gasteiger (Ed.), Handbook of Fuel Cells – Fundamentals, Technology, and Application. Wiley, New York, West Sussex, 2003.
[133] A. Capon, R. Pearson, J Electroanal. Chem 44 (1973) 1.
[134] X. Xia, T. Iwasita, J. Electrochem. Soc. 140 (1993) 2559.
[135] J. Willsau, J. Heitbaum, Electrochim. Acta 31 (1986) 943.
[136] O. Wolter, J. Willsau, J. Heitbaum, J. Electrochem. Soc. 236 (1985) 95.
[137] J. Wojtowicz, in J. Bockris, B. Conway (Ed.), Modern Aspects of Electrochemistry. Butterworths, London, 1073, p. 47.
[138] J.L. Hudson, T.T. Tsotsis, Chem. Eng. Sci. 49 (1994) 1493.
[139] M.T.M. Koper, Adv. Chem. Phys. 92 (1996) 161.
[140] P. Strasser, Interface 9 (2000) 46.
[141] P. Strasser, J. Christoph, W.F. Lin, M. Eiswirth, J.L. Hudson, J. Phys. Chem. A 104 (2000) 1854.
[142] P. Strasser, M. Eiswirth, G. Ertl, J.Chem.Phys. 107 (1997) 991.
[143] P. Strasser, M. Eiswirth, M.T.M. Koper, J. Electroanal. Chem. 478 (1999) 50.
[144] P. Strasser, M. Luebke, F. Raspel, M. Eiswirth, G. Ertl, J. Chem.Phys. 107 (1997) 979.
[145] J. Christoph, P. Strasser, M. Eiswirth, G. Ertl, Science 284 (1999) 291.
[146] J. Lee, J. Christoph, P. Strasser, M. Eiswirth, G. Ertl, J. Chem. Phys. 115 (2001) 1485.
[147] J. Lee, J. Christoph, P. Strasser, M. Eiswirth, G. Ertl, in S.G. S.R. Narayanan, T. Zawodzinski,eds. (Ed.), Direct Methanol Fuel Cells. The Electrochemical Society, Washington, 2001, p. 80.
[148] J. Lee, P. Strasser, M. Eiswirth, G. Ertl, Electrochim. Acta 47 (2001) 501.
[149] J. Lee, J. Christoph, P. Strasser, M. Eiswirth, G. Ertl, Phys. Chem. Chem. Phys. 5 (2003) 935.
[150] M.T.M. Koper, J.Chem. Soc., Faraday Trans. 94 (1998) 1369.
[151] M.T.M. Koper: Modern Aspects of Electrochemistry, Kluwer Academic Publishers/Plenum Press, New York, 2003.
[152] M. Weaver, S.C. CHang, L.W. Leung, X. Jiang, M. Rubel, M. Szklarczyk, D. Zurawski, A. Wieckowski, J. Electroanal. Chem 327 (1992) 247.
[153] H.A. Gasteiger, N. Markovic, J. Philip N. Ross, E.J. Cairns, J. Phys. Chem. 98 (1994) 617.
[154] H.A. Gasteiger, N. Markovic, P.N. Ross Jr., E.J. Cairns, Electrochimica Acta 39 (1994) 1825.
[155] H. Binder, A. Koehling, G. Sandstede, in G. Sandstede (Ed.), From Electrocatalysis to Fuel Cells. University of Washington Press, Seattle, London, 1972, p. 43.
[156] M. Watanabe, S. Motoo, J. Electroanal. Chem. 60 (1975) 267.
[157] H. Gasteiger, N. Markovic, P.N. Ross, J. Phys. Chem. 99 (1995) 16757.
[158] H. Gasteiger, N. Markovic, P.N. Ross, J. Phys. Chem. 99 (1995) 8945.
[159] H. Gasteiger, N. Markovic, P.N. Ross, Catal. Lett. 36 (1996) 1.
[160] P. Liu, Á. Logadóttir, J.K. Nørskov, Electrochim. Acta 48 (2003) 3731.
[161] P. Stonehard, J. Appl. Electrochem. 22 (1992) 995.
[162] P. Stonehart, M. Watanabe, N. Yamamoto, Electrocatalyst 1993.
[163] J.H. Cho, W.J. Roh, D.K. Kim, J.B. Yoon, J.H. Choy, H.S. Kim, J. Chem. Soc.–Faraday Transactions 94 (1998) 2835.
[164] W.C. Choi, Y.J. Kim, S.I. Woo, Catalysis Today 74 (2002) 235.
[165] K.L. Ley, R. Liu, C. Pu, Q. Fan, N. Leyarovska, C. Segre, E.S. Smotkin, J. Electrochem. Soc. 144 (1997) 1543.
[166] G. Tamizhmani, G.A. Capuano, J. Electrochem. Soc. 41 (1994) 968.
[167] B. Gurau, E. Reddington, S. Sarangpani, J. of Phy. Chem. B (1998) 9997.
[168] J. Shim, D.-Y. Yoo, J. Lee, Electrochim. Acta 45 (2000) 1943.
[169] A. Seo, J. Lee, K. Han, H. Kim, Electrochim. Acta (2006) in press.
[170] T. He, E. Kreidler, L. Xiong, J. Luo, C.J. Zhong, J. Electrochem. Soc. 153 (2006) A1637.

[171] P. Strasser, S. Gorer, M. Devenney, in S.G. S.R. Narayanan, T. Zawodzinski,eds. (Ed.), Direct Methanol Fuel Cells. The Electrochemical Society, Washington, 2001, p. 191.
[172] J.K. Nørskov, P. Liu, Anode Catalyst Materials for Use in Fuel Cells US 2002/0146614 A1, 2002.
[173] K.-W. Park, J.-H. Choi, S.-A. Lee, C. Pak, H. Chang, Y.-E. Sung, J. of Catalysis 224 (2004) 236.
[174] D. Cao, G.-Q. Lu, A. Wieckowski, S.A. Wasileski, M. Neurock, J of Phys. Chem. B 109 (2005) 11622.
[175] E.A. Batista, G.R.P. Malpass, A.J.Motheo, T. Iwasita, J. of Electroanal. Chem. 571 (2004) 273.
[176] M.A.A. Rahim, R.M.A. Hameed, M.W. Khalil, J. Power Sources 135 (2004) 42.
[177] L. Dubau, F. Hahn, C. Coutanceau, J.-M. Leger, C. Lamy, J. of Electroanal. Chem. 554–555 (2003) 407.
[178] E.R. Fachini, R. Diaz-Ayala, E. Casado-Rivera, S. File, C.R. Cabrera, Langmuir 19 (2003) 8986.
[179] E.A. Batista, G.R.P. Malpass, A.J. Motheo, T. Iwasita, Electrochem. Commun. 5 (2003) 843.
[180] T. Kamo, Electrochemistry 70 (2002) 915.
[181] O.A. Khazova, A.A. Mikhailova, A.M. Skundin, E.K. Tuseeva, A. Havranek, K. Wippermann, Fuel Cells 2 (2002) 99.
[182] J. Jiang, A. Kucernak, J. of Electroanal. Chem. 533 (2002) 153.
[183] A.V. Tripkovic, K.D. Popovic, B.N. Grgur, B. Blizanac, P.N. Ross, N.M. Markovic, Electrochim. Acta 47 (2002) 3707.
[184] J.M. Leger, Journal of Applied Electrochemistry 31 (2001) 767.
[185] J.C. Dubois, Actualite Chimique 12 (2001) 58.
[186] G. Goekagac, J.-M. Leger, F. Hahn, C. Lamy, Carbon Supported Pt, Pt/W and Pt/Mo Electrocatalysts for Methanol Oxidation: Electrochemical and IR Spectroscopic Characterisation, Electrochemical Society Proceedings, 2001, p. 174.
[187] Z. Jusys, R.J. Behm, J. Phys. Chem. B 105 (2001) 10874.
[188] A. Lima, C. Coutanceau, J.M. Leger, C. Lamy, Journal of Applied Electrochemistry 31 (2001) 379.
[189] X.M. Ren, P. Zelenay, S. Thomas, J. Davey, S. Gottesfeld, J. of Power Sources 86 (2000) 111.
[190] T. Iwasita, H. Hoster, A. John-Anacker, W.F. Lin, W. Vielstich, Langmuir 16 (2000) 522.
[191] S. Sriramulu, T.D. Jarvi, E.M. Stuve, J. Electroanal. Chem 467 (1999) 132.
[192] G. Burstein, C. Barnett, A. Kucernak, K. Williams, Catal. Today 38 (1997) 425.
[193] A. Hamnett, Catal. Today 38 (1997) 445.
[194] P.S. Kauranen, E. Skou, J. Munk, J. Electroanal. Chem. 404 (1996) 1.
[195] N.M. Markovic, H.A. Gasteiger, P.N.R. Jr, X. Jiang, I. Villegas, M.J. Weaver, Electrochim. Acta 40 (1995) 91.
[196] E. Herrero, K. Franaszczuk, A. Wieckowski, J. Phys. Chem. B 98 (1994) 5074.
[197] H.A. Gasteiger, N. Markovic, J. Philip N. Ross, E.J. Cairns, J. Phys. Chem. 97 (1993) 12020.
[198] P.N. Ross, Electrochim. Acta 36 (1991) 2053.
[199] A.B. Anderson, E. Grantscharova, S. Seong, J. Electrochem. Soc. 143 (1996) 2075.
[200] H.A. Gasteiger, N. Marcovic, P.N. Ross, E.J. Cairnes, J. Electrochem. Soc. 141 (1994) 1795.
[201] K. Park, J. Choi, K. Ahn, Y. Sung, J. Phys. Chem. 108 (2004) 5989.
[202] T. Kim, M. Takahashi, M. Nagai, K. Kobayashi, Chemistry Letters 33 (2004) 478.
[203] A. Neto, M. Giz, J. Perez, E. Ticianelli, E. Gonzalez, J. Electrochem. Soc. 149 (2002) A272.
[204] W.S. Li, L.P. Tian, Q.M. Huang, H. Li, H.Y. Chen, X.P. Lian, J. Power Sources 104 (2002) 281.
[205] G.L. Troughton, A. Hamnett, Bulletin of Electrochemistry 7 (1991) 488.
[206] B.J. Kennedy, A.W. Smith, J. Electroanal. Chem. 293 (1990) 103.
[207] T. Iwasita, F.C. Nart, W. Vielstich, Ber. Bunsenges. Phys. Chem 94 (1990) 1030.

[208] Z. Jusys, T.J. Schmidt, L. Dubau, K. Lasch, L. Jorissen, J. Garch, R.J. Behm, J. of Power Sources 105 (2002) 297.

[209] W.C. Choi, Y.J. Kim, S.I. Woo, W.H. Hong, *Science and Technology in Catalysis*. Kadansha Ltd, chapter 86, 2002, p. 395.

[210] E. Reddington, J.-S. Yu, B.C. Chan, A. Sapienza, G. Chen, T.E. Mallouk, B. Gurau, R. Viswanathan, R. Liu, E.S. Smotkin, S. Sarangapani, in H. Fenniri (Ed.), *Combinatorial Chemistry*. Oxford University Press, Oxford, 2000, p. 401.

[211] E. Reddington, J.-S. Yu, A. Sapienza, B.C. Chan, B. Gurau, Mat. Res. Soc. Symp. Proc. 549 (1999) 231.

Chapter 7

Geochemistry of Mineral Surfaces and Factors Affecting Their Chemical Reactivity

Gordon E. Brown, Jr.[1,2] Thomas P. Trainor[3] and Anne M. Chaka[4]

[1]Surface & Aqueous Geochemistry Group, Department of Geological & Environmental Sciences, Stanford University, Stanford, CA 94305-2115, USA; [2]Stanford Synchrotron Radiation Laboratory, SLAC, 2575 Sand Hill Road, MS 69, Menlo Park, CA 94025, USA; [3]Department of Chemistry and Biochemistry, University of Alaska Fairbanks, Fairbanks, AK 99775-6160, USA; [4]Optical Technology Division, Physics Laboratory, National Institute of Standards and Technology, Gaithersburg, MD 20899-8440, USA

1. Introduction

We live in a world dominated by interfaces, as shown by a view of Earth from Apollo 17 (Figure 7.1). The fluid envelope or hydrosphere of Earth, which covers some 70 % of Earth's surface, comes in contact with the solid Earth at the interfaces between water and rocks, sediments, or soils. The gaseous envelope or atmosphere interacts with the solid Earth over the continents. Earth's lithosphere, or outer 10–100 km of rock, is broken by active faults and static fractures far too numerous to record. Meteoric water (i.e., that originating from rain or snow on Earth's surface) and formation water (i.e., that occurring in pores of deeply buried sedimentary rocks) percolate along these faults and cracks, reacting with the minerals lining the flow path. When heated by shallow magma bodies or hot rocks, such fluids are even more effective at altering or dissolving minerals and transporting dissolved ions. Seawater also plays a major role in 'water-rock' interactions, resulting in alteration of enormous volumes of submarine sediments as well as fresh basaltic rocks at mid-ocean ridges. These reactions, together with emissions of gases such as CO_2, SO_2, and H_2O from volcanoes and hot springs, have had a major influence on the chemical evolution of Earth's atmosphere and oceans over geologic time (roughly 4.5 billion

Figure 7.1. View of Earth from Apollo 17 showing the African continent, the Southern Ocean and portions of the Atlantic and Indian Oceans, Antarctica, and extensive cloud cover. It emphasizes interfaces between continents and oceans (solid–liquid), continents and atmosphere (solids and gases), and oceans and atmosphere (liquids and gases). From NASA (http://visibleearth.nasa.gov/view-rec.php?id=12907).

years) [1]. In addition, catalytic reactions involving CO and CH_3SH at 3-d transition metal sulfide-aqueous solution interfaces are believed by some to have resulted in the formation of prebiotic molecules early in Earth's history [2].

From this general description of the Earth's solid-fluid-gaseous envelope, it is apparent that reactions governing the chemistry of the environment are dominated by those at environmental interfaces. The importance of such reactions is well summarized in a 1987 quotation from the late Werner Stumm [3]: 'Almost all of the problems associated with understanding the processes that control the composition of our environment concern interfaces, above all the interfaces of water with naturally occurring solids.'

This chapter focuses on mineral surfaces and some of the factors affecting their chemical reactivity with water and aqueous species. Such surfaces are terminated by reactive sites or functional groups that participate in acid-base, ligand-exchange, and electron-transfer reactions, which are the three most important classes of reactions in geochemistry [4–6]. Mineral surface reactions play a dominant role in many geochemical processes, including mineral dissolution and precipitation, heterogeneous nucleation and growth of solids, pH buffering, cation and anion sorption and desorption, heterogeneous redox reactions, and heterogeneous catalytic reactions, which in turn affect (or control) chemical weathering of minerals [7] and the development of soils [8], water quality [9,10], colloid stability [11], contaminant and nutrient sequestration and release [12,13], soil rheological properties [14], acid mine drainage [15], the respiratory cycle of many microorganisms [16], and the biogeochemical cycling of elements [17]. The surfaces of aerosol mineral particles in the troposphere are also thought to play key roles in heterogeneous atmospheric chemistry [18], including the formation of cloud condensation nuclei [19,20] and the conversion of nitrous oxide to nitrous acid [21,22], which affect climate and acid rain production, respectively. Mineral surface reactions also play a major role in mineral separation methods such as froth flotation that underpin the world's mineral industries (e.g., [23]) and in the ceramic processing industry (e.g., [24]), which produces ceramics for a variety of industrial, technological, and medical applications. Thus, a fundamental knowledge of mineral surfaces is vital to many areas of science and technology.

Understanding the reactivity of mineral surfaces under environmental conditions is difficult. This is the case because most mineral particles at Earth's surface or in the troposphere are in contact with a fluid phase (typically an aqueous solution), atmospheric or soil gas, organic matter, and/or biological organisms. Not only are the structure and composition of these interfaces poorly understood at a microscopic level, but it is also difficult to isolate the individual effects of aqueous solutions, gases, organic matter, and microbial organisms on mineral surface structure and reactivity. Another complicating factor is the presence of defects on mineral surfaces. While it is thought that such defects may dominate interface reactivity, little is known about the nature and density of such features.

Over the past 40 years, surface science studies have revealed a great deal about clean, compositionally and structurally homogeneous surfaces under controlled conditions [25–27]. However, such surfaces exist only in ultra-high vacuum, so the results of these studies are not generally applicable to 'real' interfacial systems, i.e., those in which surfaces are in contact with atmospheric gases, reactor gases and fluids in controlled catalytic reactions, aqueous solutions, organic matter, fungi, or microbial biofilms. Real surfaces have structures and reactivities that may be altered substantially by interactions with the environment. In addition, when liquid water is present, it is not likely that the geometric or electronic structures of surfaces will be the same as under UHV conditions, particularly for redox-sensitive surfaces or anhydrous bulk solids.

Mineral-liquid or mineral-gas interfaces under reactive conditions cannot be studied easily using standard UHV surface science methods. To overcome the 'pressure gap' between ex situ UHV measurements and the in situ reactivity of surfaces under atmospheric pressure or in contact with a liquid, new approaches are required, some of which have only been introduced in the last 20 years, including scanning tunneling microscopy [28,29], atomic force microscopy [30,31], non-linear optical methods [32,33], synchrotron-based surface scattering [34–38], synchrotron-based X-ray absorption fine structure spectroscopy [39,40], X-ray standing wave spectroscopy [41–45], synchrotron-based X-ray emission spectroscopy [46,47], X-ray reflection phase contrast microscopy [48], synchrotron-based resonant anomalous X-ray reflectivity [49], and quantum chemical methods [50,51]. Because of the enormous complexity of environmental interfaces, no single experimental or theoretical approach is sufficient to characterize their geometric and electronic structures or composition or to gain a fundamental understanding of their reactivity. Instead, a reductionist approach is required in which simplified model systems are studied using a variety of ex situ and in situ surface-sensitive methods coupled, where possible, with high-level theoretical calculations of surface/interface structure, thermodynamics, and reactivity [6]. Once the simplest model systems are understood, additional complexity can be introduced in a carefully controlled manner, with the ultimate aim of approaching the complexity of real interfaces. Initial characterization studies of simplified analogs of complex surfaces and interfaces usually include basic UHV surface science measurements, such as LEED and photoemission spectroscopy, on the clean surfaces prior to and following reaction with environmental species such as water. However, these ex situ measurements must be followed by in situ measurements under reactive conditions to achieve as complete an understanding as possible of the complex interfaces between solids and liquids, organic matter, gases, or microbial cells. Adapting existing UHV methods to allow spectromicroscopy studies of chemical processes at environmental interfaces under reactive conditions using differentially pumped sample chambers (e.g., [52,53]) or in real-time (e.g., [54]) represents an exciting challenge for surface and interface science.

The purpose of this chapter is to present a brief overview of the geochemistry of mineral surfaces, including their (1) dissolution mechanisms, (2) development of electrical charge when in contact with aqueous solutions, and (3) uptake of aqueous cations and anions, and to discuss some of the factors that control their chemical reactivity, including (1) defect density, (2) cooperative effects among adsorbates, (3) intrinsic differences in surface properties such as isoelectric points, and (4) differences in surface structure under hydrated conditions. Although less studied, we shall also consider interfaces between minerals and organic matter or microbial organisms and the effect of organic matter and biofilm surface coatings on the reactivity of metal-oxide surfaces with respect to heavy metal ions. Our overall aim is to provide a rudimentary understanding of how differences in atomic and electronic surface structure and composition determine properties such as chemical reactivity. Our main focus will be on metal-oxide and metal-(oxy)hydroxide surfaces because these are arguably the most reactive surfaces in the real environment [6,55,56]. Past UHV surface science studies of clean metal oxide surfaces (see, e.g., [25,57]) provide some of the background needed for understanding their changes when exposed to reactive environments. Recent structural and theoretical studies of model metal-oxide surfaces in contact with gas or liquid water under ambient conditions provide a basis for understanding factors affecting chemical reactivity in a broader range of systems.

We begin with a discussion of the most common minerals present in Earth's crust, soils, and troposphere, as well as some less common minerals that contain common environmental contaminants. Following this is (1) a discussion of the nature of environmentally important solid surfaces before and after reaction with aqueous solutions, including their charging behavior as a function of solution pH; (2) the nature of the electrical double layer and how it is altered by changes in the type of solid present and the ionic strength and pH of the solution in contact with the solid; and (3) dissolution, precipitation, and sorption processes relevant to environmental interfacial chemistry. We finish with a discussion of some of the factors affecting chemical reactivity at mineral/aqueous solution interfaces.

2. Environmental interfaces

In the next three subsections, we discuss the most important minerals in Earth's crust, soils, and atmosphere, and explore some of the basic concepts that help guide our thinking about their interactions with water, aqueous metal ions, organic matter, microbial organisms, and atmospheric and soil gases.

2.1. Common minerals in Earth's crust, soils, and atmosphere, weathering mechanisms and products, and less common minerals that contain or adsorb environmental contaminants

Among the 4000+ mineral species identified and characterized, only a small subset are common in Earth's lithosphere, biosphere, hydrosphere, and atmosphere. Table 7.1 lists the most common minerals occurring in Earth's crust, which is defined as Earth's outer 2–70 km, with the thinnest regions underlying the oceans at mid-ocean ridges and the thickest regions underlying the highest mountains on continents. The order of minerals in Table 7.1 is from the least resistant to chemical weathering (top) to the most resistant (bottom). For silicate minerals, this order roughly follows the increasing extent of polymerization of SiO_4 tetrahedra, ranging from the least resistant, olivine, an orthosilicate with isolated SiO_4 tetrahedra, to the most resistant, quartz, a tectosilicate with complete tetrahedral polymerization (i.e., all corners of all silicate tetrahedra are linked to other silicate tetrahedra). Also listed in Table 7.1 are the types of weathering reactions at low pH and the most common weathering products of the various primary minerals. The dissolution of quartz provides an example of *congruent dissolution*, which results in the stoichiometric release of one mole of silicon and two moles of oxygen to solution for each mole of quartz dissolved. *Incongruent dissolution* involves dissolution and reprecipitation (or selective leaching), resulting in one or more solid phases of different composition than the original mineral plus ions released to solution in a concentration ratio that does not match the stoichiometry of the original mineral. Albite feldspar is

Table 7.1
Common minerals in Earth's crust, types of weathering reactions, activation energies, and weathering products.

Mineral	Chemical Formula	Volume %[1]	Weathering Reaction (low pH)[2]	ΔE(act)[3]	Weathering Products
Olivines	$(Mg,Fe^{2+})_2SiO_4$	3	Congruent dissolution, Fe^{2+} oxidation	38	Hematite Goethite Si(aq)
Amphiboles	$(Na,K)_{0-1}$ $(Ca,Mg,Fe^{2+},Na)_2$ $(Mg,Fe,Al)_5(Si,Al)_8(OH)_2$	5	Congruent dissolution, Fe^{2+} oxidation	–	Ions in solution
Pyroxenes	(Ca,Mg,Fe^{2+}) $(Mg,Fe^{2+,3+},Al)$ $(Si,Al)_2O_6$	11	Congruent dissolution, Fe^{2+} oxidation	45–80	Ions in solution
Plagioclase Feldspar	$(Na_xCa_{1-x})Al_{2-x}Si_{2+x}O_8$	39	Incongruent dissolution ?	35–88	Smectite Kaolinite Gibbsite
Potassium Feldspar	$KAlSi_3O_8$	12	Incongruent dissolution ?	82	Kaolinite Gibbsite
Biotite-Muscovite	$K(Mg,Fe,Al)_{2-3}$ $(AlSi_3)O_{10}(OH)_2$	5	Incongruent dissolution, Fe^{2+} oxidation	–	Vermiculite Kaolinite Gibbsite
Quartz	SiO_2	12	Resistant to dissolution (congruent)	77	Si(aq)
Clays	Kaolinite: $Al_2Si_2O_5(OH)_4$	5	Congruent dissolution	–	Gibbsite Boehmite
	Smectite: $(Ca,Na)_{0.33}$ $(Al,Fe^{2+}, Fe^{3+},Mg)_{2-3}$ $(Si,Al)_4O_{10}(OH)_2 \cdot nH_2O$		Incongruent dissolution		Kaolinite Gibbsite Goethite
Magnetite	$(Fe^{3+})_2Fe^{2+}O_4$		Resistant to dissolution	–	Hematite Goethite
Ilmenite	$FeTiO_3$				Anatase Goethite
Apatite Carbonates	$Ca_5(PO_4)_3(OH,F,CO_3)$ $(Ca,Mg,Fe^{2+})CO_3$	8	Congruent dissolution		Ions in solution

[1]From Ronov and Yaroshevsky [62], [2]After Berner and Berner [58], [3]Activation energy (kJ/mole) of dissolution reactions (from Schott and Petit [59]).

thought by some (e.g., [58,59]) to undergo incongruent dissolution, as indicated by the following reaction of albite with carbonic acid in water:

$$3NaAlSi_3O_8 + 2H_2CO_3 + 12H_2O = NaAl_3Si_3O_{10}(OH)_2 + 6H_4SiO_4 + 2Na^+ + 2HCO_3^-$$

The results of this reaction are precipitation of smectite clay and release of silicic acid, sodium ions, and bicarbonate ions to solution. The actual dissolution reaction is significantly more complex than indicated here, as discussed in Section 2.4.

Table 7.2 lists the most common non-silicate minerals in soils. Not included are the common silicate minerals, which are the same as those listed in Table 7.1, with clays

Table 7.2
Common non-silicate minerals in soils (after shulze [63]).

Mineral Class	Mineral Name	Chemical Formula	Weathering Reaction
Halides	Halite	$NaCl$	Congruent dissolution in H_2O
Sulfates	Gypsum Anhydrite Jarosite	$CaSO_4 \cdot 2H_2O$ $CaSO_4$ $K(Al,Fe)_3(SO_4)_2(OH)_6$	Congruent dissolution in H_2O
Sulfides	Pyrite	FeS_2	Oxidation of Fe and S
Carbonates	Calcite Dolomite	$CaCO_3$ $CaMg(CO_3)_2$	Congruent dissolution by acids
Al-Oxides & Hydroxides	Gibbsite Bayerite Boehmite Diaspore	γ-$Al(OH)_3$ α-$Al(OH)_3$ γ-$AlOOH$ α-$AlOOH$	Highly resistant to dissolution
	Corundum	α-Al_2O_3	Resistant to dissolution
	Hematite Goethite Lepidocrocite	α-Fe_2O_3 α-$FeOOH$ γ-$FeOOH$	Highly resistant to dissolution
Fe-Oxides & Hydroxides	Maghemite Ferrihydrite	γ-Fe_2O_3 $5Fe_2O_3 \cdot 9H_2O$	Resistant to dissolution
	Magnetite	$(Fe^{3+})_2Fe^{2+}O_4$	Highly resistant to dissolution, Fe^{2+} oxidation
	Pyrolusite	β-MnO_2	Resistant to dissolution, Mn^{4+} reduction
Mn-Oxides & Hydroxides	Manganite Birnessite Todorokite Rutile	$MnOOH$ $NaMn_2O_4 \cdot 1.5H_2O$ $(Na,Ca,K)(Mn,Mg)_6O_{12}$ TiO_2	Resistant to dissolution
Ti- Oxides	Anatase Ilmenite	TiO_2 $FeTiO_3$	Highly resistant to dissolution

becoming more dominant as a soil becomes more mature. The Jackson–Sherman weathering stages of a soil [60,61] provide a convenient way of estimating the extent of weathering based on indicator minerals. For example, the presence of gypsum and carbonates in soils indicates very low concentrations of water and organic matter and an early stage of weathering. The presence of silicates such as olivines, pyroxenes, and amphiboles indicates reducing environments and also an early stage of weathering. Intermediate stages of weathering are indicated by the presence of quartz, micas, chlorite, and smectites. Advanced stages of weathering are indicated by the presence of kaolinite, gibbsite, goethite, hematite, and titanium oxides.

Table 7.3 lists examples of common mineral aerosol particles in the Earth's troposphere based on estimates by Leinen et al. [64] and Claquin et al. [65]. The surfaces of these particles interact with water vapor and atmospheric gases (e.g., SO_2 and NO_x), and thus serve as cloud condensation nuclei, which can have dramatic effects on global temperatures [66].

The minerals listed in these three tables act as substrates on which chemical reactions with aqueous solutions, atmospheric and soil gases, organic matter, and microbial organisms occur. Some are resistant to chemical weathering and therefore persist in soils (e.g., hematite, rutile), where they can act as adsorbents of environmental contaminants and pollutants, whereas others are so reactive to natural organic and inorganic acids in solution (e.g., carbonates) or are so sensitive to oxidizing atmospheres (e.g., olivine, pyroxene) that they undergo rapid chemical weathering and are converted through congruent and incongruent dissolution processes to aqueous species and other minerals. In addition, minerals in soils and

Table 7.3
Common mineral aerosols in the Earth's troposphere[1].

Mineral	Chemical Formula
Illite	$(K,H_3O)(Al,Mg,Fe)_2(Si,Al)_4O_{10}[(OH)_2,H_2O]$
Kaolinite	$Al_2Si_2O_5(OH)_4$
Smectite	$(Ca,Na)_{0.33}(Al,Fe^{2+},Fe^{3+},Mg)_{2-3}(Si,Al)_4O_{10}(OH)_2 \cdot nH_2O$
Feldspar	$(Na_xCa_{1-x})(Al_{2-x}Si_{2+x})O_8$ and $KAlSi_3O_8$
Quartz	α-SiO_2
Chlorite	$(Al,Fe^{2+},Fe^{3+},Mg)_{4-6}(Si,Al)_4O_{10}(OH)_8$
Calcite	$CaCO_3$
Halite	NaCl
Hematite	α-Fe_2O_3
Gypsum	$CaSO_4 \cdot 2H_2O$

[1]Table compiled using data in Leinen et al. [64] and Claquin et al. [65], which are based, respectively, on sampling clay-size and silt-size particles over the Northern Pacific Ocean, and different soil types in arid regions consistent with the Food and Agricultural Organization Soil Map of the World. The minerals are listed in approximate order of abundance for these localities, with the most abundant at the top and least abundant at the bottom.

aquatic systems are sometimes coated by secondary mineral phases (e.g., Fe(III)- and Mn(III, IV)-oxides and (oxy)hydroxides), natural organic matter, or microbial biofilms, which can inhibit further dissolution or enhance dissolution, including reductive (oxidative) dissolution in which redox-sensitive cations (and anions) in the near-surface region of a mineral are reduced (or oxidized) by abiotic or biotic processes and released into solution (see, e.g., [67–72]). Dissolution of mineral surfaces will be considered in more detail below.

Most of the heavy metal contaminants associated with mining or agricultural practices are released from less common mineral phases, some of which are primary (i.e., formed initially by an igneous, metamorphic, or sedimentary process) and some of which are secondary (i.e., formed by chemical alteration of the primary minerals). Table 7.4 lists some of the more common contaminant ions and the mineral phases with which they are associated in economic mineral deposits or weathered zones associated with these deposits.

Table 7.4
Examples of less common minerals that contain common environmental contaminants and their relative solubilities or ease of removal of adsorbed species.

Mineral	Chemical Formula	Contaminant	Relative Solubility[1]/Ease of Removal[2]
Arsenopyrite	$FeAsS$	As	Soluble
Arsenian Pyrite	$(Fe,As)S_2$	As	Soluble
Hoernesnite	$Mg_3(AsO_4)_2{\bullet}8H_2O$	As	Insoluble
Scorodite	$FeAsO_4{\bullet}2H_2O$	As	Insoluble
Orpiment	As_2S_3	As	Insoluble
Realgar	AsS	As	Insoluble
Adsorbed As^{3+}	Fe,Mn-(oxyhydr)oxides	As	Easily removed at high pH
Adsorbed As^{5+}	Fe,Mn-(oxyhydr)oxides	As	Easily removed at high pH
Cinnabar	HgS (hex)	Hg	Highly Insoluble
Metacinnabar	HgS (cub)	Hg	Highly Insoluble
Corderoite	$Hg_3S_2Cl_2$	Hg	Soluble
Eglestonite	$Hg_3Cl_3O_2H$	Hg	Highly Soluble
Montroydite	HgO	Hg	Highly Soluble
Mercuric Chloride	$HgCl_2$	Hg	Highly Soluble
Schuetteite	$Hg_3O_2SO_4$	Hg	Highly Soluble
Terlinguite	Hg_2OCl	Hg	Highly Soluble
Elemental Mercury	Hg^0	Hg	Insoluble, Volatile
Adsorbed Hg^{2+}	Fe,Mn-(oxyhydr)oxides	Hg	Easily removed at low pH
Anglesite	$PbSO_4$	Pb	Soluble

(*Continued*)

Table 7.4 (Continued)

Mineral	Chemical Formula	Contaminant	Relative Solubility[1]/Ease of Removal[2]
Cerrusite	$PbCO_3$	Pb	Soluble
Hydrocerrusite	$Pb_3(CO_3)_2(OH)_2$	Pb	Insoluble at high pH
Galena	PbS	Pb	Insoluble
Laurionite	PbOHCl	Pb	Soluble
Leadhillite	$Pb(SO_4)(CO_3)_2(OH)_2$	Pb	Soluble
Litharge	PbO (yellow)	Pb	Soluble
Massicot	PbO (red)	Pb	Soluble
Plumbogummite	$PbAl_3(PO_4)_2(OH)_5 \cdot H_2O$	Pb	Highly Insoluble
Pyromorphite	$Pb_3(PO_4)_2Cl$	Pb	Highly Insoluble
Vanadinite	$Pb_3(VO_4)_3Cl$	Pb	Insoluble
Adsorbed Pb^{2+}	Fe,Mn-(oxyhydr)oxides	Pb	Easily removed at low pH
Gypsum	$Ca(S,Se)O_4 \bullet 2H_2O$	Se	Highly Soluble
Adsorbed Se^{4+}	Fe,Mn-(oxyhydr)oxides	Se	Easily removed at high pH
Adsorbed Se^{6+}	Fe,Mn-(oxyhydr)oxides	Se	Easily removed at all pH values
Hydrozincite	$ZnCO_3(OH)H_2O$	Zn	Relatively Insoluble
Smithsonite	$ZnCO_3$	Zn	Soluble
Sphalerite	ZnS	Zn	Highly Insoluble
Zincite	ZnO	Zn	Relatively Insoluble
Zn/Al-Hydrotalcite	$Zn_xAl_y(OH)_2(CO_3)_{2y} \cdot nH_2O$	Zn	Relatively Insoluble
Adsorbed Zn^{2+}	Fe,Mn-(oxyhydr)oxides	Zn	Easily removed at low pH

[1] The relative solubilities reported are very crude estimates based on equilibrium solubility products. These estimates do not take into account variations in solubility as a function of pH, ionic strength, activities of various solution species (e.g., HCO_3^-), redox state, particle size, surface defect types and concentrations, the concentration of various types of adsorbates, including natural organic matter, on mineral surface, or the presence of different types of bacteria or microbial biofilms on mineral surfaces.

[2] The estimates of ease of removal (desorption) of adsorbed species are qualitative and do not take into account ionic strength effects, the effects of competing sorbates such as PO_4^{3-}, and other variables that can affect desorption.

2.2. Solubilities of Al- and Fe(III)-oxides and Al and Fe(III)-(oxy)hydroxides

When immersed in aqueous solutions, the surfaces of metal oxides are expected to be hydroxylated (see Section 3.1). However, metal oxides may also react with aqueous solutions leading to dissolution, which can effectively be thought of as mass transfer from the solid to aqueous phase. The rate and extent of dissolution reactions depend on a number of factors, including solution pH, acid-base properties of oxo groups on the metal oxide surface, types of ligands present in solution, metal

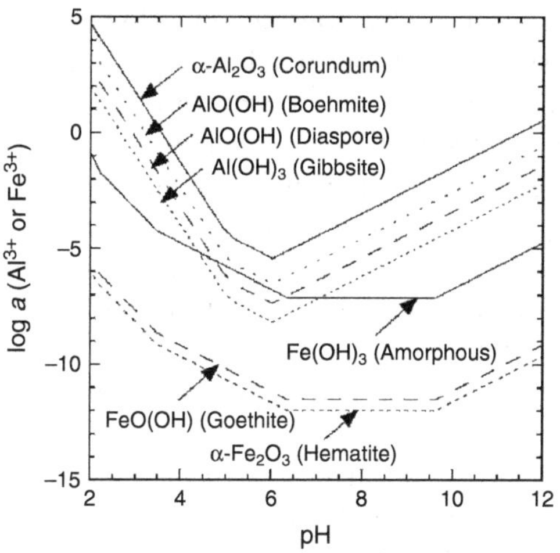

Figure 7.2. Plot of log activity of Al^{3+} or Fe^{3+} in solution vs. pH for the common Al- and Fe(III)-oxides and oxy(hydroxides). Calculations were made using The Geochemist's Workbench.

oxygen vs. metal-ligand bond strengths, etc. The dependence of solubilities of the common Al- and Fe(III)-oxides and -(oxy)hydroxides on pH is shown in Figure 7.2. These types of phases typically show minima in solubilities at pH values ranging from 6–8, with significant increases in solubilities at low and high pH due to the amphoteric nature of the aqueous Fe and Al hydrolysis products [73]. In contrast, quartz has no such minimum at intermediate pH values but instead has a relatively constant solubility over most of the acidic and neutral pH range and a steep increase in solubility at pH > 10. Although the detailed mechanisms of dissolution of metal oxides are generally unknown, protons at low pH and hydroxo ligands at high pH participate in the dissolution process.

Casey and co-workers (e.g., [74–76]) have advocated a similarity of mechanisms for the release of metal ions from dissolving metal oxide surfaces and ligand exchange around aqueous metal ions. Both reactions are controlled by the strength of bonds between the metal ion and the surrounding first-shell ligands. In the case of ligand exchange in the first hydration shell of metal ions in solution, the rates are generally thought to correlate directly with the metal-oxygen bond strength. For alkali and alkaline Earth metals and elements of the sp-block of the periodic table, solvent exchange rates correlate inversely with the ionic potential (charge/radius), whereas rates of solvent exchange for the first-row transition metals vary inversely with crystal field stabilization energy [74]. In addition, more ionically bonded metal ions tend to be more reactive than covalently bonded metal ions. The same correlations should generally hold for the exchange of oxo or hydroxo ligands at mineral surfaces, although the rate of ligand exchange for a specific type of metal ion at a metal oxide surface is generally considerably slower than the rate for solvent exchange around the same metal ion in solution [74]. Likewise, the time scale of adsorption of protons or hydroxo ions on mineral surfaces in contact with aqueous solutions is generally thought to be considerably slower (0.001 to 10 s) [77] than protonation of metal ion complexes in solution ($<10^{-6}$ s) [76], although this conclusion has been questioned for hematite [78].

The analogy between solution reactions and reactions at mineral surfaces is not perfect, however, as illustrated by the effect of proton loss from the first hydration shell of a metal ion in solution versus at a metal oxide surface on ligand-exchange rate. The loss of a proton from a water molecule in the first hydration shell of a cation in solution results in an increase in the rate of proton exchange for other water molecules in the first coordination shell of the cation by a factor of 100 to 1000 [76]. For example, the rate of ligand exchange for the first hydrolysis product of aqueous Fe(III) ($Fe(H_2O)_5(OH)^{2+}$) is 750 times faster than the fully aquated $Fe(III)(H_2O)_6^{3+}$ ion [79]. While the dissolution rates of metals oxides at low pH (where surface hydroxyl groups are expected to be fully protonated) show good correlations with the water exchange rates (c.f. dissolution rate vs. ligand exchange rate correlations presented by Casey et al. [76]), there is generally a reduction in the dissolution rate as pH increases and surface hydroxyl groups deprotonate. This apparent contrast is likely due to the weakening of multiple metal-oxygen bonds when surface hydroxyl groups are protonated, thus making the rupture of framework (hydr)oxy functional groups and removal of an aquated metal ion more probable [80]. At high pH, hydroxo ions become effective competitors for the coordination of metals at mineral surfaces, and thus there is a general increase in the rates of dissolution as the association of hydroxo ligands at the mineral surface may rupture framework (i.e., bridging) bonds in the surface and allow removal of metal-hydroxo complexes.

In apparent contrast, loss of a proton from a water molecule on a metal oxide surface with increasing pH is correlated with a reduction in dissolution, as shown in Figure 7.2, indicating a strengthening of M–O surface bonds. Reaction of hydroxo ions with metal oxide surfaces at high pH facilitates dissolution (Figure 7.2). A possible explanation for this behavior at high pH is an associative ligand-exchange reaction involving the hydroxo ligand, resulting in the formation of a transition state complex with increased coordination of the metal ion, thus a weakening of M–O bonds in the complex.

Using the reasoning of Casey [74], one might predict that the rate of dissolution of α-Al_2O_3 should be significantly slower than that of isostructural α-Fe_2O_3 at a given pH, based on the observed five orders of magnitude difference in mean residence time of water molecules in the first hydration shells of $Al(H_2O)_6^{3+}$ ($\approx$10 s) and $Fe(H_2O)_6{}^{3+}$ ($\approx 10^{-4}$ s) [81]. This difference in exchange rate should reflect differences in Al-O and Fe^{3+}-O bond strengths in these aqueous complexes (see [82]).

Because differences in M–O bond strengths should also be related to differences in solubility of isostructural metal oxides, one might anticipate that the analogy between the coordination chemistry of metal ions in solution and at mineral surfaces could be used to help explain differences in solubility of corundum and hematite as a function of pH. As shown in Figure 7.2, the minimum solubility of α-Fe_2O_3 in aqueous solution as a function of pH is about eight orders of magnitude lower than that of α-Al_2O_3. The average M−O bond length in corundum (1.91 Å) [83] is shorter than the average M−O bond length in hematite (2.03 Å) [84], which is not consistent with the much lower solubility of hematite. However, Al−O bonds at the hydrated corundum surface are significantly lengthened relative to the bulk structure [37]. In contrast, Fe^{3+}−O bonds at the hydrated hematite surface are contracted relative to those in the bulk structure [85]. These differences, coupled with the higher electronegativity of Fe^{3+} (1.96) relative to Al^{3+} (1.71) [86], which makes $^{VI}Fe^{3+}$−O bonds more covalent than Al−O bonds, should stabilize surface $^{VI}Fe^{3+}$−O bonds more relative to surface $^{VI}Al^{3+}$−O bonds. This $^{VI}Fe^{3+}$−O bond stabilization is manifested in the greater stability of hematite relative to corundum under ambient conditions. Another related factor is the lower acidity of surface hydroxyls on α-Al_2O_3 vs. α-Fe_2O_3, which suggests that the former should retain protons to higher pH values, thus should dissolve faster. It is important to point out that most dissolution models focus on dissolution rates far from equilibrium, thus they are not easily correlated with difference in free energies of the solid phases but do correlate with the lability of M−O bonds and on protonation states of bridging oxygens. This issue will be revisited in our discussion of the differences in reactivity of corundum vs. hematite to aqueous metal ions (see Section 3.3).

2.3. Dissolution mechanisms at feldspar–water interfaces

Feldspars are the most abundant minerals in Earth's crust (Table 7.1), and thus their dissolution has a major influence on the cycling of Si, Al, and alkali and alkaline Earth metals in the biosphere; atmospheric CO_2 levels; the composition of natural waters; and soil formation [87]. Despite extensive investigations of feldspar dissolution reactions over the past several decades, the nature of these reactions remains controversial (see, e.g., [88]). As alluded to in Section 2.1, the overall dissolution reaction of alkali feldspars involves a number of elementary reactions such as initial H_2O hydrolysis and H_3O^+ or OH^- catalysis of Si−O−Al linkages in feldspar, as proposed by Xiao and Lasaga [89,90] and Lasaga [91], based on quantum chemical modeling studies of silicate dissolution mechanisms and kinetics, replacement of Na^+ ions by H_3O^+, and analysis of changes in solution compositions during dissolution [88]. Potential mechanisms for the H_3O^+ and OH^- catalyzed hydrolysis reactions of feldspar and quartz have been proposed by Xiao and Lasaga [89,90]. In the case of H_3O^+ catalysis, one possible reaction step involves the attack of the oxygen ion bridging between aluminate and silicate tetrahedra, resulting in a preferential lengthening and weakening of the Al−O bond, and leading to preferential release of Al^{3+} ions, which is consistent with observations. Although this model involves only one water molecule and, therefore, ignores cooperative effects between adjacent water molecules, which should reduce the lengthening of Si−O and Al−O bonds in feldspar relative to the single water molecule model (see [90]), the calculated activation energy for this reaction step at the MP2/6-31G* computational level (15.95 kcal/mole) is reasonably close to measured proton catalyzed feldspar activation energies (17–29 kcal/mole). The reaction mechanism for OH^- catalyzed dissolution of quartz is envisioned by Xiao and Lasaga [89] to include four steps. The highest activation energy step (18.9 kcal/mole) is the one required to form the transition state structure involving 5-coordinated Si, which results in a lengthening and weakening of the bridging Si−O−Si bonds. A similar mechanism can be envisioned for the OH^- catalyzed attack on $O_3{\equiv}Al{-}O{-}Si{\equiv}O_3$ linkages in albite feldspar, which should result in preferential addition of OH^- to the AlO_4 tetrahedron and weakening of the Al−O bond. A similar dissolution mechanism has been proposed for (001) and (111) β-cristobalite surfaces [92] and neutral silica surfaces [93] based on quantum chemical studies.

Three classes of hypotheses have been proposed to account for experimental observations of feldspar dissolution – (1) the surface reaction hypothesis, (2) the armoring precipitate hypothesis, and (3) the leached layer hypothesis [59]. The most popular of these hypotheses currently is the third one. It involves preferential leaching of weakly bonded ions, such as Na^+ and K^+ in the case of alkali feldspars, from the surface region of silicates, which precedes the leaching of more strongly bonded ions such as Al^{3+} and Si^{4+} at acidic pH values (see, e.g., [59,88,94]), resulting in an altered, near-surface zone that can be tens of Å's [59] to several microns thick at the lowest pH values [95]. It was originally proposed that as the altered layer increases in thickness, the rate of dissolution decreases until a steady state rate is achieved. This diffusional rate control is now generally thought to be incorrect as feldspar dissolution rates generally increase with decreasing pH, which in turn results in a thicker leached layer [94]. Instead, it is now hypothesized that hydrolysis of the Al−O−Si bridging bond between aluminate and silicate tetrahedra is the rate-limiting reaction in feldspar dissolution at acidic pH values and that steady state is reached when the diffusion of Al^{3+} from within the leached layer is equivalent to the (slower) hydrolysis of remaining Si−O−Si bonds [94]. This scenario should not be generalized to all feldspars under all conditions, however. For example, at near-neutral to basic pH values, the leached layer, if present, is very thin and the dissolution rate is slower than at low pH [96].

More recent studies of alkali feldspar dissolution have found variations on the hypotheses listed above. For example, Nugent et al. [97] have proposed that naturally weathered albite feldspar surfaces are sodium and aluminum depleted, as found for laboratory-dissolved samples, suggesting similar dissolution mechanisms for albite dissolution in acidic soils and in acidic solutions in the laboratory. However, using atomic force microscopy, they detected a thin, hydrous, patchy coating of amorphous and crystalline aluminosilicate on natural albite surfaces, which may partially inhibit dissolution, and thus may help explain the consistently slower dissolution rates of natural albite samples in soils (up to four orders of magnitude slower) (see also [98]). In another recent study, Hellmann et al. [99] used high resolution, energy-filtered transmission electron microscopy to study the leached layer of labradorite feldspar altered under acidic conditions. They proposed that the near-surface altered zone is the result of dissolution-reprecipitation and is not due to preferential leaching of interstitial cations (Na^+ and Ca^{2+}) and framework cations (Al^{3+} and Si^{4+}). They also suggest that the intrinsic dissolution process of labradorite feldspar under acidic conditions is stoichiometric and congruent. This is an interesting but controversial hypothesis that awaits more observations before it should be generalized to other feldspar compositions (however, see below).

Additional observations on the dissolution of alkali feldspars come from studies by Fenter et al. [100–102] and Teng et al. [103] of orthoclase dissolution on the (001) and (010) cleavage surfaces, which used atomic force microscopy and synchrotron X-ray reflectivity (Figure 7.3(a)). These studies found that the measured X-ray reflectivity did not decrease monotonically with time during dissolution (Figure 7.3(a)), as would be characteristic of random dissolution (e.g., where all exposed tetrahedral sites dissolve at the same rate). Instead, the X-ray reflectivity data exhibit an oscillatory pattern at both acidic and alkaline pH values. This observation implies that two distinct reactive sites (e.g., terrace and step sites) are found in each pH regime, and that the relative reactivities of these sites differ substantially at the two extreme pH values. Dissolution at alkaline pH (pH 12.9) was found to be fully stoichiometric and dominated by lateral dissolution processes producing layer-by-layer dissolution, while dissolution at acidic pH (pH 1.1) was found to be only minimally non-stoichiometric (i.e., limited to one unit cell depth) and a more random process in which the orthoclase surface is substantially disrupted and roughened. The altered or leached layer discussed earlier based on other types of measurements of dissolving feldspars at low pH was found to be attributable to the formation of a gel-like coating. Crystal truncation rod diffraction data taken on an orthoclase feldspar (001) cleavage surface after reaction for 3 hr at pH 12.9 (Figure 7.3(b)) supports the above conclusions. The only significant change was an increase in step-edge density. Reaction of an (001) cleavage surface at pH 2.0 and 95°C for short time periods (hours) resulted in stoichiometric dissolution to a depth of one unit cell (6.4Å), whereas dissolution under these conditions for 11 months resulted in the formation of a 1300 Å-thick boehmite (γ-AlOOH) layer. These new X-ray scattering results provide a more quantitative picture of feldspar dissolution that refutes some of the older models.

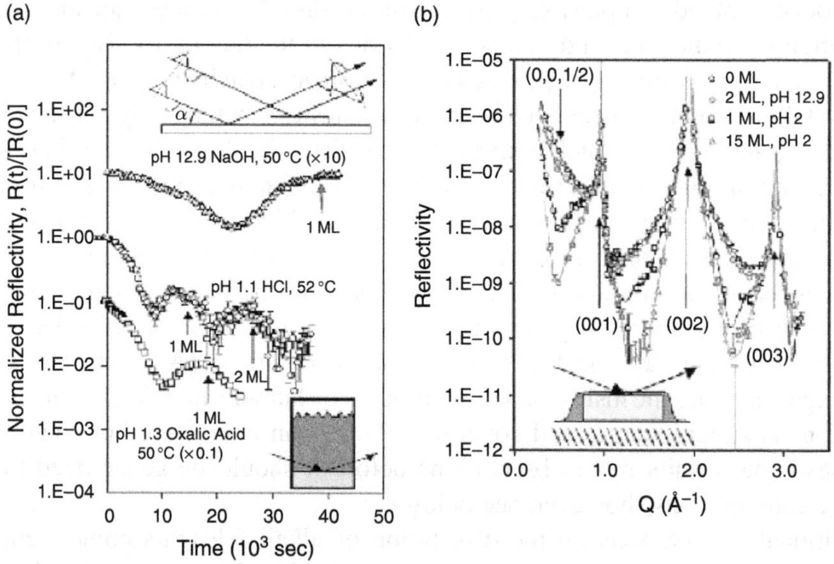

Figure 7.3. (a) In situ X-ray reflectivity vs. time (measured at the anti-Bragg condition, shown in inset at top) during dissolution of orthoclase feldspar, $KAlSi_3O_8$, (001) cleavage surface at extreme pH values. The removal of successive monolayers (ML) is noted for each set of data. (after [100]); **(b)** in situ crystal truncation rod diffraction profiles for a freshly cleaved orthoclase (001) surface (circles) and after reaction at pH = 2.0 (1 and 15 ML dissolved) (diamond and square) and pH = 12.9 (2 ML dissolved) (triangle) (after [103]). (Figures provided by P. Fenter.)

2.4. The nature of metal oxide-aqueous solution interfaces – some basics

Many models used to interpret and predict aqueous ion adsorption at metal oxide-aqueous solution interfaces are based on assumptions about the structure of surface functional groups, adsorption complexes, and the diffuse layer in the aqueous electrolyte solution adjacent to the charged metal oxide surface (e.g., [104–106]). The most commonly applied family of models is based on the hypothesis that the structure of the oxide-water interface consists of layers: a surface layer of strongly bound inner-sphere complexes including surface hydroxyls, one or more layers of weakly bound outer-sphere complexes, and a diffuse region of ions electrostatically attracted to the charged interface (e.g., [104,106] and Chapter 6) (Figure 7.4). The type of interaction of an aqueous pollutant and nutrient species with the solid surface determines its mobility, a fundamental property that affects pollutant and nutrient transport in aquifers and soils. The charge on a metal-oxide or metal-hydroxide surface, and thus the surface potential, depends on the degree of protonation of surface hydroxyl groups and the extent of adsorption of charged inner-sphere species in the inner Helmholtz planes (Figure 7.4). The potential at the planes of weakly bound

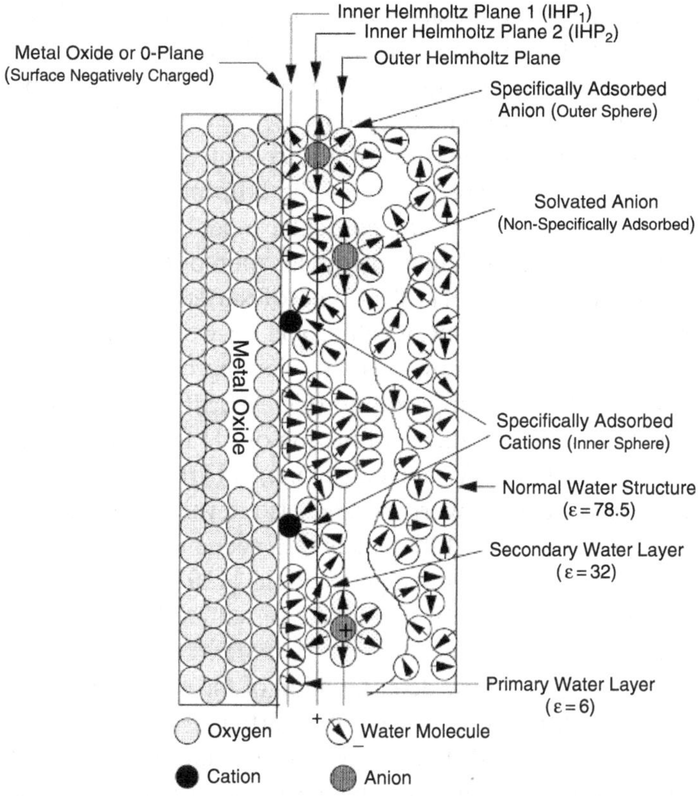

Figure 7.4. Schematic model of the Electrical Double Layer (EDL) at the metal oxide-aqueous solution interface showing elements of the Gouy–Chapman–Stern–Grahame model, including specifically adsorbed cations and non-specifically adsorbed solvated anions. The zero-plane is defined by the location of surface sites, which may be protonated or deprotonated. The inner Helmholtz plane, or β-plane, is defined by the centers of specifically adsorbed anions and cations. The outer Helmholtz plane, d-plane, or Stern plane corresponds to the beginning of the diffuse layer of counter-ions and co-ions. Cation size has been exaggerated. Estimates of the dielectric constant of water, ε, are indicated for the first and second water layers nearest the interface and for bulk water (modified after [6]).

ions (the outer Helmholtz plane or Stern plane) is related to the surface potential through constant capacitance terms, where the capacitance is a function of interlayer spacing and a dielectric constant [104]. In the diffuse layer, ions are expected to follow a Boltzmann distribution in which the potential field decays away from the outermost plane of specifically adsorbed species in a near exponential fashion according to the Gouy–Chapman theory [107–109]. The decay length of the potential is dependent upon the ionic strength of the solution and varies from $\approx$10 Å for a 100 mM solution to $\approx$1000 Å for a 0.01 mM solution (for a 1:1 electrolyte, based on Gouy–Chapman theory). This ionic strength dependence of the ion distribution and the pH dependence of the metal-oxide or metal-hydroxide surface charge relate to the total amount of charge that accumulates in the diffuse layer. The ionic structure of this interface also affects the binding constants of sorbates by altering the electrostatic potential in the interface region.

Variations in binding constants are often accounted for by using electrostatic corrections in the mass-law expressions of the adsorption reactions [106]. These corrections are calculated based on the particular model hypothesized for the structure of the interface, and, in particular, on the location of the adsorbing species with respect to the surface (Figure 7.4). This type of surface complexation model has been extended by Hiemstra, van Riemsdijk, and co-workers [110–113], who considered additional bonding constraints using a Pauling bond-valence approach. The charge distribution multi-site complexation (CD-MUSIC) model they developed is now widely used to fit uptake data as a function of pH for cations (e.g., [114,115]), anions (e.g., [116]), small organic molecules (e.g., [117]), and natural organic matter (e.g., [118]). Sverjensky and co-workers have also developed a variant of the triple-layer surface complexation model that takes into account the electrostatics of water dipole desorption during ligand exchange reactions [119]. This extended triple-layer model has recently been used to fit the uptake of sulfate [120], selenate [120], and arsenite [121] on goethite.

Macroscopic experiments allow determination of the capacitances, potentials, and binding constants by fitting titration data to a particular model of the surface complexation reaction [105,106,110–121]; however, this approach does not allow direct microscopic determination of the inter-layer spacing or the dielectric constant in the inter-layer region. While discrimination between inner-sphere and outer-sphere sorption complexes may be presumed from macroscopic experiments [122,123], direct determination of the structure and nature of surface complexes and the structure of the diffuse layer is not possible by these methods alone [40,124]. Nor is it clear that ideas from the chemistry of isolated species in solution (e.g., outer- vs. inner-sphere complexes) are directly transferable to the surface layer or if additional short- to mid-range structural ordering is important. Instead, in situ (in the presence of bulk water) molecular-scale probes such as X-ray absorption fine structure spectroscopy (XAFS) and X-ray standing wave (XSW) methods are needed to provide this information (see Section 3.4). To date, however, there have been very few molecular-scale experimental studies of the EDL at the metal oxide-aqueous solution interface (see, e.g., [125,126]).

One approach commonly used to study how strongly an aqueous ion binds to a solid surface is measurement of adsorption isotherms as a function of pH, ionic strength, and total metal-ion concentration, in the presence or absence of other ions or organic coadsorbents. It is often assumed that significant inhibition of adsorption with increasing ionic strength at a given pH indicates that the sorption complexes are dominantly of the weakly bound, outer-sphere type [123]. In contrast, when there is little or no dependence of metal-ion sorption on ionic strength, the sorption complex is assumed to be more strongly bound as an inner-sphere type complex. Even when there is little ionic strength dependence of ion sorption, the pH_{ads} relative to the point of zero charge (pH_{PZC}) of the oxide can be used to infer differences in relative strength of binding as a function of pH (pH_{ads} is defined as the pH value at 50% uptake of the metal ion). Strong uptake of a co-ion (e.g., sorption of a cation below the pH_{PZC}) implies that the free energy of chemical bond formation is significant with respect to the electrostatic terms, thus infers inner-sphere (and likely covalent) interaction.

To illustrate the effect of different metal oxides on the sorption of a specific metal ion, as well as how adsorption isotherms may be used to provide qualitative insights about the relative binding of an adsorbate, we show uptake data for aqueous Co(II) on α-SiO_2 (quartz), TiO_2 (rutile), α-Fe_2O_3 (hematite), and α-Al_2O_3 as a function of pH at a relatively constant ionic strength in Figure 7.5. The vertical bands representing pH_{PZC} values for the four oxides show the range of values reported in the literature. The Co(II) uptake data have been corrected for differences in surface area and estimated number of reactive surface sites of the respective solids, resulting in a relatively constant ratio of Σ Co(II):Σ sites of 0.002–0.006. The pH_{PZC} values of these surfaces differ significantly, ranging from about pH 2–3 for α-SiO_2 to pH 8.5 for α-Al_2O_3 [127,128]. Below the pH_{PZC}, the solid surface is positively charged, indicating an excess of surface protons, whereas above the

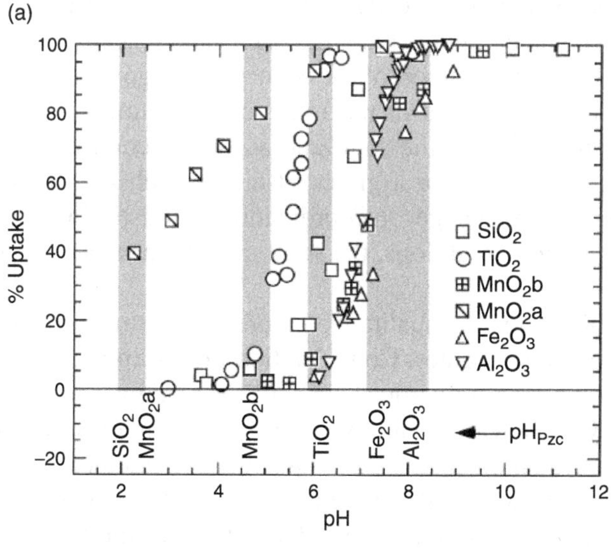

Figure 7.5. Percentages uptake of Co^{2+} vs. pH on selected metal oxides. The vertical bars indicate the range of pH_{PZC} values for the different oxides. (from [6])

pH_{PZC}, the surface is negatively charged, indicating a proton deficiency and an excess of (hydr)oxo groups.

This variation in surface charge as a function of pH can be explained using the following simplified equilibria [5]:

$$S{-}OH_2^+ \overset{K_1}{\leftrightarrow} S{-}OH \overset{K_2}{\leftrightarrow} S{-}O^-$$

where S represents a surface metal atom and K_1 and K_2 represent the first and second acidity constants, respectively, of the surface hydroxyl groups. Similarly, metal ions sorb on organic materials (e.g., humic matter and bacterial cell surfaces) via protolysis reactions involving amino (NH_3^+) and carboxyl (COOH) functional groups, among others, as shown by the following equilibria [5]:

$$R{<}\begin{matrix}COOH\\ \\ NH_3^+\end{matrix} \overset{K_1}{\leftrightarrow} R{<}\begin{matrix}COO^-\\ \\ NH_3^+\end{matrix} \overset{K_2}{\leftrightarrow} R{<}\begin{matrix}COO^-\\ \\ NH_2,\end{matrix}$$

where *R* represents a large organic molecule or a bacterium with functional groups.

Aqueous Co(II) does not sorb significantly on the α-SiO_2 surface until well above the pH_{PZC} of α-SiO_2, where the surface is negatively charged. In contrast, uptake of Co(II) on α-Al_2O_3 and α-Fe_2O_3 is essentially complete below their respective pH_{PZC} values, where the surfaces are positively charged. These differences suggest that Co(II) binds more strongly to the α-Al_2O_3 and α-Fe_2O_3 surfaces than to the α-SiO_2 surface. The uptake behavior of Co(II) on TiO_2 (rutile) is intermediate between these two cases, with uptake beginning below the pH_{PZC} and reaching 100 % above pH_{PZC}. In sum, differences in the location of the adsorption edge of a single adion relative to the pH_{PZC} reflect differences in the relative contributions of different bonding forces. The properties of both the solid and adsorbate contribute to the magnitude of these bonding forces, but for a given adsorbate such as aqueous Co(II) which does not show significant changes in coordination chemistry in the aqueous solution as a function of solution conditions, the solid must be primarily responsible for the differences.

This example illustrates the qualitative nature of information that can be gleaned from macroscopic uptake studies. Consideration of adsorption isotherms alone cannot provide mechanistic information about sorption reactions because such isotherms can be fit equally well with a variety of surface complexation models assuming different reaction stoichiometries. More quantitative, molecular-scale information about such reactions is needed if we are to develop a fundamental understanding of molecular processes at environmental interfaces. Over the past 20 years in situ XAFS spectroscopy studies have provided quantitative information on the products of sorption reactions at metal oxide-aqueous solution interfaces (e.g., [39,40,129–138]. One finding from these studies is that many metal oxide and silicate surfaces in contact with aqueous solutions are dynamic, undergoing dissolution during adsorption reactions which can result in co-precipitation of new phases such as mixed-metal layered double hydroxides [135–138]. Many of these XAFS spectroscopy studies of adsorption complexes have been reviewed recently, and the reader is referred to references [12] and [139] for reviews of such studies through 2002.

The example discussed above shows the effect of different solid substrates on the adsorption of a single type of aqueous cation [Co(II)]. The uptake vs. pH curves in Figure 7.6 show the differences in uptake behavior as a function of pH for a variety of aqueous cations and anions on one type of substrate – ferric hydroxide particles [5]. Some of the cations included in Figure 7.6(a), like Pb^{2+} and Cr^{3+}, sorb at low pH values, whereas others, like Zn^{2+} and Ni^{2+}, do not sorb until higher pH values are reached. At low pH values, metal oxide and hydroxide surfaces tend to

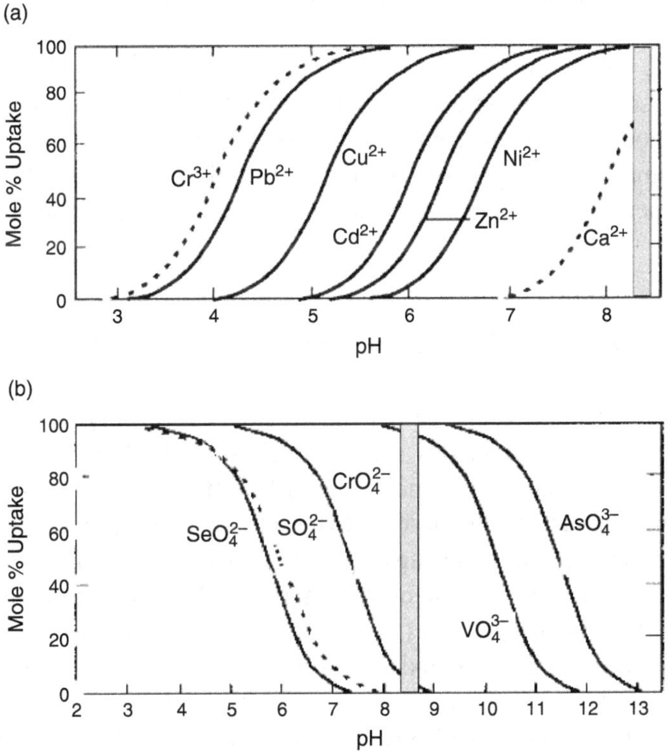

Figure 7.6. (a) Mole % uptake of cations vs. pH for various cations adsorbing at the hydrous ferric oxide – aqueous solution interface. The vertical bar at pH 8.5 represents the pH_{PZC} of hydrous ferric oxide. (b) Mole % uptake of anions vs. pH for various oxoanions adsorbing at the hydrous ferric oxide – aqueous solution interface as a function of pH. The vertical bar at pH 8.5 represents the pH_{PZC} of hydrous ferric oxide. (after [5]).

be positively charged, with an excess of protons bound to the surface, thus these surfaces tend to repel positively charged metal ions and attract negatively charged ions. At some intermediate pH value, the surface becomes electrically neutral. At the pH_{PZC}, electrostatic repulsion of a positively or negatively charged ion would be minimized. At pH values higher than the pH_{PZC}, the surface becomes negatively charged because of the predominance of hydroxo (OH^-) or oxo (O^{2-}) groups on the surface. Under these conditions, a positively charged metal ion in solution would be attracted to the surface, but a negatively charged ion would be repelled. Anions show the opposite behavior, with strong electrostatic attraction to metal oxide particle surfaces at low pH values and repulsion at high pH values (Figure 7.6(b)).

The pH_{PZC} of ferric hydroxide surfaces is about 8 [127], so aqueous Pb^{2+} should be electrostatically repelled from these surfaces at pH values less than 8. However, as seen in Figure 7.6(a), the Pb^{2+} present in this aqueous solution is sorbed essentially completely to ferric hydroxide surfaces at pH 6. This behavior suggests that Pb^{2+} forms direct chemical bonds to these surfaces in order to overcome the repulsive electrostatic forces below the pH_{PZC} of ferric hydroxide. This conclusion based on macroscopic uptake data has been confirmed by direct spectroscopic observation using X-ray absorption fine structure (XAFS) spectroscopy under in situ conditions (i.e., with aqueous solution in contact with α-FeOOH surfaces at ambient temperature and pressure) [133,134]. These studies showed that the aquated Pb(II) ion forms dominantly inner-sphere, bidentate complexes on α-FeOOH surfaces.

3. Factors affecting the chemical reactivity of mineral surfaces

A variety of interrelated factors affect the chemical reactivity of mineral surfaces with respect to water and aqueous species, including (1) defect density, (2) cooperative effects among adsorbate molecules, (3) differences in intrinsic properties of different mineral surfaces, including different isoelectric points, (4) solution pH, (5) types of surface functional groups, (6) surface structural differences of different mineral surfaces, and (7) the effect of organic and biofilm coatings on mineral surface reactivity. Factors (3) and (4) have already been discussed in Section 2.4. Also important is the effect of oxidation–reduction reactions on these variables. For example, when the fresh surface of an Fe(II)-containing oxide (e.g., Fe_3O_4) or sulfide (e.g., FeS_2) reacts with the oxidized form of chromium, $Cr(VI)O_4{}^{2-}$, in pH 7, slightly oxidizing solution, it is likely that an electron transfer reaction will occur in which Cr(VI) is reduced to Cr(III) and surface Fe(II) ions are oxidized to Fe(III) (see, e.g., [151–154]). When this reaction occurs, the solid surface changes in composition, structure, and reactivity, and the surface is likely to become passivated by a non-reactive layer after a certain time period, depending on reaction kinetics.

Many of the minerals considered in Section 2, such as alkali feldspar, have complex structures and compositions and complex dissolution behavior in water. In spite of the great insights provided by some of the studies discussed above, the pathways responsible for reactions of these complex minerals with acidic or basic solutions are not understood at a fundamental mechanistic level. In order to focus this discussion on atomic-level factors controlling surface reactivity, here we consider simplified metal oxide model systems under carefully controlled conditions. Such systems can be studied at the molecular level using a variety of modern surface science methods, many of which use intense X-rays from synchrotron radiation sources and can probe surfaces and surface adsorbates under reactive (in situ) conditions (i.e., with liquid water present at ambient temperature and pressure). The model metal oxides chosen include the following single-crystal surfaces: MgO(100), α-Al_2O_3(0001), α-Al_2O_3(1–102), α-Fe_2O_3(0001), and α-Fe_2O_3(1–102).

3.1. The reaction of water vapor with metal oxide surfaces – surface science and theoretical studies of simplified model systems illustrating effects of defect density and adsorbate cooperative effects

The interaction of liquid water with the surfaces of natural solids is one of the most fundamental chemical reactions occurring in nature. This reaction strongly influences the degree of surface hydroxylation, which in turn influences the type(s) of functional groups exposed at mineral surfaces. Thus hydration reactions are extremely important in determining the reactivity of metal oxide surfaces. Even though there have been hundreds of studies of the interaction of water with clean metal and metal oxide surfaces over the past 30 years using a variety of surface science methods [140,141], there is little fundamental understanding of how liquid water or water vapor reacts with mineral surfaces except in the simplest cases [142]. For example, the interaction of water vapor with the clean MgO (100) surface has been studied extensively (see [141,143] for a review of some of these studies). Here we focus on synchrotron radiation-based photoemission spectroscopy studies of MgO(100)-water vapor reactions as a function of p(H_2O) at 298 K [143,144]. Each surface was prepared by cleaving a MgO single-crystal in a UHV preparation chamber (base pressure of 5×10^{-11} torr), exposing it sequentially to different p(H_2O) levels, with an exposure time at each step of 3 min, transferring the sample to a UHV analysis chamber between sequential doses of water vapor, and collecting oxygen 1s, 2s, and valence band (VB) spectra.

Figure 7.7(a) shows the O 1s photoemission spectra of MgO(100) following reaction with different concentrations of water vapor. Liu et al. [143] interpreted these spectra as indicating that water vapor reacts primarily with defect sites on the Mg(100) surface at p(H_2O) $\leq 10^{-4}$ torr and with terrace sites at p(H_2O) $>10^{-4}$ torr.

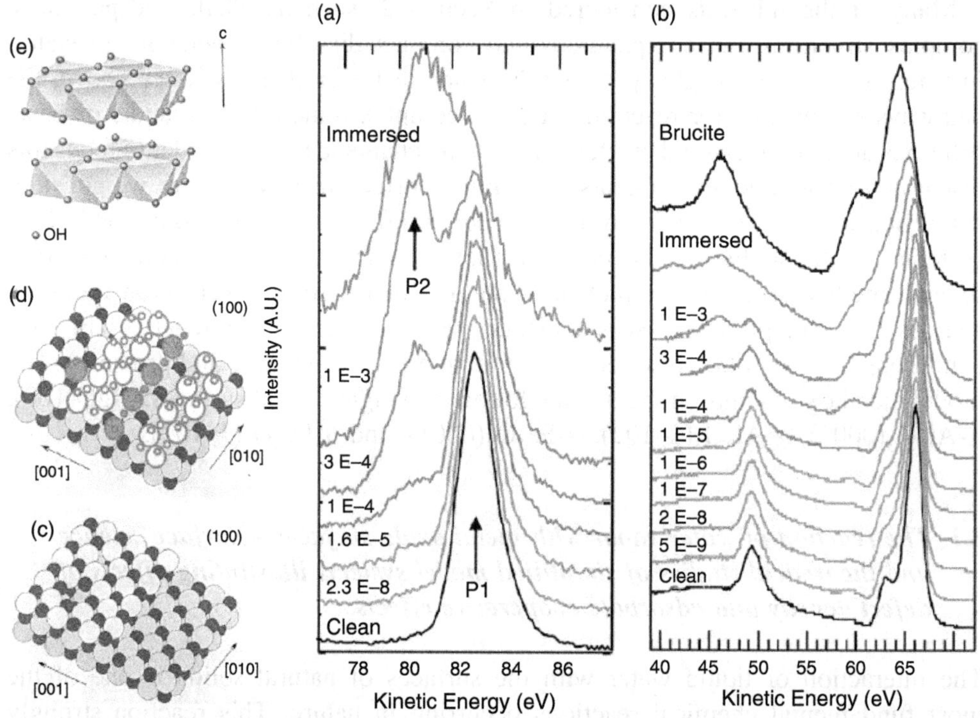

Figure 7.7. (a) Oxygen 1s synchrotron X-ray photoemission spectra of the clean MgO (100) surface and the same surface after sequential 3 min. exposures to water vapor at the p(H_2O) values indicated (in torr) (taken at 620 eV incident X-ray energy). The top spectrum was taken after immersion of MgO(100) in bulk water for 30 min. Peaks labeled P1 are due to near-surface lattice oxygens, and those labeled P2 are due to oxygens in surface hydroxyl groups. (b) Oxygen valence band and 2s spectra of the clean MgO (100) surface and the same surface after sequential exposures to water vapor at the p(H_2O) values indicated (in torr) (taken at 80 eV incident X-ray energy). The top spectrum is from the mineral brucite ($Mg(OH)_2$) (after [143]). (c) Drawing of the MgO (100) surface (black circles represent magnesium ions and open circles represent oxygen ions) showing a step edge and an oxygen vacancy defect (from [25]). (d) Schematic drawing of the interaction of water molecules with this surface, with dissociative chemisorption of water molecules occurring at the step edge and at the oxygen vacancy. (e) Drawing of the brucite structure.

Evidence for this conclusion is the appearance of a low intensity feature at a kinetic energy of ≈80.6 eV in the O 1s spectra (Figure 7.7(a)) and at ≈60 eV in the O 2s + VB spectra (Figure 7.7(b)), which are caused by the dissociation of water and the formation of OH^- groups on the MgO (100) surface. The main O 1s feature centered at ≈83 eV kinetic energy and the main VB feature centered at ≈66 eV kinetic energy are due to lattice oxygens. The nature of these water-surface interactions is shown schematically in the structural drawings in Figures 7.7(c) and (d), which depict two MgO (100) surfaces, one showing terrace sites, a step defect, and an oxygen vacancy defect prior to exposure to water vapor (Figure 7.7(c)), and the second showing hydroxyl groups on the surface following exposure at $p(H_2O) > 3 \times 10^{-4}$ torr (Figure 7.7(d)). At exposure pressures less than this 'threshold pressure', water chemisorbs dissociatively on these types of defects, which are thought to be more reactive than terrace sites on MgO(100) [145,146]. This low kinetic energy shoulder does not grow appreciably in intensity as $p(H_2O)$ is increased from 10^{-9} to 10^{-4} torr, even after prolonged exposures (several hours) of the surface to water vapor (up to 1.8×10^4 Langmuirs (L), where 1 L corresponds to 10^{-6} torr-sec). However, when the defect density of these surfaces was increased (from ≈5% to 35%) by various means, this low kinetic energy shoulder increased significantly in intensity at $p(H_2O) < 10^{-3}$ torr [144]. This effect is discussed in more detail below. In addition, when $p(H_2O)$ was raised to 3×10^{-4} torr with an exposure time of 3 min (corresponding to an exposure of 5.4×10^4 L), the dissociation reaction was rapid as indicated by the rapid increase in intensity of the low kinetic energy feature, which continued to grow in intensity with exposure of the surface to higher water pressures [143].

These photoemission results are at variance with earlier high-level quantum chemical calculations, which modeled the interaction of one water molecule per surface unit cell and indicated that the interaction of a water molecule with a terrace site on MgO(100) is endothermic [145,146]. However, more recent density functional calculations [147,148] in which a more realistic 3×2 ordered array of water molecules was placed on the MgO (100) surface, predict that water should dissociate on terrace sites, in agreement with the photoemission results. An important outcome of these calculations is that a cooperative effect associated with hydrogen bonding between adjacent water molecules on the surface appears to be required for dissociation of water on MgO(100) terrace sites. Apparently, when a critical coverage of water molecules is reached (≈50% monolayer coverage) on these surfaces, these hydrogen bonding interactions, coupled with the surface interactions, destabilize the water molecule, leading to dissociation and surface hydroxylation. Also shown in Fig. 7.7(b) are O 1s and O 2s + VB spectra of MgO(100) immersed in liquid water and of brucite ($Mg(OH)_2$). The spectral features indicative of hydroxyl groups on the MgO(100) surfaces are amplified in these "high water pressure" spectra. Prolonged exposure of MgO to water will result in the formation of brucite, whose structure is shown in Fig. 7.7(e).

The important role played by defect density in chemical reactivity can be seen in Figure 7.8, which shows O 1s photoemission spectra of a vacuum-cleaved, clean MgO (100) surface in comparison with (1) the O 1s spectrum of a vacuum-cleaved MgO (100) surface exposed to 2.3×10^{-8} torr $p(H_2O)$ for 3 min, (2) the spectrum of an MgO (100) surface purposely cleaved off angle in vacuum then exposed to $p(H_2O) = 3 \times 10^{-8}$ torr for 3 minutes, and (3) the spectrum of an MgO (100) surface that was Ar^+ ion sputtered and then exposed to $p(H_2O) = 2.5 \times 10^{-8}$ torr

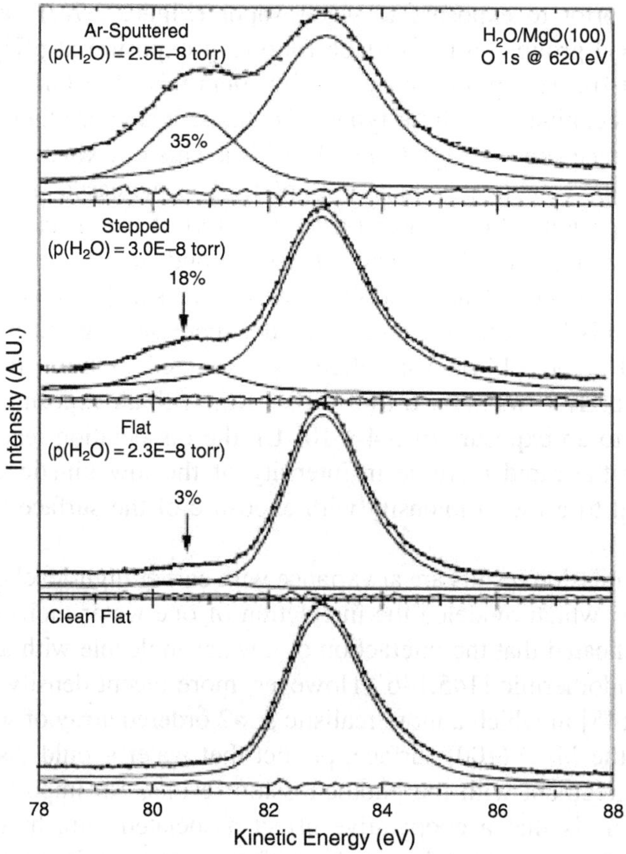

Figure 7.8. O 1s spectra of vacuum-cleaved, clean flat MgO(100) and various water-dosed MgO(100) surfaces (3 min exposures at the pressures indicated). One (clean surface) and two component (water-dosed surfaces) fits are shown, along with the fit residual. The incident photon energy was 620 eV. (from [144])

for 3 min. The low kinetic energy peaks in the three top spectra represent surface hydroxyls. At these low pressures, water should react only with defect sites (see earlier discussion of the reaction of water vapor with low defect MgO(100)). Their intensities increase in proportion to the defect density on MgO(100), indicating that dissociation of water is greatest on the most defective surfaces.

Another example of the interaction of water with a relatively simple metal oxide surface is provided by the water vapor/α-Al_2O_3(0001) system (Figure 7.9(a)). Oxygen 1s synchrotron radiation photoemission results indicate that significant dissociative chemisorption of water molecules does not occur below ≈1 torr p(H_2O) [149]. However, following exposure of the alumina (0001) surface to water vapor above this 'threshold p(H_2O)', a low kinetic energy feature in the O 1s spectrum grows quickly,

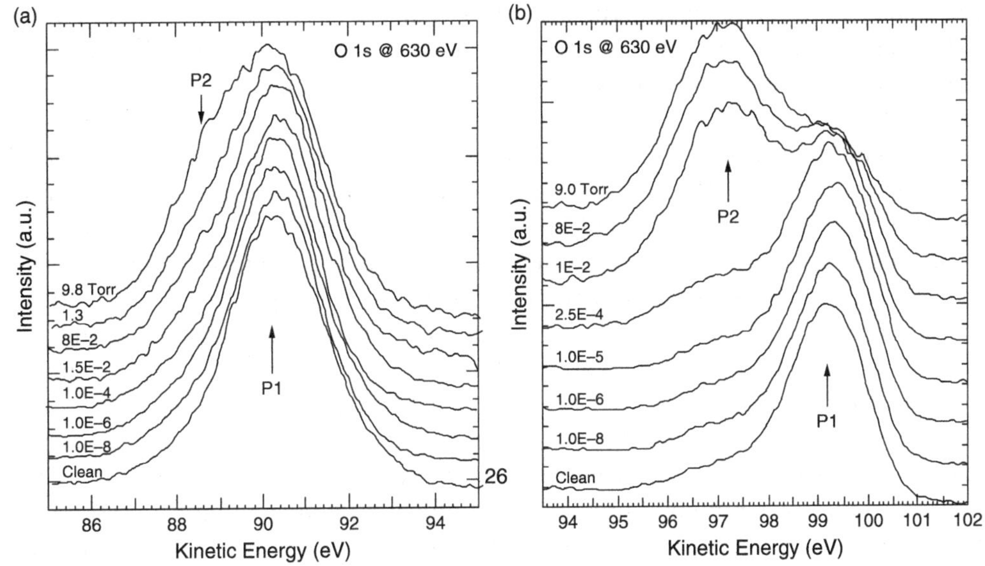

Figure 7.9. (a) Oxygen 1s synchrotron X-ray photoemission spectra of the clean α-Al_2O_3 (0001) surface (bottom spectrum) and the same surface after sequential interactions with water vapor at various p(H_2O) values (as indicated in Torr). (b) Oxygen 1s synchrotron X-ray photoemission spectra of the clean α-Fe_2O_3 (0001) surface (bottom spectrum) and the same surface after sequential interactions with water vapor at various p(H_2O) values (as indicated in Torr). P1 indicates the O 1s spectrum of lattice oxygen, and P2 indicates the O 1s spectrum of oxygen in surface hydroxyl groups. (after [149]).

indicating increasing levels of dissociative chemisorption of water on terrace sites. The results of recent density functional calculations of the interaction of water on α-Al_2O_3(0001) as a function of water coverage [50,51] are in substantial agreement with the photoemission results of Liu et al. [149]. Similar photoemission experiments on α-Fe_2O_3 (0001) indicate a 'threshold p(H_2O)' of 10^{-4} torr, above which water dissociates on terrace sites (Figure 7.9(b)) [149]. These observations raise the interesting question as to why the 'threshold p(H_2O)' values of isostructural corundum and hematite differ by about five orders of magnitude.

The factors contributing to the differences in 'threshold p(H_2O)' for isostructural corundum and hematite can be examined using theoretical methods. The free energy of a mineral surface can be calculated as a function of the concentration of species in the surrounding environment by linking quantum mechanics and experimental thermo-dynamic data in a method known as '*ab initio* thermodynamics' [51,150]. The 0 K binding energy of an atom in any given surface stoichiometry or environmental species, such as a water or oxygen molecule, can be calculated using density functional theory (DFT). The change in enthalpy and entropy as a function of temperature and pressure is determined through the calculation of harmonic vibrational frequencies and tabulated experimental data. Thus it is possible to determine the p(H_2O) at which the dissociation of water on the mineral surface and formation of a hydroxylated structure becomes exothermic, relative to a clean surface in equilibrium with water vapor. For corundum and hematite, the *ab initio* thermodynamics calculations indicate that as water vapor pressure is increased, the formation of fully hydroxylated (0001) surfaces likely occurs in a stepwise fashion, beginning with dissociation of one water molecule per exposed metal cation [A.M. Chaka, in prep.]. This first dissociation is exothermic for both corundum and hematite and does not require the presence of defects to occur. The subsequent stages of hydroxylation correspond to dissociation of a second, then a third, water molecule per exposed cation until the octahedral coordinations of the surface Al and Fe atoms are restored. The thermodynamic pH_2O threshold for dissociation of three water molecules per surface cation, which corresponds to the fully hydroxylated surface observed experimentally, is calculated to be 0.1 torr for α-Al_2O_3 and 10^{-9} torr for α-Fe_2O_3, a difference of eight orders of magnitude, reflecting a dramatic difference in the strength of the Fe−O bonds compared to Al−O. Most of the difference in the calculated thermodynamic pH_2O thresholds compared to the experimentally observed values is likely due to kinetic factors. The higher pH_2O values observed experimentally may reflect the need for additional physisorbed waters to be present to facilitate a cooperative dissociation mechanism to lower the activation barrier, as was indicated by the Car–Parinello dynamics simulations performed by Haas and coworkers on α-Al_2O_3 (0001) [50].

3.2. *Grazing incidence EXAFS spectroscopic studies of Pb(II)aq adsorption on metal oxide surfaces – effect of differences in surface functional groups on reactivity*

Additional information about intrinsic differences or similarities in reactivity of metal oxide surfaces can be derived from grazing-incidence (GI)-EXAFS spectroscopy studies of metal ions adsorbed on these surfaces. Here we focus on in situ GI-EXAFS studies of Pb(II)aq adsorption on α-Al_2O_3 (0001), α-Al_2O_3 (1–102), α-Fe_2O_3 (0001), and α-Fe_2O_3 (1–102) single crystal surfaces [155–157]. In a grazing-incidence EXAFS spectroscopy experiment, the incident X-ray beam strikes the single-crystal surface at an angle below the critical angles of these substrates, which are 0.16 – 0.185° at 14 keV. Below the critical angle, total external reflection of the X-ray beam should occur, resulting in up to a factor of four-times greater sensitivity to adsorbates relative to bulk EXAFS methods [158].

The GI-EXAFS study of Bargar et al. [155] found that Pb(II)aq forms dominantly outer-sphere complexes at the α-Al_2O_3(0001)/aqueous solution interface after reaction of the substrate for a minimum of 90 min with a 0.25 mM $Pb(NO_3)_2$ solution

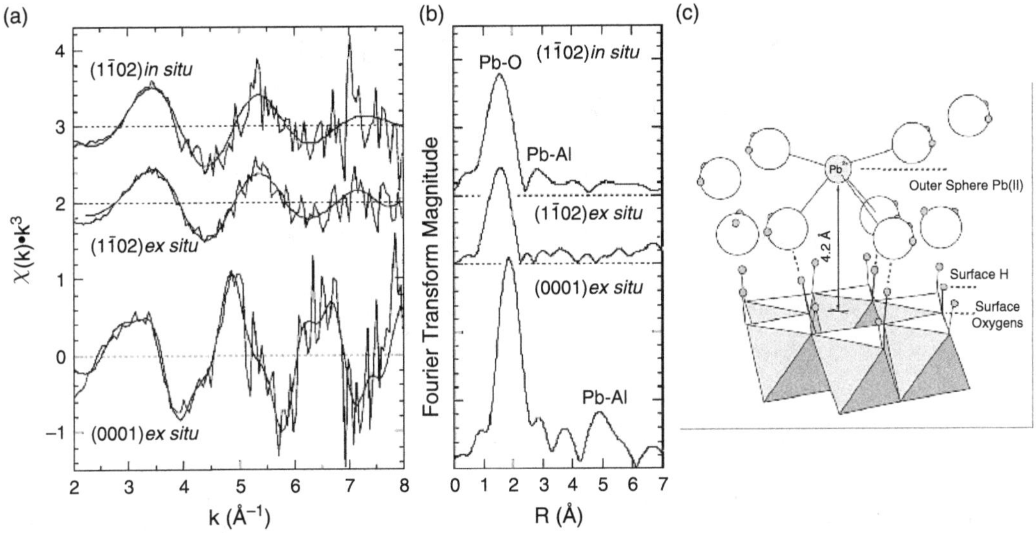

Figure 7.10. (a) Normalized k^3-weighted GI-EXAFS spectra of Pb(II)/α- Al_2O_3(0001) and Pb(II)/α-Al_2O_3(1–102) samples under in situ (i.e., with water vapor present and a relative humidity of 50–80%) and ex situ (i.e., no water vapor) conditions. (b) Fourier transforms of the EXAFS data showing Pb−O pair correlations (features at 1.6–2.0 Å, uncorrected for phase shift) in all three FT's and a Pb−Al pair correlation (feature at about 5 Å, uncorrected for phase shift) in the FT of the Pb(II)/α-Al_2O_3(0001) sample. (c) Schematic drawing of an outer-sphere $Pb(H_2O)_4{}^{2+}$ complex at the α-Al_2O_3(0001) surface (from [155]).

at pH 7 and 21°C. Fits of the Pb L_3-edge EXAFS data (Figure 7.10(a)) yielded a Pb−O distance of 2.51(±0.02) Å, which is consistent with Pb^{2+} bonded to water molecules. If Pb^{2+} were bonded to oxo ligands, the Pb−O distance would fall in the range 2.20–2.31 Å, whereas if bonded to hydroxo ligands, the Pb−OH distance would be 2.25–2.32 Å. In addition, the Pb−Al distance was found to be 5.78(±0.04) Å for the (0001) alumina surface, which is consistent with the distance expected for a fully aquated, outer-sphere Pb(II) complex at the α-Al_2O_3(0001)/aqueous solution interface (Figure 7.10(c)). Under similar solution conditions, Pb(II)aq was found to form dominantly inner-sphere complexes on the α-Al_2O_3(1–102) surface [155]. XPS data on similar samples showed that the sorption density of Pb^{2+} on the (1–102) surface is $1.7 \pm 0.1\,\mu$moles/m^2, whereas a sorption density of $0.05\,\mu$moles/m^2 was found for the (0001) surface [155]. These results indicate that the α-Al_2O_3(1–102) surface is significantly more reactive to aqueous Pb^{2+} than the (0001) surface.

In a parallel study, Co(II)aq was found to form dominantly inner-sphere complexes on both surfaces of α-Al_2O_3 [156]. This difference in reactivity between the (0001) and (1–102) surfaces with respect to aqueous Pb(II) and Co(II) was explained by

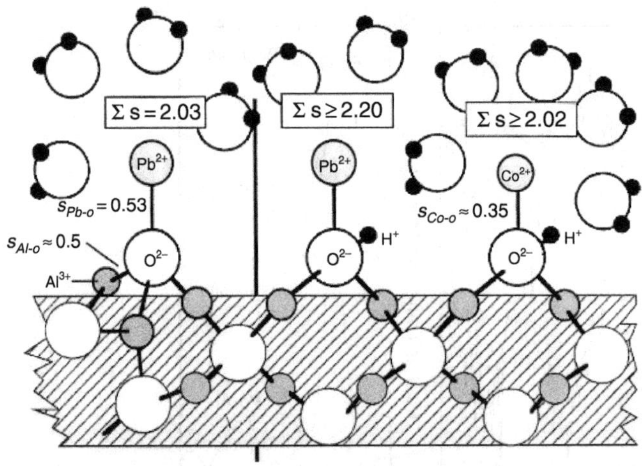

Figure 7.11. Schematic drawing of functional groups on the (a) α-Al_2O_3(1–102) surface, which is predicted to have both singly and triply coordinated surface oxygens, and (b) α-Al_2O_3(0001) surface, which is predicted to have only doubly coordinated surface oxygens. It shows that 4-coordinated Pb^{2+}, with a Pb−O bond strength of 0.53 valence units (v.u.), can bond to surface oxygens bonded to three 6-coordinated Al^{3+} ions on the (1–102) surface, resulting in a bond strength sum of 2.03 v.u., which obeys Pauling's bond valence sum rule for oxygen. In contrast, 4-coordinated Pb^{2+} is not predicted to bond to surface oxygens that are bonded to two 6-coordinated Al^{3+} ions and a proton on the (0001) surface because this coordination results in a bond valence sum of 2.20 v.u. However, 6-coordinated Co^{2+}, with a Co−O bond strength of 0.35 v.u., can bond to these types of oxygens on the (0001) surface. The bond strength of the O−H bond to these oxygens is about 0.8 v.u. (see Figure 7.12) (from [156]).

Bargar et al. [156] as being due to differences in surface functional groups on the two surfaces (Figure 7.11). Assuming a perfect termination of the bulk structure along the (0001) and (1–102) planes and no relaxation or reconstruction, the (0001) surface is predicted to have only oxygens bonded to two 6-coordinated Al^{3+} ions, while the (1–102) surface is predicted to have oxygens bonded to one 6-coordinated Al^{3+} ion and oxygens bonded to three 6-coordinated Al^{3+} ions [156]. These proposed surface structures are best guesses, and, as pointed out by Bargar et al. [156], may not be correct when the surfaces are in contact with water. Singly and triply coordinated oxygens are predicted to be significantly more reactive than doubly coordinated oxygens, thus the predicted presence of dominantly inner-sphere Pb(II) on the α-Al_2O_3 (1–102) surface and dominantly outer-sphere Pb(II) on the (0001) alumina surface is consistent with the EXAFS results [155,156].

The number of protons bonded to surface oxygens cannot be determined directly from EXAFS spectroscopy and must be inferred by some other means. One approach is to add protons at an O−H distance needed to satisfy Pauling's bond strength sum to surface oxygens. Bond length and bond strength curves for O−H bonds

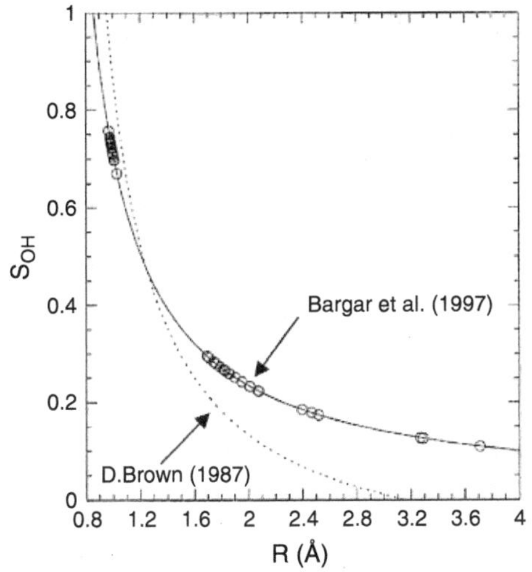

Figure 7.12. Bond length vs. bond valence curves for O−H bonds. The solid line is a fit of O−H bond lengths from a variety of structures containing water molecules or hydroxyl groups using the following equation: $s_{O-H} = 0.241/(R_{O-H} - 0.677)$, where RO−H is the experimentally determined O−H distance [156]. The dotted line is from Brown [159]. (Figure from [156].)

are required for this approach and have been derived by Bargar et al. [156] and Brown [159] (Figure 7.12). Two populations of O−H distances are evident, one corresponding to hydroxyl O−H bonds (with OH distances ranging from 0.8 to 1.0 Å and bond strengths ranging from 0.76 to 0.67 v.u.) and one corresponding to H-bonds in water molecules (with O−H distances ranging from 1.6 to >3 Å and bond strengths ranging from 0.3 to 0.1 v.u.).

Because α-Al_2O_3 and α-Fe_2O_3 are isostructural, they would be expected to have very similar surface structures along the same crystallographic plane, such as (0001) or (1-102), and similar types of metal ion adsorption complexes on a specific surface. This prediction was found to be incorrect for the α-Fe_2O_3 (0001) surface in a GI-EXAFS study of aqueous Pb(II) adsorption at the α-Fe_2O_3(0001)/water and α-Fe_2O_3(1–102)/water interfaces (Figure 7.13) [157]. Instead, Pb^{2+} was found to form dominantly inner-sphere complexes on both surfaces, with the (0001) surface having a higher coverage of Pb^{2+} than the (1–102) surface based on XPS results [157]. This finding suggests that assumptions about similarity of the structures of hydrated α-Fe_2O_3 (0001) and α-Al_2O_3 (0001) are incorrect. The next section explores this suggestion.

To summarize, the order of reactivity of the α-Fe_2O_3 and α-Al_2O_3 single-crystal surfaces with respect to both water vapor and aqueous Pb(II) is as follows: $\alpha\text{-}Fe_2O_3(0001) > \alpha\text{-}Fe_2O_3(1-102) \approx \alpha\text{-}Al_2O_3(1-102) >> \alpha\text{-}Al_2O_3(0001)$.

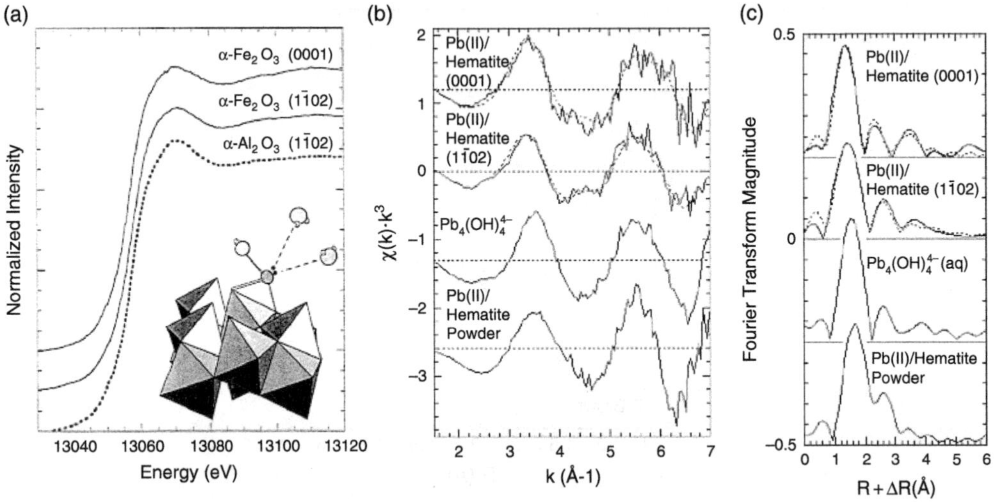

Figure 7.13. (a) Pb L_3-XANES spectra for Pb(II)/α-Fe_2O_3(0001)/water, Pb(II)/α-Fe_2O_3(1–102)/water, and Pb(II)/α-Al_2O_3(1–102)/water systems, showing a schematic model of Pb^{2+} inner-sphere complexes on these surfaces. (b) Background-subtracted, k^3-weighted Pb L_3-EXAFS spectra of Pb(II)/α-Fe_2O_3(0001)/water and Pb(II)/α-Fe_2O_3(1–102)/water systems compared with Pb L_3-EXAFS spectra of aqueous $Pb_4(OH)_4^{4-}$ and Pb(II) sorbed on powdered hematite samples. (c) Fourier transforms of the EXAFS spectra in Figure 7.13(b) showing Pb–O and Pb–Fe pair correlations. (from [157].)

3.3. The structure of hydrated metal oxide surfaces from X-ray diffraction studies

X-ray scattering techniques have been widely used for investigation of the atomic scale structure of (crystalline) surface terminations [36,38,160]. As opposed to traditional ultra-high vacuum surface characterization techniques based on electron or soft X-ray spectroscopies, hard X-ray scattering can be used to investigate surface and interface structure in the presence of liquid water films. Therefore, the influence of hydration or hydroxylation and the liquid water structure adjacent to a mineral surface can be probed in situ, thus avoiding the need to extrapolate UHV surface characterization to high pressure conditions. This direct analysis is important to building models that accurately reflect surface structure and composition, and hence mineral surface reactivity, under conditions that reflect those encountered in aqueous environments.

Among the most common surface X-ray scattering techniques used to probe mineral-fluid interface structure is the measurement of crystal truncation rods (CTRs). CTRs are diffuse streaks of intensity connecting bulk reciprocal lattice (Bragg) points in the direction perpendicular to a surface, and arise as a natural consequence of the presence of a sharp termination of the crystal [34,161]. The CTR scattering is a (in-phase) summation of the bulk and surface contributions, where the net scattering factor can be expressed as

$$F_T \propto F_{bc}(Q)F_{CTR}(Q_\perp)+F_{sc}(Q) \qquad (1)$$

where Q is the conventional scattering vector, $Q_\perp$ is the component of Q parallel to the surface normal direction, F_{bc} is the bulk unit cell structure factor, F_{ctr} is the CTR form factor, and F_{sc} is the structure factor of the surface unit cell [161–163]. The measured CTR intensity is proportional to $|F_T^2|$ and shows characteristic $1/4\sin^2(\pi L)$ variation between Bragg points, where L is the reciprocal lattice index in the surface normal direction (Figure 7.14). Modifications in the structure and/or composition of the surface with respect to the bulk unit cell modifies the F_{sc} term; therefore, the interference between the scattering from the surface layer and the bulk results in structure along the rod profiles different from that expected for the bulk termination. The most surface sensitive parts of the rods are located at the mid-points between bulk Bragg reflections, since this is where F_{CTR} approaches a minimal value. CTRs are highly sensitive to atomic positions at the terminating surface and can provide detailed information on the three-dimensional structure of surfaces. This includes the identification of the terminating surface atoms (i.e., surface functional group coordination information) and relaxations of near surface atomic positions, as well as the structure of surface reconstructions through the analysis of fractional order

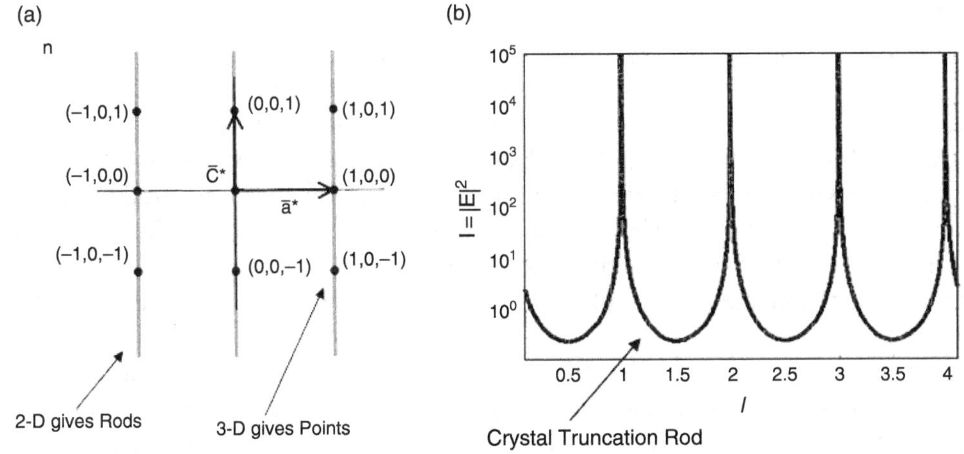

Figure 7.14. (a) Reciprocal lattice, showing discrete points (hkl) representing diffraction maxima from 3-dimensional X-ray scattering and rods between the reciprocal lattice points representing 2-dimensional scattering from surfaces. (b) plot of scattered X-ray intensity vs. reciprocal lattice direction l, showing Bragg maxima and crystal truncation rods between the diffraction maxima.

rods [34,36,161]. CTR's are also highly sensitive to surface roughness, providing a measure of the degree of surface imperfection, and in certain cases can provide information about the long-range correlation of surface features.

A comprehensive review of mineral-fluid interface systems studied by surface X-ray scattering has recently been published by Fenter and Sturchio [38]. While the number of such studies is still limited, they have provided detailed information that has begun to allow some generalizations regarding the factors dictating the structure and reactivity of hydrous interfaces. Under UHV conditions, relaxation and reconstruction of the surface is driven by the need to maintain a charge-neutral metal-oxygen stoichiometry and minimization in the net surface dipole [36,164]. In hydrous systems, dissociative adsorption of water leads to significantly different surface structures than observed under UHV conditions. From a coordination chemistry perspective, three factors appear to be dominant in dictating the experimentally observed hydroxylated surface structures: (1) reaction of metal oxide surfaces with water results in complete coordination shells of near-surface metal ions, potentially resulting in non-stoichiometric metal-oxygen ratios in the surface layer, (2) the apparent charge associated with a non-stoichiometric surface is compensated for by variable hydrogen ion stoichiometry, and (3) relaxations of surface atoms are generally much smaller than for 'clean' surfaces, likely driven primarily by the relaxation of unfavorable polyhedral arrangements present in the bulk structure, as evidenced by reduction of metal-oxygen bond length distortions or removal of metal centers associated with unfavorable polyhedral linkages. Furthermore, numerous experimental and theoretical studies indicate that physically adsorbed water adjacent to the hydrated surface is generally associated through H-bonding, and the alteration of the liquid water structure generally only extends a few molecular layers, depending on the degree of hydrogen bonding association [38].

Here we discuss the results of surface diffraction studies on hydroxylated low-index faces of corundum-types oxides (e.g., α-Al_2O_3, α-Fe_2O_3, α-Cr_2O_3) including α-Al_2O_3(0001) [37] and (1–102) [165], and α-Fe_2O_3 (0001) [85] and (1–102) [173]. Numerous experimental and theoretical investigations have been focused on understanding the structure of UHV clean surfaces (e.g., [166,167]). More recently, work has focused on understanding the affinity of these surfaces towards the binding of water and the effects of surface hydroxylation on surface structure and reactivity.

The surfaces used in our CTR diffraction studies were prepared using a chemical-mechanical polishing procedure, generally producing surface roughness of less than 10Å rms, followed by a mild acid etching procedure to ensure the surfaces were fully hydroxylated. The hydroxylated α-Al_2O_3 (0001) surface (Figure 7.15B) takes on a structural configuration that appears intermediate between the ideal oxygen termination of the bulk crystal lattice and a gibbsite (γ-$Al(OH)_3$) structure [37]. The oxygen atoms exposed at the (0001) surface are hydroxyl groups coordinated

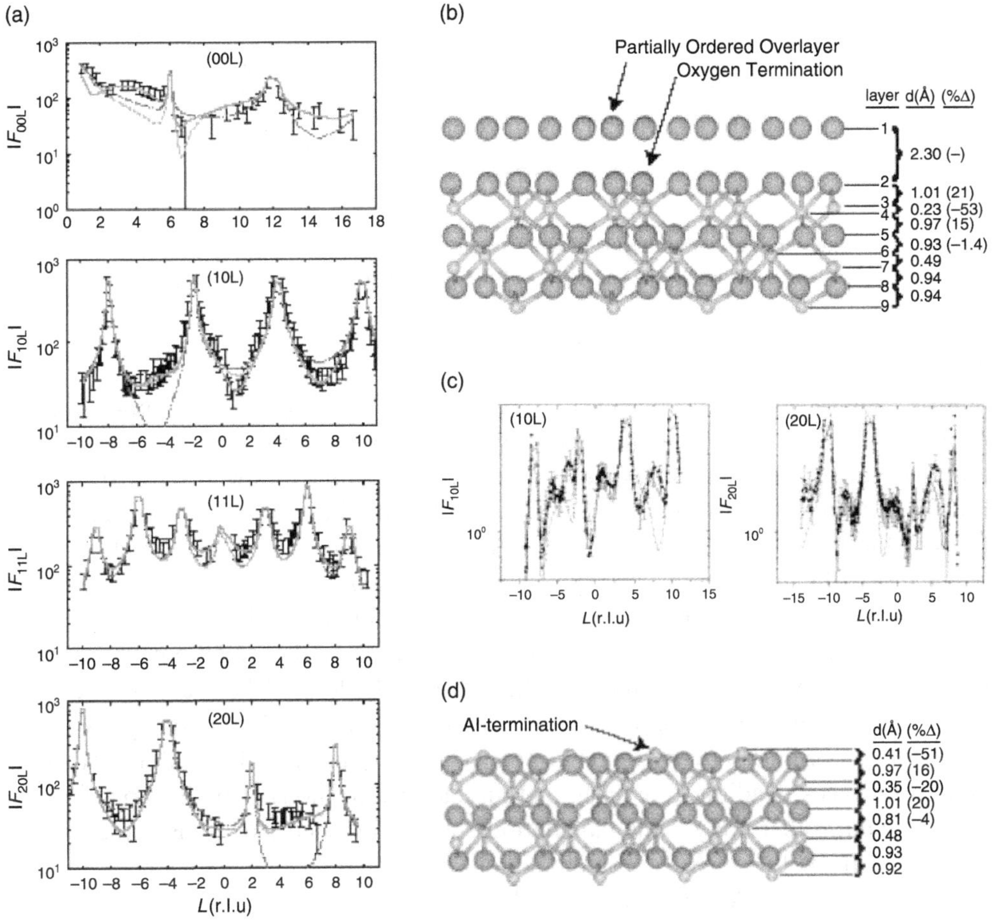

Figure 7.15. (A) Crystal truncation rods along the (00L), (10L), (11L), and (20L) zones, showing results of modeling [38]; (B) Best-fit model for the α-Al_2O_3 (0001)/water interface [38]; (C) CTR diffraction data for the UHV clean α-Al_2O_3 (0001) surface [168]; (D) Best-fit model of the UHV-clean α-Al_2O_3 (0001) surface showing that the surface is terminated by Al atoms [168].

by two octahedrally coordinated bulk Al atoms, resulting in a particularly stable hydrogen bonding surface configuration, where both in-plane hydrogen bonding and hydrogen bonding to physically adsorbed water molecules may result [50,51]. This configuration is in sharp contrast to the vacuum-terminated α-Al_2O_3 (0001) surface, which was found to have Al atoms in three-fold coordination with oxygen [168] (Figure 7.15D) and results in dramatic reactivity differences when compared to the hydroxylated α-Al_2O_3 (0001) surface [169].

The structure of the hydrated α-Al_2O_3 (1–102) surface differs substantially from the hydrated (0001) surface, having surface (hydr)oxo functional groups

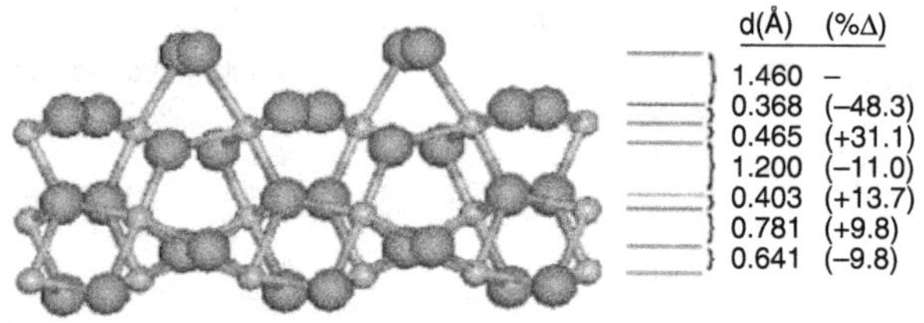

Figure 7.16. Side-view of the α-Al_2O_3 (1–102) surface, showing the missing Al positions, the d-spacings perpendicular to the (1–102) surface and the percentage changes relative to the bulk lattice spacings (from [165]).

that are singly, doubly, and triply coordinated by octahedrally coordinated bulk Al atoms [165]. In this study the surface was initially prepared in an ultra-high vacuum system by Ar^+ ion sputtering and oxygen annealing followed by dosing with H_2O. The results suggest that the 5-fold coordinated Al atoms that would terminate the stoichiometric surface model have been lost from the surface, presumably being balanced by the addition of 3 H^+ per lost Al^{3+} (Figure 7.16). The DFT-calculated relaxations for this structure with three protons replacing each lost Al^{3+} are in good agreement with the CTR data, whereas structures based on the stoichiometric surface model are not (A. M. Chaka, in prep.). In addition, the loss of the 5-coordinate Al^{3+} is consistent with a lowering of the surface energy due to minimizing unfavorable polyhedral coordination environments (e.g., the near-surface AlO_6 polyhedra share octahedral faces, leading to a strong cation-cation repulsion). This is supported by recent *ab initio* thermodynamic calculations, which indicate the surface free energy is reduced from 126 meV/Å^2 for the stoichiometric surface to 20 meV/Å^2 for the protonated 'missing-Al' structure. (A. M. Chaka, in prep.)

The structure of the hydrated α-Fe_2O_3 (0001) surface [85] differs significantly from that of the α-Al_2O_3 (0001) surface. Analysis of the CTR data suggests that the α-Fe_2O_3 (0001) surface is hydroxyl-terminated with ~1/3 mono-layer (based on crystallographic site densities) hydroxylated-iron species exposed at the surface (Figure 7.17, Table 7.5). There is a significant relaxation of the surface atoms, primarily a contraction of the double-iron layer below the terminating oxygen plane, similar to the α-Al_2O_3 (0001) surface [37], as well as an expansion of the terminal Fe position away from the surface oxygen layer. Comparison of the CTR-derived structural model with results of DFT calculations for a variety of surface stoichiometries (Figure 7.18, Table 7.5) suggests that the terminal Fe is coordinated by hydroxyl groups resulting from a dissociative water adsorption mechanism ($(HO)_x$–Fe–H_xO_3–R) because this is the only case which correctly predicts the directions of all surface relaxations. If the surface is considered to consist of two distinct chemical domains, as has been observed in previous STM studies [170,171], the CTR results are consistent with a surface that is roughly 34 % hydroxylated

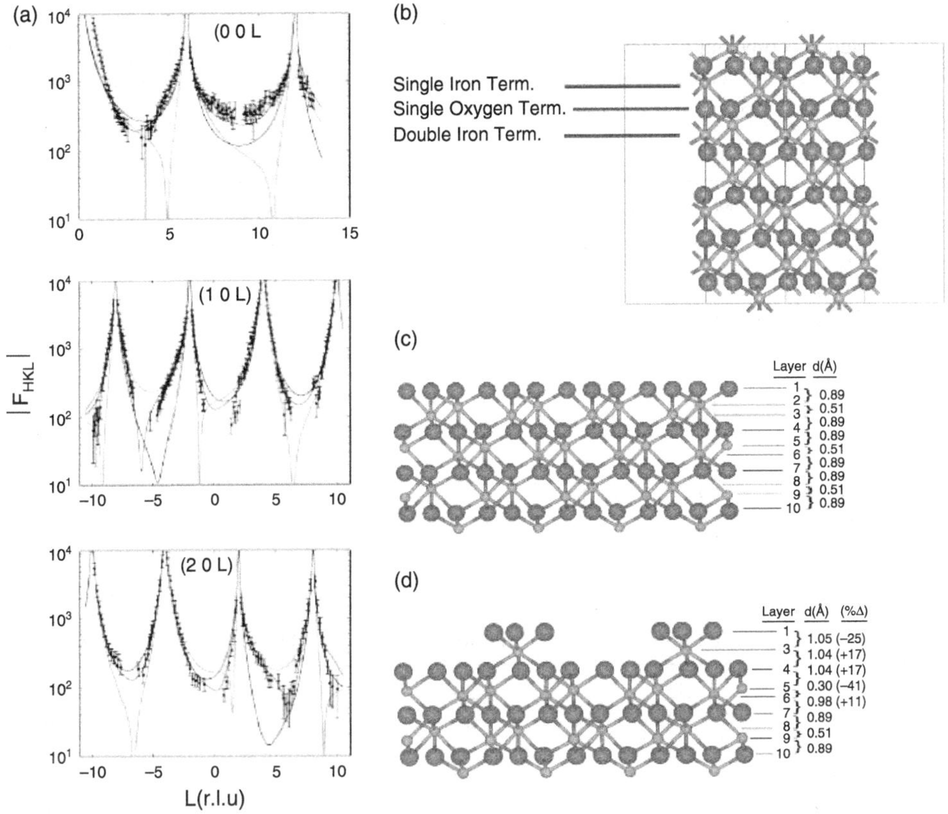

Figure 7.17. (A) CTR data for the hydrated α-Fe_2O_3 (0001) surface [85]; (B) Side view [(10–11) plane] of the α-Fe_2O_3 structure, showing three possible terminations of the bulk structure parallel to (0001); (C) Ideal oxygen termination of the (0001) surface showing the d-spacings of the various atom layers; (D) Side-view of the best-fit model of the (0001) surface of the hydrated α-Fe_2O_3 (0001) surface, showing the d-spacings of the various atom layers and the percentage difference relative to the d-spacings in the bulk structure (from [85]).

Fe-termination and 67 % hydroxyl termination; the latter termination being very similar to that found on α-Al_2O_3 (0001) while the former is consistent with water reaction at the three-coordinated Fe site terminating the clean stoichiometric surface. These findings agree well with the results of the *ab initio* thermodynamic calculations (Table 7.5) that predict the lowest energy surfaces under conditions of high water activity and low oxygen activity are the hydroxylated Fe-terminations, $(HO)_x{-}Fe{-}H_xO_3{-}R$, and the hydroxyl termination, $(HO)_3{-}Fe{-}Fe{-}R$. The thermodynamic calculations predict that a single surface should be most stable; however, the experimental observation of a mixed termination suggests that meta-stable domains are kinetically hindered from further dissolution in aqueous solution under circum-neutral pHs.

Table 7.5
Calculated and experimental percent relaxations for various (0001) terminations of α-Fe_2O_3. The DMol calculations were done according to the method outlined in [85]. LAPW refers to linear augmented plane wave calculations.

Layer		$Fe-O_3-Fe-R$			$O_3-Fe-Fe-R^{\dagger}$		$(HO)_3-Fe-Fe-R^{\dagger}$
		DMol3	LAPW[a]	Expt[b]	DMol3	LAPW[a]	DMol3
1–2	O_3-Fe	–	–	–	–	–	–
2–3	$Fe-Fe$	–	–	–	–	–	–
3–4	$Fe-O_3$	−65	−57	−59	–	–	–
4–5	O_3-Fe	7	7	17	−3	−1	35
5–6	$Fe-Fe$	−26	−33	−16.7	−78	−79	−41
6–7	$Fe-O_3$	13	15	41	38	37	12
7–8	O_3-Fe	5	5		−7	−6	2
8–9	$Fe-Fe$	−4	−3		12	+16	6
9–10	$Fe-O_3$	2	1		−4	−4	−4
10–11	O_3-Fe	0	4		−3		0

Layer		$(H_2O)_3-Fe-O_3-R$	$(HO)_3-Fe-O_3-R$	$(HO)_2-Fe-O_3-R$	$HO-Fe-O_3-R$
		DMol3	DMol3	DMol3	DMol3
1–3	O_3-Fe	32	−29	−14	18
3–4	$Fe-O_3$	−40	27	5	−36
4–5	O_3-Fe	9	−9	−8	10
5–6	$Fe-Fe$	−57	−69	−56	−72
6–7	$Fe-O_3$	22	31	22	21
7–8	O_3-Fe	0	−20	−4	−1
8–9	$Fe-Fe$	−25	8	−24	12
9–10	$Fe-O_3$	2	0	−2	−4
10–11	O_3-Fe	4	1	1	−1

Layer		$(HO)_3-Fe-H_3O_3-R$	$(HO)-Fe-HO_3-R$	$O_x-Fe_y-O_3-R$	$(HO)_3-Al-Al-R^{\dagger}$
		DMol3	DMol3	Expt.[c]	Expt[d]
1–3	O_3-Fe	−28	26	−12	–
3–4	$Fe-O_3$	20	−13[††]	15	–
4–5	O_3-Fe	26	13	15	21
5–6	$Fe-Fe$	−69	−23	−37	−53
6–7	$Fe-O_3$	15	9	17	15
7–8	O_3-Fe		2	−3	−1.4
8–9	$Fe-Fe$		1	−5	
9–10	$Fe-O_3$		0	1	
10–11	O_3-Fe		−1	3	

[a] Wang et al. [171]; [b] Chambers and Yi [172]; [c] From the single-domain structural analysis of CTR data (see [85]); [d] Eng et al. [37]; † The relaxations for these models are shifted to start at layer 4 to facilitate comparison. †† Average of three values: (−21.7, −7.1, −11.3).

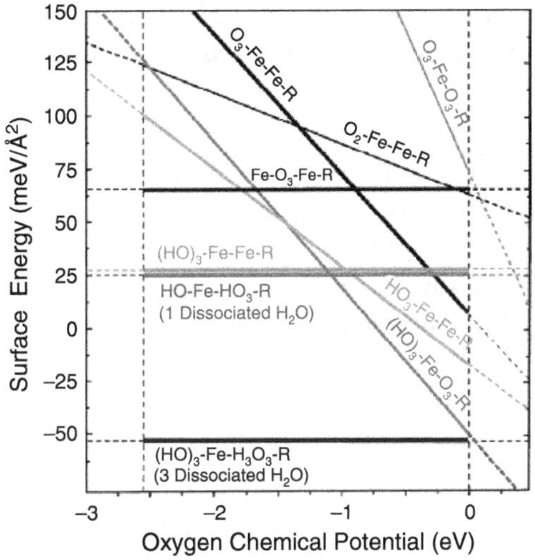

Figure 7.18. Plot of surface energy vs. oxygen chemical potential for different terminations of the α-Fe_2O_3 (0001) from *ab initio* thermodynamic calculations. (from [85]).

These results provide strong evidence that surface hydroxylation results in the presence of multiple surface oxygen coordination environments on the low-index faces of α-Fe_2O_3 and α-Al_2O_3. As discussed in Section 3.3, these findings provide a structural basis for understanding the differences in reactivity among these substrates with respect to the adsorption of aqueous metal ions. These results, in combination with other similar studies [38], underscore the fact that the local coordination environments of the surface functional groups at a hydrated metal-oxide surface may be substantially different than expected based on ideal terminations of the bulk crystal and with clean or UHV-prepared surfaces. This also provides a structural interpretation for understanding the difference in reactivity between hydroxylated metal oxide surfaces that are predominantly terminated with Lewis acid hydroxyl sites, and clean metal-oxide surfaces that are terminated with Lewis acid oxygen sites and Lewis base sites from under-coordinated surface metals.

Recently, Tanwar et al. [173] completed a CTR study of the hydrated α-Fe_2O_3 (1–102) surface and found that it is very similar to that of the hydrated α-Al_2O_3 (1-102) surface, with equal proportions of singly, doubly, and triply coordinated oxygens. In addition, Lo *et al.* [174] carried out an *ab initio* thermodynamic study of this surface in the presence of 20 torr H_2O, and found that the predicted surface structure is the same as that determined experimentally. Thus, this surface is expected to have a reactivity intermediate between those of α-Fe_2O_3 (0001) (most reactive) and α-Al_2O_3 (0001) (least reactive).

3.4. X-ray standing wave studies of the electrical double layer at solid-aqueous solution interfaces and in situ measurements of surface reactivity

The nature of the electrical double layer (EDL) at metal (hydr)oxide surfaces will be reviewed next, including a discussion of the classical models of this interfacial region, drawing from Refs. [105–111] and its variation with solution conditions, particularly pH and ionic strength.

Experimental investigations of EDL structure and interface reactivity require the ability to probe solid/solution ion partitioning and interfacial ion distribution functions at length scales from less than one Å to greater than 1000 Å. A number of previous studies have shown that X-ray standing wave (XSW) techniques can provide detailed information on the partitioning of aqueous metals and metalloids to surfaces and can examine the EDL distribution of charged species or investigate the distribution of metals in a variety of organic surface coatings (see below).

Long-period XSW fluorescent-yield methods in particular, have been found to provide good sensitivity to ion partitioning and interfacial elemental distributions. The long-period XSW technique is based on constructive interference of incident and specularly reflected beams, when the incidence angle is in the region of the critical angle for total external reflection, leading to the formation of a standing wave field above the mirror surface [175,176]. The position and spacing of the XSW antinodes are controlled by adjusting the angle of incidence; if the incident energy is above the absorption edge for an element of interest, there will be a characteristic modulation in the fluorescent-yield profile depending on the functional form of the distribution function (see [177,178] for recent reviews of the experimental details).

An example application of the long-period XSW technique, and the coupling of complementary interfacial analysis methods, is provided by the study of metal-ion sorption on α-Al_2O_3 and α-Fe_2O_3 single-crystal surfaces. As discussed above, the low-index surfaces of α-Al_2O_3 and α-Fe_2O_3 are important model systems for investigating the influence of surface structure and composition on the overall reactivity of metal-oxide surfaces. The surface-specific reactivity of α-Al_2O_3 and α-Fe_2O_3 with respect to the adsorption of aqueous metals (Pb(II), Co(II), Zn(II), Cu(II)) has been examined in a number of studies using surface-specific x-ray spectroscopic and scattering techniques.

The in situ partitioning of Pb(II) and Pb(II)/SeO_4^{2-} on the α-Al_2O_3 (0001) and (1–102) surfaces was studied using long-period XSW fluorescent yield methods [179]. Thin films of Pb(II)/Se(VI) bearing solutions (1–3 μm) were placed in contact with the substrates, and encapsulated using a 2 μm polypropylene film, and the experimental Pb L_α and Se K_α fluorescent-yield profiles were measured as a function of incidence angle. The fluorescent-yield profiles were analyzed using a 3-layer model consisting of the 2 μm polypropylene film, a solution layer of variable thickness, and the α-Al_2O_3 substrate. The distribution model used for the calculation of the fluorescent-yield profile was similar in form to the Stern-Grahame model for interfacial distribution of charge species,

$$N(z) = \begin{cases} N_{ad} & 0 \leq z \leq z_{ad} \\ N_o \exp\left(\dfrac{-ZF\Psi(z)}{RT}\right) & z_{ad} < z < d_{soln} \end{cases} \tag{2}$$

where N_o is the bulk solution concentration, N_{ad} is the adsorbed layer concentration, F is Faraday's constant and Z is the charge of the ion in solution. The region near the surface is termed the 'compact layer' consisting of directly bound adsorbates, and the region $z > z_{ad}$ is the diffuse accumulation/depletion of dissolved ions balancing the net surface charge. Ψ is the interfacial electrostatic potential (Volts) from the Gouy-Chapman model of electrified interfaces. At low surface potentials the form of the potential function is generally modeled using a simple exponential decay given by

$$\Psi = \Psi_o e^{-(z-z_{ad})/L_d} \tag{3}$$

where L_d is the decay length which is a function of the ionic strength (I) of the solution, $1/L_d = 0.329I^{1/2}$ [109,179]. When $\Psi_o = 0$ (i.e., no surface charge), equation (2) reduces to a simple two-component model, with a single compact adsorbed layer and no diffuse layer.

Calculations of the variations expected in the fluorescent-yield (FY) profiles as a function of the distribution model parameters are shown in Figure 7.19. When the species of interest resides predominantly at the solid surface, the FY profile shows a maximum at the critical angle for total external reflection. As the ratio of the surface-bound species to the total number of species in the solution volume adjacent to the surface decreases, the FY distribution broadens at the low angles. A similar effect is noted when a diffuse layer accumulation arises due to an interfacial electrostatic potential.

Pb Lα fluorescent-yield profiles measured at total solution concentrations of 6, 60, and 600 μM Pb(II) (pH = 4.5) on the α-Al_2O_3 (1–102) surface are displayed in Figure 7.20(a). The relatively sharp peak of the profiles near the critical angle (~0.16°) indicates a high ratio of surface-bound to solution species in each case. Fits to the data provide the compact layer adsorption density as a function of the solution layer concentration, thus allowing determination of an in situ crystal face-specific adsorption isotherm (Figure 7.20(b)). Similar measurements on the α-Al_2O_3 (0001) surface suggested sorption densities were approximately 2.5× lower, confirming previous ex situ observation of reactivity differences among the substrates [155,156]. Furthermore, no diffuse layer component was required to fit the data. Based on our previous studies of the structure of the hydroxylated surfaces [37,165], we proposed

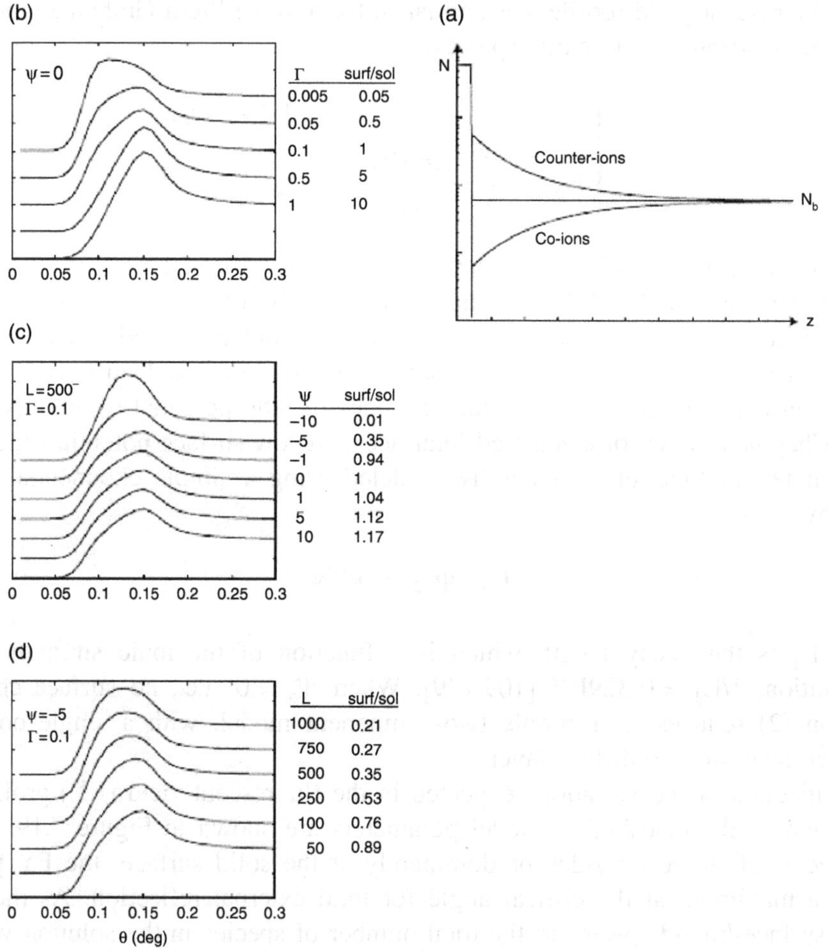

Figure 7.19. (A) Schematic of the distribution function, showing the compact layer, and the diffuse concentration tail corresponding to co-ions and counter-ions. (B–D) Model calculations of the fluorescence-yield intensity in the grazing-angle geometry. Calculations were performed using an α-Al_2O_3 substrate, a 1 μm thick solution layer and a 2 μm polypropylene cap layer at 14 keV. The thickness of the compact layer is fixed at 5 Å and the bulk solution concentration was fixed at 100 μM. The calculated profiles were convoluted with a Gaussian of 30 mdeg FWHM to account for experimental broadening. γ is the adsorption density in the compact layer (μmoles/m^2), Ψ* is the reduced potential of the interface and L is the potential decay length (Å) (from [179]).

that the protonation of the surface hydroxyl groups at pH 4.5 would result in a net positive surface charge, which is consistent with the lack of a diffuse layer accumulation. However, analysis of Pb(II)/Se(VI) solutions suggested that a diffuse layer accumulation of Pb(II) resulted from high surface loading of Se(VI), indicating a reversal of the surface charge due to adsorption of SeO_4^{2-} oxoanions.

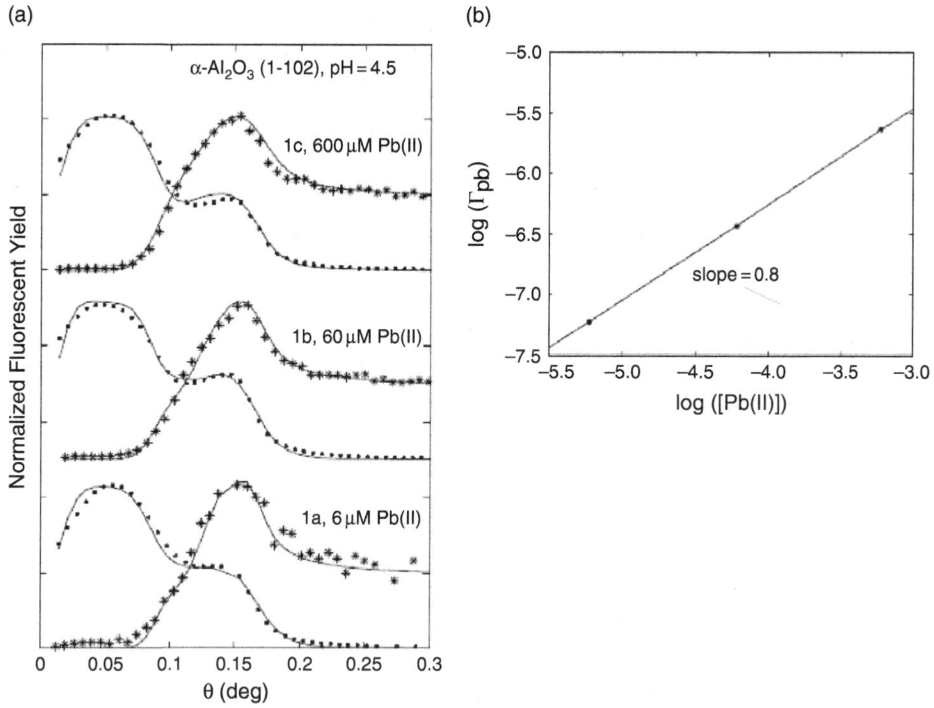

Figure 7.20. (a) Reflectivity (·) and Pb Lα fluorescent-yield (FY) (*) data and corresponding best fits (solid lines) for Pb(II)-bearing solutions in contact with the α-Al_2O_3 (1–102) surface (pH 4.5). (b) Pb adsorption isotherm corresponding to the best-fit adsorption densities from the FY data in Figure 7.20(a) (from [179]).

3.5. Effect of organic coatings and microbial biofilms on metal oxide surface reactivity – X-ray standing wave studies of metal ion partitioning between coating and surface

Natural organic matter (NOM) and microbial biofilms are widespread in soils and often are found as coatings on mineral surfaces. Biofilms in particular form microenvironments in which aqueous chemical conditions may differ from those of the bulk environment. Reactive functional groups in NOM and on bacterial surfaces and in exopolysaccharides provide a large array of binding sites for metals, posing the question of whether or not NOM and bacterial biomass plays a dominant role in controlling metal ion migration in soils and aquifers. Such processes could have major implications for ground water quality, as the migration and toxicity of heavy metal contaminants in the environment are controlled by interactions between the metal solutes, aqueous solutions, and soil materials. Another important issue concerns the extent to which NOM and biofilm coatings on mineral surfaces change the intrinsic reactivity of the coated mineral surface. Templeton et al. [180] used XSW fluorescence-yield measurements, coupled with XAFS, to probe the distributions of Pb(II) within mineral-biofilm-water systems. The main purpose of this study was to determine how biofilm coatings change the reactivity of metal oxide surfaces. In situ XSW measurements were performed on *Berkholderia cepacia* biofilms grown on single-crystal α-Al_2O_3 (0001) and (1–102) and α-Fe_2O_3 (0001) surfaces, subsequently reacted for 2 h with aqueous solutions containing various Pb concentrations (ranging from 1 to 150 μM) at pH 6 and an ionic strength of 0.005 M. At a Pb concentration of 1 μM, the FY intensity for the α-Al_2O_3 (1–102) and α-Fe_2O_3 (0001) surfaces peaks at the critical angles of the two substrates (≈160 and 185 mdeg, respectively, at 14 keV), indicating that Pb(II) is located primarily at the corundum or hematite surfaces at this concentration (Figure 7.21). With increasing

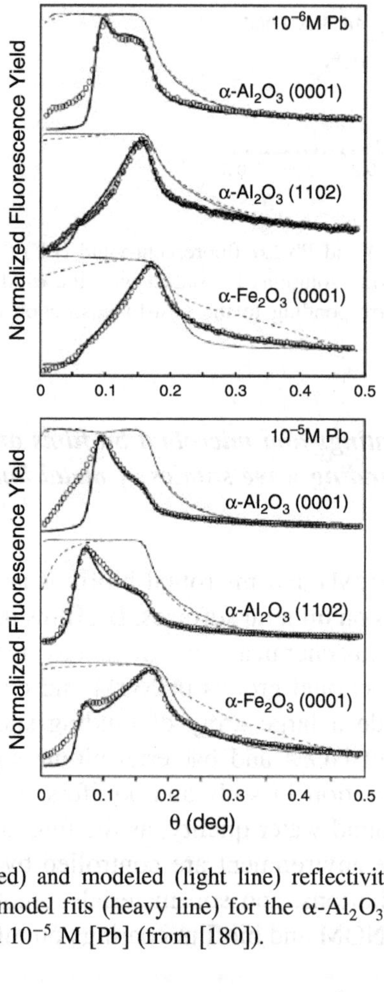

Figure 7.21. Measured (dashed) and modeled (light line) reflectivity (Log Io/I1) profiles and Pb Lα FY profiles (circles) with model fits (heavy line) for the α-Al_2O_3 (0001), α-Al_2O_3 (1-102), and α-Fe_2O_3 (0001) at 10^{-6} M and 10^{-5} M [Pb] (from [180]).

Pb concentration, there are two FY peaks, one occurring at the critical angle of each substrate and one occurring at ≈60 mdeg for the coated α-Al_2O_3 (1–102) surface and at about 85 mdeg for the coated α-Fe_2O_3 (0001) surface. The growing intensity at the lower incidence angles with increasing [Pb] indicates that Pb^{2+} is also binding to sites in the *B. cepacia* biofilm coating. At all Pb concentrations studied, the XSW-FY-data indicate that Pb^{2+} binds primarily to functional groups in the biofilm on the α-Al_2O_3 (0001) sample.

Tables 7.6 and 7.7 compare the binding affinities of sites on the α-Al_2O_3 (0001), α-Al_2O_3 (1–102), and α-Fe_2O_3 (0001) surfaces and those in various types of bacteria, natural organic matter, and polyacrylic acid (PAA). The metal oxide substrates and humic acid are characterized by two binding sites, one strong and one weak, for Pb(II). The stability constant for the PAA-Pb(II) complex (i.e., ML_2 species) is typically larger than those of higher affinity mineral surface binding sites (i.e., M_1).

Table 7.6
Comparison of log binding site concentrations and Pb(II) binding affinities (log $K_{apparent}$, pH 6 in 0.005M $NaNO_3$) for the substrates (after [180] and [181]).

Mineral substrates	Type of binding site	Site concentration (log n)	Binding affinities (log K_{app})
α-Al_2O_3 (0001)	M_1	−6.25	5.35
	M_2	−5.05	2.7
α-Al_2O_3 (1–102)	M_1	−6.25	6.0
	M_2	−5.15	3.55
α-Fe_2O_3 (0001)	M_1	−5.85	6.65
	M_2	−5.1	3.5

Table 7.7
Comparison of Pb(II) binding affinities (log stability constants) for various microorganisms, natural humic acid, and polyacrylic acid (PAA) (from [181]).

Type of organic matter		Stability constant
Cyanobacteria	log K (cell wall)	4.67
	log K (sheath)	5.07
Enterobacteriaceae	log $K_{carboxyl}$	3.9
	log $K_{phosphoryl}$	5.0
B. subtilis	log $K_{carboxyl}$	4.2
	log $K_{phosphoryl}$	5.6
B. licheniformis	log $K_{carboxyl}$	4.7
	log $K_{phosphoryl}$	5.7
Humic Acid	log K_{Pb-S1}	3.40
	log K_{Pb-S2}	8.75
PAA	log β_{102}	6.75–7.00

Moreover, the stability constant for the higher affinity Pb(II)-humate complex (i.e., log $K_{Pb\text{-}S2}$) is 2.1 to 3.4 units larger than those of higher affinity mineral surface binding sites. Therefore, in the case of a mineral substrate coated by PAA or HA, it is possible that binding sites of PAA or HA will 'outcompete' mineral surface binding sites.

The results of the XSW study by Templeton et al. [180] show that Pb^{2+} binds initially to reactive sites on the α-Al_2O_3 (1–102) and α-Fe_2O_3 (0001) surfaces even with a biofilm coating that covers essentially the entire mineral surface, as shown by SEM and confocal microscopy studies. The order of reactivity of these biofilm-coated surfaces for Pb(II) [α-Fe_2O_3 (0001) > α-Al_2O_3 (1-102) >> α-Al_2O_3 (0001)] is the same as that observed in uptake and EXAFS studies of Pb(II) sorption on biofilm-free alumina and hematite surfaces (Figure 7.22) (see [155–157]).

The study by Templeton et al. [180] also shows that the *B. cepacia* biofilm does not block all reactive sites on the alumina and hematite surfaces and that sites on the α-Al_2O_3 (1–102) and α-Fe_2O_3 (0001) surfaces 'outcompete' functional groups in the biofilm (including the exopolysaccharide exudate) at low Pb concentrations. Although it is not appropriate to generalize these findings to the interaction of aqueous heavy metal ions with all NOM- or biofilm-coated mineral surfaces, they do raise questions about the generalization that NOM or biofilm coatings change the adsorption characteristics of mineral surfaces (cf., [182,183]), and they are also inconsistent with the suggestion that NOM blocks reactive sites on mineral surfaces (cf., [184]).

In a related study, Yoon et al. [181] used XSW-FY methods to study the partitioning of aqueous Pb(II) between metal oxide substrates and polyacrylic acid (PAA)

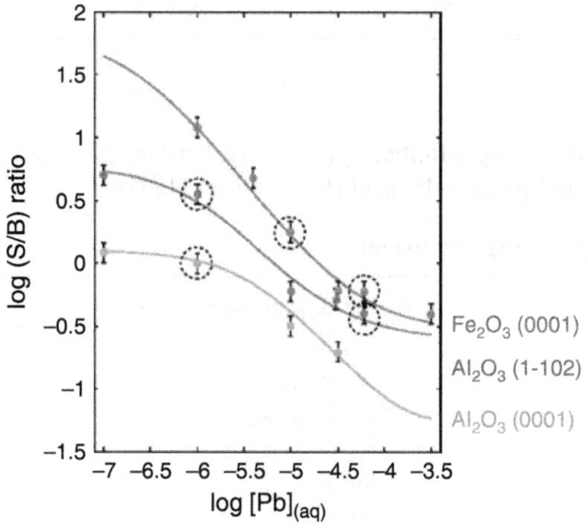

Figure 7.22. Plot of log (Pb at the metal oxide surface/Pb in biofilm) vs. log Pb concentration for the α-Al_2O_3 (0001), α-Al_2O_3 (1-102), and α-Fe_2O_3 (0001) surfaces coated by a monolayer biofilm of the Gram(-) bacterium *Burkholderia cepacia*. The α-Fe_2O_3 (0001) surface is most reactive to aqueous Pb(II) and the α-Al_2O_3 (0001) surface is the least reactive. (from [180])

coatings that were about 500 Å thick. The solution pH in these experiments was 4.5, and the PAA is a surrogate for natural organic matter, with carboxyl functional groups at a site density of 1.4×10^{-2} mol/g dry weight. As shown in Figure 7.23, the order of reactivity of the PAA-coated metal oxide surfaces with respect to aqueous Pb(II) is α-Fe_2O_3 (0001) > α-Al_2O_3 (1-102) >> α-Al_2O_3 (0001) [181].

These results emphasize the importance of mineral surfaces as sinks for heavy metals in natural environments, even when coated by NOM or biofilms, and raise questions about conventional assumptions that biofilm and NOM coatings on mineral surfaces can change the way in which aqueous heavy metals interact with these surfaces.

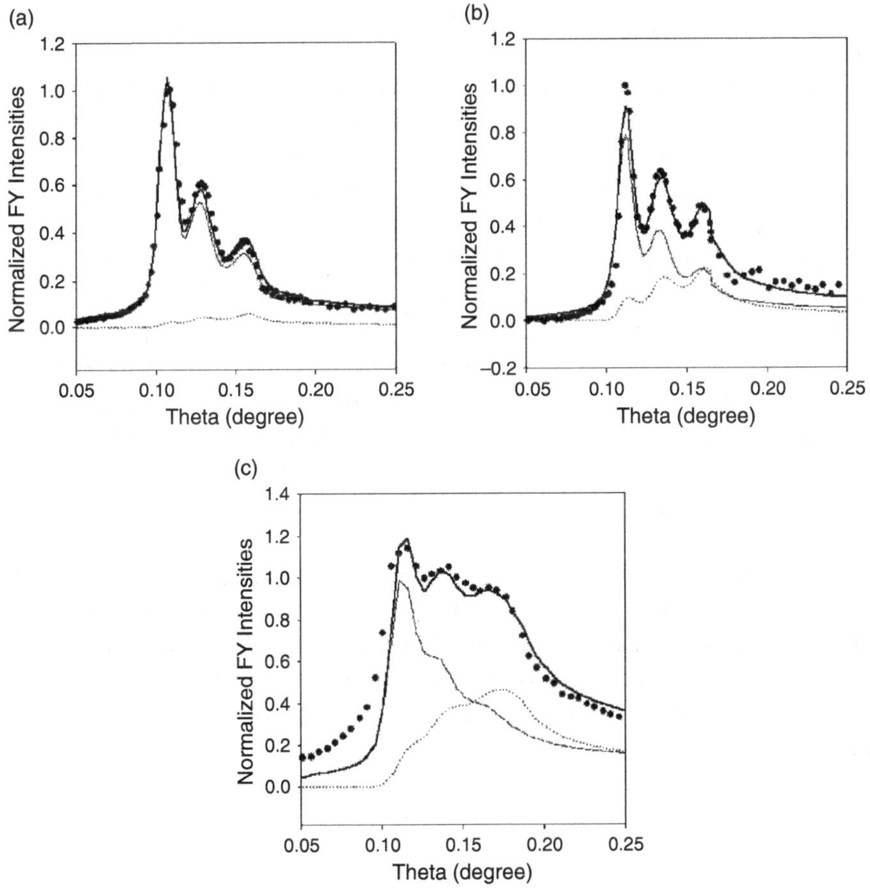

Figure 7.23. XSW-FY profiles of (a) PAA/α-Al_2O_3(0001), (b) PAA/α-Al_2O_3(1-102), and (c) PAA/α-Fe_2O_3(0001) following interaction with Pb(II) ions at 2×10^{-5} M, taken at an X-ray excitation energy of 14 keV. Filled circles: experimental data, solid line: best-fit results, dotted line: XSW-FY components from adsorbed species, dashed line: XSW-FY components from PAA film bound species (from [181]).

4. Conclusions

The studies of the α-Al_2O_3 (0001), α-Al_2O_3 (1-102), and α-Fe_2O_3 (0001) surfaces discussed above indicate that they differ significantly in reactivity with respect to water and aqueous Pb(II), with the following order of reactivity: α-Fe_2O_3 (0001) > α-Al_2O_3 (1-102) >> α-Al_2O_3 (0001). They also show that microbial biofilms or NOM coatings don't alter this order of reactivity in the systems studied. The key factor that explains the relative reactivities of these surfaces is difference in surface structure of the substrates. Based on crystal-chemical arguments, the doubly coordinated hydroxyl groups exposed at the α-Al_2O_3 (0001) surface are thought to be highly stable (i.e., non-proton-titratable) over a broad range of pH [112,165]. These surface functional groups are also not likely to form strong complexes with Pb(II), since this would result in an over saturation of the surface oxygen valence [156]. The hydroxylated α-Al_2O_3 (1-102) surface, however, is terminated by surface hydroxyl groups that are singly, doubly, and triply coordinated by Al^{3+} [165] (Figure 7.16). Again the doubly coordinated hydroxyls are likely low reactivity sites, while the singly and triply coordinated hydroxyls likely have much higher proton lability and affinity towards the binding of Pb(II) [112,156,165].

The surface structural analysis of α-Fe_2O_3 (0001) discussed above suggests a significantly more complicated structure than that of either of the α-Al_2O_3 surfaces considered or the α-Fe_2O_3 (1-102) surface. The postulated two-domain structure presents surface hydroxyl groups that are doubly coordinated by bulk Fe(III), similar to the α-Al_2O_3 (0001) surface, as well as domains dominated by singly and triply coordinated hydroxyls. Crystal-chemical arguments similar to those used for the α-Al_2O_3 surfaces apply and predict that the domain consisting of doubly coordinated hydroxyls will be relatively low in reactivity, while the domain dominated by singly and triply coordinated hydroxyls will readily bind Pb(II). Further, it is likely that a surface dominated by nano-scale domains would result in a high proportion of domain edges, which would also be reactive towards Pb(II) binding. This observation is consistent with Templeton et al.'s suggestion of a greater proportion of highly reactive "defect-sites" on the α-Fe_2O_3 (0001) surface [179], as well as with recent GI-XAFS results on arsenate adsorption on the α-Fe_2O_3 (0001) surface [185].

Acknowledgments

The work reported in this chapter has been supported by the National Science Foundation (Grants CHE-0431425 and BES-0404400), the Department of Energy, Basic Energy Sciences, Geoscience Program (Grant DE-FG03-93ER14347-A007), and the DOE Environmental Management Science Program (Grants DE-FG07-98ER14842, DE-FG07-99ER15022, and DE-FG07-99ER15024. We thank the staff of the Stanford Synchrotron Radiation Laboratory, particularly Dr. John Bargar, and the staff of GSE-CARS (Sector 13) at the Advanced Photon Source, particularly Dr. Peter Eng, for technical assistance. Much of the spectroscopic work reported in this chapter was carried out at SSRL on beamline 11.2, and all of the crystal truncation rod diffraction studies reported here were carried out at the APS on beamline ID-13-C. SSRL is supported by the Department of Energy (Divisions of Chemical Science and Materials Science and the Office of Health and Environmental Research) and by the National Institutes of Health. GSECARS is supported by the National Science Foundation (Earth Sciences Division) and the Department of Energy (Division of Chemical Sciences, Geosciences, and Biosciences). We also wish to acknowledge support of computer resources at the Arctic Region Supercomputer Center (ARSC) and the Helix System at the National Institutes of Health, Bethesda, MD. ARSC is supported by the Department of Defense, and the Helix System is supported by the Center for Information Technology, NIH, and the National Institute of Standards and Technology.

References

[1] H. D. Holland, *The Chemical Evolution of the Atmosphere and Oceans*, (Princeton University Press, Princeton, NJ, 1984).

[2] C. Huber and G. Wachtershauser, Science **276**, 245 (1997).

[3] W. Stumm, B. Wehrli, and E. Wieland, Croatica Chemica Acta **60**, 429 (1987).

[4] W. Stumm and J. J. Morgan, *Aquatic Chemistry*, 3rd Ed. (Wiley-Interscience, New York, 1996).

[5] W. Stumm, *Chemistry of the Solid-Water Interface – Processes at the Mineral-Water and Particle-Water Interface in Natural Systems*, pp. 1–8 (John Wiley & Sons, New York, 1992).

[6] G. E. Brown Jr. et al., Chem. Rev. **99**, 77 (1999).

[7] A. F. White and S. L. Brantley (eds.), *Chemical Weathering Rates of Silicate Minerals*. Vol. 31, *Reviews in Mineralogy* (Mineralogical Society of America, Washington, DC, 1995).

[8] F. C. Ugolini and H. Spaltenstein, in *Global Biogeochemical Cycles*, edited by S. S. Butcher, R. J. Charlson, G. H. Orians, and G. V. Wolfe, pp. 123–153 (Academic Press, New York, 1992).

[9] W. J. Deutsch. *Groundwater Geochemistry – Fundamentals and Applications to Contamination* (Lewis Publishers, Boca Raton, FL, 1997).

[10] E. A. Laws. *Aquatic Pollution*, 2nd Ed. (John Wiley & Sons, New York, 1993).

[11] L. M. Mosley, K. A. Hunter, and W. A. Ducker, Environ. Sci. Technol. **37**, 3303 (2003).

[12] G. E. Brown, Jr. and G. A. Parks, Internat. Geol. Rev. **43**, 963 (2001).

[13] D. L. Sparks, *Environmental Soil Chemistry*. 2nd Ed. (Academic Press, San Diego, 2002).

[14] N. Guven and R. M. Pollastro (eds.), *Clay-Water Interface and Its Rheological Implications*, *Clay Minerals Society Workshop Lectures*, Vol. 4 (The Clay Minerals Society, Boulder, CO, 1992).

[15] D. K. Nordstrom and C. N. Alpers, in *The Environmental Geochemistry of Mineral Deposits. Part A: Processes, Techniques, and Health Issues*, edited by G. S. Plumlee and M. J. Logsdon, *Reviews in Economic Geology*, Vol. 6A, pp. 133–160 (Society of Economic Geologists, Inc., Littleton CO, 1999).

[16] D. R. Lovley (ed.), *Environmental Microbe–Metal Interactions* (American Society for Microbiology Press, Washington, DC, 2000).
[17] S. S. Butcher et al. (eds.), *Global Biogeochemical Cycles* (Academic Press, San Diego, CA, 1992).
[18] C. R. Usher, A. E. Michel, and V. H. Grassian, Chem. Rev. **103**, 4883 (2003).
[19] P. R. Buseck and M. Posfai, Proc. Nat. Acad. Sci. USA **96**, 3372 (1999).
[20] B. J. Finlayson-Pitts and J. N. Pitts, Jr., *Chemistry of the Upper and Lower Atmosphere – Theory, Experiments, and Applications* (Academic Press, San Diego, CA, 2000).
[21] V. H. Grassian, Internat. Rev. Phys. Chem. **20**, 467 (2001).
[22] S. Wang et al., Geophys. Res. Lett. **30**, (2003).
[23] S. B. Johnson et al., Internat. J. Min. Processing **58**, 267 (2000).
[24] J. A. Lewis, J. Am. Ceram. Soc. **83**, 2341 (2000).
[25] V. E. Henrich and P. A. Cox. *The Surface Science of Metal Oxides*, (Cambridge University Press, Cambridge, UK, 1994).
[26] C. B. Duke (ed.), *Surface Science: The First Thirty Years*, Surf. Sci. (1994).
[27] C. B. Duke and E. W. Plummer (eds.), *Frontiers in Surface and Interface Science*, Surf. Sci. **500** (2002).
[28] G. Binnig and H. Rohrer, Surf. Sci. **126**, 236 (1983).
[29] U. Becker and M. F. Hochella, Jr., Geochim. Cosmochim. Acta **60**, 2413 (1996).
[30] G. Binnig et al., Surf. Sci. **189**, 1 (1987).
[31] A. G. Stack, S. R. Higgins, and C. M. Eggleston, Geochim. Cosmochim. Acta **65**, 3055 (2001).
[32] G. L Richmond, Chem. Phys. Lett. **106**, 26 (1984).
[33] S. Ong, X. Zhao, and K. B. Eisenthal, Chem. Phys. Lett. **191**, 327 (1992).
[34] I. K. Robinson, Phys. Rev. B **33**, 3830 (1986).
[35] M. F. Toney et al., Nature **368**, 444 (1994).
[36] G. Renaud, Surf. Sci. Rept. **32**, 1 (1998).
[37] P. J. Eng et al., Science **288**, 1029 (2000).
[38] P. Fenter and N. C. Sturchio, Prog. Surf. Sci. **77**, 171 (2004)
[39] K. F. Hayes et al., Science **238**, 783 (1987).
[40] G. E. Brown, Jr., Rev. Mineral. **23**, 309 (1990).
[41] M. J. Bedzyk et al., Science **248**, 52 (1990).
[42] H. D. Abruna (ed.), *Electrochemical Interfaces – Modern Techniques for In-Situ Interface Characterization* (VCH Publishers, Inc., New York, 1991).
[43] C. Malgrange and D. Ferret, Nucl. Instrum. Meth. Phys. Res. A **314**, 285 (1992).
[44] Y. Qian et al., Science **265**, 1555 (1994).
[45] L. Cheng et al., Phys. Rev. Lett. **90**, 255503/1 (2003).
[46] A. Nilsson, J. Elect. Spectros. Rel. Phenom. **126**, 3 (2002).
[47] A. Nilsson et al., Phys. Rev. Lett. **78**, 2847 (1997).
[48] P. Fenter et al., Nature Physics **2**, 700 (2006).
[49] C. Park and P. A. Fenter, J. Appl. Crystallog., in press (2006).
[50] K. C. Hass et al., Science **282**, 265 (1998).
[51] X.-G. Wang, A. M. Chaka, and M. Scheffler, Phys. Rev. Lett. **84**, 3650 (2000).
[52] D. F. Ogletree et al., Rev. Sci. Instrum. **73**, 3872 (2002).
[53] H. K. Bluhm et al., J. Elec. Spectros. Rel. Phenom. **150**, 86 (2006).
[54] A. Baraldi et al., Surf. Sci. Repts. **49**, 169 (2003).
[55] G. Sposito, *The Surface Chemistry of Soils* (Oxford University Press, New York, 1984).
[56] H. A. Alk-Abadeleh and V. H. Grassian. Surf. Sci. Repts. **52**, 63 (2003).
[57] H-J. Freund, Faraday Discuss **114**, 1 (1999).

[58] E. K. Berner and R. A. Berner, *Global Environment: Water, Air, and Geochemical Cycles* (Prentice Hall, Upper Saddle River, NJ, 1996).
[59] J. Schott and J-F. Petit, in *Aquatic Surface Chemistry*, edited by W. Stumm, pp. 293–315 (John Wiley & Sons, New York, 1987).
[60] M. L. Jackson and G. D. Sherman, Adv. Agron. **5**, 219 (1953).
[61] M. L. Jackson, Soil Sci. **99**, 15 (1965).
[62] A. B. Ronov and A. A. Yaroshevsky, Am. Geophys. Union Monograph **13**, 50 (1969).
[63] D. G. Schulze. in *Minerals in Soil Environments,* 2nd Ed., J. B. Dixon and S. B. Weed (eds.), pp. 1–34 (Soil Science Society of America, Madison, WI, 1989).
[64] M. Leinen et al., J. Geophys. Res. D **99**, 21017 (1994).
[65] T. Claquin, M. Schulz, and Y. J. Balkanski, J. Geophys. Res. D **104**, 22243 (1999).
[66] P. R. Buseck et al., Internat. Geol. Rev. **42**, 577 (2000).
[67] G. Furrer and W. Stumm, Geochim. Cosmochim. Acta **50**, 1847 (1986).
[68] A. T. Stone and J. J. Morgan, in *Aquatic Surface Chemistry*, edited by W. Stumm, pp. 221–254 (Wiley-Intersciences, New York, 1987).
[69] W. Stumm and G. Furrer, in *Aquatic Surface Chemistry*, edited by W. Stumm, pp. 197–219 (Wiley-Intersciences, New York, 1987).
[70] A. T. Stone, Rev. Mineral. **35**, 309 (1997).
[71] R. G. Arnold, T. J. DiChristina, and M. R. Hoffman, Biotech. Bioeng. **32**, 1081 (1988).
[72] D. R. Lovley, Microbiol. Rev. **55**, 1472 (1991).
[73] C. F. Baes, Jr. and R. E. Mesmer, *The Hydrolysis of Cations* (Robert F. Krieger Pub. Co., Malabar, FL, 1986).
[74] W. H. Casey, in *Mineral Surfaces*, edited by D. J. Vaughan and R. A. D. Pattrick, pp. 185–217, *The Mineralogical Society Series 5* (Chapman & Hall, London, 1995).
[75] W. H. Casey and C. Ludwig, Nature **381**, 506 (1996).
[76] W. H. Casey and C. Ludwig, Rev. Mineral. **31**, 87 (1995).
[77] R. D. Astumian et al., J. Phys. Chem. 3832 (1981).
[78] S. D. Samson and C. M. Eggleston, Geochim. Cosmochim. Acta **64**, 3675 (2000).
[79] A. E. Merbach and J. W. Akitt, NMR Basic Principles and Progress **24**, 190 (1990).
[80] W. Stumm and E. Weiland, in *Aquatic Chemical Kinetics*, edited by W. Stumm, pp. 367–400 (Wiley-Interscience, New York, 1990).
[81] J. Burgess, *Ions in Solution: Basic Principles of Chemical Interactions* (Ellis-Horwood Ltd, Chichester, UK 1988).
[82] M. Magini et al., *X-ray Diffraction of Ions in Solutions: Hydration and Complex Formation* (CRC Press, Boca Raton, FL, 1988).
[83] R. E. Newnham and Y. M. deHaan. Zeitschrift Kristallogr. **117**, 235 (1962).
[84] R. L. Blake et al., Am. Mineral. **51**, 123 (1966).
[85] T. P. Trainor et al., Surf. Sci. **573**, 204 (2004).
[86] A. L. Allred, J. Inorg. Nucl. Chem. **17**, 215 (1961).
[87] R. A. Berner, A. C. Lasaga, and R. M. Garrels, Am. J. Sci. **283**, 641 (1983).
[88] A. E. Blum and L. L. Stillings, Rev. Mineral. **31**, 291 (1995).
[89] Y. Xiao and A. C. Lasaga, Geochim. Cosmochim. Acta **58**, 5379 (1994).
[90] Y. Xiao and A. C. Lasaga, Geochim. Cosmochim. Acta **60**, 2283 (1996).
[91] A. C. Lasaga. Rev. Mineral. **31**, 23 (1995).
[92] A. Pelmenschikov et al., J. Phys. Chem. B **104**, 5779 (2000).
[93] A. Pelmenschikov, J. Leszczynski, and L. G. M. Pettersson, J. Phys. Chem. A **105**, 9528 (2001).
[94] S. L. Brantley and L. L. Stillings, Am. J. Sci. **296**, 101 (1996).
[95] W. H. Casey et al., Chem. Geol. **78**, 205 (1989).
[96] L. Chou and R. Wollast, Am. J. Sci. **285**, 963 (1984).

[97] M. A. Nugent et al., Nature **395**, 588 (1998).
[98] A. F. White and S. L. Brantley, Chem. Geol. **202**, 479 (2003).
[99] R. Hellmann et al., Phys. Chem. Minerals **30**, 192 (2003).
[100] P. Fenter et al., Geochim. Cosmochim. Acta **64**, 3663 (2000).
[101] P. Fenter et al., Geochim. Cosmochim. Acta **67**, 197 (2003).
[102] P. Fenter et al., Geochim. Cosmochim. Acta **67**, 4267 (2003).
[103] H. H. Teng et al., Geochim. Cosmochim. Acta **65**, 3459 (2001).
[104] J. A. Davis, R. O. James, and J. O. Leckie, J. Colloid Interface Sci. **63**, 480 (1978).
[105] J. Westall and H. Hohl, Adv. Colloid Interface Sci. **12**, 265 (1980).
[106] J. A. Davis and D. B. Kent, Rev. Mineral. **23**, 177 (1990).
[107] A. W. Adamson and A. Gast, *Physical Chemistry of Surfaces*, 6th Ed. (John Wiley & Sons, Inc, New York, 1990).
[108] S. L. Carnie and G. M. Torrie, Adv. Chem. Phys. **56**, 141 (1984).
[109] R. J. Hunter, *Foundations of Colloid Science*, Vol. 1. (Clarendon Press, Oxford, 1987).
[110] T. Hiemstra, W. H. van Riemsdijk, and G. H. Bolt, J. Colloid Interface Sci. **133**, 91 (1989).
[111] T. Hiemstra, J. C. M. De Wit, and W. H. van Riemsdijk, J. Colloid Interface Sci. **133**, 105 (1989).
[112] T. Hiemstra and W. H. van Riemsdijk, J. Colloid Interface Sci. **179**, 488 (1996).
[113] T. Hiemstra, P. Venema, and W. H. van Riemsdijk, J. Colloid Interface Sci. **184**, 680 (1996).
[114] P. Venema, T. Hiemstra, and W. H. van Riemsdijk, J. Colloid Interface Sci. 181, 45, (1996).
[115] B. Cances et al., Am. Inst. Phys. Conf. Proc., 13th Int. XAFS Conf. **882**, 217 (2007).
[116] R. P. J. J. Rietra, T. Hiemstra, and W. H. van Riemsdijk, J. Colloid Interface Sci. 218, 511 (1999).
[117] S. B. Johnson et al., Langmuir **20**, 11480 (2004).
[118] L. Weng et al., Environ. Sci. Technol. **40**, 7494 (2006).
[119] D. A. Sverjensky and K. Fukushi, Environ. Sci. Technol. **40**, 263 (2006).
[120] K. Fukushi and D. A. Sverjensky, Geochim. Cosmochim. Acta **71**, 1 (2007).
[121] D. A. Sverjensky and K. Fukushi, Geochim. Cosmochim. Acta **70**, 3778 (2006).
[122] G. A. Parks, in *Mineral Water Interface Geochemistry*, edited by M. F. Hochella, Jr. and A. F. White, *Reviews in Mineralogy,* Vol. **23**, pp. 133–169 (Mineralogical Society of America, Washington, DC, 1990).
[123] K. F. Hayes and J. O. Leckie, J. Colloid Interface Sci. **115**, 564 (1987).
[124] G. Sposito, Rev. Mineral. **23**, 262 (1990).
[125] Z. Zhang et al., Langmuir **20**, 4954 (2004).
[126] Z. Zhang et al., Surf. Sci., in press (2007).
[127] G. A. Parks, Chem. Rev. **65**, 177 (1965).
[128] D. A. Sverjensky, Geochim. Cosmochim. Acta **58**, 3123 (1994).
[129] G. E. Brown, Jr., G. A. Parks, and C. J. Chisholm-Brause, Chimia **43**, 248 (1989).
[130] C. J. Chisholm-Brause et al., Nature **348**, 528 (1990).
[131] L. Charlet and A. Manceau, J. Colloid Interface Sci. **148**, 443 (1992).
[132] G. E. Brown, Jr., G. A. Parks, and P. A. O'Day, in *Mineral Surfaces*, edited by D. J. Vaughan and R. A. D. Pattrick, pp. 129–183 (Chapman & Hall, London, 1995).
[133] J. R. Bargar, G. E. Brown, Jr., and G. A. Parks, Geochim. Cosmochim. Acta **61**, 2617 (1997).
[134] J. R. Bargar, G. E. Brown, Jr., and G. A. Parks, Geochim. Cosmochim. Acta **61**, 2639 (1997).
[135] J. B. d'Espinose de la Caillerie, M. Kermarec, and O. Clause, J. Am. Chem. Soc. **117**, 11471 (1995).
[136] S. N. Towle et al., J. Colloid Interface Sci. **187,** 62 (1997).
[137] A. M. Scheidegger, G. M. Lamble, and D. L. Sparks, J. Colloid Interface Sci. **186**, 118 (1997).
[138] H. A. Thompson, G. A. Parks, and G. E. Brown, Jr., Geochim. Cosmochim. Acta **63**, 1767 (1999).

[139] G. E. Brown, Jr. and N. C. Sturchio, Rev. Mineral. Geochem. **49**, 1 (2002).
[140] P. A. Thiel and T. F. Madey, Surf. Sci. Rep. **7**, 211 (1987).
[141] M. A. Henderson, Surf. Sci. Rep. **285**, 1 (2002).
[142] G. E. Brown, Jr., Science **294**, 67 (2001).
[143] P. Liu et al., Surf. Sci. **412/413**, 287 (1998).
[144] P. Liu, T. Kendelewicz, and G. E. Brown, Jr., Surf. Sci. **412/413**, 315 (1998).
[145] W. Langel and M. Parinello, Phys. Rev. Lett. **73**, 504 (1994).
[146] C. A. Scamehorn, N. M. Harrison, and M. I. McCarthy, J. Chem. Phys. **101**, 1547 (1994).
[147] L. Giordano, J. Goniakowski, and J. Suzanne, Phys. Rev. Lett. **81**, 1271 (1998).
[148] M. Odelius, Phys. Rev. Lett. **82**, 3919 (1999).
[149] P. Liu et al., Surf. Sci. **417**, 53 (1998).
[150] K. Reuter and M. Scheffler, Phys. Rev. B **104**, 6195 (2000).
[151] M. L. Peterson et al., Environ. Sci. Technol. **31**, 1573 (1997).
[152] T. Kendelewicz et al., Surf. Sci. **424**, 219 (1999).
[153] T. Kendelewicz et al., Surf. Sci. **469**, 144 (2000).
[154] C. S. Doyle et al., Geochim. Cosmochim. Acta **68**, 4287 (2004).
[155] J. R. Bargar et al., Geochim. Cosmochim. Acta **60**, 3541 (1996).
[156] J. R. Bargar et al., J. Colloid Interface Sci. **185**, 473 (1997).
[157] J. R. Bargar et al., Langmuir **20**, 1667 (2004).
[158] S. M. Heald, H. Chen, and J. M. Tranquada, Phys. Rev. B **38**, 1016 (1987).
[159] I. D. Brown, Phys. Chem. Minerals **15**, 30 (1987).
[160] I. K. Robinson and D. J. Tweet, Rep. Prog. Phys. **55**, 599 (1992).
[161] E. Vlieg et al., Surf. Sci. **210**, 301 (1989).
[162] T. P. Trainor, P. J. Eng, and I. K. Robinson, J. Appl. Crystallogr. **35**, 696 (2002).
[163] P. A. Fenter, Rev. Mineral. Geochem. **49**, 149 (2002).
[164] C. Noguera, J. Physics: Condensed Matter **12**, R367 (2000).
[165] T. P. Trainor et al., Surf. Sci. **496**, 238 (2002).
[166] N. G. Condon et al., Surf. Sci. Lett. **310**, L609 (1994).
[167] N. G. Condon et al., Phys. Rev. Lett. 75, 1961 (1995).
[168] P. Guenard et al., Surf. Rev. Lett. **5**, 321 (1998).
[169] S. A. Chambers et al., Science **297**, 827 (2002).
[170] C. A. Eggleston et al., Geochim. Cosmochim. Acta **67**, 985 (2003).
[171] X. G. Wang et al., Phys. Rev. Lett. **81**, 1038 (1998).
[172] S. A. Chambers and S. I. Yi, Surf. Sci. **439**, L785 (1999).
[173] K. S. Tanwar et al., Surf. Sci. **601**, 460 (2007).
[174] C. S. Lo, K. S. Tanwar, A. M. Chaka, and T. P. Trainor, Phys. Rev. B 75, 075425 (2007).
[175] M. J. Bedzyk, G. M. Bommarito, J. S. Schildkraut, Phys. Rev. Lett. **62**, 1376 (1989).
[176] D. K. G. de Boer, Phys. Rev. B **44**, 498 (1991).
[177] T. P. Trainor, A. S. Templeton, and P. J. Eng, J. Electron Spectros. Related Phenomena **150**, 66 (2006).
[178] M. J. Bedzyk and L. Cheng, Rev. Mineral. Geochem. **49**, 221 (2002).
[179] T. P. Trainor et al., Langmuir **18**, 5782 (2002).
[180] A. S. Templeton et al., Proc. Nat. Acad. Sci. U.S.A. **98**, 11897 (2001).
[181] T.H. Yoon et al., Langmuir **21**, 4503 (2005).
[182] R. A. Neihof and G. I. Loeb, Limnology and Oceanography **17**, 7 (1972).
[183] R. A. Neihof and G. I. Loeb, J. Marine Res. **32**, 5 (1974).
[184] J. A. Davis, Geochim. Cosmochim. Acta **48**, 679 (1984).
[185] G. A. Waychunas et al., Analytical and Bioanalytical Chemistry **383**, 12 (2005).

Index

图书在版编目（CIP）数据

表面和界面的化学键合 ： 英文 / （瑞典） 安德斯·尼尔松，（瑞典） 拉尔斯·G.M.彼得松，（丹） 延斯·K.内斯科乌编. -- 影印本. -- 北京 ： 世界图书出版有限公司北京分公司，2025. 8. -- ISBN 978-7-5232-2345-1

Ⅰ. O647.11

中国国家版本馆CIP数据核字第2025KT2076号

中文书名	表面和界面的化学键合
英文书名	Chemical Bonding at Surfaces and Interfaces
著　　者	［瑞典］安德斯·尼尔松（Anders Nilsson）
	［瑞典］拉尔斯·G.M. 彼得松（Lars G.M. Pettersson）
	［丹麦］延斯·K. 内斯科乌（Jens K. Nørskov）
责任编辑	陈　亮　刘叶青
出版发行	世界图书出版有限公司北京分公司
地　　址	北京市东城区朝内大街 137 号
邮　　编	100010
电　　话	010-64038355（发行）　64033507（总编室）
网　　址	http://www.wpcbj.com.cn
邮　　箱	wpcbjst@vip.163.com
销　　售	新华书店
印　　刷	北京建宏印刷有限公司
开　　本	710mm × 1000mm　1/16
印　　张	33.5
字　　数	532 千字
版　　次	2025 年 8 月第 1 版
印　　次	2025 年 8 月第 1 次印刷
版权登记	01-2022-1889
国际书号	ISBN 978-7-5232-2345-1
定　　价	119.00 元
